The Teeth of Non-Mammalian Vertebrates

The Teeth of Non-Mammalian Vertebrates

Barry Berkovitz
Emeritus Reader in Dental Anatomy,
King's College London, United Kingdom
Visiting Professor, Oman Dental College,
Mina Al Fahal, Oman
Honorary Curator, Odontological Collection,
Hunterian Museum, Royal College of Surgeons
of England, London, United Kingdom

Peter Shellis
Department of Preventive,
Restorative and Pediatric Dentistry,
University of Bern, Switzerland

AMSTERDAM • BOSTON • HEIDELBERG • LONDON • NEW YORK • OXFORD • PARIS
SAN DIEGO • SAN FRANCISCO • SINGAPORE • SYDNEY • TOKYO

Academic Press is an imprint of Elsevier

Academic Press is an imprint of Elsevier
125 London Wall, London EC2Y 5AS, United Kingdom
525 B Street, Suite 1800, San Diego, CA 92101-4495, United States
50 Hampshire Street, 5th Floor, Cambridge, MA 02139, United States
The Boulevard, Langford Lane, Kidlington, Oxford OX5 1GB, United Kingdom

Notices
Knowledge and best practice in this field are constantly changing. As new research and experience broaden our understanding, changes in research methods, professional practices, or medical treatment may become necessary.

Practitioners and researchers must always rely on their own experience and knowledge in evaluating and using any information, methods, compounds, or experiments described herein. In using such information or methods they should be mindful of their own safety and the safety of others, including parties for whom they have a professional responsibility.

To the fullest extent of the law, neither the Publisher nor the authors, contributors, or editors, assume any liability for any injury and/or damage to persons or property as a matter of products liability, negligence or otherwise, or from any use or operation of any methods, products, instructions, or ideas contained in the material herein.

Library of Congress Cataloging-in-Publication Data
A catalog record for this book is available from the Library of Congress

British Library Cataloguing-in-Publication Data
A catalogue record for this book is available from the British Library

ISBN: 978-0-12-802850-6

For information on all Academic Press publications
visit our website at https://www.elsevier.com/

  Working together
to grow libraries in
developing countries

www.elsevier.com • www.bookaid.org

Publisher: Sara Tenney
Acquisition Editor: Kristi Gomez
Editorial Project Manager: Pat Gonzalez
Production Project Manager: Edward Taylor
Designer: Matthew Limbert

Typeset by TNQ Books and Journals

Front cover: Skull of rattlesnake. Courtesy of Dr. R Hardiman (Figure 7.25A in book, Micro CT scan of the dentition of a rattlesnake (*Crotalus* sp.)).

Contents

Preface

Teeth appeared about 450 million years ago (Rücklin et al., 2012; Benton, 2015), soon after the origin of hinged jaws during the evolution of the vertebrates. Teeth are homologous with **odontodes**: the denticles found on the surface of the dermal skeleton of ancient, jawless vertebrates and the placoid scales of chondrichthyans (Sire and Huysseune, 2003). Teeth and odontodes have a base of dentine, a mineralized tissue formed by mesenchymal odontoblasts (Chapter 11), that is usually covered by an outer layer of hypermineralized tissue (Chapter 12). Teeth, placoid scales, and presumably odontodes in the dermal skeleton of extinct vertebrates are formed by an interaction between the epithelium and the underlying mesenchyme. The classical theory of the origin of teeth was that the competence of the skin to develop odontodes was extended into the mouth, enabling the formation of teeth by interaction between oral epithelium and mesenchyme. The classical view was challenged by Smith and coworkers, who suggested that teeth appeared before the origin of jaws, from odontodes associated with pharyngeal endoderm (the "inside-out" hypothesis) (Smith and Coates, 1998), and, moreover, that teeth evolved independently in placoderms, chondrichthyans, and osteichthyans (Smith and Johanson, 2003). These suggestions have been very fruitful in that they stimulated extensive research and have led to hypotheses that combine elements of the classical and inside-out theories. The "revised outside-in" hypothesis of Huysseune et al. (2010) proposed that formation of pharyngeal teeth depends on ectoderm invading the oropharyngeal cavity through the gill slits. Fraser et al. (2010) suggested that odontodes or teeth will form (in the skin or the oropharyngeal cavity, respectively) whenever epithelial and neural crest gene regulatory networks interact (the "inside and out" theory), but they do not address the phylogenetic relationships between odontodes and teeth. The balance of opinion supports the classical view: teeth evolved from dermal odontodes in jawed vertebrates and they evolved only once (Benton, 2015; Witten et al., 2014; Donoghue and Rücklin, 2016).

Although there is partial or total loss of teeth in birds, turtles, and a significant number of other taxa (Davit-Béal et al., 2009), for most vertebrates the possession of teeth considerably increases the efficiency of the jaws in feeding. Since their first appearance, the teeth of non-mammalian vertebrates have become adapted to a great variety of functions: grasping, piercing, cutting, chopping, crushing, and even grinding. Heterodonty, the coexistence of different tooth forms within the same dentition, has further expanded the possibilities for efficient acquisition and processing of food.

The greater part of this book is devoted to descriptions of the morphological diversity of the dentition and its relationship to function (Chapters 1–8). This subject has always attracted the attention of anatomists. For example, John Hunter's first book was 'The natural history of the human teeth' (1771), whereas Richard Owen's 'Odontography' (1840–1845) described the morphology of teeth in all vertebrate classes. Until the latter part of the 20th century, comparative dental anatomy formed part of the education of dental students, and the subject featured in most textbooks of dental anatomy, from Charles Tomes' 'A manual of dental anatomy, human and comparative' (1876–1904) onward, but was last included in the 1978 edition of 'A color atlas and textbook of oral anatomy' (B K B Berkovitz, G R Holland, B J Moxham). There is now no readily available textbook detailing the dentition of non-mammalian vertebrates. Our primary purpose in writing this book was to remedy this deficiency, using high-quality photographs and images prepared by radiography, and other techniques, such as micro-CT scanning. The value of a comprehensive survey of this kind is that it provides not only an overview of the adaptation of the dentition to a range of diets, but also a basis for identification of dental material in archeological and other contexts. To discuss the dentitions of both extant and extinct non-mammalian vertebrates would be an enormous task and, in this book, we have restricted the scope to living representatives of the group. Even so, the range of morphological specializations is very large.

To complement the morphological descriptions of teeth and dentitions, we summarize the functional morphology of feeding in the chapter devoted to each major taxon. In addition, we review some special topics: tooth development, ontogeny of the dentition and tooth replacement, and biology of the dental hard tissues.

Teeth provide important models for the formation of organs that develop through a series of epithelial–mesenchymal interactions. The development of a dentition involves a complex interplay between the processes controlling formation of individual teeth and factors controlling the number, position, and replacement of teeth (Stock, 2001). The study of tooth development has expanded through the availability of sophisticated immunohistochemical techniques that allow the interactions between gene networks during odontogenesis to be elucidated (Chapter 9). Similarly, studies of the signaling mechanisms that control patterning of the dentition and tooth replacement have increased (Chapter 10). Research on growth of teeth and patterning of the dentition have, up to now, focused on the dentitions of a few mammals, but attention is now being given to the more varied dentitions of non-mammalian vertebrates to offer new insights. Finally, the evolutionary relationships between the constituent hard tissues of teeth have been clarified by studies of the distribution of matrix proteins and by new insights into the evolution of the genes that code for these proteins (Chapters 11 and 12). We are aware that a separate monograph on each of these chapter topics, and even for topics within in each chapter, could easily be written, but we hope that we have provided introductions that will enable interested readers to explore these topics in greater depth while enhancing their understanding of tooth biology.

BKB Berkovitz and RP Shellis
2016/17

A NOTE ON MAGNIFICATION

We have used various methods to indicate the size of a specimen. Some images were supplied with scale bars or rules, so needed no further attention. In most cases, we supply the "image width," ie, the width of the field of view before magnification. In the case of a few images, unfortunately, we have no information on the magnification.

REFERENCES

Benton, M.J., 2015. Vertebrate Palaeontology, fourth ed. Wiley-Blackwell, Chichester.

Davit-Béal, T., Tucker, A.S., Sire, J.Y., 2009. Loss of teeth and enamel in tetrapods: fossil record, genetic data and morphological adaptations. J. Anat. 214, 477–501.

Donoghue, P.C.J., Rücklin, M., 2016. The ins and outs of the evolutionary origin of teeth. Evol. Dev. 18, 19–30.

Fraser, G.J., Cerny, R., Soukup, V., Bronner-Fraser, M., Streelman, J.T., 2010. The odontode explosion: the origin of tooth-like structures in vertebrates. BioEssays 32, 808–817.

Huysseune, A., Sire, J.Y., Witten, P.E., 2010. A revised hypothesis on the evolutionary origin of the vertebrate dentition. J. Appl. Ichthyol. 26, 152–155.

Rücklin, M., Donoghue, P.C.J., Johanson, Z., Trinajstic, K., Marone, F., Stampanoni, M., 2012. Development of teeth and jaws in the earliest jawed vertebrates. Nature 491, 748–752.

Sire, J.Y., Huysseune, A., 2003. Formation of dermal skeletal and dental tissues in fish: a comparative and evolutionary approach. Biol. Rev. 78, 219–249.

Smith, M.M., Coates, M., 1998. Evolutionary origins of the vertebrate dentition: phylogenetic patterns and developmental evolution. Eur. J. Oral Sci. 106 (Suppl. 1), 482–500.

Smith, M.M., Johanson, Z., 2003. Separate evolutionary origins of teeth from evidence in fossil jawed vertebrates. Science 299, 1235–1236.

Stock, D.W., 2001. The genetic basis of modularity in the development and evolution of the vertebrate dentition. Phil. Trans. R. Soc. 356, 1633–1653.

Witten, P.E., Sire, J.Y., Huysseune, A., 2014. Old, new and new-old concepts about the evolution of teeth. J. Appl. Ichthyol. 30, 636–642.

Acknowledgments

This book could not have been written without the help and encouragement of a very large number of people. All of the people who provided images for our book are acknowledged in the legends, but we offer them here our grateful thanks. It was a humbling experience for us to write to colleagues worldwide who usually did not know us, to find that they immediately responded so quickly and so positively, by offering us the use of such important material. Their encouragement ensured that we completed this project.

Special thanks are extended to the following museum staff, who provided help beyond the call of duty:

. Hunterian Museum, Royal College of Surgeons of England: Dr. S. Alberti, C Phillips, M Cooke, M Farrell, S Morton, K Hussey.
. Horrniman Museum & Gardens: Dr. P Viscardi.
. University College London, Grant Museum of Zoology; and Oxford University Museum of Natural History: Dr. M Carnall.
. Museum of Life Sciences, King's College London: Dr. G Sales.
. Natural History Museum: Dr. M Wilkinson and Dr. P Kitching.
. Hunterian Zoology Museum, University of Glasgow: Dr. M Reilly.
. Australian Museum Research Institute: Dr. M McGrouther.

In the figure legends, the names of three museums are given in abbreviated form. RCSOM = Royal College of Surgeons (of England) Odontological Museum; MoLSKCL = Museum of Life Sciences King's College London. In "UCL Grant Museum of Zoology," UCL = University College London.

We are most grateful to Dr. Maisano from Digimorph.org (Digital Morphology Library, University of Texas at Austin) for providing many high-quality micro-CT scans.

We express our thanks to the following colleagues who provided considerable expertise that we lacked in several topics: Dr. R Britz (Natural History Museum, London), Dr. R Cerny (Department of Zoology, Charles University, Prague), Prof. S E Evans (Research Department of Cell and Developmental Biology, University College London), Dr. A Konings (Cichlid Press), Prof. M M Smith (Dental Institute, King's College London), Prof. A S Tucker (Department of Craniofacial Development and Stem Cell Biology, King's College London), Dr. C Underwood (Department of Earth and Planetary Sciences, Birkbeck, University of London). Drs. Evans, Smith, Tucker, and Underwood read chapters and provided invaluable criticisms.

Dr. Shellis thanks Professor A Lussi (Department of Restorative Dentistry, University of Bern) for support and encouragement over many years.

We are much indebted to J Carr, S Franey, and M Simon for photographic assistance.

Acknowledgments

Chapter 1

Cyclostomes

The cyclostomes (class Agnatha, subclass Cyclostomata) are eel-like fishes that have sucker-like mouths (Fig. 1.1) and comprise the **lampreys** (Petromyzontiformes) and the **hagfishes** (Myxiniformes). They are the only extant representatives of the **Agnatha**, fishes that lack jaws, in contrast to the **Gnathostomata**, the jawed vertebrates that include all other extant vertebrates. In addition, the hagfishes lack vertebrae. Neither lampreys nor hagfishes have mineralized skeletal tissues or true teeth, and they are thought to have evolved before hard tissues had appeared. The taxonomic and phylogenetic positions of the cyclostomes have been a matter of debate (Nicholls, 2009). Phenotypic analyses have supported a closer relationship of lampreys with gnathostomes than with hagfishes, thus excluding hagfishes from the vertebrates. In contrast, molecular analyses have consistently indicated that the cyclostomes are monophyletic (Near, 2009; Heimberg et al., 2010).

Cyclostomes possess tooth-like structures that are similar in structure and composition to other ectodermally derived organs, such as claws and horns. The teeth appear at metamorphosis into the adult stage and are then continuously replaced throughout life. The skin and oral mucosa of cyclostomes secrete mucus, so these structures are specializations hardened by localized keratinization. The teeth are not mineralized. Cyclostome teeth consist of cones of keratinized cells that are replaced by new cones that form beneath the functional teeth (Dawson, 1969; Uehara, 1983; Yokoyama and Ishiyama, 1998; Alibardi and Segalla, 2011). Although they bear no structural resemblance to true teeth, we nevertheless refer to them as teeth (as we do for the keratinized tooth-like structures of frog tadpoles in Chapter 5), because there is no other available term.

Cyclostome teeth are composed mainly of acidic keratins bound tightly together by higher amounts of nonkeratin proteins (Alibardi and Segalla, 2011). One reason for including cyclostome teeth in this book is that it has been suggested that enamel protein antigens are present in the teeth of the **Pacific hagfish** (*Eptatretus stoutii*). At first, these antigens were identified by electrophoresis as enamelins (Slavkin et al., 1983), but Slavkin et al. (1991) later referred to them as amelogenins and also reported immunoreactivity to an antibody against an amelogenin peptide. The reactivity was localized to the pokal cells, a group of cells that participate in formation of replacement tooth cones (Dawson, 1969). However, immunohistochemical investigations by Yokoyama and Ishiyama (1998) detected neither enamelin nor amelogenin in teeth of the **brown hagfish** (*Paramyxine atami*). This agrees with genomic analysis that shows that enamel matrix protein genes are absent from cyclostomes (Venkatesh et al., 2014). Yokoyama and Ishiyama also concluded that the pokal cells were keratinocytes, and not secretory cells, as Slavkin et al. (1991) had suggested.

MYXINIFORMES

Hagfishes comprise a single family of about 70 species, divided into seven genera. They are eel-like scavengers that live on ocean and riverbeds. They feed on invertebrates, such as polychaete worms and on the flesh of dead and dying fish. A hagfish has two pairs of crescent-shaped serrated teeth carried on a cartilaginous **dental plate** (Fig. 1.2) that is hinged in the midline. The dental plate rests on the dorsal aspect of the **basal plate**, a complex of longitudinally orientated bars of cartilage. During feeding, the dental plate is protracted, so that it slides forward and then rotates around the anterior crest of the basal plate and is exposed to the exterior. At rest, the dental plate is folded along the midline, but in the protracted position it opens to an angle of almost 180°, so that the teeth are exposed and can be applied to the surface of the prey. When the dental plate is retracted the teeth close, thereby grasping a morsel of flesh that is then drawn into the mouth as the dental plate is pulled caudally along the basal plate (Clark and Summers, 2007; Clark et al., 2010). Modeling suggests that the dental apparatus of hagfishes can exert large bite forces (Clark and Summers, 2007). However, operation of this feeding mechanism is significantly slower than that of the jaws in fishes and tetrapods that are thus better adapted to capturing active prey (Clark and Summers, 2007).

The Teeth of Non-Mammalian Vertebrates. http://dx.doi.org/10.1016/B978-0-12-802850-6.00001-1

FIGURE 1.1 Body of European river lamprey (*Lampetra fluviatilis*) showing sucker-like mouth. *Courtesy Wikipedia.*

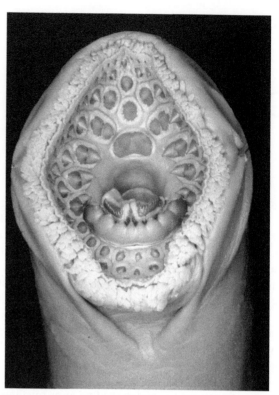

FIGURE 1.3 Mouth of lamprey showing horny dental teeth and plates. Note the two lateral lingual plates on the tongue in the center, with the tip of the transverse lingual plate just visible in front. *From Berkovitz, B.K.B., 2013. Nothing but the Tooth, Elsevier, London.*

FIGURE 1.2 Mouth of hagfish showing two horny dental plates. *Courtesy Professor A Gorbman.*

Differences in the morphologies of the tooth plates help to identify the different species or hagfish. For example, the **white-headed hagfish** (*Myxine ios*) has a total cusp count of 44–51, whereas the **Atlantic hagfish** (*Myxine glutinosa*) has a total cusp count of 32–36.

PETROMYZONTIFORMES

There are about 50 species of lamprey divided into three families. Some species in each family are parasitic. They attach themselves to fish and aquatic mammals by a sucker-like mouth and rasp away the skin to obtain body fluids and

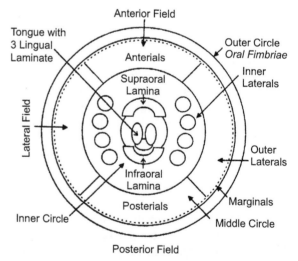

FIGURE 1.4 Diagram showing disposition of horny dental plates of a lamprey. *From Smith, R.E., Butler, V.L., 2008. Towards the identification of Lamprey (Lampetra spp.) in archaeological contexts. J. Northwest Anthropol. 42, 131–142. Courtesy Editors of Journal of Northwest Anthropology.*

tissue by means of horny plates on the tongue. Other species, which tend to be freshwater species, do not feed after metamorphosis, but survive on food reserves accumulated during the larval phase.

Petromyzontidae

The **sea lamprey** (*Petromyzon marinus*) is an example of a parasitic species. Its teeth are hollow and more numerous than those of the hagfish (Fig. 1.3). The sea lamprey has three horny lingual plates containing "teeth," one central (transverse lingual) plate and two lateral (longitudinal lingual) plates, and numerous other tooth-like structures surround the mouth.

As in hagfishes, lamprey species can be distinguished on the basis of the number and distribution of teeth (Hubbs and Potter, 1971: Renaud, 2011). A generally accepted method for standardizing the distribution of teeth is shown in Fig. 1.4 (Smith and Butler, 2008), whereas Fig. 1.5 (Maitland, 1972; Igoe et al., 2004) and Table 1.1 (Wydoski and Whitney, 2003; Smith and Butler, 2008) illustrate the differences in distribution of the teeth between three species of lampreys. **Pacific lamprey** (*Lampetra tridentata*) has 17–21 cusps on the transverse lingual lamina, whereas the **Ohio lamprey** (*Ichthyomyzon bdellium*) has 21–32 cusps.

It has been estimated that the horny teeth of lampreys are replaced 20–40 times between metamorphosis and spawning.

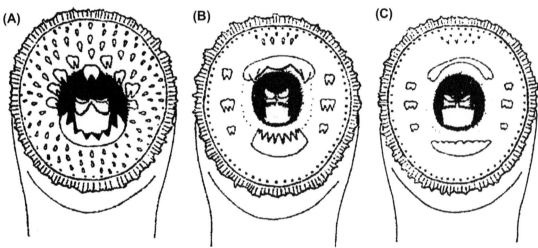

FIGURE 1.5 Mouth structures in three different lampreys. (A) Sea lamprey (*Petromyzon marinus*); (B) river lamprey (*Lampetra fluviatilis*); (C) brook lamprey (*Lampetra planeri*). *From Igoe, F., Quigley, D.T.G., Marnell, E., Meskell, E., O'Connor, W., Byrne C., 2004. The sea lamprey* Petromyzon marinus *(L), river lamprey* Lampetra fluviatilis *(L) and brook lamprey* Lampetra planeri *(Bloch) in Ireland: general biology, ecology, distribution and status with recommendations for conservation. Biol. Env. Proc. R. Ir. Acad. 2004;104B:43–56. Courtesy Editors of Proceedings of the Royal Irish Academy.*

TABLE 1.1 Distinguishing Features of Dentition in Adult Sea, River, and Brook Lampreys

Name	Supra-Oral Lamina	Infra-Oral Lamina	Lateral Teeth	Posterior Teeth
Pacific lamprey (*Lampetra tridentata*) Maximum length *ca.* 760 mm	3 cusps	5 cusps	4 pairs	Present
River lamprey (*Lampetra ayresi*) Ca. 280–300 mm	2 cusps	≥6 cusps	3 pairs	Absent
Western brook lamprey (*Lampetra richardsoni*) Less than 180 mm	Very small, weakly developed, rounded teeth			

From Smith, R.E., Butler, V.L., 2008. Towards the identification of Lamprey (Lampetra spp.) in archaeological contexts. J. Northwest Anthropol. 42, 131–142.

REFERENCES

Alibardi, L., Segalla, A., 2011. The process of cornification in the horny teeth of the lamprey involves proteins in the keratin range and other keratin-associated proteins. Zool. Stud. 50, 416–425.

Berkovitz, B.K.B., 2013. Nothing but the Tooth. Elsevier, London.

Clark, A.J., Summers, A.P., 2007. Morphology and kinematics of feeding in hagfish: possible functional advantages of jaws. J. Exp. Biol. 210, 3897–3909.

Clark, A.J., Maravilla, E.J., Summers, A.P., 2010. A soft origin for a forceful bite: motor patterns of the feeding musculature in Atlantic hagfish, *Myxine glutinosa*. Zoology 113, 259–268.

Dawson, J.A., 1969. The keratinised teeth of *Myxine glutinosa*. A histological, histochemical, ultrastructural and experimental study. Acta Zool. 50, 35–68.

Heimberg, A.M., Cowper-Sallari, R., Sémon, M., Donoghue, P.C.J., Peterson, K.J., 2010. microRNAs reveal the interrelationships of hagfish, lampreys, and gnathostomes and the nature of the ancestral vertebrate. Proc. Natl. Acad. Sci. 107, 19379–19383.

Hubbs, C.L., Potter, I.C., 1971. Distribution, phylogeny and taxonomy. In: Hardisty, M.W., Potter, I.C. (Eds.), The Biology of Lampreys. vol. 1. Academic Press, London, pp. 127–206.

Igoe, F., Quigley, D.T.G., Marnell, E., Meskell, E., O'Connor, W., Byrne, C., 2004. The sea lamprey *Petromyzon marinus* (L), river lamprey *Lampetra fluviatilis* (L) and brook lamprey *Lampetra planeri* (Bloch) in Ireland: general biology, ecology, distribution and status with recommendations for conservation. Biol. Env. Proc. R. Ir. Acad. 104B, 43–56.

Maitland, P.S., 1972. A Key to British Freshwater Fishes. Fresh Water Biological Association, Cumbria.

Near, T.J., 2009. Conflict and resolution between phylogenies inferred from molecular and phenotypic data sets for hagfish, lampreys, and gnathostomes. J. Exp. Zool. Mol. Dev. Evol. 312B, 749–761.

Nicholls, H., 2009. Evolution: mouth to mouth. Nature 461, 164–166.

Renaud, C.B., 2011. Lampreys of the World: An Annotated and Illustrated Catalogue of Lamprey Species Known to Date. Food and Agricultural Organisation of the United Nations, Rome.

Slavkin, H.C., Graham, E., Zeichner-David, M., Hildemann, W., 1983. Enamel-like antigens in hagfish: possible evolutionary significance. Evolution 37, 404–412.

Slavkin, H.C., Kreisa, R.J., Fincham, A., Bringas, P., Santos, V., Sassano, Y., Snead, M.L., Zeichner-David, M., 1991. Evolution of enamel proteins: a paradigm for mechanisms of biomineralization. In: Suga, S., Nakahara, H. (Eds.), Mechanisms and Phylogeny of Mineralization in Biological Systems. Springer, Berlin, pp. 383–389.

Smith, R.E., Butler, V.L., 2008. Towards the identification of Lamprey (*Lampetra* spp.) in archaeological contexts. J. Northwest Anthropol. 42, 131–142.

Uehara, K., 1983. Fine structure of the horny teeth of the lamprey, *Entosphenus japonicus*. Cell Tiss. Res. 231, 1–15.

Venkatesh, B., Lee, A.P., Ravi, V., Maurya, A.K., Lian, M.M., Swann, J.B., Ohta, Y., Flajnik, M.F., Sutoh, Y., Kasahara, M., Hoon, S., Gangu, V., Roy, S.W., Irimia, M., Korzh, V., Kondrychyn, I., Lim, Z.W., Tay, B.-H., Tohari, S., Kong, K.-W., Ho, S., Lorente-Galdos, B., Quilez, J., Marques-Bonet, T., Raney, B.J., Ingham, P.W., Tay, A., Hillier, L.W., Minx, P., Boehm, T., Wilson, R.K., Brenner, S., Warren, W.C., 2014. Elephant shark genome provides unique insights into gnathostome evolution. Nat. Lond. 505, 174–179.

Wydoski, R.S., Whitney, R.R., 2003. Inland Fish of Washington. University of Washington Press.

Yokoyama, M., Ishiyama, M., 1998. Histological and immunohistochemical studies of the horny teeth in the hagfish. Odontology 85, 674–688 (Japanese, English summary and tables).

Chapter 2

Chondrichthyes 1: Sharks

CHONDRICHTHYES

Chondrichthyes are fishes that lack bone. Instead, the skeleton is composed of cartilage that is partly calcified. The group comprises two subclasses: Elasmobranchii and Holocephali. The elasmobranchs include sharks and rays of which there are more than 800 species, whereas the Holocephali (chimaeras) is a much smaller group of about 40 species.

The classification used here (Table 2.1) can be accessed from the Website www.sharksrays.org and is based on DNA analyses of a very large number of sharks and rays. Rays are not considered to be derived sharks. Thus, there are two subclasses: Elasmobranchii and Holocephali.

Elasmobranchs

In sharks and rays, the upper jaw (the palatoquadrate cartilage) is not fused to the chondrocranium, but rather articulates with it. The posterior articulation is mediated by the hyomandibula, the dorsal element of the first gill arch posterior to the jaws, that connects the jaw joint with the chondrocranium. There are four forms of jaw suspension among extant elasmobranchs (Wilga et al., 2007; Motta and Huber, 2012), distinguished by the presence and location of the anterior connections to the chondrocranium (Fig. 2.1) (Wilga, 2005). In **amphistyly** (Hexanchiformes), the palatoquadrate has a postorbital articulation and also an articulation in the orbital region. In **orbitostyly** (Squalea except for Hexanchiformes), only the orbital articulation is present. In **hyostyly** (galeomorph sharks), the palatoquadrate articulates with the ethmoid (preorbital) region of the chondrocranium. In **euhyostyly** (batoids), an anterior articulation is lacking and the hyomandibula supplies the only articulation between the chondrocranium and palatoquadrate.

Amphistylic suspension allows limited movement of the palatoquadrate relative to the chondrocranium, but in the other forms of suspension, the jaw assembly is free to move to a greater or lesser extent. Euhyostylic suspension confers the greatest freedom of movement because of the lack of an anterior articulation, whereas the mobility in orbitostyly

and hyostyly depends on the extensibility of the ligament forming the ethmoid or orbital articulation.

The mobility of the jaw assembly is of great importance in the feeding mechanics of elasmobranchs, especially because it allows the upper jaw to be protruded during food capture (Wilga et al., 2001; Wilga, 2002) (Fig. 2.2). The first phase of feeding usually involves the elevation of the head and opening of the mouth through depression of the lower jaw. The upper jaw is then protruded by being drawn forward and downward along the chondrocranium through muscular action and, simultaneously, the lower jaw closes on the food object. Finally, the upper jaw is retracted and the swimming position of the jaws is reestablished. The details of this cycle vary according to species and feeding requirements. For example, in species that feed on large prey, such as the **great white shark** (*Carcharodon carcharias*), the head is lifted high and the mouth opened as far as possible, thereby allowing an anteriorly directed bite. In species that take food located below them, the head is lifted much less or not at all.

Comprehensive reviews of feeding among elasmobranchs have been published by Wilga et al. (2007) and, in greater detail, by Motta and Huber (2012). Among sharks and rays there are three basic modes of food capture: **biting, suction,** and **filtering**. Predatory and filter-feeding elasmobranchs use ram feeding, whereby forward motion enables a predator to overtake prey or creates a flow of water into the mouth of a filter feeder. Biting is considered to be the ancestral mode of food acquisition. Biting is supplemented in many groups by suction, helping to draw food toward the mouth. Suction is created by expansion of the buccal and hyoid cavities to create negative pressure as the mouth opens (Wilga et al., 2007; Wilga and Sandford, 2008). In some groups, such as heterodontid and orectolobid sharks, and in many batoids, suction may be the exclusive method of acquiring food.

The jaws of elasmobranchs are armed with rows of teeth and no species is edentulous, although there is a small number of filter-feeding species with vestigial dentitions. In general, sharks are predatory carnivores and have basically triangular, sharp-edged teeth adapted to cutting, piercing, or grasping prey. Rays, most of which live near the sea

TABLE 2.1 Classification of Chondrichthyes
(www.sharksrays.org)

Subclass Elasmobranchii

Superorder Selachii (sharks)

Division Galeomorphi

Order Heterodontiformes (bullhead sharks)

Order Orectolobiformes (carpet sharks, wobbegongs)

Order Lamniformes (mackerel sharks)

Order Carchariniformes (ground sharks)

Division Squalea

Order Hexanchiformes (frilled and cow sharks)

Order Echinorhiniformes (bramble sharks)

Order Squaliformes (dogfishes, sleeper, and kitefin sharks)

Order Squatiniformes (angel sharks)

Order Pristiophoriformes (saw sharks)

Superorder Batoidea (rays, skates)

Order Rajiformes (skates)

Order Torpediniformes (electric rays)

Order Rhinopristiformes (guitarfish, sawfishes)

Order Myliobatiformes (stingrays, eagle rays, mantas)

Subclass Holocephali

Order Chimaeriformes (chimaeras, ratfishes)

bed, feed on benthic invertebrates as well as on fishes, and their teeth tend to be relatively small and close set. A few sharks (eg, *Heterodontus, Sphyrna tiburo, Chiloscyllium, Mustelus*) and many rays are durophagous and possess teeth adapted for crushing.

The size of the teeth varies within each dentition. Most commonly the size decreases from the midline symphysis toward the jaw joint. Whereas a small number of sharks and rays are close to **homodont**, showing no marked variation in tooth shape within the jaw, most show some degree of **heterodonty**. Tooth morphology may vary along the length of the same jaw, differ between upper and lower jaws, change during ontogeny, or differ within adult males and females of the same species.

Unlike most mammals, which have a fixed dental formula (DF), the number of teeth in the mouths of elasmobranchs varies considerably between species. The number and size of the teeth can increase during the fish's life as part of the growth process, although this is usually restricted to the early stages of ontogeny. Usually, the dentition is bilaterally symmetrical, ie, there are equal numbers of teeth on opposite sides of the midline, but the

symmetry can be altered by the presence of one or more teeth at the symphysis. This arrangement is most clearly seen in the myliobatid rays in which the largest teeth straddle the midline.

The teeth in all elasmobranchs are attached via a basal plate to a sheet of fibrous tissue running over the surface of the jaw cartilage, a type of attachment that allows varying degrees of tooth mobility. Forces generated by an unknown mechanism within the sheet of tooth-bearing connective tissue carry the teeth over the crest of the jaw, like a conveyor belt, and into a functional position, after which they are lost at the front of the jaws. Therefore, in any specimen several replacement teeth will be present at each tooth position (see, eg, in Figs. 2.3 and 2.6). In describing elasmobranch dentitions, we follow the recommendation by H F Mollet and J A Bourdon (www.elasmo.com) and use the term **row** to describe the line of teeth along the jaw and **file** to describe a marginal tooth plus its successors, whether these are also functional or still in the process of development. In a dried specimen, several replacement teeth in addition to those in functional positions can often be observed (Figs. 2.3 and 2.6), but in a live or fresh specimen some of these replacement teeth will be concealed beneath oral tissue. The number of rows of teeth visible in the mouth of a living elasmobranch varies. Among sharks, only a few rows of teeth (see Fig. 2.16) or even only one row (see Fig. 2.36) are exposed, in conformity with the cutting or grasping function. Among rays, the teeth are used to grasp or crush the prey, and many rows, forming a pavement, are typically exposed.

In a few species of sharks (eg, *Isistius*), the whole row of functional teeth is shed simultaneously but, in most species, only a few teeth are in the process of replacement at any given time, so that there are few gaps in the tooth row (Springer, 1960; Strasburg, 1963). This can be illustrated by the range in tooth number as a proportion of the maximum tooth number as seen in eight carcharhinids, varying from 4% to 26% (mean 13%) (Springer, 1960). Strasburg (1963) ascribed this variation to differences in how the teeth were set in the row. In tooth rows where the teeth alternate or form a row without overlap, the most anterior teeth in a file can be shed without hindrance, whereas in an imbricated (closely overlapping) row, each tooth can prevent its neighbor from being shed, so that the whole row must be shed together. The teeth of batoids always seem to be arranged in an alternating array, so teeth will be shed at the same time from alternate tooth positions (see Chapter 10).

The rate of tooth replacement varies widely, eg, from 18 days per row to 5 weeks per row in **nurse shark** (*Ginglymostoma cirratum*); the replacement rate in nurse sharks also varies with the season and showed that teeth were replaced at increasingly extended intervals as the

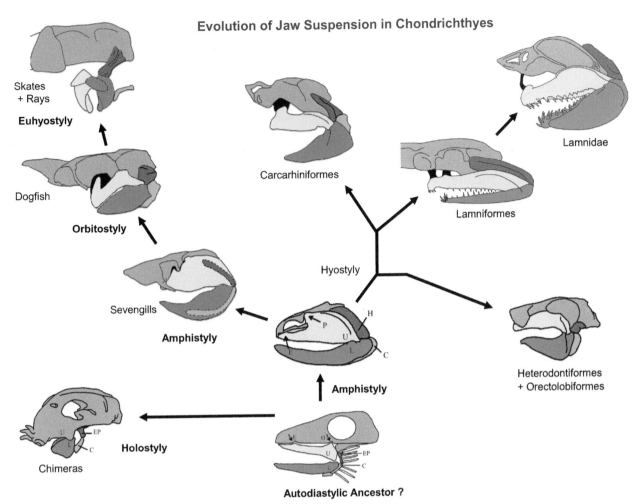

FIGURE 2.1 Types of jaw suspension among elasmobranchs. Black=ethmopalatine and postorbital ligaments; blue=Meckel's cartilage; gray=chondrocranium; green=ceratohyal; red=hyomandibula, yellow=palatoquadrate or upper jaw. The suggested evolutionary sequence assumes a phylogeny that differs from that adopted in this book. The derivation of the euhystylic suspension of batoids (top left) from an orbitostylic condition would not be tenable if batoids form a separate clade from selachians (Table 2.1). *From Wilga, C.D., 2005. Morphology and evolution of the jaw suspension in lamniform sharks. J. Morphol. 265, 102–119. Image by courtesy Dr. C Wilga. ©John Wiley and Sons, reproduced by permission. Copyright and Photocopying: VC 2016 Wiley Periodicals, Inc. All rights reserved. No part of this publication may be reproduced, stored or transmitted in any form or by any means without the prior permission in writing from the copyright holder. Authorization to photocopy items for internal and personal use is granted by the copyright holder for libraries and other users registered with their local Reproduction Rights Organisation (RRO), eg, Copyright Clearance Center (CCC), 222 Rosewood Drive, Danvers, MA01923, USA (www.copyright.com), provided the appropriate fee is paid directly to the RRO. This consent does not extend to other kinds of copying such as copying for general distribution, for advertising or promotional purposes, for creating new collective works or for resale. Special requests should be addressed to: permissions@wiley.com.*

water temperature decreased (Luer et al., 1990). Bruner (1998) in an abstract reported much slower rates in great white sharks: 106 days per row in the upper jaw and 114 days per row in the lower jaw of young specimens, with corresponding rates of 226 and 242 days per row in old specimens. There are no data on tooth replacement rate among batoids. Tooth replacement is discussed further in Chapter 10.

Chimaeras

Chimaeras live near the sea bed and feed on mollusks and crustaceans. The upper jaw is fused to the cranium (**holostyly**), in contrast to the more mobile forms of suspension in elasmobranchs. Moreover, chimaeras lack typical individual teeth. Instead, they possess three pairs of cutting tooth plates.

The rays, together with the chimaeras, are discussed in the Chapter 3. The rest of the chapter describes the dentition of sharks.

DENTITIONS OF SHARKS

There are more than 450 shark species, divided into the Galeomorphi and Squalea.

Sharks have five gill slits at the sides of the head, except for the Hexanchiformes and one species of

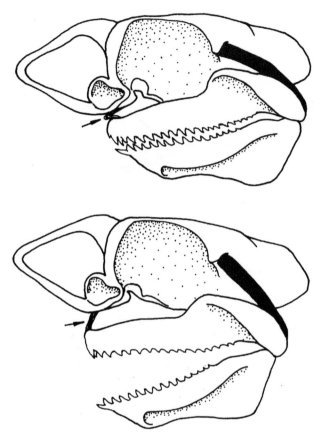

FIGURE 2.2 Jaw protrusion in a hyostylic shark. Top: Jaws closed. Palatoquadrate apposed to chondrocranium, suspended posteriorly by the hyomandibula (*black*) and anteriorly by the ethmoid ligament (*arrow*). Bottom: Jaws open and protruded as the jaws start to close on the prey. Jaw protrusion is accomplished by the palatoquadrate sliding forward, accompanied by rotation of the hyomandibula, until the extension of the ethmoid ligament prevents further forward movement.

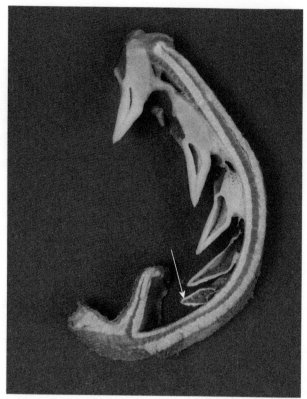

FIGURE 2.3 Cross section of jaw of a sand tiger shark (*Carcharias taurus*), showing five mineralized sets of replacing teeth, the youngest and most lightly mineralized at the bottom (*arrow*). Image width = 5.5 cm. *Courtesy RCSOMA/ 434.6.*

sawshark, which have six or seven. The mouth is curved and lies just below and behind the tip of the snout. Nearly all sharks are high-level predators (Cortés, 1999; Wetherbee and Cortés, 2012). In aggregate, the major component of the diet is teleost fish, with other prey including cephalopods, invertebrates (such as mollusks and crustaceans), and other elasmobranchs. A few species include mammals or birds in their diet, and three genera consume planktonic organisms by filter feeding.

Among biting sharks, the mouth is ventral and the width of the gape is enlarged by lifting the head as the shark strikes its prey. Sharks sometimes ingest prey whole, but more often take bites from prey that is too large to take in whole. This is possibly facilitated by upper jaw protrusion, and biting is often accompanied by shaking of the head to complete the detachment of the food morsel. Many sharks that rely on biting to capture prey use a small degree of suction (mean subambient

buccal pressure ca. −2 kPa). However, in many sharks that feed near the seabed (eg, orectolobids and heterodontids), prey is acquired exclusively using much greater suction (mean subambient buccal pressure ca. −20 to −25 kPa), which is generated by expansion of the buccal and pharyngeal cavities as the mouth opens. The mouth in these sharks has a terminal or near-terminal position. As it opens, its margins are supported by the labial cartilages, so that the gape is circular and directed forward. This arrangement enhances the pressure exerted on the prey. The distance over which suction can draw prey into the mouth is limited to a few centimeters. It is therefore necessary for suction-feeding sharks to reduce the distance from free-swimming prey by stalking or ambush. The suction is more effective in confined spaces, so can be used to extract prey from crevices or prey that is buried in mud or sand.

Considering interspecific variation, the overall bite force exerted by shark jaws scales isometrically with body mass. Average anterior bite force ranges from 8 N in the **spiny dogfish** (*Squalus acanthias*) (0.39-kg body mass) to 2400 N in the **great hammerhead** (*Sphyrna mokarran*) (581-kg body mass) (Habegger et al., 2012).

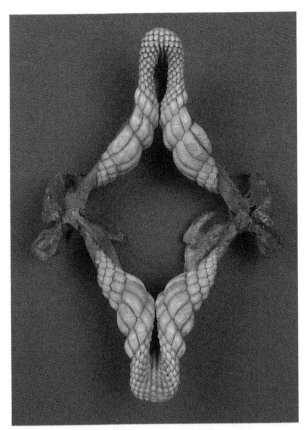

FIGURE 2.4 Dentition of Port Jackson shark (*Heterodontus portusjacksoni*). In both jaws, the anterior teeth are single pointed and the posterior teeth are large, rounded, and adapted to crushing. Tooth size reaches a maximum in the middle of the crushing region. Image width=14cm. *Courtesy RCSOMA/ 438.1.*

FIGURE 2.5 Anterior three-pointed teeth in lower jaw of zebra bullhead shark (*Heterodontus zebra*). *Courtesy Dr. C Underwood.*

Filter-feeding sharks use suction, ram (water intake driven by forward motion), engulfment, or a combination of these mechanisms to take in seawater containing planktonic

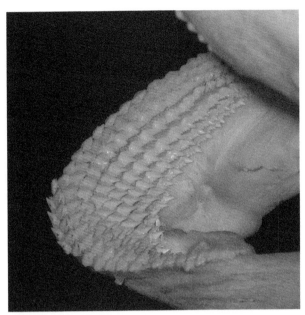

FIGURE 2.6 Teeth of nurse shark (*Ginglymostoma cirratum*), each with a large central cusp and smaller flanking cusps. Note the lack of overlap at the bases, which allows teeth to be replaced individually. *Courtesy Wikipedia.*

food organisms (Motta and Huber, 2012). The mechanism by which the small prey is filtered from the water also varies between species.

The teeth of many sharks are relatively widely spaced and have basal plates with a simple shape. This allows some anteroposterior rotation that provides a shock-absorbing mechanism. In addition, as the jaws open, tension within the tooth-bearing, connective tissue sheet can erect the anterior row(s) of teeth (Shellis, 1982). However, several species (eg, *Squalus acanthias* and members of the Dalatiidae and Centrophoridae) possess rows of sectorial teeth that are set very close together and that seem to be immobile on the jaw margin.

As will become apparent, there is a great variety of tooth form among sharks. In general, sharks have one of three basic tooth shapes: pointed, blade like, or rounded. Pointed teeth vary in form from squat cones to highly elongated spikes, which are frequently curved toward the mouth. They tend to be rounded in cross section, but may possess blunt edges on their mesial and distal surfaces. Besides the main cusp, subsidiary lateral cusps may be present. Blade-like teeth are thin relative to their mesio-distal width and have one or more cutting edges, which may lie parallel with the line of the jaw or at an oblique angle to it. These cutting edges may be smooth or serrated, and in a few species the teeth resemble saws because of the presence of a row of sharp cusps along the edges. Finally, sharks such as the Heterodontidae that have a diet that includes a large proportion of hard-shelled prey may possess crushing teeth with rounded surfaces.

Extensive descriptions and illustrations of tooth form among sharks are available in the series of publications by Herman et al. (1987–1993) and at several websites (http://www.elasmo.com; http://homepage2.nifty.com/megalodon/index.htm).

Pointed teeth are considered to be adapted to grasping small prey or piercing and tearing larger prey, whereas blade-like teeth are suited for shearing through skin and muscle. Experimental studies (Whitenack and Motta, 2010) suggest that symmetrical sharply pointed teeth puncture prey more easily than blade-like teeth with obliquely oriented tips, which require greater force to penetrate prey. However, when drawn laterally through flesh, cutting edges with different morphologies (smooth, serrated, or multicusped) seem to cut with similar efficiencies. In several sharks with blade-like teeth, notably the Dalatiidae, the crowns of neighboring teeth overlap, so that the tooth row effectively presents a single, serrated cutting edge.

Shark dentitions frequently display heterodonty. The Heterodontidae, as the name suggests, show the greatest variation in tooth form along the line of the jaw, with pointed anterior teeth and rounded, crushing posterior teeth. However, anterior-posterior variations occur in other species [eg, **snaggletooth shark** (*Hemipristis elongata*), which possesses elongated, sharply pointed anterior teeth that grade posteriorly into blade-like teeth]. A common form of heterodonty is the occurrence of markedly different tooth forms in upper and lower jaws. Usually, one jaw is furnished with pointed teeth and the other with blade-like teeth. Sharks with exclusively pointed teeth are considered to feed by grasping (clutching) or tearing, whereas those with exclusively blade-like teeth feed by cutting flesh from their prey. Feeding by sharks with heterodont dentitions combines different functions, eg, grasping and crushing in the **bullhead sharks** (Heterodontidae) or grasping and cutting (Dalatiidae).

As teeth are frequently the only preserved organs from extinct forms, knowledge of tooth morphology becomes vital in understanding shark evolution and in the precise identification of living and extinct species. Features on individual teeth that help in classification include the number of cusps, the curvature and angulation of cusps, the presence of serrations on edges, and the degree of notching on the distal margin where the tooth meets the attachment base. However, identification of teeth is further complicated by other factors already referred to, such as:

1. Differences in tooth size and morphology along the tooth row.
2. Presence of additional small teeth in the region of the midline symphysis.
3. Morphological differences between teeth in the upper and lower jaws.

4. Age-related changes reflecting a change in diet during life. This may result in a loss or gain of serrations and cusps, narrowing or broadening of the crown and changes in size.
5. Sexual dimorphism. For example, the teeth of **bigeye thresher shark** (*Alopias superciliosus*) females are broader than those of males. The upper teeth of large males of the **copper shark** (*Carcharhinus brachyurus*) are more slender and more curved (hooked) than in females. There may also be differences in serration patterns.

In addition to identifying the species of shark from tooth shape, it is often necessary to locate the position of an isolated tooth along the tooth row. This is no mean feat when, as in the case of the **spotted cat shark** (*Scyliorhinus canicula*), up to nine different tooth morphologies may be present in the upper and lower jaws (see Fig. 2.27). It may be necessary to run through a list of about 50 morphological features on a shark tooth before identifying its exact position along the tooth row of a specific shark.

In the descriptions that follow for sharks, the numbers of tooth files in the upper and lower dentitions are indicated after the Linnaean name. These are drawn mainly from the data of J A Bourdon and W Heim (www.elasmo.com) and from the Florida Museum of Natural History (www.flmnh.ufl.edu/fish/Sharks/sharks.htm), with a few modifications that reflect observations on the specimens available to us. Each DF provides, first, the number of teeth in one-half of the upper jaw, separated by "/" from, second, the number in one-half of the lower jaw. If symphysial teeth are present, this is indicated by the letter "S" before the number for the relevant jaw. It must be stressed that these data, obtained from a limited number of specimens (the ages and sexes of which were probably not known), are provided solely to give an idea of the number of teeth. Elasmobranchs do not, like mammals, have a fixed DF. Tooth number may vary with the age and size of the animal and can differ between geographic populations.

GALEOMORPH SHARKS

HETERODONTIFORMES

Heterodontidae

The bullhead sharks are represented by nine living species in a single genus, *Heterodontus*. All are relatively small bottom feeders living on a diet of shelled benthic mollusks and crustaceans.

Bullhead sharks derive their common name from the short, blunt head with high ridges above the eyes. Their generic name is related to the presence of teeth of different shapes along the jaws, exemplified by the dentition of the **Port Jackson shark** (*Heterodontus portusjacksoni*: DF 15/14) (Fig. 2.4). The anterior teeth are pointed and used for grasping, whereas the rhomboidal teeth in the back half of the jaws provide a convex crushing surface. The most anterior four to five files of crushing teeth are small and are succeeded by two files of much larger crushing teeth located midway along the jaw. Tooth size then decreases in the succeeding portion of the jaw. The prey is grasped by anterior teeth and then transferred to the back teeth, where it is crushed before being swallowed. Reflecting this diet, the jaws are robust and mineralized and there are large jaw adductor muscles. However, in *Heterodontus* the midline symphyses of the upper and lower jaws are not solidly mineralized, as they are in the durophagous myliobatid rays (see page 31).

Similar to the Port Jackson shark, the **zebra bullhead shark** (*Heterodontus zebra*: DF 12/S13) is also heterodont in both the upper and lower jaws. However, it can be distinguished by its carinated crushing posterior teeth and by its three-pointed anterior teeth (Fig. 2.5).

ORECTOLOBIFORMES

This order contains more than 40 species in seven families and is very diverse, ranging from the whale shark to small carpet and bamboo sharks.

Ginglymostomatidae

Nurse sharks (eg, *G. cirratum*: DF 17-18/16-17) have sharp pointed teeth with accessory cusps at each side. Three rows of teeth may be seen at the front of the lower jaw. The teeth are used for grasping prey, including teleosts and crustaceans, that are captured by suction as the shark cruises over the seabed. As the bases of the teeth do not overlap very much, the teeth may be replaced individually (Fig. 2.6).

Hemiscyllidae

This family comprises small, bottom-living sharks. The **brown-banded bamboo shark** (*Chiloscyllium punctatum*: DF 14-15/13-14) feeds on shrimp, scallops and squid plus small fishes. The teeth have an exceedingly simple, conical shape, forming a pavement used for grasping the prey (Fig. 2.7). The feeding mechanics in the related **white-spotted bamboo shark** (*Chiloscyllium plagiosum*) have been extensively studied (Motta and Huber, 2012).

Rhincodontidae

The **whale shark** (*Rhincodon typus*) is the sole living member of this family. It is the largest of all fishes, reaching lengths of up to 20 m. It may live for 60 years or more. The whale shark is a slow-moving fish that feeds on plankton. As it swims, large amounts of seawater flow into the huge mouth cavity

(Fig. 2.8), propelled partly by the fish's forward motion and partly by suction generated in the buccal and pharyngeal cavities. Plankton is recovered by filtration through 20 specialized pads, described in detail by Motta and Huber (2012). The pads fill the pharyngeal openings and water must therefore traverse the filters, which have openings of about 1 mm, before passing over the gills and leaving the body through the gill slits. In the mouth, whale sharks possess about 300 files of small teeth in each jaw, each with a hooked, single-point cusp (Fig. 2.9). The function of these teeth is unknown, but they may be used to help grasp hold of another individual during mating.

LAMNIFORMES

This group comprises 17 living species in seven families.

Lamnidae

The Lamnidae (mackerel or white sharks) include some well-known, fast-swimming sharks.

The **great white shark** (*Carcharodon carcharias*: DF 12-13/11-13) has a most fearsome reputation as it is the species responsible for most attacks on humans (Nambier et al., 1991). Females reach 4.5–5.0 m in length and the smaller males 3.5–4.0 m. The prey of the great white shark includes not only teleost and elasmobranch fishes, but also a relatively high proportion of large sea mammals, such as dolphins, sea lions, and seals. Only one row of teeth is visible in the upper jaw, but in the center of the lower jaw two or three rows may be visible (Fig. 2.10). The teeth, which may reach a length of 8 cm, are triangular with serrated margins (Fig. 2.11).

Young great white sharks have narrower teeth than older individuals. In addition, the teeth of very young great white sharks have a small but distinct basal cusplet on either side of the main blade. This morphology is suited to grasping slippery-bodied, small prey that can be swallowed whole.

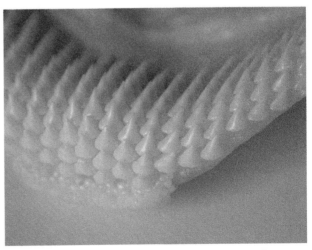

FIGURE 2.7   Teeth of brown-banded bamboo shark (*Chiloscyllium punctatum*). *Courtesy Dr. C Underwood.*

FIGURE 2.8 Whale shark (*Rhincodon typus*), showing the large, terminal mouth opening, through which water passes to the pharynx, where suspended plankton and other organisms are collected on filters attached to the gills. ©Izenbar/Dreamstime.com.

FIGURE 2.9 Numerous small, pointed teeth of in the dentition of whale shark (*Rhincodon typus*). *Courtesy Stuart Humphries. ©Australian Museum.*

The broader teeth of older great white sharks are adapted to slicing bite-sized gobbets of flesh from creatures too large to be swallowed whole. Side-to-side movements of the head help the shark excise large lumps of flesh from its prey and it seems that, when great white sharks feed on large prey such as seals, the bite is oblique, using the anterolateral teeth rather than head-on, using the anterior teeth (Martin et al., 2005).

In the next two species illustrated, the teeth are pointed rather than blade-like and lack cutting edges. They are more suited for use in feeding by grasping and tearing prey than by slicing through flesh.

In **porbeagle shark** (*Lamna nasus*: DF 14/11-14), the teeth are spaced apart and each tooth has a main narrow central point with small cusps on either side at the base (Fig. 2.12). The variations in tooth size along the jaws are extreme.

The anterior two to three files of teeth of **shortfin mako shark** (*Isurus oxyrinchus*: DF 13-14/12-14) are highly elongated, with a double curve (Fig. 2.13). Two or three rows of anterior teeth are in function at once. The posterior teeth are shorter and more blade like with smooth edges. This is the fastest shark and is capable of reaching a speed of 32 km/h. The method of feeding is to launch a high-speed lunge at the prey (mainly cephalopods and large bony fishes) and to tear off lumps of flesh.

Alopiidae

Thresher sharks can be easily recognized by the long upper lobe of the caudal fin, which they use as a weapon in catching fish and cephalopods. There are three living species. The teeth of the **common thresher** (*Alopias vulpinus*: DF 20-21/16-21) are triangular and blade like, and they are adapted for cutting, like those of the great white shark, although they are not serrated. The upper teeth are curved and point distally, whereas the lower teeth increasingly approach this morphology posteriorly (Figs. 2.14 and 2.15).

Odontaspididae

The **grey nurse shark** (*Carcharias taurus*: DF 12-13/11-13) has prominent anterior teeth with a narrow central point and small accessory cusps. More than one tooth row is visible

FIGURE 2.10   Great white shark (*Carcharodon carcharias*) showing multiple rows of teeth in the lower jaw. *Courtesy Wikipedia.*

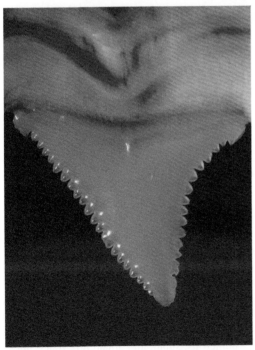

FIGURE 2.11   Single tooth of great white shark (*Carcharodon carcharias*). Note the serrated margins. *Courtesy NEFSC/NOAA.*

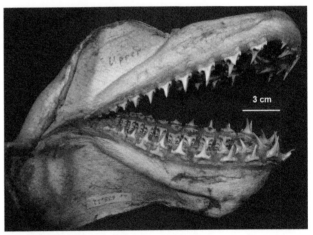

FIGURE 2.12   Dentition of porbeagle shark (*Lamna nasus*), showing extreme gradation of tooth size from the front to the back of the jaws. Note the spacing of the teeth and the small accessory cusps at the base. *Courtesy RCSOMA/ 433.11.*

Mitsukurinidae

The **goblin shark** (*Mitsukurina owstoni*: DF 26/24) is a deep-sea shark and is the only living species of this family. It possesses an elongated, flattened snout that bears numerous ampullae of Lorenzini and is thought to detect the weak electric fields emitted by prey (Fig. 2.17). The highly protrusible jaws can be extended forward rapidly. This generates considerable suction and facilitates capture of prey. The prey is then immobilized by the widely spaced, needle-like teeth.

in the mouth (Fig. 2.16).The grey nurse shark is ovoviviparous, and several embryos start to develop in each uterus. As the embryos can move around and their teeth erupt in utero, the result is cannibalism and only one embryo survives in each uterus (**embryophagy**).

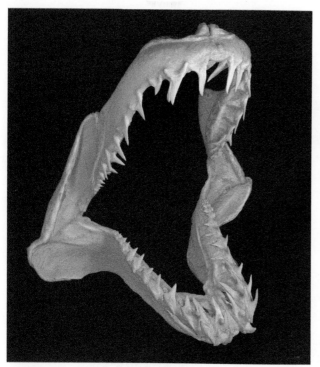

FIGURE 2.13 Dentition of mako shark (*Isurus oxyrinchus*). *Courtesy Wikipedia.*

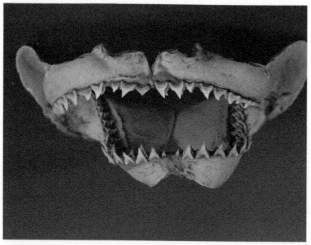

FIGURE 2.14 Dentition of common thresher shark (*Alopias vulpinus*). Image width = 18 cm. *Courtesy RCSOMA/ 433.9.*

Cetorhinidae

The **basking shark** (*Cetorhinus maximus*) is the only living species of this family. It is the second largest fish in the ocean after the whale shark and, like that species, it feeds on microscopic planktonic animals and plants. The mouth can be opened very wide, and water intake is driven by the fish's forward motion. Prey is filtered from the water by densely packed gill rakers, spaced a little less than 1 mm apart. A large number of very small teeth are found on the jaws (Fig. 2.18), but their function, if any, is unknown.

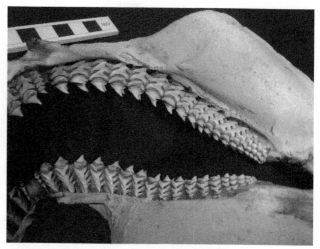

FIGURE 2.15 Jaws of common thresher shark (*Alopias vulpinus*), viewed from the medial side, showing one to two functional teeth and numerous replacement teeth on the inside of the jaw. Marker bar in 1 cm divisions. *Courtesy RCSOMA/ 433.9.*

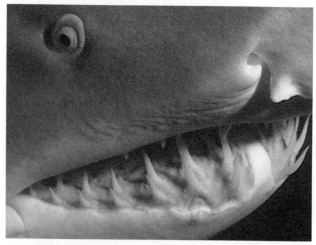

FIGURE 2.16 Mouth of grey nurse shark with two to three functional teeth visible at each tooth position. *Courtesy Richard Vevers.*

Megachasmidae

The **megamouth shark** (*Megachasma pelagios*), the only living species of this family, was first described in 1975. It is a deep-water filter feeder that uses a combination of ram, suction, and engulfment to fill its buccopharyngeal cavity with water. This water is then forced out between closely spaced gill rakers and then over the gills. Megamouth sharks possess numerous, small, and apparently functionless teeth (Fig. 2.19) (Yabumoto et al., 1997; Yano et al., 1997). It is the smallest of the filter-feeding sharks.

CARCHARHINIFORMES

The ground sharks, containing more than 270 species in eight families, are the largest order of sharks.

FIGURE 2.17 Head of goblin shark (*Mitsukurina owstoni*), showing the elongated snout. The jaws, armed with curved, needle-like teeth, are in the protruded position. *Courtesy Wikipedia.*

Hemigaleidae

The **snaggletooth shark** (*Hemigaleus elongata*: DF 14-15/17-18) has a heterodont dentition in which tooth shape varies both within the jaws and between upper and lower jaws. Near the symphysis of the upper jaw there are a few, small, pointed teeth, whereas the rest of the jaw is occupied by distally curved, blade-like teeth with serrated edges (Fig. 2.20A). In the lower jaw, the anterior region is occupied by several rows of elongated, pointed teeth with poorly developed mesial and distal edges and with small lateral cusps. Posteriorly, these teeth grade into blade-like teeth through a process of height reduction, labiolingual compression, and accentuation of the edges (Fig. 2.20B).

Sphyrnidae

There are about nine species of hammerhead sharks; these species are characterized by the unusual shape of the head, with the eyes mounted on the sides of the "hammer." Variations in tooth form between different species reflect differences in diet. The teeth of hammerheads are typically triangular. Teeth are unserrated and relatively homodont in the **smooth hammerhead shark** (*Sphyrna zygaena*: DF 13-15/12-14)) (Figs. 2.21 and 2.22), an active predator on bony fishes and invertebrates (especially cephalopods). The **great hammerhead** (*S. mokarran*: DF 17/16-17) preys extensively on rays and other sharks and has teeth with serrated edges.

Bonnethead shark (*Sphyrna tiburo*: DF S12-14/S12-13), the smallest of the hammerhead sharks, is distinguished

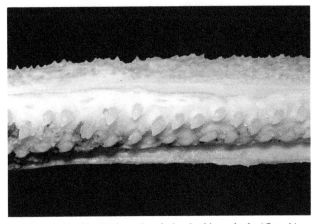

FIGURE 2.18 Vestigial teeth of the basking shark (*Cetorhinus maximus*). *Courtesy Dr. L. Marshall, Sciences Department, Museum of Victoria, Australia.*

from other members of the family by the flattened, bonnet-shaped head that is rounded between the eyes rather than being hammer shaped. Unlike the majority of sharks that eat fish, bonnetheads eat mainly crabs. Their dentition reflects this diet as the teeth are not serrated and show differences in shape along the length of the jaw (Fig. 2.23). The front teeth are symmetrical and have a sharp central cusp. In the middle of the tooth row the cusp is obliquely inclined distally, whereas at the back of the tooth row the teeth form flattened blades for crushing the shells of the prey. However, the bite force is much lower than in other durophagous sharks (Mara et al., 2010).

(A)

(B)

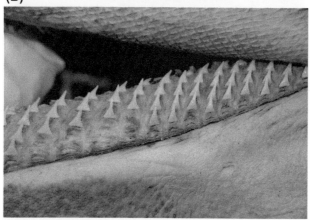

FIGURE 2.19 Dentition of megamouth shark (*Megachasma pelagios*). (A) Dentition; (B) Lower teeth. *Courtesy Dr. G. Moore, Western Australia Museum. http://commons.wikimedia.org/wiki/File:Megamouth shark preservation tank WAMM.jpg.*

(A)

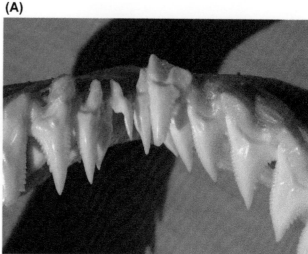

(B)

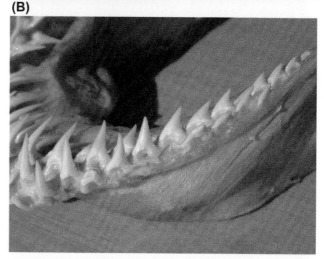

FIGURE 2.20 Snaggle-tooth shark (*Hemipristis elongatus*). (A) Teeth in the symphyseal region of the upper jaw. (B) Teeth in the lower jaw. Note the more blade-like appearance of the tooth lying posteriorly at the right side of the image. *Courtesy Dr. C Underwood.*

Carcharinidae

The requiem sharks are medium-to-large, active predators, capable of dealing with prey of considerable size. Some have been known to attack humans. The dentition of the **bull shark** (*Carcharhinus leucas*: DF S12-14/S12-13) is typical for the group, consisting of broad, triangular, and heavily serrated upper teeth (Figs. 2.24 and 2.25) and narrow, triangular lower teeth with fine serrations and broad bases. Bull sharks are regarded as having the highest mass-specific bite force (Habegger et al., 2012), even exceeding that of the great white shark and the great hammerhead.

The **sandbar shark** (*Carcharinus plumbeus*: DF S13-15/S12-15) has triangular, serrated, blade-like teeth in both jaws, although those in the upper jaw are broader and more deeply serrated than those in the lower jaw. Similar to grey nurse sharks (*Carcharias*), sandbar sharks are ovoviviparous, but only one embryo survives in utero cannibalism.

Tiger shark (*Galeocerdo cuvier*: DF 11/10), so named because of the series of dark spots and vertical lines visible on the body, is a voracious hunter taking

a wide variety of prey: crustaceans, elasmobranchs, and tetrapods, including turtles. Both upper and lower jaws surrounding the wide mouth are furnished with distinctive teeth that incorporate shredding and sawing components. The teeth are broad with highly serrated cutting edges and with tiny, secondary serrations overprinting the larger serrations. The main (mesial) blade is triangular and slopes distally. It is separated from a round subsidiary distal blade by a notch. The serrations on the mesial blade become finer toward the tip, whereas the edge of the distal blade has fewer, but much more pronounced serrations (Fig. 2.26).

Scyliorhinidae

Cat sharks are small, benthic sharks that feed on a variety of fishes and invertebrates. Many species seem to specialize in certain foods, such as crustaceans or cephalopods.

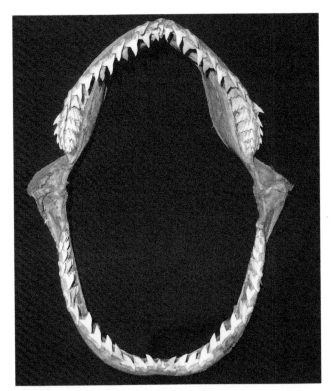

FIGURE 2.21 Dentition of smooth hammerhead shark (*Sphyrna zygaena*). Image width = 16 cm. *Courtesy RCSOMA/ 435.1.*

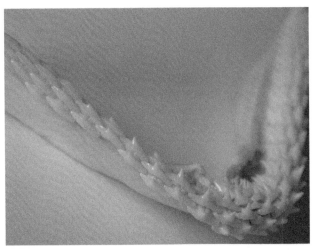

FIGURE 2.23 Part of lower dentition of bonnet-head shark (*Sphyrna tiburo*), showing pointed, grasping, anterior teeth grading into low-cusped, posterior teeth that form a crushing pavement. *Courtesy Dr. C Underwood.*

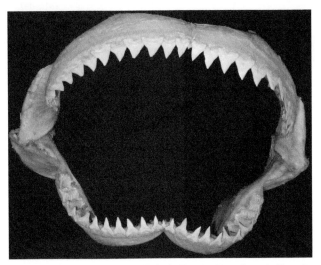

FIGURE 2.24 Dentition of bull shark (*Carcharhinus leucas*). Note the broad, triangular upper teeth and the narrow triangular lower teeth. Image width = 36 cm. *Courtesy Dr. C Underwood.*

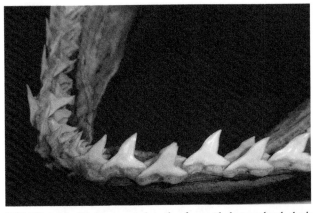

FIGURE 2.22 Weakly serrated teeth of smooth hammerhead shark (*Sphyrna zygaena*). Image width = 15 cm. *Courtesy RCSOMA/ 435.1.*

The **spotted cat shark** (*S. canicula*: DF S21-23/S20-24) has several visible rows of teeth that are used to grasp and tear prey. Tooth morphology varies extensively, both within and between the jaws (Fig. 2.27) and between males and females (Fig. 2.28). In total, nine different tooth types have been recorded, ranging from uni- or tricuspid with a prominent central cusp to low crowned with five cusps of similar size. The teeth near the midline are small, narrow, and triangular. The adjacent teeth are the largest in the dentition and have a tall central cusp. Posteriorly, tooth size decreases, but additional cusps may be added. Sexual dimorphism of

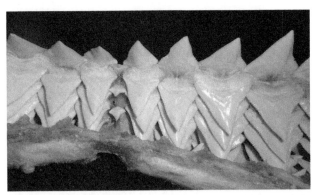

FIGURE 2.25 Higher power view of the upper jaw dentition of the bull shark (*Carcharhinus leucas*) viewed medially and demonstrating the replacing teeth. Note the serrated edges. Image width = 8 cm. *Courtesy Dr. C Underwood.*

the adult dentition seems to be related to the fact that males grasp females with their teeth during copulation, and not to dietary differences (Ellis and Shackley, 1995; Crooks, 2011). Males have longer, narrower mouths than females. In males the upper teeth are mostly unicuspid, with a small additional cusp on some posterior teeth, whereas in the lower jaw tooth form varies from uni- or bicuspid anteriorly to pentacuspid posteriorly. In females all the teeth have four or five cusps. The largest anterior teeth are taller in males.

Triakidae

This family consists of about 40 species in nine genera. Most smooth-hound sharks include a large proportion of crustaceans, such as crabs and lobsters, in their diets. The dentition of the **gummy shark** (*Mustelus antarcticus*: DF

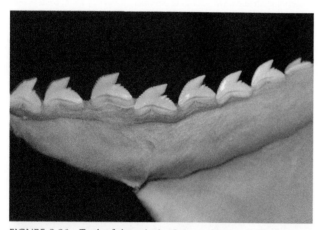

FIGURE 2.26 Teeth of tiger shark (*Galeocerdo cuvier*). The teeth are broad with highly serrated, cutting edges. The main mesial (anterior) blade is triangular and is separated from a round subsidiary distal blade by a notch. *Courtesy Coolwater Photos.*

31-32/33-35) reflects this diet, as the low-cusped teeth form a crushing pavement (Fig. 2.29), like the dentitions of rays that eat similar organisms (see Chapter 3). The dentitions are similar in upper and lower jaws. In contrast, the **whiskery shark** (*Furgaleus macki*: DF S15/19), which feeds mainly on octopus, is heterodont. The lower teeth have a single low, sharp point (Fig. 2.30A), whereas the upper teeth, apart from three files of single-pointed teeth in the symphysial region, possess a smooth oblique mesial edge and a serrated distal margin bearing four to six cusps (Fig. 2.30B).

SQUALEAN SHARKS
HEXANCHIFORMES

This order contains the most primitive sharks, with two families and six living species. They possess six or seven gill slits, in contrast to the five slits in most sharks, and they have only one dorsal fin.

Chlamydoselachidae

There are two species of frilled sharks. The **six-gill frilled shark** (*Chlamydoselachus anguineus*: DF 16/13) is a deep-water species that feeds predominantly on cephalopods. Its dentition consists of spaced, three-cusped teeth in both jaws (Figs. 2.31 and 2.32).

Hexanchidae

There are four species of cow sharks. The **seven-gill shark** (*Notorynchus cepedianus*: DF 7/S6) feeds on both teleost and elasmobranch fishes and possibly also on mammals. There are considerable differences between the teeth at the front and back of the jaws, as well as differences between the upper and lower teeth. In the upper jaw, there are small

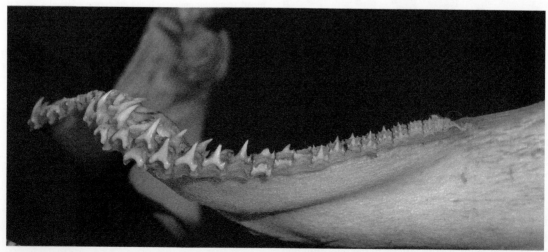

FIGURE 2.27 Teeth in the lower jaw of spotted cat shark (*Scyliorhinus canicula*). Narrow, triangular teeth are present near the midline, whereas further back in the tooth row a variable number of low accessory points are seen. Actual image size = 6 cm. *Courtesy MoLS KCL.*

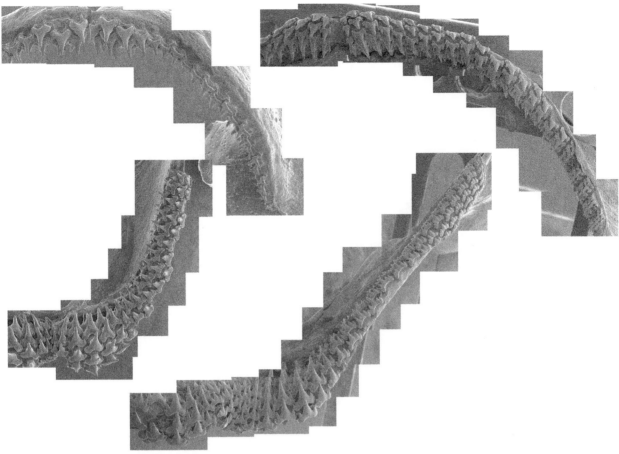

FIGURE 2.28 Montages of scanning electron micrographs showing the dentition of spotted cat shark (*Scyliorhinus canicula*). Left, male; right, female. Upper jaws to top; lower jaws to bottom. *Courtesy Dr. C. Underwood.*

parasymphysial teeth, followed by seven teeth that gradually increase in breadth, with the addition of further cusps distally (Fig. 2.33), then by a suite of very small, low teeth at the back of the jaw. In the lower jaw, the symphysial tooth is large and bilaterally symmetrical, whereas the following six teeth are broad, saw-like teeth with five points of descending size (Fig. 2.34). In both jaws, the teeth at the back of the tooth row are very small and flattened (Fig. 2.35); these small teeth do not appear in the DF quoted here. **Bluntnose six-gill shark** (*Hexanchus griseus*: DF 10/S6) has a similar dentition to the seven-gill shark, except that the six very broad, saw-like teeth at the front of the lower jaw may have 10 or more sharp cusps in a graded series (Fig. 2.36).

ECHINORHINIFORMES

This order contains just two species—prickly sharks and bramble sharks—in a single family. These sharks feed on teleost and elasmobranch fishes.

Echinorhinidae

In young **prickly shark** (*Echinorhinus cookei*: DF 21-25/20-27) the teeth have a single oblique, triangular blade. In

FIGURE 2.29 Low-cusped teeth of the gummy shark (*Mustelus antarcticus*), forming a close-set pavement. Actual length of tooth rows approximately 3 cm. *Courtesy Dr. C Underwood.*

adults, the teeth acquire a saw-like form by addition of smaller cusps on both sides of the main blade (Fig. 2.37). The dentition of the **bramble shark** (*Echinorhinus brucus*: DF 11/11) is very similar.

(A)

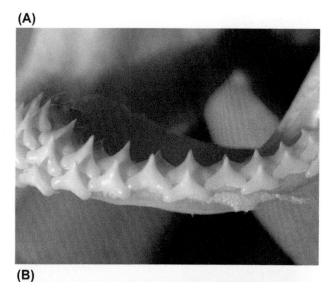

(B)

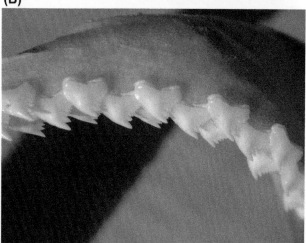

FIGURE 2.30 Whiskery shark (*Furgaleus macki*). (A) Lower teeth with a single low, sharp point. (B) Upper teeth possess a smooth oblique mesial edge and a serrated distal margin bearing four to six cusps. *Courtesy Dr. C Underwood.*

FIGURE 2.31 Head of frilled shark (*Clamydoselachus anguineus*). Note the spaced three-cusped teeth in both jaws. *Courtesy Carl Bento.* ©*Australian Museum.*

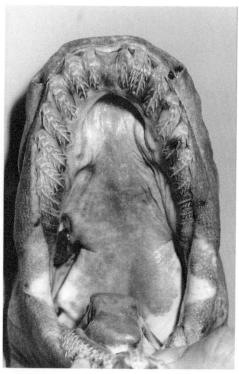

FIGURE 2.32 Ventral view of upper jaw of frilled shark (*Clamydoselachus anguineus*). Note the spaced three-cusped teeth. *Courtesy Carl Bento.* ©*Australian Museum.*

FIGURE 2.33 Fourth tooth (left side) to the seventh (right side) in the upper jaw of the seven-gill shark (*Notorynchus cepedianus*). Note the increase in size and number of cusps. *Courtesy Dr. C Underwood.*

SQUALIFORMES

This order of sharks contains more than 125 species in six families.

Dalatiidae

Kitefin sharks of the family Dalatiidae comprise seven genera, of which five are monotypic, representing the highest

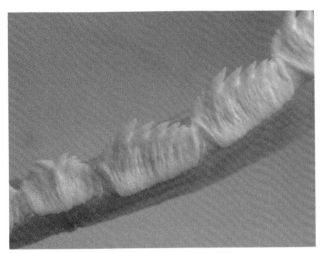

FIGURE 2.34 Symmetrical symphysial tooth (left) and the first two large teeth (right) in the lower jaw of the seven-gill shark (*Notorynchus cepedianus*). *Courtesy Dr. C Underwood.*

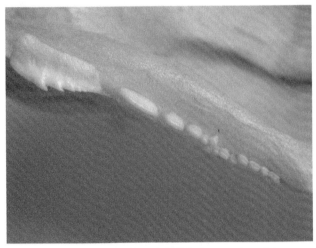

FIGURE 2.35 Broad seventh tooth (left) in the upper jaw of the seven-gill shark (*Notorynchus cepedianus*), posterior to which is a row of small teeth. *Courtesy Dr. C Underwood.*

percentage of monotypic genera for any family in the Order Squaliformes.

Kitefin shark (*Dalatias licha*: DF 9/9) is a small- to medium-sized, short-snouted shark preying mainly on fish, but also on a variety of invertebrates. The dentition is heterodont, in that the teeth in the upper and lower jaws exhibit different morphologies (Fig. 2.38). In the upper jaw the teeth have a single narrow, distally curved point with smooth edges (Fig. 2.39). The lower jaw is massive and bears a row of teeth that are very large in relation to the body size. The teeth are erect and triangular with serrated edges and are so closely interlocked at their bases (Fig. 2.40) that the tooth row functionally presents a single, continuous, serrated edge reminiscent of the dentition in the lower jaw of piranhas (see page 64). The close interlocking of the lower teeth means that they cannot be replaced individually, but only as a complete row.

The **cookie-cutter shark** (*Isistius brasiliensis*: DF 15-19/12-16) is a small relative of the kitefin shark, but it feeds in the manner of an ectoparasite. The dentition (Fig. 2.41) resembles that of the kitefin shark, but the mouth is surrounded by suctorial lips. Prey is attracted by numerous photopores on the ventral surface of the shark. After applying its suctorial lips to the skin of the prey, the shark bites and then, by rotating the body through 360°, cuts out a cone of flesh that is retained by the lips and by the pointed upper teeth. The skin of even large whales is attacked in this way (Fig. 2.42). A similar species, the **large cookie-cutter shark** (*Isistius plutodus*) has fewer teeth (DF 15/10) but, relative to body size, these teeth are said to be larger than in any other shark.

The exfoliated tooth rows of *Isistius* seem to be swallowed rather than lost into the ocean, perhaps allowing the minerals in the teeth to be recycled.

The diminutive **pocket shark** (*Mollisquama parini*) can be considered the rarest of all sharks, as only two specimens have ever been found. It is related to *Isistius* and *Dalatias*. The holotype, a larger female, was caught in the southeast Pacific Ocean, and the second pocket shark was a very small juvenile male from the Gulf of Mexico. The second specimen has been described in detail by Grace et al. (2015). The pocket shark derives its name from the presence of a conspicuous villi-lined internal pocket located just above each pectoral fin base. The juvenile specimen is just more than 8 cm in length. The dental formula is S9/S15 (Fig. 2.43A). The dentition is heterodont. The less numerous teeth on the upper jaw are more slender and shorter than the broader and longer teeth on the lower jaw (Fig. 2.43B and C). The upper teeth have a basal plate with posteriorly directed wings, whereas in the lower teeth the basal plate is indistinguishable from the body of the tooth. These morphological differences seem to be correlated with greater tooth mobility in the lower jaw (Grace et al., 2015). The teeth become smaller toward the back of the jaw. The lower teeth have high triangular crowns that curve increasingly backwards from the midline toward the mouth corners. The teeth are not serrated but irregularly spaced, and shallow notches are present on the crowns of the lower teeth. Whether the juvenile is a separate species awaits further clarification.

Oxynotidae

There are five species in this family, commonly known as rough sharks. Like the cookie-cutter shark, the **spiny or prickly dogfish** (*Oxynotus bruniensis*: DF 9-10/S7) has a fleshy-lipped mouth, small upper teeth with narrow upright cusps, and larger lower teeth that are broad, triangular, and knife-like (Fig. 2.44).

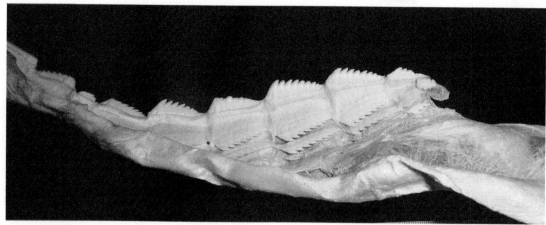

FIGURE 2.36 Multicusped, saw-like teeth viewed medially on the right side of the lower jaw of bluntnose six-gill shark (*Hexanchus griseus*). Symphysial tooth arrowed. Image width=18 cm. *Courtesy Dr. C Underwood.*

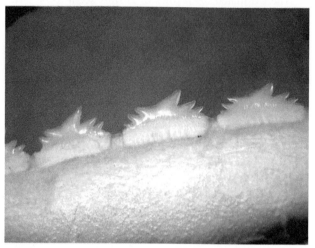

FIGURE 2.37 Teeth of prickly shark (*Echinorhinus cookei*). Note the presence of small cusps on both sides of the main blade. *Courtesy Wikipedia.*

FIGURE 2.39 Pointed teeth in upper jaw of kitefin shark (*Dalatias licha*). *Courtesy Dr. C Underwood.*

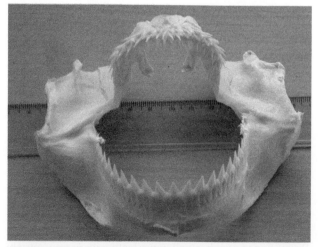

FIGURE 2.38 Jaws of kitefin shark (*Dalatias licha*). Note the different size and morphologies between upper and lower teeth. Background scale in 1-mm divisions. *Courtesy Dr. C Underwood.*

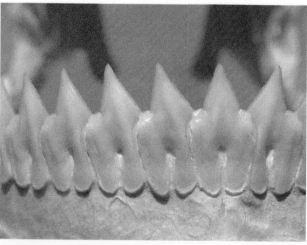

FIGURE 2.40 Close-set triangular teeth in lower jaw of kitefin shark (*Dalatias licha*). *Courtesy Dr. C Underwood.*

FIGURE 2.41 Jaws of a cookie-cutter shark. *Courtesy Dr. M. J. Miller.*

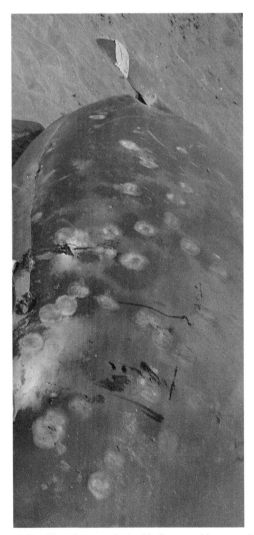

FIGURE 2.42 Bites from attacks by kitefin or cookie-cutter shark on the skin of Gray's beaked whale (*Mesoplodon grayi*). *Courtesy Wikipedia.*

Etmopteridae

This family contains the small lantern sharks, so-called because of the presence of light-producing photophores on their bodies. There are 45 species in five genera. The diet of most species is dominated by teleosts and cephalopods. Most species have a strongly heterodont dentition rather similar to that of the dalatiids, but there are exceptions.

The **viper dogfish** (*Trigonognathus kabeyai*: DF S10/S10) was first described in 1986 and is localized mainly to the deep seas around Japan (Mochizuki and Ohe, 1990: Shirai and Okamura, 1992). The teeth are distinctively fang like and widely spaced (Fig. 2.45). When the mouth is closed, the upper symphyseal tooth overlaps the lower, whereas the upper and lower lateral teeth interdigitate. The viper dogfish feeds primarily on teleosts. Its teeth are adapted to grasp rather than cut up prey, which is swallowed whole. The jaws can protrude and open unusually widely for a shark (Fig. 2.45), allowing it to swallow prey up to 40% of its own length.

Centrophoridae

This family contains about 15 species. The **gulper shark** (*Centrophorus granulosus*: DF 17/S14) has closely set teeth in both jaws that form interconnected blades (Fig. 2.46). In the upper jaw, the triangular teeth are symmetrical and upright in front (Fig. 2.47), but slope slightly obliquely further back in the tooth row. In the lower jaw the teeth are broader, with distally sloping triangular main blades and rounded distal subsidiary blades. The mesial part of the main blade overlaps the distal blade of the tooth in front (Fig. 2.48), an arrangement very similar to that in piranhas (page 64).

SQUATINIFORMES

This order contains a single family, Squatinidae, with 23 species of angel sharks.

Squatinidae

Angel sharks (*Squatina*: DF 20/18) have a flattened body and large pectoral fins, which makes them resemble rays in general form (Fig. 2.49A). They are ambush predators and feed near the seabed, particularly on flatfishes, cephalopods, and crustaceans. The teeth are spaced apart and each has a broad base and a narrow, smooth point (Fig. 2.49B).

PRISTIOPHORIFORMES

Sawsharks have the general body shape of a shark with the gills at the side of the head, but they resemble sawfishes (Chapter 3) in that they have a long, toothed snout, which is used in feeding near the ocean floor. There are nine species in a single family.

(A)   **(B)**   **(C)**

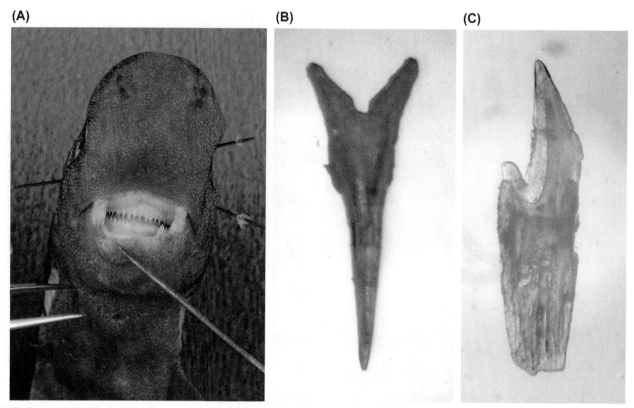

FIGURE 2.43   Pocket shark (*Mollisquama parini*). (A) Ventral view, showing upper and lower dentitions. Image width = 1.3 cm *(Courtesy Dr. M. H. Doosey.)*. (B) Isolated unicuspid upper tooth. Wings of basal plate form an acute angle. Image width = 8 mm *(Courtesy Dr. M. A. Grace.)*. (C) Flattened, blade-like lower tooth, with main cusp and subsidiary cusp. Basal plate not distinguishable from body of tooth. Image width = 10 mm *(Courtesy Dr. M. A. Grace.)*.

FIGURE 2.44   Dentition of spiny or prickly dogfish (*Oxynotus bruniensis*). *Courtesy NORFANZ Carl Bento: AMS1.39544-001.*

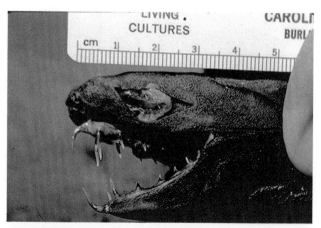

FIGURE 2.45   Viper shark (*Trigonognathus kabeyai*), with open mouth. *Courtesy Wikipedia.*

Pristiophoridae

The rostrum of sawsharks is armed on each side with a row of "teeth" that are really modified dermal spines (Figs. 2.50 and 2.51). The ventral surface of the rostrum is furnished with electro-receptors (ampullae of Lorenzini) and a pair of long, nasal barbels that, together, are used to sense prey

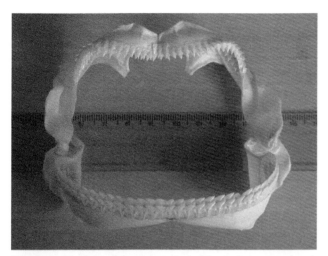

FIGURE 2.46 Jaws of the gulper shark (*Centrophorus granulosus*). Background scale in 1-mm divisions. *Courtesy Dr. C Underwood.*

FIGURE 2.47 Teeth in the upper jaw of the gulper shark (*Centrophorus granulosus*). *Courtesy Dr. C Underwood.*

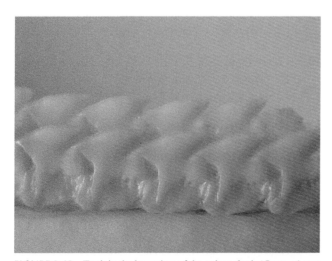

FIGURE 2.48 Teeth in the lower jaw of the gulper shark (*Centrophorus granulosus*). *Courtesy Dr. C Underwood.*

(A)

(B)

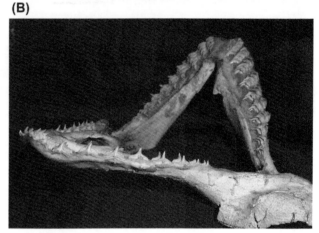

FIGURE 2.49 (A) Japanese angel shark (*Squatina japonica*), showing a flattened ray-like body form *(Courtesy Wikipedia.)*; (B) Jaws of angel shark (*Squatina squatina*). Image width = 14 cm. *(Courtesy RCSOMA/ 440.1.).*

buried in the sand. The rostrum is used to stir up the seabed to expose and to stun or impale the prey, which consists of small fish and squid. The sawshark then takes the prey into its mouth.

The rostral teeth vary in size. The largest teeth seem reasonably regularly spaced, between which are medium- and small-sized teeth (Welten et al., 2015). They are replaced throughout life and increase in number with age from about 23 pairs in the newborn to as many as 80 pairs in old, large specimens. This contrasts with the continuously growing rostral teeth of sawfishes (Chapter 3) which are much fewer in number, are of similar size and are not replaced (Slaughter and Springer, 1968). The rostral teeth of sawsharks develop parallel to the long axis of the rostrum and are directed posteriorly, but they flip up on erupting to become perpendicular to the long axis (Fig. 2.51) (Welten et al., 2015). The oral teeth are small and numerous and possess a single cusp (Fig. 2.52).

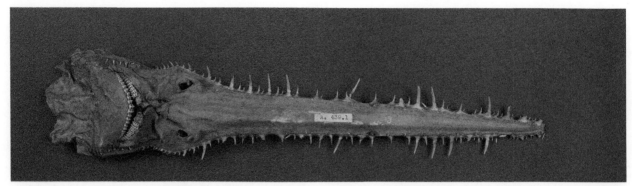

FIGURE 2.50 Rostrum of sawshark (*Pristiophorus* sp.), showing gradation in size of the rostral teeth. Image width=31 cm. *Courtesy RCSOMA/ 439.1.*

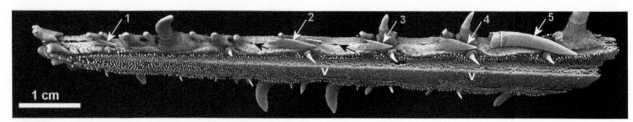

FIGURE 2.51 *Pristiophorus cirratus* adult. Lateral view of rostrum: computed tomography scan with three-dimensional rendering. Replacements for five large denticles (numbers, *white arrows*) at different stages of formation of crown and base. Replacement 1 (most distal) is the least advanced and replacement 5 (most proximal) is the most advanced. Note pore openings (*black arrows*) at the base of the hollows in the jaw cartilage marking the tooth loci. V=denticle of the ventral series. CU=unregistered specimen. *From Welten, M., Smith, M.M., Underwood, C., Johanson, Z., 2015. Evolutionary origins and development of saw-teeth on the sawfish and sawshark rostrum (Elasmobranchii; Chondrichthyes). R. Soc. Open Sci. 2, 150189, http://dx.doi. org/10.1098/rsos.150189, Courtesy Prof M. M. Smith and Editors Royal Society London.*

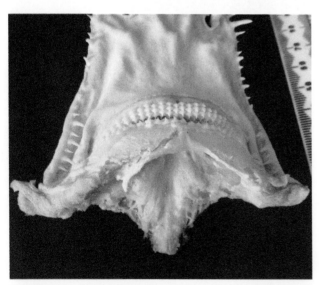

FIGURE 2.52 Oral teeth of sawshark (*Pristiophorus* sp.). Marker bar in 1-mm divisions. *Courtesy RCSOMA/ 439.1.*

The number of teeth on the saw of the sawshark differs in different species (Lange et al., 2015). In the **longnose shark** (*Pristiophorus cirratus*), there are nine teeth in front of the barbels and nine teeth behind. In the **Japanese sawshark** (*Pristiophorus japonicus*), there are 15–26 teeth in front of the barbels and 9–17 teeth behind. In the **shortnose sawshark** (*Pristiophorus nudipinnus*), there are 13 teeth in front of the barbels and 6 behind, whereas in the **Bahamas sawshark** (*Pristiophorus schroderi*), there are 13–14 teeth in front of the barbels and 9–10 teeth behind.

ONLINE RESOURCES

Elasmo.com: http://www.elasmo.com/selachin/slides/ss_dentition.html?sp=sharks. (Images and numerical data on dentitions and tooth morphology of numerous shark species.)

Elasmodiver Shark and Ray Field Guide: http://www.elasmodiver.com/sharks_and_rays.html

Florida Museum of Natural History: www.flmnh.ufl.edu/fish/Sharks/sharks.htm.

J-elasmo: http://homepage2.nifty.com/megalodon/index.htm. (Good-quality photographs of the dentitions of a number of sharks.)

Animal Diversity Web (University of Michigan Museum of Zoology): http://animaldiversity.ummz.umich.edu/accounts/Chondrichthyes/classification/

Reefquest Centre for Shark Research: http://www.elasmo-research.org/index.html

Sharksrays.org (for up-to-date classification)

Shark Trust: www.sharktrust.org

REFERENCES

Bruner, J.C., 1998. Tooth Replacement Rate of *Carcharodon carcharias* (Linnaeus, 1758). Abstract 305, AES 14th annual meeting. Program and Abstracts. Guelph, Ontario, Canada. pp. 200

Cortés, E., 1999. Standardised diet composition and trophic levels of sharks. ICES J. Mar. Sci. 56, 707–717.

Crooks, N., 2011. Sexual and Seasonal Dimorphisms in the Dermal, Dental and Ampullary Structures of the Lesser-spotted Catshark, *Scyliorhinus canicula* (Ph.D. thesis). University of Portsmouth.

Ellis, J.R., Shackley, S.E., 1995. Ontogenic changes and sexual dimorphism in the head, mouth and teeth of the lesser spotted dogfish. J. Fish. Biol. 47, 155–164.

Grace, M.A., Doosey, M.H., Bart, H.L., Naylor, G.J.P., 2015. First record of *Mollisquama* sp. (Chondrichthyes: Squaliformes: Dalatiidae) from the Gulf of Mexico, with a morphological comparison to the holotype description of *Mollisquama parini* Dolganov. Zootaxa 3948, 587–600.

Habegger, M.L., Motta, P.J., Huber, D.R., Dean, M.N., 2012. Feeding biomechanics and theoretical calculations of bite force in bull sharks (*Carcharhinus leucas*) during ontogeny. Zoology 115, 354–364.

Herman, J., Hovestadt-Euler, M., Hovestadt, D.C., 1987. Contributions to the study of the comparative morphology of teeth and other relevant ichthyodolurites in living supraspecific taxa of chondrichthyan fishes. In: Stehmann, M. (Ed.), Bulletin de l'Institut Royal des Sciences Naturelles de Belgique, Biologie 57, pp. 43–56 1988; 58:99–126, 1989; 59:101–157, 1990; 60:181–230, 1991; 61:73–120, 1992; 62:193–254, 1993; 63:185–256.

Lange, T., Brehm, J., Moritz, T., 2015. A practical key for the identification of large fish rostra. Spixiana 38, 145–160.

Luer, C.A., Blum, P.C., Gilbert, P.W., 1990. Rate of tooth replacement in the nurse shark, *Ginglymostoma cirratum*. Copeia 182–191.

Mara, K.R., Motta, P.J., Huber, D.R., 2010. Bite force and performance in the durophagous bonnethead shark, *Sphyrna tiburo*. J. Exp. Zool. A Ecol. Genet. Physiol. 313, 95–105.

Martin, R.A., Hammerschlag, N., Collier, R.S., Fallows, C., 2005. Predatory behaviour of white sharks (*Carcharodon carcharias*) at Seal Island, South Africa. J. Mar. Biol. Assoc. U.K. 85, 1121–1135.

Mochizuki, K., Ohe, F., 1990. *Trigonognathus kabeyai*, a new genus and species of the squalid sharks from Japan. Jap J. Ichthyol. 36, 385–390.

Motta, P.J., Huber, D.R., 2012. Prey capture behaviour and feeding mechanics of elasmobranchs. In: Carrier, J.C., Musick, J.A., Heithaus, M.R. (Eds.), Biology of Sharks and Their Relatives, second ed. CRC Press, Boca Raton, pp. 153–209.

Nambier, P., Bridges, T.E., Brown, K.A., 1991. Allometric relationships of the dentition of the Great White Shark, *Carcharodon carcharias*, in forensic investigations of shark attacks. J. Forensic Odontostomatol. 1991 (9), 1–16.

Shellis, R.P., 1982. Comparative anatomy of tooth support. In: Berkovitz, B.K.B., Moxham, B., Newman, H.N. (Eds.), The Periodontal Ligament in Health and Disease. Pergamon, Oxford, pp. 3–24.

Shirai, S., Okamura, O., 1992. Anatomy of *Trigonognathus kabeyai*, with comments on feeding mechanism and phylogenetic relationships (Elasmobranchii, Squalidae). Jap J. Ichthyol. 39, 139–150.

Slaughter, B., Springer, S., 1968. Replacement of rostral teeth in sawfishes and sawsharks. Copeia 499–506.

Springer, S., 1960. Natural history of the sandbar shark *Eulamia milberti*. U.S. Fish. Wildl. Serv. Fish. Bull. 61, 1–38.

Strasburg, D.W., 1963. The diet and dentition of *Isistius brasiliensis*, with remarks on tooth replacement in other sharks. Copeia 33–40.

Welten, M., Smith, M.M., Underwood, C., Johanson, Z., 2015. Evolutionary origins and development of saw-teeth on the sawfish and sawshark rostrum (Elasmobranchii; Chondrichthyes). R. Soc. Open Sci. 2, 150189. http://dx.doi.org/10.1098/rsos.150189.

Wetherbee, B.M., Cortés, E., 2012. Food consumption and feeding habits. In: Carrier, J.C., Musick, J.A., Heithaus, M.R. (Eds.), Biology of Sharks and Their Relatives, second ed. CRC Press, Boca Raton, pp. 225–246.

Whitenack, L.B., Motta, P.J., 2010. Performance of shark teeth during puncture and draw: implications for the mechanics of feeding. Biol. J. Linn. Soc. 100, 271–286.

Wilga, C.D., 2002. A functional analysis of jaw suspensions in elasmobranchs. Biol. J. Linn. Soc. 75, 483–502.

Wilga, C.D., 2005. Morphology and evolution of the jaw suspension in lamniform sharks. J. Morphol. 265, 102–119.

Wilga, C.D., Hueter, R.E., Wainwright, P.C., Motta, P.J., 2001. Evolution of upper jaw protrusion mechanisms in elasmobranchs. Am. Zool. 41, 1248–1257.

Wilga, C.D., Motta, P.J., Sanford, C.P., 2007. Evolution and ecology of feeding in elasmobranchs. Integr. Comp. Biol. 47, 55–69.

Wilga, C.D., Sanford, C.P., 2008. Suction generation in white-spotted bamboo sharks *Chiloscyllium plagiosum*. J. Exp. Biol. 211, 3128–3138.

Yabumoto, Y., Goto, M., Yano, K., Uyeno, T., 1997. Dentition of a female megamouth, *Megachasma pelagios*, collected from Hakata Bay, Japan. In: Yano, K., Morrissey, J.F., Yabumoto, Y., Nakaya, K. (Eds.), Biology of the Megamouth Shark. Tokai University Press, Tokyo, Japan, pp. 63–75.

Yano, K., Yabumoto, Y., Ogawa, H., Hasegawa, T., Naganobu, K., Matumura, S., Misuna, Y., Matumura, K., 1997. X-ray observations on vertebrae and dentition of a megamouth shark, *Megachasma pelagios*, from Hakata Bay, Japan. In: Yano, K., Morrissey, J.F., Yabumoto, Y., Nakaya, K. (Eds.), Biology of the Megamouth Shark. Tokai University Press, Japan, pp. 21–29.

FURTHER READING

Castro, J.I., 1983. The Sharks of North American Waters. Texas A&M University Press, College Station, TX, pp. 35–36.

Compagno, L.J.V., 1984a. Sharks of the World. An Annotated and Illustrated Catalogue of Shark Species Known to Date, vol. 4. Hexanchiformes to Lamniformes, Rome, FAO. Part 1.

Compagno, L.J.V., 1984b. Sharks of the World. An Annotated and Illustrated Catalogue of Shark Species Known to Date, vol. 4. Carcharhiniformes, Rome, FAO. Part 2.

Last, P.R., Stevens, J.D., 2009. Sharks and Rays of Australia, second ed. Harvard University Press, Cambridge, MA, pp. 179–180.

Shimada, K., 2001. Teeth of embryos in lamniform sharks (Chondrichthyes: Elasmobranchii). Env. Biol. Fishes 63, 309–319.

Wydoski, R.S., Whitney, R.R., 2003. Inland Fish of Washington. University of Washington Press, Seattle.

REFERENCES

FURTHER READING

Chapter 3

Chondrichthyes 2: Rays and Chimaeras

BATOIDEA

The Batoidea, or rays, are classified into four orders: Myliobatiformes (including stingrays, manta rays, and eagle rays), Rajiformes (skates and guitarfish), Pristiformes (sawfish), and Torpediniformes (electric rays). Rays are flat-bodied and have ventrally located gill slits. Their paired pectoral fins are greatly enlarged and merged with the head and body. In Myliobatiformes and skates, the tail is slender, and the dorsal and caudal fins are reduced, while the large pectoral fins form a diamond-shaped pair of "wings" that undulate and provide the propulsive force in swimming. Torpediniformes, Pristiformes, and guitarfish, however, swim using their muscular tail, like sharks.

DIET AND FEEDING

The large pelagic manta and mobula rays are filter feeders, with the mouth terminal or subterminal. Most batoids, however, have a ventrally directed mouth and feed on demersal fish and invertebrates. To capture their prey the rays adopt a variety of strategies, including: stunning prey electrically (electric rays); ambush predation; trapping prey beneath the pectoral fins; biting off protruding parts of benthic prey; exposing buried prey by flapping the pectoral fins or blowing water from the mouth to remove covering sand; and burrowing for prey beneath the surface of the ocean floor.

As noted in Chapter 2 (Fig. 2.1), the euhyostylic mode of jaw suspension in rays confers exceptional mobility on the jaws. The lack of an anterior connection to the chondrocranium means that the jaws can be highly protrusible: in the **Atlantic guitarfish** (*Rhinobatos*) by 26% of head length and in the **lesser electric ray** (*Narcine*) by about 100% of head length, assisted by flexibility at the symphysis, which allows the jaws to be brought closer together. Although in some sharks the jaws are protruded by about 30% of head length, more usually the value is 7–18% (Dean and Motta, 2004).

The protrusibility of the jaws probably contributes to the generation of suction, which is a very important element of feeding among rays. Some groups, such as skates and

guitarfish, use a combination of biting and suction to capture prey (Wilga and Motta, 1998), whereas others, such as electric rays and at least some durophagous myliobatids, use suction as the main component of feeding (Wilga et al., 2007; Motta and Huber, 2012). Rays such as *Narcine* can suck worms and similar prey from beneath the substrate surface. The extreme protrusibility of the jaws of *Narcine* allows the mouth to be pushed deep into the substrate to forage for prey. Capture of prey from beneath the surface entails ingestion of sand and other extraneous material, and strategies, such as the use of repeated suction or blowing water from the pharynx are used to clean the prey before swallowing.

In most elasmobranchs, the jaws open and close in the sagittal plane but at least one species of electric ray (*Narcine*) has the unique ability to protrude its jaws asymmetrically, by up to 60 degree from the sagittal plane, through asymmetric rotation of the hyomandibulae, and utilizes this ability in cleaning prey of debris (Dean and Motta, 2004).

The range of tooth form in batoids is narrower than in sharks, and dentitions seem to be adapted for either grasping or crushing prey. Grasping teeth are furnished with sharp points, while crushing teeth present a flat or rounded surface to the prey. In both types of dentition, the number of rows of exposed functional teeth is much higher than the one or two rows generally observed in shark dentitions (Fig. 3.1). In both grasping and crushing dentitions, the teeth are set close together in an alternating array to form a pavement. The most highly developed crushing dentitions are found in Myliobatidae. These rays possess only a few files of large, block-like teeth that interlock to form a flat, mill-like surface. In some crushing-type dentitions, the teeth may have edges as well as rounded upper surfaces that could provide a shredding as well as a crushing function.

In grasping dentitions, the teeth have short, sharp points directed obliquely toward the interior of the mouth. The alternating array presents a field of points adapted to grasping prey without necessarily inflicting injury to a great depth, like the teeth of sharks. The teeth are also less mobile than those of sharks, movement being restricted by the

The Teeth of Non-Mammalian Vertebrates. http://dx.doi.org/10.1016/B978-0-12-802850-6.00003-5

morphology of the basal plate of the tooth and by the closer packing of the teeth. In most batoids, the basal plate possesses two lateral lobes separated by a medial groove, in the center of which is a pore allowing passage of blood vessels to the dental pulp. The block-like teeth of Myliobatiformes are stabilized by numerous anteroposteriorly oriented ridges (see later in this chapter). It is conceivable that, in some batoids, pointed teeth could be erected for grasping or lowered to form a crushing surface (by tension or relaxation of the fibrous sheet to which they are attached), as occurs in the bamboo shark (*Chiloscyllium plagiosum*) (Ramsay and Wilga, 2007).

In some species, such as thornback rays, the teeth show sexual dimorphism, with males having more pointed teeth that are used in grasping the female.

DENTITIONS OF RAYS

The following descriptions can be expanded by reading the descriptions of batoid dentitions and tooth morphology in Herman et al. (1992–2002) at www.elasmo.com (follow link to batoids) and www.flmnh.ufl.edu/fish/rays.

MYLIOBATIFORMES

This order contains over 220 species of stingrays, eagle rays, manta rays, and devil rays, classified into 10 families.

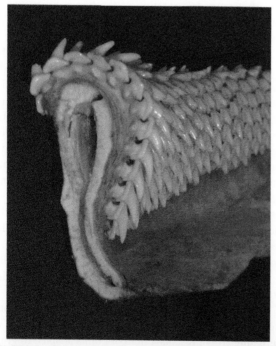

FIGURE 3.1 Cross-section of the jaw of a skate (*Raia* sp.) showing the presence of 13 successional rows of pointed teeth. *Courtesy Professor P. T. Sharpe.*

Dasyatidae

One of four families of stingrays, members of the Dasyatidae are highly electroreceptive fish that can detect the weak electric fields generated by prey. Examples of this type of ray are the **whiptail stingray** (*Dasyatis brevis*) and **bluntnose stingray** (*Dasyatis say*). Their dentitions consist of small, diamond-shaped, and slightly curved teeth that are arranged as a tessellated pavement. *D. brevis* has 21–37 upper tooth rows and 23–44 lower tooth rows (Fig. 3.2). During the reproductive season, the teeth in the males of these species develop points that curve toward the corners of the mouth and allow them to hold on to the females during copulation (Fig. 3.3).

Unlike most other stingrays, the **pelagic stingray** (*Pteroplatytrygon violacea*) has sharp-edged teeth in both males and females. These teeth are used for grasping while the ridges are used for cutting, unlike the typical crushing dentition associated with most stingrays (Fig. 3.4). There are 17 files of teeth in each row on either side of the upper jaw and 19 teeth on each side in the lower jaw, which also possesses a bicuspid symphyseal tooth.

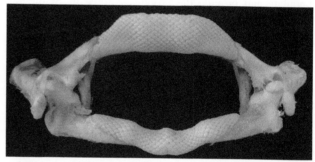

FIGURE 3.2 Dentition of whiptail stingray (*Dasyatis brevis*). Image width = 12 cm. *Courtesy Dr. C. Underwood.*

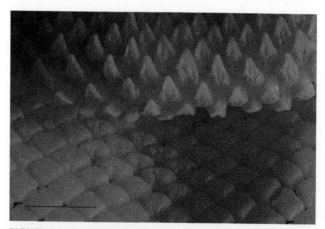

FIGURE 3.3 Dentition of bluntnose stingrays (*Dasyatis say*) comparing the more pointed teeth in the male (*above*) with the blunter teeth of the female (*below*) during the mating season. Marker bar = 5 mm. *Courtesy Dr. J. C. Seitz.*

Gymnuridae

The teeth of the **smooth butterfly ray** (*Gymnura micrura*) have one short, sharp, conical point on a somewhat swollen base (Fig. 3.5).

Myliobatidae

This family contains over 40 species in three subfamilies and includes the bat, eagle, manta, devil, and cow-nose rays. They are free-swimming rays whose pectoral fins make up broad, powerful "wings" that can measure over 6m from tip to tip. The Myliobatidae includes two types of rays with quite distinct feeding habits and dentitions. Eagle and cow-nose rays feed on molluscs and have dentitions adapted to crushing (durophagy). In contrast, the devil and manta rays (*Mobula, Manta*) are large, open-water filter feeders with a reduced dentition.

The dentitions of the durophagous myliobatids show a number of specializations in the jaws and teeth related to their

diet (Summers, 2000). The cartilaginous jaws are strengthened by calcified struts (trabeculae), and the palatoquadrate and mandibular symphysis are fused. Strong ligaments connecting the upper and lower jaws restrict the gape. The strong adductor muscles can be asynchronously activated.

In eagle and cow-nose rays, the dentition consists of fewer and larger teeth than in any other ray and the shape and structure of the teeth are unique. The teeth have the form of blocks arranged in anteroposterior files packed closely together in an alternating array to form an almost gap-free pavement. Where there are multiple files, teeth forming the central portion of the pavement have hexagonal outlines, while those forming the edges have pentagonal, square, or triangular outlines. The body of the tooth consists of osteodentine, supporting an overlying layer of orthodentine. At the functional surface, the teeth are covered with a layer of enameloid (Fig. 3.6). At the base of the tooth the basal plate is divided into numerous anteroposteriorly oriented ridges (Fig. 3.7) instead of the two lobes, which occur in most batoids; these increase the surface area for fibrous attachment to the jaw. The tooth pavement is stabilized by a variety of mechanisms. First, the polygonal shapes of the alternating teeth in most myliobatids interlock. Second, the vertical surfaces bear ridges and grooves which interdigitate with those on neighboring teeth. Third, in *Myliobatis* and *Aetobatus*, the anteroposterior ridges of the basal plate extend from the posterior margin of the tooth; these interdigitate with those of the succeeding tooth and also form a shelf on which the body of the neighboring tooth rests (Herman et al., 2000 and http://www.elasmo.com/refs/slides/ss_myliobatid.html).

FIGURE 3.4 Dentition of the pelagic stingray (*Pteroplatytrygon violacea*) composed of sharp, pointed teeth. Note the bicuspid symphysial tooth. Marker bar = 5mm. *Courtesy Dr. J. C. Seitz.*

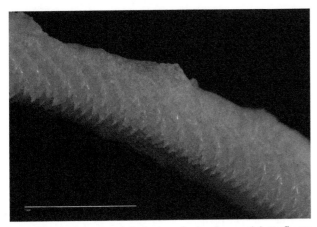

FIGURE 3.5 Pointed teeth in the jaw of a female smooth butterfly ray (*Gymnura micrura*). Marker bar = 5mm. *Courtesy Dr. J. C. Seitz.*

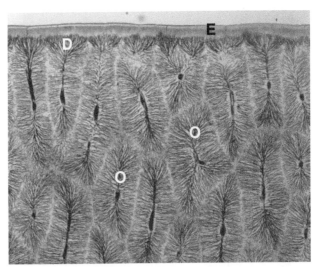

FIGURE 3.6 Vertical section of tooth of *Myliobatis* sp. Functional surface composed of orthodentine (D), covered with enameloid (E), supported by trabeculae of osteodentine (O). Image width = 4.28mm. Museums at the Royal College of Surgeons, Tomes slide collection. Catalog no. 123.

The dentition of the **bat ray** (*Myliobatis californica*) is made up of a series of seven files of crushing teeth. The central hexagonal plate is very wide, taking up about half the width of the occlusal surface and it is flanked by three lateral files of smaller teeth on each side (Fig. 3.8), the outermost being pentagonal. The crushing surface formed by the teeth of the upper jaw is more curved than that of the lower jaw.

Cow-nose rays (*Rhinoptera javanica*) are specialized suction feeders (Sasko et al., 2006), which open and close their jaws to generate water movements that are used to excavate buried prey. Food capture is achieved by suction and the prey is then cleaned by actions similar to those used in excavation. As in the bat ray, there are seven files of teeth. However, although the central file of crushing teeth is the widest, the adjacent two files are nearly as wide. Together, these three files occupy about three-quarters of the jaw width and the rest is taken up by two lateral files of small teeth on each side, the more medial being hexagonal and the most lateral being pentagonal (Fig. 3.9A). Fig. 3.9B illustrates a cow-nose ray with a mutation resulting in an additional row of

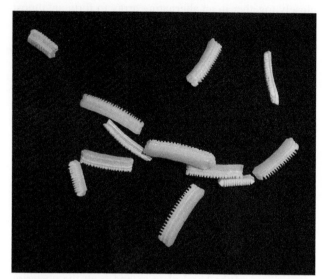

FIGURE 3.7 Isolated teeth of *Myliobatis* sp., showing smooth grinding surfaces and corrugated basal surfaces, the projections of which enhance the stability of attachment. *Courtesy Dr. C. Underwood.*

(A)

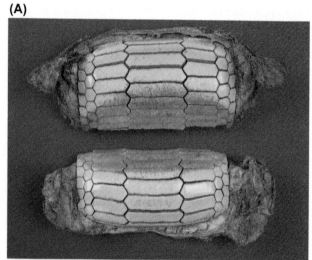

(B)

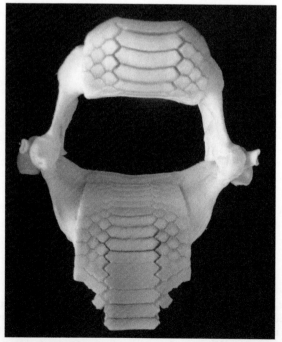

FIGURE 3.8 Dentition of bat ray (*Myliobatis californica*). Image width = 16 cm. *Courtesy Dr. C. Underwood.*

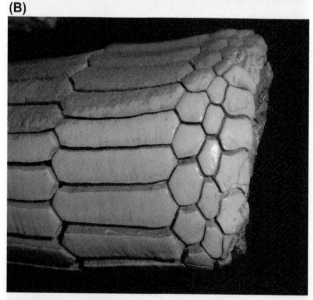

FIGURE 3.9 (A) Dentition of cow-nose ray (*Rhinoptera javanica*). Image width = 18 cm. *(Courtesy RCSOMA/ 448.1.)* (B) Dentition of cow-nose ray (*R. javanica*) with an extra file of pentagonal teeth (*arrow*) on one side. *(Courtesy RCSOMA/ 448.2.)*

pentagonal teeth on one side. It would appear that the extra file is the middle of the three lateral ones and has a squashed hexagonal form.

In contrast to the tooth plates in *Myliobatis* and *Rhinoptera*, those of the **spotted eagle ray** (*Aetobatus narinari*) are each made up of a single file of teeth. Those in the upper jaw are gently curving oblongs (Fig. 3.10), while those in the lower jaw are chevron-shaped (Fig. 3.11). The upper plate is broader than the lower, the two plates occupying about 80% and 60% of the width of the mouth, respectively. The upper plate is less deep than the lower and the three most anterior rows of the lower plate project beyond the upper tooth plate when the mouth is closed.

Manta rays (*Manta birostris*) swim along slowly through the water with their terminally placed mouths open, so that water passively enters the mouth. This process is assisted by a pair of cephalic lobes that help deflect water into the mouth. The small planktonic animals and plants suspended in the water are collected on plates of sponge-like tissue located between the gill bars. Manta rays have numerous, small, flat, apparently functionless teeth in the lower jaw.

There are two species of manta rays, both with teeth limited to the lower jaw. The **giant Pacific manta ray** (*Manta birostris*) has small cusped teeth. In the lower jaw there are 220–250 files of teeth, with 12–16 rows, giving a total tooth count of 3000–4000 (Fig. 3.12A). In the smaller **inshore manta ray** (*Manta alfredi*), there are 142–182 files of teeth in six to eight rows, giving a total tooth count of 9000–15,000 (Fig. 3.12B). (Marshall et al., 2009).

The **devil fish** or **giant devil rays** (*Mobula hypostoma*) resemble manta rays but have subterminal mouths (Fig. 3.13A). Their teeth are also small and seemingly functionless (Fig. 3.13B). The teeth show sexual dimorphism. Female teeth are diamond-shaped, square, or rectangular, sometimes with all variations present simultaneously and with a range of sizes. The teeth of the male are more crowded, with usually one or three (but sometimes up to four or five) long, slender, blunt points facing inward. In both sexes, there are approximately 10 rows of functional teeth in each jaw, each row of teeth overlapping the next.

RAJIFORMES

This group includes skates and guitarfish, with over 270 species distributed among five families. The skates, but not the guitarfish, use their enlarged pectoral fins in swimming (rajiform locomotion).

Rajidae

The skate family contains about 200 species within 14 genera. The dentition of the **undulate ray** (*Raja undulata*), which consists of a mosaic of small teeth, is typical. It also shows sexual dimorphism in that, while the teeth of females are rounded (Fig. 3.14A), those of the males are pointed (Fig. 3.14B).

The **common skate** (*Dipturus betis*), like the undulate ray, also exhibits sexual dimorphism, with the female possessing flattened rows of teeth (Fig. 3.15A), while those of the male are more pointed (Fig. 3.15B). It is now known that there are two species of common skate: the rarer **blue skate** (*Dipturus flossada*) and the larger **flapper skate** (*Dipturus intermedia*). Among the distinguishing features are the narrower bases of the teeth in the blue skate (Fig. 3.16; Iglesias et al., 2010).

Rhinobatidae

Guitarfish share some bodily features with both rays and sharks. Their bodies are flattened dorsoventrally but they retain a muscular tail with two functional dorsal fins. The

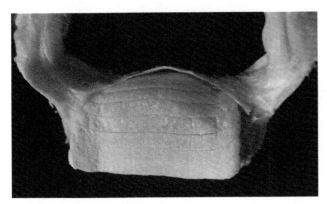

FIGURE 3.10 Upper dentition of spotted eagle ray (*Aetobatus narinari*). Image width = 6 cm. *Courtesy Dr. C. Underwood.*

FIGURE 3.11 Lower dentition of spotted eagle ray (*Aetobatus narinari*). Image width = 8 cm. *Courtesy Dr. C. Underwood.*

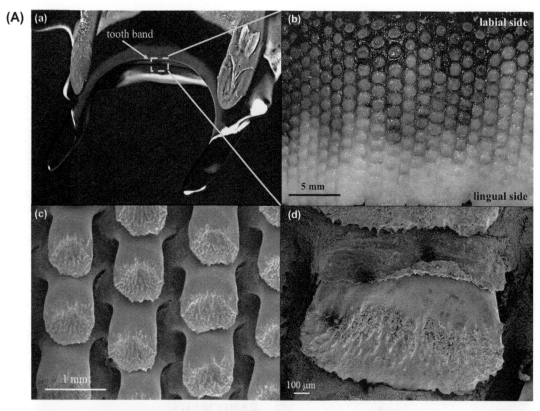

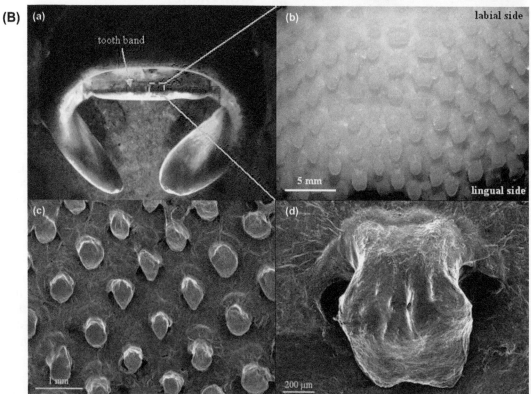

FIGURE 3.12 (A) Dentition and tooth morphology in the giant Pacific manta ray (*Manta birostris*): (a) lower jaw with elongated tooth band; (b) section of teeth mid-band; (c) embedded teeth of male ray; (d) view of single embedded female tooth. (B) Dentition and tooth morphology in the inshore manta ray (*M. birostris*): (a) lower jaw with elongated tooth band; (b) section of teeth mid-band; (c) embedded teeth of male ray; (d) view of single embedded female tooth. *From Marshall, A.D., Compagno, L.J.V., Bennett, M.B., 2009. Redescription of the genus* Manta *with resurrection of* Manta alfredi. *Zootaxa 2301, 1–28. Courtesy Editors of Zootaxa.*

FIGURE 3.13   Giant devil ray (*Mobula hypostoma*). (A) Swimming devil ray, showing cephalic lobes flanking mouth opening. *(Courtesy Wikipedia.)* (B) Teeth retained on the jaw of a devil ray. Image width=7 cm. *(Courtesy Dr. C. Underwood.)*

diet of these fish consists of benthic organisms including molluscs and crustaceans, and also small fish. The dentition is adapted to crushing. The **giant guitarfish** (*Rhyncobatis djiddensis*) has a dentition consisting of numerous, rectangular, blunt teeth arranged in about 10 rows in the upper jaw and in slightly fewer rows in the lower jaw (Figs. 3.17 and 3.18A). The teeth surfaces are slightly roughened (Fig. 3.18B). The teeth of the **bowmouth guitarfish** (*Rhina ancylostoma*) are rounded and have ridged surfaces (Figs. 3.19 and 3.20). The roughened surfaces of the teeth of *Rhynchodon* and *Rhina* presumably increase grip during the crushing of prey.

PRISTIFORMES

Sawfish are characterized by their projecting "saw" or rostrum, which they use to search out prey (Fig. 3.21). The rostrum is armed with "teeth" (modified dermal denticles) set in sockets perpendicular to the long axis (Fig. 3.22). In contrast to the rostral teeth of the sawshark (Chapter 2), those of sawfish are of almost uniform size, grow continuously, and are not replaced. The rostral teeth are composed of osteodentine, with the denteons aligned parallel with the long axis of the tooth (Fig. 3.23). This structure may allow dentine to be abraded away from the sides of the tooth more easily than from the tip, so that the sharpness of the tooth tip is maintained (Shellis and Berkovitz, 1980).

Pristidae

According to recent classification, sawfish form a single family that includes five species in two genera. Four species belong to *Pristis*: *Pristis clavata*, *Pristis pectinata*, *Pristis pristis*, *Pristis zijsron*. One species belongs to *Anoxypristis*: *Anoxypristis cuspidata*. The morphology of the rostrum and teeth, and the tooth number, can be used in species identification (Wueringer et al., 2009; Whitty et al., 2014; Lange et al., 2015).

Pristis

The common name of the **dwarf (Queensland) sawfish** (*P. clavata*) comes from a previous assumption that the species reached only a relatively small size (ca. 2.5 m total length). This assumption has since been disproven by the capture of specimens greater than this size. Indeed, individuals do not attain sexual maturity until reaching 2.3 m or more. The dwarf sawfish can be differentiated from all other members of the genus by the incomplete posterior groove on the rostral teeth (complete in all other members) that does not reach the base of the tooth. Rostral teeth number 18–27 per side. It can also be distinguished from the smalltooth sawfish and green sawfish, which look similar to it, by the relatively short rostrum. The dwarf sawfish has the smallest distribution of any living sawfish species, as it appears to be restricted to the tropical waters of Australia and Papua New Guinea.

(A)

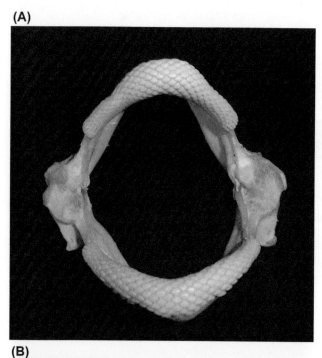

(B)

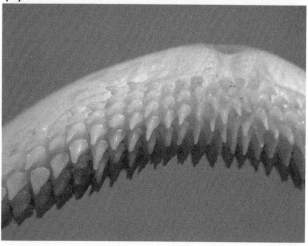

FIGURE 3.14 Undulate ray (*Raja undulata*). (A) Dentition of female. Image width = 7 cm. (B) Teeth in the upper jaw of male. *Both courtesy Dr. C. Underwood.*

(A)

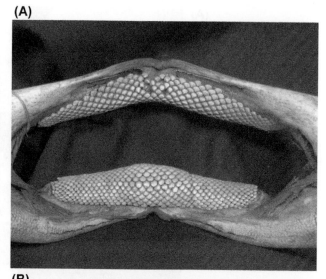

(B)

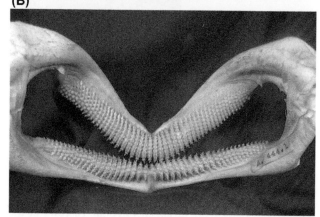

FIGURE 3.15 Common skate (*Dipturus betis*). (A) Dentition of a female. Image width = 12 cm. RCSOMA/441.11. (B) Dentition of a male, viewed from inner aspect of jaw. Actual image size = 16 cm. RCSOMA/441.1.

The rostrum of the **smalltooth sawfish** (*P. pectinata*) (Fig. 3.22A) resembles that of the green sawfish (*P. zijsron*) (Fig. 3.22B), although: (1) its rostrum has a slight but noticeable taper; (2) it generally has a lower average number of rostral teeth (23–30 but rarely as low as 20) per side than the green sawfish (23–37); and (3) it has a first dorsal fin positioned directly over, or slightly anterior to, the pelvic fins. The western Atlantic population of small-tooth sawfish has a higher average number of teeth per side than does the eastern Atlantic population, with averages of 26.3 and 23.7, respectively. There is an anterior–posterior gradient in the spacing of the rostral teeth: the gap between the posterior-most two teeth (on the same side) is 2–4 times that of the anterior-most two teeth. The

posterior edge of the tooth is grooved. The tips of the teeth may become blunted over time.

The **largetooth sawfish** (*P. pristis*) resembles the smalltooth sawfish, but it can usually be distinguished from it by having a robust rostrum with a wide base and a greater amount of tapering, and typically having a fewer number of teeth per side [14–23 compared with 23–30 (rarely as low as 20); Fig. 3.22C]. The rostrum of male largetooth sawfish has on average two more teeth per side than that of females.

Owing to its great length, the **green sawfish** (*P. zijsron*) has the longest rostrum of any living species of sawfish (to at least 1.6 m; Fig. 3.22B). There are between 23 and 37 teeth per side, often with more on one side of the rostrum. As with the smalltooth sawfish, the spacing between the teeth increases from anterior to posterior, but the gradient is more marked: the space between the posterior-most teeth is about 4–8 times that of the anterior-most two teeth. The Indian Ocean population has a higher number of teeth

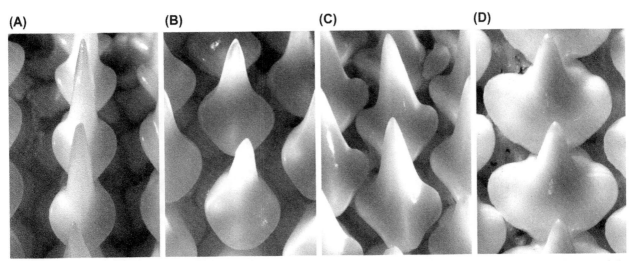

FIGURE 3.16 Images of skate teeth from the fifth row from the symphysis in the lower jaw. (A) *Dipturus* cf. *flossada*, adult male. (B) *Dipturus* cf. *flossada*, adult female. (C) *Dipturus* cf. *intermedia*, adult male. (D) *Dipturus* cf. *flossada*, adult female. Image widths = 5.5 mm. *From Iglesias, S.P., Toulhaut, L., Sellos, D.Y., 2010. Taxonomic confusion and market mislabelling of threatened skates: important consequences for their conservation status. Aquat. Conserv. Mar. Freshw. Ecosyst. 20, 319–333. Courtesy Editors Aquatic Conservation, Marine, and Freshwater Ecosystems.*

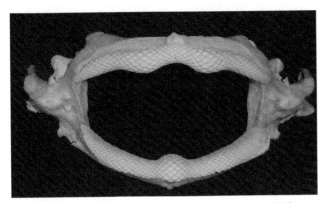

FIGURE 3.17 Dentition of giant guitarfish (*Rhyncobatis djiddensis*), viewed from the front. Image width = 16 cm. *Courtesy Dr. C. Underwood.*

(A)

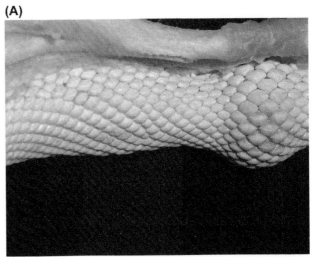

(B)

FIGURE 3.18 (A) Dentition of giant guitarfish (*Rhyncobatis djiddensis*) viewed from behind to show replacement teeth. Image width = 4.5 cm. (B) High power view of teeth in lower jaw to show roughened surfaces. Image width = 2.5 cm. *Both courtesy Dr. C. Underwood.*

per side than does the western Pacific population, with averages of 30.2 and 27.0, respectively. The rostrum is rather narrow and has little or no noticeable tapering. This species has particularly sharply pointed rostral teeth.

Anoxypristis

The **knifetooth sawfish** (*A. cuspidata*) can be identified by its triangular, blade-like, rostral teeth that are broad at the base (Fig. 3.22D), and which number between 16 and 33 per side. Specimens from the Indian Ocean typically have a higher number of rostral teeth per side than do specimens from the western Pacific, averaging 25.6 (range 22–29) and 21.2 (range 17–30), respectively. The basal quarter of the rostrum is devoid of teeth. The rostrum is rather small and narrow compared to other sawfish and seldom reaches 0.8 m.

The oral teeth of sawfish consist of many diagonal rows of tiny blunt teeth with rounded cusps and smooth surfaces (Fig. 3.24).

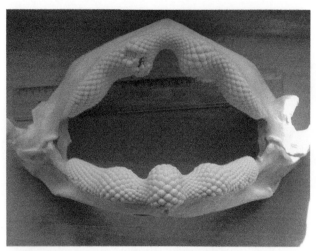

FIGURE 3.19 Dentition of the bowmouth guitarfish (*Rhina ancylostoma*). Top background scale in 1 mm divisions. *Courtesy Dr. C. Underwood.*

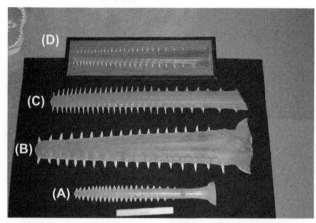

FIGURE 3.22 Rostra of various sawfish. (A) Smalltooth sawfish (*Pristis pectinata*); (B) green sawfish (*Pristis zijsron*); (C) largetooth sawfish (*Pristis pristis*); (D) knifetooth sawfish (*Anoxypristis cuspidata*). Scale = 30 cm. *Courtesy Dr. J. C. Seitz.*

TORPEDINIFORMES

Electric rays have the ability to stun their prey by generating an electrical charge, which is concentrated as the ray wraps its disk around its prey. This charge can measure up to 50 V in a large animal and, with low internal organ resistance, can produce a power output of 1 kW. There are 69 species in four families.

Torpedinidae

The teeth of electric rays are small as their prey of fish is killed or stunned prior to eating and therefore put up minimal resistance. As exemplified by the **Pacific electric ray** (*Torpedo californica*), the dentition consists of numerous, small, curved, pointed teeth with approximately seven rows of teeth exposed (Fig. 3.25). There are 25–28 files of teeth on the upper jaw and 19–26 on the lower jaw.

FIGURE 3.20 High power view of teeth in the lower jaw of bowmouth guitarfish (*Rhina ancylostoma*) showing ridging on the surface. Image width = 4.5 cm. *Courtesy Dr. C. Underwood.*

FIGURE 3.21 Sawfish (*Pristis* sp.), showing the rostrum. *Tamara Bauer/Dreamstime.com.*

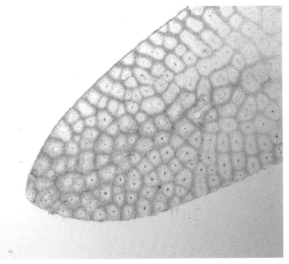

FIGURE 3.23 Cross-section of rostral tooth of *Pristis pristis*. The tooth consists of uniform units of osteodentine oriented parallel with the long axis of the tooth and held together by transverse network of transverse mineralized collagen fibers. Image width=4.28 mm. © *Museums at the Royal College of Surgeons, Tomes slide collection. Catalog no. 104.*

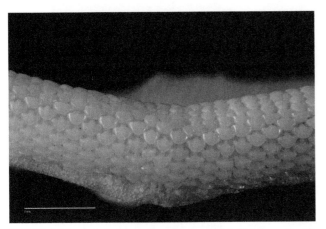

FIGURE 3.24 Oral dentition of smalltooth sawfish (*Pristis pectinata*). Marker bar=5 mm. *Courtesy Dr. J. C. Seitz.*

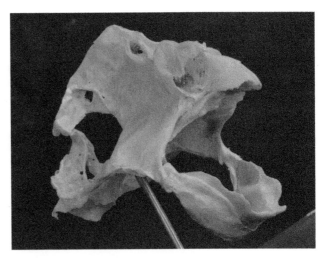

FIGURE 3.26 Tooth plates of ratfish (*Hydrolagus colliei*). *Courtesy Osteology Collection, Wesleyan College.*

DENTITIONS OF CHIMAERAS

This subclass of cartilaginous fish is characterized by the fusion of the upper jaw to the cranium (Fig. 3.26). There are three living families: Callorhinchidae (plough-nose chimaeras), Chimaeridae (short-nosed chimaeras), and Rhinochimaeridae (long-nose chimaeras), with over 30 species.

The dentitions are specialized and are reduced to three pairs of cutting/shearing plates adapted for processing their diet of invertebrates, such as molluscs, shrimp, crabs, polychaete worms, and small fish. There are two tooth plates on each side of the upper jaw and one on the lower.

Chimaeridae

The **spotted ratfish** (*Hydrolagus colliei*) has one elongated mineralized plate in the lower jaw and two plates in the upper jaw on each side (Fig. 3.26). Each tooth plate represents a single tooth family, usually with two members. The

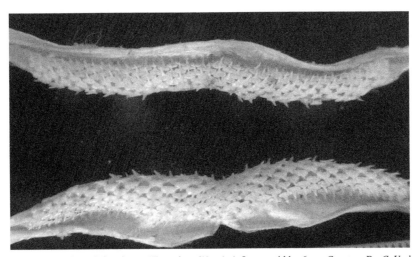

FIGURE 3.25 Dentition of electric ray (*Torpedo californica*). Image width=6 cm. *Courtesy Dr. C. Underwood.*

FIGURE 3.27   Elephantfish (*Callorhinchus callorynchus*). *Courtesy Wikipedia.*

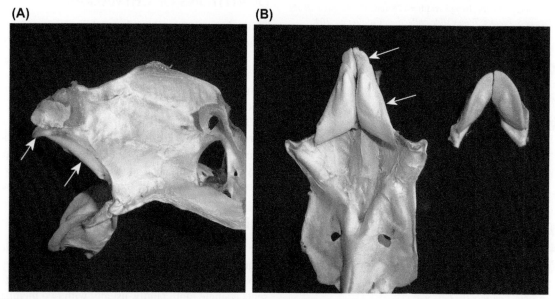

FIGURE 3.28   Dentition of the elephantfish (*Callorhinchus callorynchus*). (A) Side view. Image width=9 cm; (B) occlusal view of upper and lower jaws. Note the two tooth plates (*arrows*) on each side of the upper jaw. Image width=10 cm. *Courtesy Dr. C. Underwood.*

tooth has a wide base at which new tooth tissue is continuously formed, allowing it to grow throughout the life of the animal. The jaws of *H. colliei* are not reinforced by mineralization. Nevertheless, this species has a large mass-specific bite force, like *Heterodontus* and other durophagous fish (Huber et al., 2008).

Callorhinchidae

The **elephantfish** (**Australian ghost shark**) (*Callorhinchus callorynchus*) is named after the protuberance on the snout (Fig. 3.27). It has the typical three dental plates on each side (Fig. 3.28A), two of which can be seen in an occlusal view of the upper jaw (Fig. 3.28B).

ONLINE RESOURCES

Elasmo.com: http://www.elasmo.com/refs/slides/ss_myliobatid.html. (Detailed account of the tooth plates of myliobatids.)

Florida Museum of Natural History: http://www.flmnh.ufl.edu/fish/.

Shark Trust of Great Britain: http://www.sharktrust.org/en/skate_and_ray_taxonomy

Elasmo.com: www.elasmo.com. (Detailed descriptions of dentitions of numerous species of sharks and other topics.)

Reefquest Center for Shark Research: http://www.elasmo-research.org/education/shark_profiles/batoids.htm

Florida Fish and Wildlife Conservation Commission: http://myfwc.com/research/saltwater/fish/sawfish/. (This Website gives an up-to-date list of references concerning sawfish.)

REFERENCES

Dean, M.N., Motta, P.J., 2004. Feeding behavior and kinematics of the lesser electric ray, *Narcine brasiliensis* (Elasmobranchii: Batoidea). Zoology 107, 171–189.

Herman, J., Hovestadt-Euler, M., Hovestadt, D.C., Stehmann, M., 1992–2002. Contributions to the study of the comparative morphology of teeth and other relevant ichthydolurites in living supraspecific taxa of chondrichthyan fishes. In: Stehmann, M. (Ed.), Bulletin of the Royal Institute of Natural Sciences of Belgium, Biologie 1992, 62, 193–254; 1994, 64, 165–207; 1995, 65, 237–307; 1997, 67, 107–162; 1998, 68, 145–197; 1999, 69, 161–200; 2000, 70, 5–67; 2001, 71, 5–35; 2002, 72, 5–45.

Huber, D.R., Dean, M.N., Summers, A.P., 2008. Hard prey, soft jaws and the ontogeny of feeding mechanics in the spotted ratfish *Hydrolagus colliei*. J. R. Soc. Interface 5, 941–952.

Iglesias, S.P., Toulhaut, L., Sellos, D.Y., 2010. Taxonomic confusion and market mislabelling of threatened skates: important consequences for their conservation status. Aquat. Conserv. Mar. Freshw. Ecosyst. 20, 319–333.

Lange, T., Brehm, J., Moritz, T., 2015. A practical key for the identification of large fish rostra. Spixiana 38, 145–160.

Marshall, A.D., Compagno, L.J.V., Bennett, M.B., 2009. Redescription of the genus *Manta* with resurrection of *Manta alfredi*. Zootaxa 2301, 1–28.

Motta, P.J., Huber, D.R., 2012. Prey capture behaviour and feeding mechanics of elasmobranchs. In J.C. Carrier, J.A. Musick, M.R. Heithaus (Eds.), *Biology of Sharks and their Relatives*, second ed. CRC Press, Boca Raton, pp. 153–209.

Ramsay, J.B., Wilga, C.D., 2007. Morphology and mechanics of the teeth and jaws of white-spotted bamboo sharks (*Chiloscyllium plagiosum*). J. Morphol. 268, 664–682.

Sasko, D.E., Dean, M.N., Motta, P.J., Hueter, R.E., 2006. Prey capture behavior and kinematics of the Atlantic cownose ray, *Rhinoptera bonasus*. Zoology 109, 171–181.

Shellis, R.P., Berkovitz, B.K.B., 1980. Dentine structure in the rostral teeth of the sawfish *Pristis* (Elasmobranchii). Arch. Oral Biol. 25, 339–343.

Summers, A.P., 2000. Stiffening the stingray skeleton – an investigation of durophagy in myliobatid stingrays (Chondrichthyes, Batoidea, Myliobatidae). J. Morphol. 243, 113–126.

Whitty, J.M., Phillips, N.M., Thorburn, D.C., Simpfendorfer, C.A., Field, I., Peverell, S.C., Morgan, D.L., 2014. Utility of rostra in the identification of Australian sawfishes (Chondrichthyes: Pristidae). Aquat. Conserv. Mar. Freshw. Ecosyst. 24, 791–804.

Wilga, C.D., Motta, P.J., 1998. Feeding mechanism of the atlantic guitarfish *Rhinobatos lentiginosus*: modulation of kinematic and motor activity. J. Exp. Biol. 201, 3167–3184.

Wilga, C.D., Motta, P.J., Sanford, C.P., 2007. Evolution and ecology of feeding in elasmobranchs. Integr. Comp. Biol. 47, 55–69.

Wueringer, B.W., Squire, L., Collin, S.P., 2009. The biology of extinct and extant sawfish (Batoidea: Sclerorhynchidae and Pristidae). Rev. Fish. Biol. Fish. 19, 445–464.

FURTHER READING

Faria, V.V., McDavitt, M.T., Charvet, P., Wiley, T.R., Simpfendorfer, C.A., Naylor, G.J.P., 2013. Species delineation and global population structure of critically endangered sawfishes (Pristidae). Zool. J. Linn. Soc. 167, 136–164.

Miller, W.A., 1974. Observations on the developing rostrum and rostral teeth of sawfish: *Pristis perotteti* and *Pristis cuspidatus*. Copeia 311–318.

Slaughter, B.H., Springer, S., 1968. Replacement of rostral teeth in sawfishes and sawsharks. Copeia 499–506.

Tricas, T.C., Deacon, K., Last, P.R., McCosker, J.C. (Eds.), 1997. Sharks and Rays: The Ultimate Guide to Underwater Predators. Harper-Collins, London.

Chapter 4

Osteichthyes

CLASSIFICATION

There are roughly 30,000 species of bony fishes (Osteichthyes), about 60% of which live in the sea and 40% in fresh water. They account for approximately half of all species of vertebrates. Given such a large group, the classification of bony fishes has often been revised and remains under continuous review. Here, we use a phylogenetic classification (Betancur et al., 2013, 2014) based mostly on molecular evidence. As the following summary of the classification shows, the bony fishes can be divided into two superclasses, the vastly more numerous ray-finned fishes (Actinopterygii) and the lobe-finned fishes (Sarcopterygii) comprising the coelacanths and lungfishes:

Megaclass Osteichthyes
Superclass Actinopterygii
Class Cladistia (bichirs and reedfishes)
Class Actinopteri
Subclass Chondrostei (sturgeons, paddlefishes)
Subclass Neopterygii
Infraclass Holostei (bowfin, gars)
Infraclass Teleostei (vast majority of bony fishes)
Superclass Sarcopterygii
Class Coelacanthinomorpha (= Actinistia) (coelacanths)
Class Dipnotetrapodomorpha
Subclass Dipnomorpha (lungfishes)
Subclass Tetrapodomorpha (tetrapods)

The Cladistia, Chondrostei, and Holostei have mainly cartilaginous skeletons that show some degree of ossification. The Chondrostei consists of a single order, the Acipenseriformes, and two families: the Acipenseridae (sturgeons, 25 species) and the Polyodontidae (paddlefishes, 2 species). The Chondrostei have small teeth in the larval stage, but adults are edentulous and are not considered further in this book. The Neopterygii comprises a small infraclass, the Holostei, and a much larger infraclass, the Teleostei, which is the largest group of vertebrates. A condensed classification of the teleosts is provided in Fig. 4.1.

DIET

In size, actinopterygian fishes span a very wide range, from miniature freshwater fishes less than 20-mm long to billfishes about 4-m long. They have occupied an enormous variety of ecological niches and exploit a correspondingly wide range of food sources.

Several groups of actinopterygians have undergone major adaptive radiations that have attracted considerable attention from evolutionary biologists. The most famous radiation is that of the cichlids that colonized East African lakes, but extensive evolutionary radiations also occurred among characiforms, parrotfishes, and gobies. Streelman and Danley (2003) identified three overlapping phases of adaptive radiation: colonization of major habitats, specialization in exploiting food resources, and diversification in sensory communication. These phases often, although not always, follow this sequence. Obviously, morphological adaptation of the dentition is a central component of the phase of trophic diversification.

Most actinopterygians are carnivorous and prey on fish and invertebrates, such as worms, arthropods, and mollusks. Predators generally consume prey whole, but they may bite mouthfuls of flesh from larger prey. In addition, several species exploit other animals as food sources in ways that do not kill or disable the prey. Cleaner fishes, such as the **cleaner wrasses** (*Labroides*), derive nutrients from skin parasites that they remove from other fishes: an example of mutualism. Other fishes pluck scales, or bite pieces of fins, from prey species: both scales and fin tissue have high nutrient content and regenerate within a fairly short time (days or weeks) (Nico and de Morales, 1994).

Scale eating has evolved independently at least 12 times among bony fishes (Sazima, 1983). Species in which scale eating is the sole or main source of nutrition may display morphological asymmetry of the jaws and dentition that is correlated with a preference for removing scales from one side of the prey (eg, Hata et al., 2011; Lee et al., 2012). Both scale eating and fin eating can form part of the feeding repertoire of omnivorous species and are often more common in juveniles than in adults. Scale eating could have originated in other feeding activities, such as piscivory, scraping algae, or mucus eating, or in other behaviors, such as removing ectoparasites or aggression (Sazima, 1983). Given the multiple independent occurrences of scale eating, the origin of the habit probably varies between different families (Sazima, 1983).

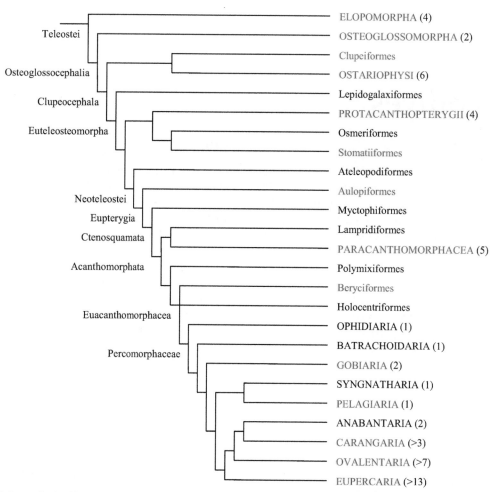

FIGURE 4.1 Phylogenetic classification of teleosts. To right, capitalized taxa include more than one order (number of contained orders in parentheses). Taxa in red type are represented in this chapter. *Condensed from Betancur, R.R., Broughton, R.E., Wiley, E.O., Carpenter, K., López, J.A., Li, C., Holcroft, N.I., Arcila, D., Sanciangco, M., Cureton II, J.C., Zhang, F., Buser, T., Campbell, M.A., Ballesteros, J.A., Roa-Varon, A., Willis, S., Borden, W.C., Rowley, T., Reneau, P.C., Hough, D.J., Lu, G., Grande, T., Arratia, G., Ortí, G., 2013. The tree of life and a new classification of bony fishes. PLoS Curr. Tree Life. Edition 1.*

Planktonic organisms may be exploited by particle feeding, in which the food organisms are simply engulfed a few at a time, or may be harvested by filter feeding, in which water is taken into the mouth and then expelled through the gills, where the suspended plankton is filtered out.

Only about 5% of bony fishes are herbivorous (Bone et al., 1995; Tolentino-Pablico et al., 2007), of which the majority inhabits tropical fresh water. Most tropical marine, herbivorous bony fishes live on seagrass or scrape algae from the surface of coral. In temperate coastal waters, blennies and gobies feed on seaweed. Freshwater herbivorous species can exploit, as well as algae, a wider range of plants, such as moss, submerged and decaying grasses, and seeds and fruit, all of which fall into the water. Algae attached to submerged surfaces harbor a variety of microscopic organisms and small invertebrates and form a layer referred to as **periphyton** (on vegetation) or **Aufwuchs** (on rocks; the biocover that grows on rocks). It is likely that the presence

of these nonplant components augments the nutrients available to herbivorous fishes (Bone et al., 1995).

A major source of nutrition for many fishes is detritus. Detritus contains a variety of dead fragments of plants and animals, together with microorganisms.

FEEDING AND FOOD PROCESSING IN ACTINOPTERYGII

Bony fishes are unique among vertebrates in that the oral jaws are augmented by a set of pharyngeal jaws, derived from the branchial arches (Nelson, 1969). The oral and pharyngeal jaws function together as a highly coordinated system for capturing food; transporting it to the esophagus; and, to a variable extent, preparing it for digestion.

In general, feeding by teleosts involves a sequence of stages: prey capture, manipulation of prey within the mouth and pharyngeal apparatus, and intrapharyngeal transport

(Vandewalle et al., 1994, 2000; Schwenk and Rubega, 2005). Intrapharyngeal transport tends to be the most prolonged phase of the feeding cycle and can include trituration of the prey.

In general, the capture of food is the main function of the oral jaws. Food processing is carried out largely by the pharyngeal jaw system, but sometimes with contributions from the oral jaws and the hyoid apparatus.

Oral Jaws

The oral jaws of bony fishes are formed largely of dermal bones, although the jaw joint is formed between two cartilage bones—the quadrate and articular—that are derived, respectively, from the palatoquadrate (upper jaw) and Meckel cartilages (lower jaw). The jaw joint is suspended from the cranium by the hyomandibula, the pterygoids, and the palatines, collectively referred to as the **suspensorium** (see Fig. 4.2). The bones forming the upper jaw are the premaxilla and maxilla. In basal actinopterygians both the premaxilla and maxilla carry teeth, but during the evolution of the group, the maxilla has become edentulous in several orders. The anterior portion of the lower jaw is formed by the dentary and the posterior portion by the angular, surangular, and coronoid bones: of these bones, the dentary is the main tooth-bearing bone. Among primitive bony fishes, additional teeth are commonly present on other bones within the mouth, such as the vomers, palatines, ectopterygoids, and parasphenoid on the roof of the mouth, and even on the tongue, ie, on pads of dermal bone associated with the basihyals. Among advanced teleosts, teeth tend to be confined to the marginal bones, and to be absent or sparse elsewhere in the mouth.

The actinopterygians are notable for the degree of movement between the bones making up the upper jaw, and in some taxa, other parts of the skull. In many groups, the moveable elements of the cranial skeleton form a system of levers that allows the upper jaw to be protruded. Protrusible jaws have evolved through a series of modifications to the

jaw skeleton, especially in the arrangement and function of the bones forming the upper jaw. Among basal actinopterygians (Schaeffer and Rosen, 1961; Lauder, 1980), the maxilla and premaxilla are not mobile in *Polypterus* and *Lepisosteus*; but in *Amia*, the premaxilla is a small bone with limited mobility, and the maxilla (the major tooth-bearing bone in this genus) is more loosely connected to the cranium and hence more mobile (Fig. 4.2). The maxilla is attached by connective tissue to the palatine bone at one end and to the lower jaw at the other end. When the jaws open, the posterior end of the maxilla is drawn downward through its connection with the lower jaw, and so swings forward to form the lateral margins of the gape [clearly seen in Figure 13 of Lauder (1980)]. The resulting occlusion of the gape increases fluid flow due to the suction that accompanies mouth opening, and it also hinders escape of prey or food particles. A jaw mechanism similar to that seen in *Amia* is also present in many other actinopterygii, eg, *Salmo* (see Fig. 4.43). However, this arrangement has been modified in many actinoptergyians through many changes (Schaeffer and Rosen, 1961; Motta, 1984a):

1. The number and size of the teeth on the maxilla are reduced and the bone finally becomes edentulous in advanced actinopterygians.
2. The ligamentous joint between the maxilla and the palatine bone is replaced by a ball-and-socket joint.
3. The premaxilla enlarges and may become the sole tooth-bearing bone in the upper jaw. It overlaps the maxilla, which is excluded from the gape. The two bones are joined by connective tissue, and a ligamentous connection may be established between the premaxilla and the lower jaw.
4. The premaxilla becomes increasingly mobile.

Consequently, in most actinopterygians, the premaxilla is rotated and braced by the maxilla during jaw opening. In protrusible upper jaws, this causes the premaxilla to be thrust forward as the lower jaw is depressed (Figs. 4.3–4.5). Protrusible jaws have evolved independently several times and the detailed mechanism varies correspondingly (Schaeffer and Rosen, 1961; Motta, 1984a; Westneat, 2004). Among Perciformes, which is the largest group to possess this feature, protrusible upper jaws have dorsal processes that can slide on the snout and ventral processes that articulate with the anterior end of the maxilla. The driving force for protrusion of the premaxilla is provided by pressure from the maxilla rotating about its anterior end (Fig. 4.3). The system of levers formed by the lower jaw, maxilla, and premaxilla has been modeled as a **four-bar linkage** (eg, Westneat, 2004; Hulsey and García de Léon, 2005). This allows the transmission of motion from the lower jaw to the maxilla to be quantified.

Typically, protrusion of the upper jaw extends the position of the mouth opening in advance of the fish by around

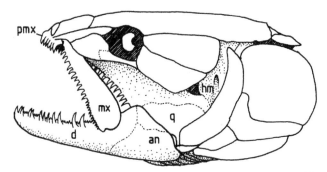

FIGURE 4.2 Diagram of skull of *Amia* with jaws open. Both premaxilla (pmx) and maxilla (mx) bear teeth. During jaw opening, premaxilla does not move, but maxilla rotates forward around maxilla-palatine connection. *an*, angular; *d*, dentary; *hm*, hyomandibula; *q*, quadrate.

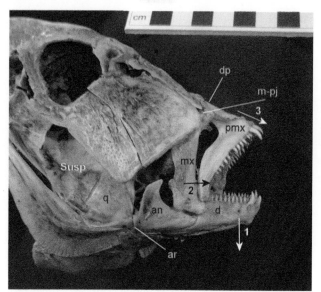

FIGURE 4.3 Skull of black bream (*Spondyliosoma cantharus*) with jaws open. Teeth borne on premaxilla, but not on maxilla, and maxilla is excluded from gape. As the lower jaw is depressed during mouth opening (white vertical *arrow, 1*), the maxilla, which is connected to the lower jaw, rotates downward, pushing the premaxilla forward (black horizontal *arrow, 2*). The dorsal processes of the premaxilla slide down grooves in the snout (white oblique *arrow, 3*), leading to protrusion of the upper jaw. *an*, angular; *ar*, articular; *d*, dentary; *dp*, dorsal processes of premaxilla; *m-pj*, maxilla-palatine joint; *mx*, maxilla; *pmx*, premaxilla; *q*, quadrate; *Susp*, suspensorium (component bones not distinguishable). *Courtesy RCSOMA/ 467.4.*

FIGURE 4.4 Head of gilt-head (sea) bream (*Sparus auratus*), with mouth closed.

10% of head length. In some species, the protrusion can be as great as 65% of head length. This is accomplished by rotation of the suspensorium through increased flexibility of various joints, depending on the species, thereby advancing the position of the quadrate and hence that of the jaw assembly (Westneat and Wainwright, 1991; Waltzek and Wainwright, 2003).

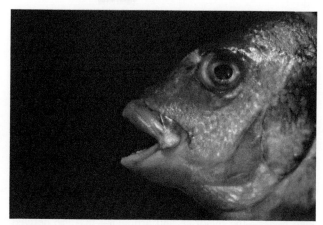

FIGURE 4.5 Head of gilt-head (sea) bream (*Sparus auratus*), with mouth open. Note the anterior displacement of the premaxilla and the rotation of the maxilla.

Investigations of the lever systems involved in operation of the jaws show that there is considerable variation in the mechanical advantage associated with opening and closing of the mouth. This reflects variation in the degree to which dentitions are adapted toward rapid jaw action or to exerting a powerful bite (Westneat, 2004).

In many groups of fishes that feed on prey attached to the substrate, particularly coral reefs, the joint between the dentary and angular bones of the lower jaw is moveable. This mobility increases the freedom of movement of the jaw apparatus and hence allows close adaptation of scraping teeth to the substrate (Price et al., 2010). These fishes have short lower jaws that confer a large mechanical advantage. The intramandibular joint can either increase or decrease the gape, depending on the direction of rotation (Konow et al., 2008). In Girellidae, Chaetodontidae, Acanthuroidei, and Scarinae, movement at the intramandibular joint increases the gape, thereby allowing more food to be taken at each bite than would be the case if the gape were limited by the short lower jaws. In the Pomacentridae, in contrast, movement at the intramandibular joint restricts the gape. These fishes have highly protrusible jaws and use them to grip and tear off food attached to the surface. A mobile intramandibular joint is also present among Distichodontidae (Characiformes)

Food Capture

Bony fishes can acquire their food in a great variety of ways. The first phase, in which the mouth is brought into contact with the food, is achieved either by **suction** that brings the prey toward the mouth or by **ram** in which the fish swims toward the food (Wainwright et al., 2015).

Suction, used in feeding by a large proportion of bony fishes, is generated by an increase in buccal volume as the fish opens its mouth, by elevation of the head, retraction and depression of the hyoid arch, and expansion of the operculum. The resulting pressure gradient at mouth opening causes flow of water into the mouth, and this exerts force on prey, whether free swimming or attached to the substratum, and draws it toward the feeding fish (Holtzman et al., 2008). Suction is maximized in fishes with a small, approximately circular, mouth opening and a deep body form (Wainwright et al., 2015). Observations on an efficient suction feeder, the **bluegill sunfish** (*Lepomis macrochirus*) (Ferry-Graham and Lauder, 2001; Day et al., 2005), show that the peak fluid flow is achieved when the mouth is almost fully open. Water is drawn into the mouth from all directions, but the fluid velocity decreases rapidly with distance from the mouth and the effect of suction extends only over a distance equal to about the diameter of the mouth opening. Fluid velocity during suction should in theory be greater for small mouths than for large mouths (Wainwright et al., 2007), but Wainwright et al. (2001) found that, in seven species of cichlids, the distance from which prey can be captured is not correlated with the gape. Wainwright et al. (2007) suggested that the advantage of greater water velocity during suction may be that prey clinging to a substrate is more easily detached by the increased force. In efficient suction feeders, prey is transported rapidly past the jaw margins, so oral teeth are not required and they might in fact impede suction by increasing friction (Gosline, 1973). Consequently, in many suction feeders, such as cyprinids, the oral dentition is reduced or lost.

Suction is most effective in relation to relatively small prey, and because it operates only over a short distance, it is usually necessary for fishes to approach the prey actively. Some predatory fishes, such as pike, salmon, and billfish, have wide, cleft-shaped mouths that are not well adapted to generating suction. These fishes are frequently strong swimmers that launch a high-speed ram attack to overtake their prey, with the cranium lifted to enlarge the gape, as in sharks. During a ram attack, large quantities of water enter the mouth opening and must be removed through the sides of the mouth and the open opercula. The direction of high-speed attacks by powerful swimmers, such as **barracudas** (*Sphyraena*) and **pikes** (*Esox*), is usually oblique or perpendicular to the axis of the prey and directed at the center of the prey fish. Weaker swimmers, such as **garfishes** (*Lepisosteus*), approach the prey by stealth and strike the prey from close quarters (Porter and Motta, 2004; Habegger et al., 2011). Observations on the **great barracuda** (*Sphyraena barracuda*) and the **king mackerel** (*Scomberomorus cavalla*) show that, in these ram feeders, the jaws are adapted for rapid water clearance and that the size-corrected bite force is lower than average (Westneat, 2004; Habegger et al., 2011; Ferguson et al., 2015).

Many other fishes use a combination of ram and suction to acquire food. For example, protrusion of the jaws not only contributes to suction, and thus exerts an attractive force on the prey, but also advances the mouth opening, so it is equivalent to an increase in the speed of approach. The improvement can be considerable in species with extremely protrusible jaws (Motta, 1984a; Westneat and Wainwright, 1989; Waltzek and Wainwright, 2003). However, it seems that extreme jaw protrusion does not enhance suction performance compared with more typical protrusion (Waltzek and Wainwright, 2003).

Biting is used to kill or capture prey, or to take tissue from other fishes: flesh, pieces of fin, or scales. Bottom-feeding fishes use biting to pick up mollusks or crustaceans, or to graze on algae and other vegetation. Cleaning fishes use biting to remove ectoparasites from other fishes. Prey capture frequently involves a combination of suction and biting, as Ferry-Graham et al. (2002) observed among benthic-feeding labrids. In capture of relatively large prey, suction alone may be insufficient to transport it completely into the mouth, and biting is required to retain the prey.

Among fishes that feed on coral, or exploit algal deposits, the dentition is adapted to removing food from solid surfaces. Although the dentition may be used to scrape, or even excavate, the surface, the feeding mechanism is usually less mechanically demanding and teeth are used to pluck deposits or to comb through them.

Suspension-feeding fishes, such as herring (*Clupea*), may draw water into the mouth by using either suction or by ram feeding (engulfing plankton by swimming forward with mouth and opercula wide open). The microscopic prey organisms are separated from the water by cross-flow filtration on the surfaces of the gill rakers and carried toward the esophagus (Sanderson et al., 2001). The principal filter-feeding bony fishes are the Clupeidae. Species such as the **menhaden** (genera *Brevoortia* and *Ethmidium*) use filter feeding exclusively, but many species, such as **herring** (*Clupea*), switch between filter feeding for small plankton and particle feeding for larger planktonic organisms (Bone et al., 1995).

Finally, in some deep-sea fishes, the teeth are highly elongated and are used to imprison prey within the oral cavity rather than to bite.

Intraoral Transport and Prey Manipulation

Prey is usually killed or at least disabled during capture and can therefore be swallowed immediately or manipulated into a suitable orientation for swallowing. Prey is reorientated by use of pressure gradients in the water within the mouth and by use of teeth on the palate and tongue.

During intraoral manipulation, the prey can be bitten repeatedly (Lauder, 1980), which perforates the integument and makes it permeable to digestive enzymes. Osteoglossomorph and salmoniform fishes possess a more efficient mechanism for lacerating prey, in the form of the so-called **tongue bite apparatus**. The "tongue" carries, attached to the basihyal, a dermal pad of bone that bears sharp, often enlarged, teeth (Figs. 4.6 and 4.43) and can be brought into contact with other teeth on the roof of the mouth (Fig. 4.44). The tongue teeth and palatal teeth can bite against each other in a "chewing" action. In addition, the tongue bite apparatus is used to lacerate and cause internal damage to prey by a process called **raking**. The prey is gripped between the marginal teeth and anteroposterior movements of the tongue and palate drag teeth through its flesh. The palatal teeth are dragged forward by lifting the snout. The tongue teeth are dragged backward by retraction of the pectoral girdle, to which the hyoid and branchial arches are connected by a strong ligament. There is significant variation in the kinematics of both chewing and raking, especially with respect to the relative contribution made to raking by protraction of the upper teeth and retraction of the lower teeth (Sanford and Lauder, 1990; Camp et al., 2009).

Osteoglossomorphs and salmoniforms are not the only group of bony fishes that use palatal teeth in combination with lower teeth to process food. Some percomorph teleosts possess teeth on the parasphenoid, which are used in chewing and transporting prey. In **leaf-fishes** (*Pristolepis*), the parasphenoid teeth engage with teeth on the tongue, whereas in anabantoids (gouramis), the parasphenoid teeth are placed posteriorly and engage with the lower pharyngeal jaw (Liem and Greenwood, 1981).

In some mollusk-eating fishes (eg, sheep's-head fish and wolffish), there is at the posterior region of the jaws a bed of rounded teeth that are used to crush prey. Such durophagous species generate above-average bite forces (Ferguson et al., 2015).

Intrapharyngeal Transport and Prey Processing

The pharyngeal jaws evolved from specific elements of the system of arches of endochondral bone that support the gills (five in teleosts)—the **branchial basket**. The lower pharyngeal jaw is derived from the fifth ceratobranchials, whereas the upper pharyngeal jaw is derived from the second, third, or fourth pharyngobranchials, depending on species. These elements carry major concentrations of teeth on pads of dermal bone (Fig. 4.7). In basal teleosts, such as the Elopiformes, sharply pointed teeth also occur on small pads attached by connective tissue to many other elements of the branchial arches besides the pharyngeal jaws, so the branchial basket is more or less lined with small teeth. These teeth assist in transport of food that, in these fishes, is the main function of the pharyngeal apparatus, accomplished chiefly by anteroposterior movements of the lower pharyngeal jaws. Prey is transported from the mouth to the esophagus with little modification.

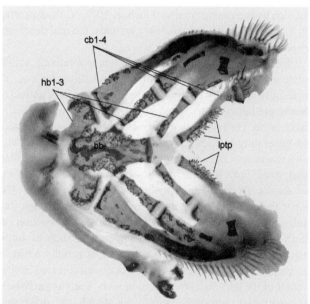

FIGURE 4.7 Branchial arches of the reedfish *Polypterus* (anterior to left): ventral elements. *Polypterus* has four branchial arches, not five as in holosteans and teleosts. Arches 1–3 are complete, being connected to the median basibranchial (bb) via hypobranchials (hb), whereas the fourth arch lacks hypobranchials (as does the fifth arch in actinopterygians). The principal lower pharyngeal tooth plates (lptp) are located on the most posterior arch (here the fourth, in teleosts the fifth), but in primitive fishes such as *Polypterus* smaller tooth plates bearing small teeth are distributed over all the elements of the arches. Cleared specimen. Cartilage stained with alcian blue, bone stained with alizarin red. Image width = 10 mm. *From Claeson, K., Bernis, W.E., Hagadom, J.W., 2007. New interpretations of the skull of a primitive bony fish* Erpetoichthys calabaricus. *J. Morphol. 268, 1021–1039. Courtesy Editors of Journal of Morphology.*

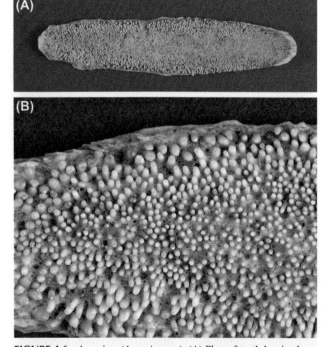

FIGURE 4.6 Arapaima (*Arapaima* sp.). (A) Plate of tooth-bearing bone attached to basihyals, forming ventral component of tongue bite apparatus. Image width = 26 cm. (B) High-power view of Fig. 4.6A. Image width = 5 cm. *Courtesy RCSOMA/ 453.1.*

In more advanced actinopterygians, the pharyngeal jaws take on the main role in processing food in addition to transport. Breaking down food requires the use of greater force than is required in simple transport, and this is facilitated by a variety of changes that enable the pharyngeal jaws to engage more effectively (Vandewalle et al., 1994, 2000). Teeth are concentrated on the tooth pads of the pharyngeal jaws. The pads become fused to the supporting endochondral bones, so that mechanical coupling is improved. Mechanical stability may be further improved by fusion of the lower tooth-bearing plates to form a single plate. The tooth pads on the upper pharyngeal jaws remain separate, but they are mechanically coupled by ligamentous connections.

A large group of percomorph fishes (cichlids, embiotocids, labrids, scarids, and pomacentrids) are described as **pharyngognathous**. They share several features that are considered to improve the efficiency of the pharyngeal jaws: development of a synovial joint between the upper pharyngeal jaw and the parasphenoid, union of the lower pharyngeal jaws by a suture or by fusion to form a single structure, and a muscular sling supporting the lower pharyngeal plate from the neurocranium (Liem, 1973; Liem and Greenwood, 1981; Wainwright et al., 2012). These characters were often considered as synapomorphies shared by the above-mentioned groups that were grouped together as Labroidei. However, subsequent work shows that pharyngognathy evolved independently 6–10 times and that the Labroidei are no longer accepted as a monophyletic group (Wainwright et al., 2012; Betancur et al., 2013, 2014). It was widely accepted (eg, Liem, 1973) that pharyngognathy was a key evolutionary innovation, in that it freed the oral jaws from food processing and promoted phylogenetic diversity. Hulsey et al. (2006) presented evidence for decoupling among cichlids, but phylogenetic studies suggest that the possession of specialized pharyngeal jaws does not seem to have been the direct cause of diversification in the radiations of cichlids and labrids (Seehausen, 2006; Alfaro et al., 2009; Price et al., 2010). However, this structural modification could have been essential for exploitation of new ecological niches (eg, coral reefs) that later resulted in diversification (Alfaro et al., 2009; Kazancıoğlu et al., 2009; Price et al., 2010).

Cypriniforms lack an upper pharyngeal jaw and the lower pharyngeal jaw occludes with a horny pad on the basioccipital region of the skull (Sibbing, 1982; Vandewalle et al., 2000; Pasco-Viel et al., 2010).

Food processing within the pharyngeal jaw apparatus is especially important to break down food items (eg, plant material or invertebrates) where access of digestive enzymes is hindered by cell walls, cuticle, or shell. In labrids, eg, mollusks are crushed between the pharyngeal teeth and the shell fragments ejected from the mouth before the soft parts are swallowed. The pharyngeal mill in halfbeaks has a shredding action (Tibbetts and Carseldine, 2003, 2004; Buddery et al., 2009). The parrotfishes have robust pharyngeal jaws that reduce plant material and ingested coral by a genuine grinding action (Carr et al., 2006): an extremely rare attribute among non-mammalian dentitions. There is considerable variation in the movements of the pharyngeal jaws among teleosts and the movements can be adapted to the physical properties of the prey. The operation of the jaws and the muscular activities involved are described in detail by Vandewalle et al. (1994, 2000).

Tooth Form

As in cartilaginous fishes, the number, and sometimes the form, of teeth vary between species and with age. However, some groups, such as piranhas and many other characins, retain the same tooth number throughout life. The dentition of bony fishes is usually homodont, and in nearly all species, the teeth are replaced continuously. Sharp, pointed teeth, usually recurved, are primitive for actinopterygians and most carnivorous actinopterygians possess a simple dentition consisting of a few rows of small conical teeth used for seizing and grasping prey, which tends to be swallowed whole. Predatory fishes, such as pike and barracuda, have mouths furnished with numerous fang-like teeth adapted to piercing and impaling prey. However, oral teeth are not essential for capturing prey, as in some successful carnivores, eg, billfishes, they are highly reduced or missing altogether. Although such relatively simple tooth forms are the rule, there is more variety of tooth form in bony fishes than in any other class of vertebrates. Oral teeth may be blade like (slicing), rounded (crushing), gouge or chisel shaped (scraping), or spatulate (plucking). Some groups have tooth plates or beaks, composite structures made up of individual teeth cemented together with bone. Some species, particularly those fishes that feed on shelled invertebrates, have heterodont dentitions: the anterior teeth are adapted to grasping shellfish and the posterior teeth to crushing them.

A similar variety of form is found in pharyngeal teeth. Among basal actinopterygians the pharyngeal jaws, like the oral jaws, are furnished with sharp, pointed teeth that provide grip for food transport. In more advanced teleosts, in which the pharyngeal jaws are responsible for most of the food processing, the pharyngeal teeth have a range of shapes, such as spatulate or gouge shaped (scraping), blunt and rounded (crushing or grinding), or blade like (shredding). Tooth form often varies within a single tooth plate or between upper and lower tooth plates. A good example is the pharyngeal mill of the carnivorous **river gar-fishes** (*Zenarchopterus*), which seems to be adapted to shredding prey. In the upper jaw, the teeth on the second pharyngobranchial pads have a simple pointed shape, whereas most of the teeth on the third pharyngobranchial are tricuspid and directed posteriorly. On the lower tooth pad, most of the teeth are also tricuspid and curved posteriorly, but the posterior margin of the pad bears larger pointed, curved teeth that point anteriorly (Tibbetts and Carseldine, 2004).

Several groups of bony fishes, such as Cyprinidae, lack oral teeth.

Tooth Attachment

The wide range of feeding methods places correspondingly variable mechanical demands on the teeth, and this is reflected in the variety of tooth attachments from rigid ankylosis to specialized hinge mechanisms. Formation of the tissues involved in tooth attachment is described in Chapter 9. From an extensive survey, Fink (1981) established the distribution of attachment modes among actinopterygians and demonstrated associations between certain types of attachment and particular lineages. He also showed that different modes of attachment can coexist in the oral dentition and that the attachment mode in the oral dentition often differs from that in the pharyngeal dentition. In the following, a combination of his classification and that of Shellis (1982) is used. Descriptive names are used in preference to the numbered categories of Fink (1981).

Ankylosis

[Type 1 of Fink (1981)]. This refers to teeth that are not divided like those in other modes of attachment and are fixed rigidly at the base to the bone of the jaws by ankylosis: fusion of tooth and jawbone by deposition of mineralized bone-like tissue (**bone of attachment**) (Fig. 4.8). Pedicellate and hinged modes of attachment also involve ankylosis between tooth and bone, but the division of the tooth into two parts confers some mobility. Fink (1981) concluded that ankylosis is the primitive form of tooth attachment among actinopterygians, as it is the almost universal mode in Cladistia, Holostei, and among osteoglossomorph teleosts. Rigidly attached teeth also occur among more advanced teleosts. Among sarcopterygians it is the mode of attachment in the tooth plates of lungfishes (Smith, 1985) and in the coelacanth (Shellis and Poole, 1978).

The rigid connection between tooth and jaw bone transmits the bite force to the functional tip of the tooth with minimal loss due to deflection. This mode of attachment is thus associated with biting dentitions (eg, garfishes, tigerfishes) and with crushing dentitions. Rigidly ankylosed teeth coexist with hinged teeth in some dentitions (eg, pikes).

Direct Fibrous Attachment

[Not in the classification of Fink (1981)]. Here, a narrow dividing zone at the base of the tooth is unmineralized, and it is traversed by collagen fibers running between the dentine of the crown portion of the tooth and the underlying bone of attachment. It has been reported in the piranha

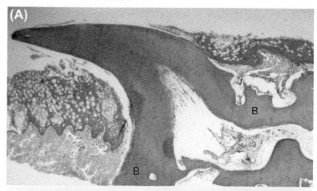
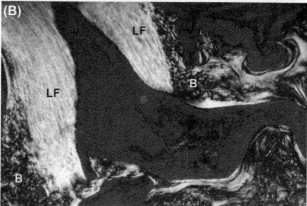

FIGURE 4.8 Ankylosis in coelacanth (West Indian Ocean *Latimeria chalumnae*). Ground section. (A) Bright field illumination. *B*, Bone. Image width=2.9 mm; (B) Higher power view of junction between tooth and bone. Polarized light: longitudinal fibers of dentine (LF) surrounded by woven bone (B), which has a checkered appearance because of the crisscrossing fibers in the tissue. Image width=450 μm.

(Fig. 4.9) (Shellis and Berkovitz, 1976) and can be observed in crushing teeth of some durophagous species. The mechanical properties of this mode of attachment are probably very similar to those of ankylosis, in view of the narrow width of the dividing zone.

Pedicellate Attachment

[Type 2 of Fink (1981)]. This mode has been referred to as indirect fibrous attachment (Shellis, 1982), but the term pedicellate is preferred, as it is widely used to describe the similar attachment in Amphibia (see Chapter 5). The distal ("crown") portion of the tooth is connected by an unmineralized, fibrous **dividing zone**[1] to a cylindrical structure of mineralized tissue, termed the **pedicel**, that is in turn ankylosed to the bone. The pedicel and the crown are developmentally a unit (see Chapter 9). In teleosts, the pedicel is shorter than the crown and is usually a short, squat cylinder (Figs. 4.10 and 4.11). Sometimes, the pedicel is incomplete, so that at one side the

1. We have adopted the term "dividing zone" from the literature on amphibian dental anatomy, for the sake of consistency, as the structure and development of the amphibian and teleost structures are very similar (see Chapters 5 and 9).

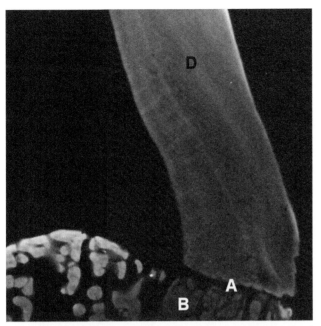

FIGURE 4.9 Piranha (*Serrasalmus rhombeus*), microradiograph of ground section, showing one side of the direct fibrous attachment of tooth shaft dentine (D) to bone of attachment (B). Site of dividing zone (A) looks dark because it is unmineralized. Image width=2.95 mm. *From Shellis, R.P., Berkovitz, B.K.B., 1976. Observations on the dental anatomy of piranhas (Characidae) with special reference to tooth structure. J. Zool. Lond. 180, 69–84. © John Wiley and Sons, reproduced by permission. See Fig. 2.1, p. 7, for copyright declaration.*

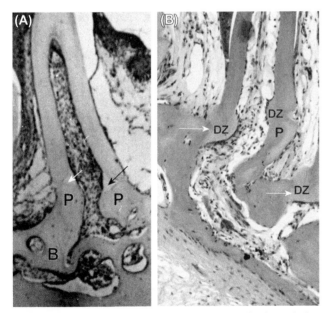

FIGURE 4.10 European eel (*Anguilla anguilla*). (A) Tooth attached to bone (B) via pedicel (P) and dividing zone (*arrows*). Image width=500 μm. (B) Tooth and part of its neighbor, with pedicels (P) that are incomplete on one side so that the dividing zone (DZ) connects directly with bone (point of connection marked by *arrows*). Hematoxylin and eosin. Image width=500 μm.

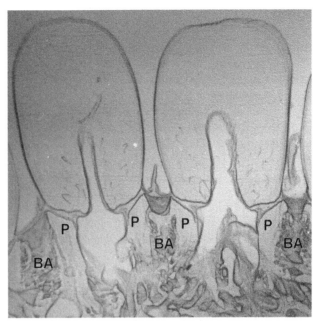

FIGURE 4.11 European plaice (*Pleuronectes platessa*). Ground section of two pedicellate teeth. The joint surface of the distal "crown" portion of each tooth is convex and rests inside the concave joint surface of the pedicel (P). The pedicels are embedded within bone of attachment (BA). Image width=1.8 mm.

fibrous attachment connects with bone rather than pedicel (Fig. 4.10B). The pedicellate attachment is derived from the rigidly ankylosed form of attachment (Fink, 1981) and is the most common mode of attachment among teleosts. It seems to be the universal mode in some groups, such as elopiforms and atheriniforms, and is widespread in many other taxa.

The fibrous joint in this mode of attachment confers mobility on the crown. Although this reduces the power of the bite, it provides a shock-absorbing mechanism. Pedicellate attachment occurs widely among teleosts with small pointed teeth and a grasping dentition, but it is not confined to this type of dentition, as it is also found in the incisiform teeth of the **pinfish** (*Lagodon rhomboides*: Sparidae) (Fink, 1981) and the **European plaice** (*Pleuronectes platessa*) (Fig. 4.11). Although the topic has not been studied systematically, variations in the morphology of the joint must affect tooth mobility and the mechanical properties of the attachment. Wide joints, with longer fibers, are presumably more mobile than equivalent but narrow joints. The shape of the two articulating surfaces must also affect the mobility. The joint surfaces are often flat or concave (sloping inward toward the pulp) and would offer little resistance to lateral force (Fig. 4.10A and B). In such joints, bending at the joint would be translated into tensile stresses in the attachment fibers. In other joints, the pedicel surface is concave and the distal surface is convex (Fig. 4.11). Such a joint may have primarily a cushioning role, with the attachment fibers being compressed rather than placed in tension.

The pedicellate attachment is an important feature of the teeth of the large number of teleosts that graze on Aufwuchs or periphyton. Grazing fishes, such as blennies (Blenniidae), mullets (Mugilidae), or blackfishes (Girellidae) have rows of numerous small teeth on their jaws. For an effective scraping action, the individual teeth need to have a flexible attachment, so that the tooth row can be applied closely to the irregular surface to which the food adheres. Usually, the width of the dividing zone is radially symmetrical, so that the degree of mobility is the same in all directions, but in some grazing fishes the pedicellate attachment can act as a hinge; it restricts mobility to one plane, albeit with a more restricted range of movement than in the hinged teeth described next. In the **opaleye** (*Girella nigricans*), eg, the hinge action is achieved by the development of a groove on the distal surface, into which fits a flange on the pedicel surface (Norris and Prescott, 1959). Restriction of movement to one plane prevents the tooth tip from twisting axially during operation (Norris and Prescott, 1959), so that the efficiency of each tooth is enhanced and neighboring teeth do not interfere with each other.

It should be noted that, as in amphibians, the dividing zone is sometimes mineralized, so the tooth is no longer mobile (Huysseune et al., 1998).

Hinged Teeth

The forms of attachment described by this term are derived from the pedicellate attachment. The term "hinged teeth" refers to teeth in which the joint between the crown and the pedicel not only restricts rotation to one plane, like the joint in *Girella* teeth, but also can be depressed almost completely in one direction while resisting depression in the opposite direction. The functions of hinged teeth are to allow prey to enter the mouth and to prevent it from leaving: the teeth are depressed by inward (lingual) movement and erected by outward labial) movement.

Hinged teeth occur in biting predatory fishes, and often coexist with rigidly ankylosed, stabbing teeth (Fink, 1981).

Anterior-Hinged Attachment

[Type 3 of Fink (1981)]. In this mode of attachment, mineralization of the crown extends into the anterior region of the fibrous attachment. This region forms a hinge point, with an anterior axis of rotation. Fink (1981) reported that teeth with this type of attachment can be depressed by 90° or more. The force restoring the teeth to the upright position is presumably provided by elastic strain energy stored in the thin mineralized anterior part of the hinge joint as the tooth is depressed.

Anterior-hinged teeth occur mainly in the deep-sea fishes belonging to the Stomiiformes (Fink, 1981). A similar type of attachment was also identified in the splenial teeth of *Polypterus* (Fink, 1981).

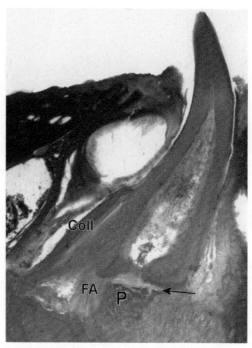

FIGURE 4.12 European hake (*Merluccius merluccius*). Slightly parasagittal longitudinal ground section of hinged tooth. *P*=pedicel; FA=fibrous attachment; Coll=collagenous strip connecting posterior side of distal portion of tooth to jaw bone. Discontinuity at anterior part of pedicel joint indicated by an *arrow*. Image width=4.28 mm. © *Museums at the Royal College of Surgeons, Tomes slide collection. Catalogue no. 362.*

Posterior-Hinged Attachment [Type 4 of Fink (1981)]. The main feature of this form of attachment is that there is no fibrous connection in the anterior region of the dividing zone; instead, there is a gap between the crown and the pedicel that thus cannot restrict depression. The remaining fibrous connection provides a hinge with a posterior axis of rotation. The mechanism restoring the tooth to its upright position varies between species. In the hinged teeth of the **northern pike** (*Esox lucius*) thick bundles of fibers in the dental pulp run between the tip of the crown and the bone of the jaw. These are stretched as the crown is depressed and their elastic recovery as the tooth is unloaded provides the restoring force. The hinged teeth of the **European hake** (*Merluccius merluccius*) possess a thick strip of collagen running from the posterior aspect of the crown to the jaw bone (Fig. 4.12). When the crown is depressed, the collagenous strip buckles and the stored elastic strain energy then provides a restorative force once the external force is removed.

This type of attachment is widespread among advanced teleosts. It occurs among the Stomiatiformes, among the Aulopiformes and Myctophiformes (Eurypterygia), among many paracanthomorphacean taxa, and among about half the clades within the Percomorphaceae.

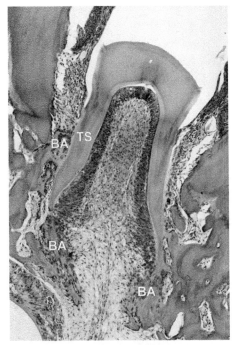

FIGURE 4.13 Thecodont attachment in Ballan wrasse (*Labrus bergylta*): beginning of ankylosis in erupted tooth. Spongy bone of attachment (BA) has been deposited between the tooth base and the adjoining bone. Bone of attachment is also starting to form basal to the tooth, especially at left, and between the lateral surfaces of the tooth shaft (TS) and the surrounding bone. Harris hematoxylin. Image width = 820 μm.

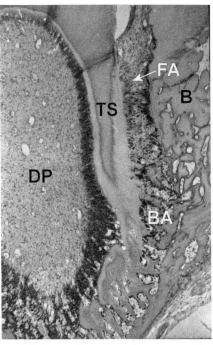

FIGURE 4.14 Ballan wrasse (*Labrus bergylta*). Junction between tooth shaft (TS: left) and surrounding jaw bone (B: right). Basally, the region between tooth and jaw bone has been filled with bone of attachment (BA) connecting the two surfaces. Apically, the tooth and bone are connected by obliquely orientated collagen fibers forming a fibrous attachment (FA), which resembles a periodontal ligament. This fibrous attachment will be replaced by spongy bone in due course. Harris hematoxylin. Image width = 380 μm.

Thecodont Attachment

In some teleosts, the teeth develop intraosseously, within crypts or troughs in the dentigerous bones and then erupt into function (Trapani, 2001). In most fishes with this type of development, such as piranhas (Berkovitz and Shellis, 1978), the teeth eventually attach to the crest of the jaw bone by their bases, but in the Labridae and Sparidae, the teeth become ankylosed by their lateral surfaces to the bone surrounding the eruption channel, and bone is also deposited basal to the tooth (Fig. 4.13). This is a form of **thecodont**, or socketed, attachment. It differs from that which occurs among crocodilians (Chapter 8), not only because the attachment is mineralized rather than fibrous, but also because the socket is not persistent: each replacement tooth erupts through its own channel that then becomes its socket. Tooth attachment in some species, such as the plaice (Fig. 4.11) and the **gilt-head (sea) bream** (*Sparus auratus*) (Shellis, 1982) combines the pedicellate and socketed modes of attachment, as the pedicel is ankylosed within the eruption channel[2]. Fink (1981) considered that thecodont attachment is rare and restricted to durophagous teleosts. However, the occurrence of socketed fang-like teeth in

some predatory teleosts (Bemis et al., 2005) suggests that thecodonty may be more widespread.

During the lateral ankylosis of the teeth of *Labrus*, a fibrous connection between bone and tooth surface is established as a first step (Fig. 4.14) (Shellis, 1982). Although it is a temporary structure that acts as a template for subsequent deposition of bone of attachment (Fig. 4.14), this fibrous connection closely resembles the periodontal ligament that supports teeth of crocodilians and mammals. Soule (1969) reported that the teeth of trigger fishes (Balistidae) are attached by a periodontal ligament. However, although the ligament is similar to that of mammals or crocodilians, it seems to connect neighboring teeth to each other rather than to the jaw bone. The teeth seem to be attached by ankylosis to shallow depressions in which the teeth rest.

"Unattached" Teeth

In Chilodontidae and Prochilodontidae (Characiformes), which feed on mud and detritus, only the first generation of teeth are attached to the jaw bones. Subsequent generations are attached only in the dermis of the lips, so the erupted tips convert the skin of the lips into a rasping surface.

---

2. Note that Fig. 19 of Shellis (1982) is incorrectly labeled: the letters a and b should be reversed, so that b = pedicel and a = bone of attachment.

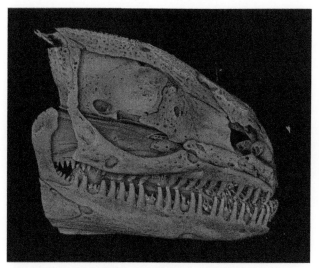

FIGURE 4.15 Senegal bichir (*Polypterus senegalus*). Micro-CT image of dentition seen in lateral view. Image width = 16 mm. *Courtesy Prof M.M. Smith.*

FIGURE 4.16 Senegal bichir (*Polypterus senegalus*). Micro-CT image of dentition seen in medial view. Image width = 18 mm. *Courtesy Prof M.M. Smith.*

DENTITIONS OF ACTINOPTERYGII

The descriptions in this chapter focus mainly on the oral dentition, because of the greater availability of specimens for illustration. However, we present detailed accounts of the pharyngeal dentitions of some taxa.

CLADISTA

The Cladista consists of only the order **Polypteriformes** that includes 12 living species of ropefishes and bichirs in two genera. They differ from the majority of Osteichthyes as they have lungs, breathe air, and do not have typical swim bladders. As the name suggests, these fishes have numerous fins.

Polypteriformes

The **Senegal bichir** (*Polypterus senegalus*) and the **reedfish** (ropefish or snakefish) (*Erpetoichthys calabaricus*) are freshwater fishes that inhabit swamps and flood plains, feeding on small vertebrates, crustaceans, and insects. In both fishes, the dentary, premaxilla, and maxilla each carry a single row of large, slightly recurved teeth (Fig. 4.15). Inside these rows of large teeth lie multiple rows of smaller teeth on the prearticular and coronoids in the lower jaw and on all bones forming the roof of the mouth (Figs. 4.16 and 4.17). Pads of teeth also occur on most of the branchial arch bones (Fig. 4.7). *Polypterus* uses suction to ingest prey from the bottom of freshwater lakes and rivers (Lauder, 1980).

NEOPTERYGII: HOLOSTEI

This infraclass contains two small orders of North American bony fishes displaying primitive characteristics. Bowfins and gars inhabit fresh and brackish water in North America

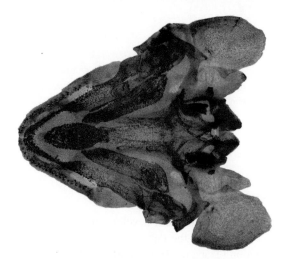

FIGURE 4.17 Reedfish (*Erpetoichthys calabaricus*). Ventral view of palate, showing numerous teeth on the premaxilla, maxilla, parasphenoid, dermopalatine, and endopterygoid. Alizarin red preparation. Image width = 27 mm. *From Claeson, K., Bernis, W.E., Hagadom, J.W., 2007. New interpretations of the skull of a primitive bony fish* Erpetoichthys calabaricus. *J. Morphol. 268, 1021–1039. Courtesy Editors of Journal of Morphology.*

and, like the polypteriforms, can obtain oxygen from both water and air.

Amiiformes

Bowfin (*Amia calva*: Amiidae) is the single representative of this order. The premaxillae, maxillae, and dentaries carry numerous strong, conical teeth (Fig. 4.18). Median tooth numbers and ranges are premaxilla 8 (6–9); maxilla 42 (29–50), and dentary 15 (10–19). Tooth numbers do not seem to increase with size (Miller and Radnor, 1973). Additional teeth, some of which are larger than on the premaxillae and maxillae, are carried on the vomers, palatines, and ectopterygoids (Fig. 4.19). Suction feeding in *Amia* is enhanced by strong forward rotation of the maxillae, which

FIGURE 4.18 Teeth of bowfin (*Amia calva*). Image width = 12 cm. *Courtesy Dr. C. Underwood.*

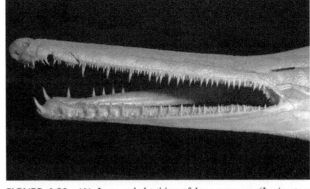

FIGURE 4.20 (A) Jaws and dentition of long-nose gar (*Lepisosteus osseus*). Image width = 16 cm. *Courtesy Dr. C. Underwood.*

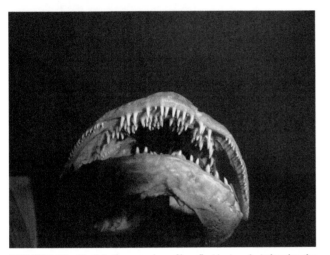

FIGURE 4.19 Teeth in the upper jaw of bowfin (*Amia calva*) showing the presence of many rows of palatine teeth. Image width = 10 cm. *Courtesy of UCL, Grant Museum of Zoology. Catalogue No. LDUCZ-V348.*

FIGURE 4.21 Teeth of alligator gar (*Atractosteus spatula*). © *Betty Wills, 2005.*

brings forward a fold of skin that limits the gape to an anteriorly directed opening (Lauder, 1980).

Lepisosteiformes

There is in this order a single family (Lepisosteidae), containing seven species of gars in two genera. *Lepisosteus* is an ambush predator that, after a slow, prolonged approach, aligns its snout alongside the prey. The ensuing strike is a rapid sideways lunge at the prey that is then seized in the middle of the body (Lauder, 1980; Porter and Motta, 2004). The long jaws are adapted for rapid closing (Westneat, 2004; Porter and Motta, 2004) and the strike, from initiation to contact with the prey, is over in 20–35 ms on average (Lauder, 1980; Porter and Motta, 2004). The **longnose gar** (*Lepisosteus osseus*) can be distinguished from other gars by having a snout that is more than twice the length of the rest of the head. Its elongated jaws are furnished with literally hundreds of small, thin, sharply pointed teeth. Larger caniniform teeth are regularly spaced between the smaller teeth

(Fig. 4.20). **Alligator gar** (*Atractosteus spatula*) has double rows of teeth on each side of the upper jaw (Fig. 4.21).

NEOPTERYGII: TELEOSTEI

The teleosts comprise nearly all living fishes, representing about 27,000 species in approximately 40 orders and 450 families. In "traditional" classifications, the teleosts have been divided into 12 main superorders: Osteoglossomorpha, Elopomorpha, Clupeomorpha, Ostariophysi, Protacanthopterygii, Stenopterygii, Cyclosquamata, Scopelomorpha, Lampridiomorpha, Polymyxiomorpha, Paracanthopterygii, and Acanthopterygii. In the classification of Betancur et al. (2014), much of the traditional terminology is retained (see Fig. 4.1).

Elopomorpha

This cohort contains two orders of fusiform, predatory fishes and three orders of eels, of which there are about 800 species. All members have flat, transparent, leptocephalus larvae.

Anguilliformes

Congridae

Conger eel (*Conger conger*) is an opportunistic predator that generally eats smaller fish. The dentition consists of numerous, closely packed, sharp teeth that are mainly arranged in one or two rows, although anteriorly and in the roof of the mouth there are many more rows (Figs. 4.22 and 4.23).

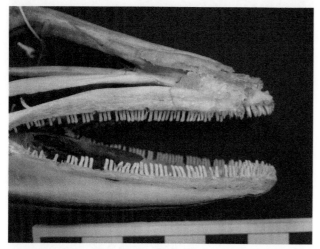

FIGURE 4.22 Jaws of conger eel (*Conger conger*). Image width=7 cm. *Courtesy RCSOMA/ 460.2*

FIGURE 4.23 Upper jaw of conger eel (*Conger conger*).Ventral view. Teeth on maxillae, premaxillae, and vomers. Note the replacement teeth on the inner surface of the maxillae. Image width=7 cm. *Courtesy RCSOMA/ 460.21*

Muraenidae

Moray eels are fearsome predators with teeth to match. In the **fangtooth moray** (*Enchelycore anatina*), the teeth may reach a length of more than 2 cm, look transparent, and are arranged in multiple rows (Fig. 4.24). In addition to marginal rows in the upper and lower jaws, teeth are also present on the vomers, as illustrated in the **fimbriated moray eel** (*Gymnothorax fimbriatus*) (Fig. 4.25). In the **Nago snake moray** (*Uropterygius nagoensis*), the teeth are arranged in multiple rows on the premaxilla, maxilla, and dentary; the teeth forming the inner rows are the largest (Hatooka, 1984)

FIGURE 4.24 Fangtooth moray (or tiger moray) (*Enchelycore anatina*), showing long, sharp teeth. © *Natursports/Dreamstime.com.*

FIGURE 4.25 Palatal teeth of fimbriated moray eel (*Gymnothorax fimbriatus*). © *Rod Andrewartha.*

Because they have narrow heads and hide in confined spaces, moray eels seem to be unable to expand their mouth and pharynx enough to generate the negative pressure required to suck prey into their mouths. In some species (eg, **reticulated moray**, *Muraena retifera*), the loss of suction is compensated by the development of unique protrusible pharyngeal jaws (Fig. 4.26). The upper pharyngeal jaw is formed by the fourth epibranchials and the lower pharyngeal jaw by the fourth ceratobranchials; both jaws bear sharp recurved teeth. Instead of processing food already delivered into the pharynx, the pharyngeal jaws are thrust forward to help grasp prey initially seized by the oral teeth. Prey is then moved into the esophagus by cooperative movements of the two sets of jaws. The pharyngeal jaws retract and the head is moved forward as the oral jaws release their grip, and this is followed by the oral jaws restraining the prey while the pharyngeal jaws are protracted to take a new grip (Mehta and Wainwright, 2007).

Muraenesocidae

The **common pike conger** (*Muraenesox bagio*), like other members of this family, has a long, narrow head and a large mouth. The pike conger feeds on benthic fishes and crustaceans. The highly elongated jaws carry numerous conical teeth (Fig. 4.27). The marginal teeth comprise an outer row of small teeth and an inner row of medium-sized teeth, of which those in the anterior three-quarters of the jaw are flattened and blade like, with mesial and distal cutting edges. Teeth in the posterior quarter of the jaw are blunt and peg shaped. At the front of the upper and lower jaws are three to five large, spaced, fang-like teeth. On the palate there is a double row of small conical teeth and a median row of very large, spaced, dagger-like teeth with sharp cutting edges.

Ostariophysi

This subcohort is the second largest grouping of teleosts. It is the dominant fish group in fresh water and contains catfishes, carps, characins, and electric eels. Ostariophysans are distinguished by the possession of the Weberian apparatus: a chain of ossicles linking the swim bladder to the inner ear. This system acts as a sound amplifier, with the swim bladder acting as a resonator that allows detection of faint sounds that would be otherwise inaudible.

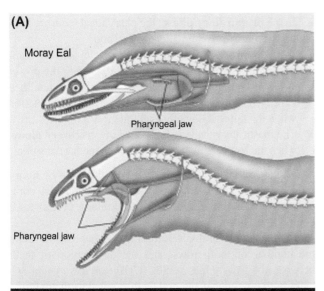

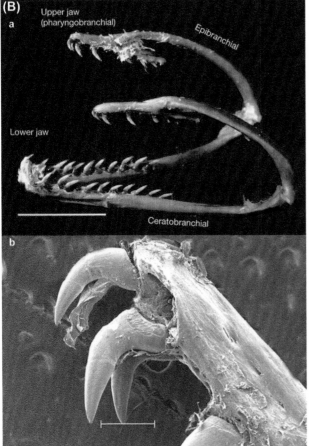

FIGURE 4.26 (A) Diagram of pharyngeal jaws in moray eels. Top: jaws at rest, with pharyngeal jaws in normal position. Bottom: oral jaws open and pharyngeal jaws protracted to sieze prey. *(Courtesy of Wikipedia.)* (B) Left lateral view of a cleared and Alizarin red–stained pharyngeal jaw apparatus, illustrating the sharp, recurved teeth on the pharyngobranchials used to grasp prey. Scale = 1 cm. (C) Scanning electron micrograph of recurved teeth at the front of the left anterior upper pharyngobranchial. Scale = 0.5 mm. *From Mehta, R.S., Wainwright, P.C., 2007. Raptorial jaws in the throat help moray eels swallow large prey. Nat. Lond. 449, 79–82. Courtesy Editors Nature, London.*

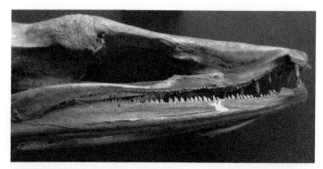

FIGURE 4.27 Jaws and dentition of common pike conger (or pike eel) (*Muraenesox bagio*). Image width = 10 cm. *Courtesy Horniman Museum & Gardens and Dr. Paolo Viscardi. Catalog no. LDHRN-NH.83.1.16.*

Within the Ostariophysi there has been widespread loss of teeth from the roof of the mouth, the basihyals, and the upper pharyngeals, although teeth occur in one or more of these sites in all orders except for Cypriniformes (Stock, 2007).

Cypriniformes

This order contains about 3700 species in three superfamilies: Cobitoidea (nine families of loaches), Catastomoidea (two families of suckers), and Cyprinoidea (two families of carp, barbs, goldfishes, and minnows).

Cypriniforms have lost all teeth except for those on the fifth ceratobranchials. Tooth loss seems to have commenced in the last common ostariophysan ancestor and to have continued in stages during later evolution of the cypriniforms. The timing of the important event of loss of the marginal oral teeth is, however, uncertain (Stock, 2007). The cypriniforms are thought to have evolved from a bottom-feeding ancestor with a simple dentition (Gosline, 1973; Stock, 2007). Many extant species continue to feed in this way, but the mode of feeding has diversified and many species feed in open water. Although the diet of some species is dominated by a particular food, eg, snails or algae, most cypriniforms are omnivorous and feed on plant material, small invertebrates, and decaying matter. Cypriniforms are suction feeders and the loss of oral teeth may have facilitated passage of food on the stream of water through the oral cavity (Gosline, 1973). Protrusible jaws, using a distinctive mechanism (Motta, 1984a), have evolved independently in this group (Westneat, 2004) and probably enhance retrieval of food from the bottom of the water as they enable suction from a more ventral direction (Gosline, 1973).

Cypriniforms lack oral teeth and rely entirely on the pharyngeal jaws for processing food. Food is taken into the mouth by suction and transported by peristaltic movements of the branchial basket to the pharyngeal dentition, where it is ground before being swallowed. Large food particles are crushed first, before being ground. The lower pharyngeal jaws have a bent shape, with the lower half directed anteroventrally, articulating with the branchial basket and the upper half orientated approximately vertically. The pharyngeal teeth are located medially, at the level of the angle between the upper and lower halves. There is no upper pharyngeal jaw. Instead, the lower pharyngeal teeth oppose a horny pad that is anchored to the basioccipital process: a structure derived from fusion of the ventral arches of the first three vertebrae with the skull.

In a detailed study of food processing in the **common carp** (*Cyprinus carpio*), Sibbing (1982) described the masticatory cycle as consisting of three phases:

- In the **preparatory phase**, the pharyngeal teeth are lowered, thereby providing access for food.
- In the **power stroke**, the teeth are elevated to compress the food against the horny pad. During the power stroke, the teeth may have a crushing action when they approach the horny pad vertically, or a grinding action when the approach is more oblique.
- The masticatory cycle is completed by a **recovery phase**, when the teeth are withdrawn to an intermediate position.

The movements of the teeth are brought about by rotary movements and abduction of the ceratobranchials, combined with rotation of the skull [see Sibbing (1982) for a comprehensive account].

There is considerable variation in the shape of the teeth, the number of tooth rows, and the distribution of teeth between the rows. One species (*Gyrinocheilus aymonieri*) lacks both pharyngeal and oral teeth, conical teeth are found among Cobitoidea and Catastomoidea, and a further five shapes occur among Cyprinoidea (Pasco-Viel et al., 2010). The six tooth shapes found among Cypriniformes are:

1. conical: simple teeth with a rounded tip (Fig. 4.28D);
2. spoon shaped: conical teeth with a concave, pointed tip and a hook (Fig. 4.28A);
3. spatulate: compressed but with the apical regions swollen and fitted together and with truncated grinding surfaces, forming a common roundish chewing area (Fig. 4.28B);
4. compressed: very broad teeth, with margins either straight or convex anteriorly and concave posteriorly (Fig. 4.28C);
5. saw shaped: compressed teeth with a grinding surface containing many protuberances (Fig. 4.28E); and
6. molariform: crushing teeth with broad grinding surfaces (Fig. 4.28F).

There are one to three tooth rows, with different numbers of teeth in each row (Cobitoidea and Catastomoidea have one row). Although many species have no more than five teeth in a row, Cobitoidea have 10–20 teeth per row, whereas some Catostomoidea possess up to 100 teeth per row. A dental formula may be written for cypriforms. For the **zebrafish** (*Danio rerio*), with three tooth rows, the formula is 2,4,5 (Fig. 4.28A), indicating there are two teeth in the dorsal row, four teeth in the intermediate row, and five teeth in the ventral row. Table 4.1 gives the dental formulae for the pharyngeal dentitions of some Cypriniformes. Where data for only one series are given, it implies that there is

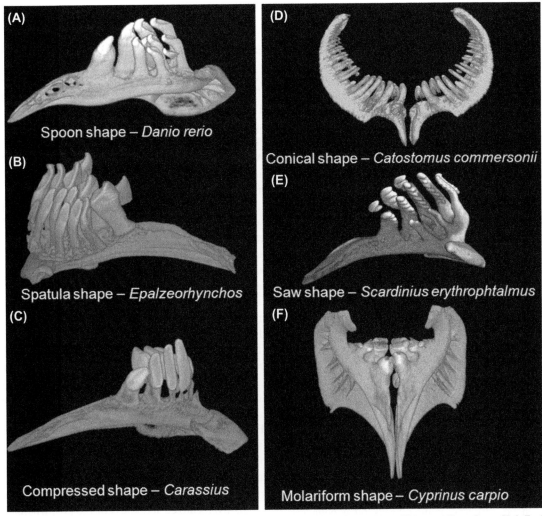

FIGURE 4.28 Pharyngeal tooth plates in Cypriniformes, showing six variations in tooth form. *Micro-CT images. From Pasco-Viel, E., Charles, C., Chevret, P., Semon, M., Tafforeau, P., Viriot, L., Laudet, V., 2010. Evolutionary trends of the pharyngeal dentition in cypriniformes (actinopterygii: Ostariophysi). PLoS One 5, e11293. Courtesy Editors of Public Library of Science. (PLoS One).*

TABLE 4.1 Shape and Distribution of Pharyngeal Teeth Among Some Cypriniforms

Family	Species	Dental formula	Tooth shape
Cobitidae–	*Misgurnus anguillicaudatus*	14–16	Conical
Catastomidae	*Carpiodes microstomus*	50	Conical
Cyprinidae–Rasborinae	*Danio rerio*	2.4.5	Spoon
	Laubuca laubuca	2.4.5–4.4.2	Spoon
	Rasbora borapettensis	4.5	Spoon
Cyprinidae–Cyprininae	*Carasssius auratus*	4	Compressed
	Cyprinus carpio	1.1.3	Molariform
	Epalzeorhynchus frenatum	2.4.5	Spatulate
Cyprinidae–Cultrinae	*Ctenopharyngodon idella*	2.4	Compressed
	Xenocypris yunnanensis	2.3.5–6.3.2	Compressed
Cyprinidae–Acheilognathinae	*Rhodeus sericeus*	5	Compressed
Cyprinidae–Gobioninae	*Gobio gobio*	3.5–5.2	Spoon

Modified from Pasco-Viel, E., Charles, C., Chevret, P., Semon, M., Tafforeau, P., Viriot, L., Laudet, V., 2010: Evolutionary trends of the pharyngeal dentition in cypriniformes (actinopterygii: Ostariophysi). PLoS One 5, e11293. Courtesy Editors Public Library of Science (PLoS One).

bilateral symmetry. However, the cyprinoid pharyngeal dentition is frequently asymmetrical. Thus, the dental formula of *Xenocypris yunnanensis* indicates that the dorsal row on the left has six teeth and that on the right side has five.

Because of the great variety of dental morphology, diet, and ecology, small cypriniforms, such as cichlids, have attracted attention as useful subjects for studying adaptation and evolution. Pasco-Viel et al. (2014) explored the evolutionary radiation of two tribes of the subfamily Cyprininae: the Poropuntini (100 species), which are confined to Southeast Asia; and the Labeonini (400 species), which occur between East Asia and Africa. The difference in distribution is thought to be due to changes in availability of migration routes between Asia and Africa. The Poropuntini retained several ancestral features of the feeding apparatus, including a terminal mouth opening and a high pharyngeal cavity. Radiation within this group involved diversification of the number, shape, and arrangement of the pharyngeal teeth in adaptation to a wide range of diets. In contrast, there occurred within the Labeonini a correlated series of morphological adaptations to feeding on small benthic foods, including algae and zooplankton adherent to the substrate: a ventrally positioned mouth, low pharyngeal cavity, and a pavement-like arrangement of pharyngeal teeth. Thus, the evolutionary radiations in the Poropuntini and Labeonini were driven by ecological diversification and specialization, respectively.

One cyprinid, the **zebrafish**, is a very important model organism in many fields of research, including development and regenerative medicine. Its genome has been fully sequenced and transgenic hybrids have been created. It is attracting increasing attention in dental research. Its simple dentition (Fig. 4.28A) offers possibilities for investigating such problems as the control of number and shape of teeth and the loss and gain of teeth during evolution. Progress in this area, and potential contributions of research on zebrafish, is reviewed by Stock (2007) and Bruneel et al. (2015).

Perhaps the most remarkable dentition among the cypriniforms belongs to the **Dracula fish** (*Danionella dracula*), a relative of the zebrafish discovered in Myanmar (Britz et al., 2009). This tiny fish—only 17-mm long—has a much reduced neurocranial and branchial skeleton. Unlike all other cypriniforms, males possess tooth-like structures (odontoid projections) on the jaws (Fig. 4.29A, C and D). Each of the jaw bones in males has 6–13 odontoid projections that decrease in size distally (Fig. 4.29E). The most anterior of the upper projections form a pair of recurved fangs that project roughly 0.3 mm beyond the epithelium, whereas the lower fangs have only the tips barely exposed. Females have only a few short projections and the fangs are rudimentary (Fig. 4.29B). The odontoid projections are not, however, true teeth, made up of dentine and enameloid; instead, they are bony structures that penetrate into the mouth (Fig. 4.29F). The

pharyngeal teeth of *D. dracula* are true teeth, with a pulp cavity and an enameloid cap (Fig. 4.29H).

Characiformes

This order is divided into 23 families with more than 1500 species and includes a vast array of fishes that live in rivers, streams, and lakes. The characins comprise a small group of African species (30 genera in 4 families) and a much larger and more diverse group of Neotropical species (250 genera in 19 families).

The characins have nonprotrusible jaws and often possess teeth on both the maxilla and premaxilla. Guisande et al. (2012) distinguished 14 tooth shapes among characins. Tooth development and succession are unusual among characins. First, although in most actinopterygians replacement teeth develop lingual to the functional teeth, those of characins, with a few exceptions, develop intraosseously beneath the functional teeth, either in a cavity within the jaw bone, as in *Astyanax* (see Fig. 4.36B) (Marinho and Lima, 2009) or in a trench, as in *Hemigrammus* (see Fig. 4.37B) (Trapani, 2001; Trapani et al., 2005; Lima et al., 2009). Second, in many characins all the teeth in each quadrant develop, erupt, and are shed almost simultaneously (Roberts, 1967; Berkovitz and Shellis, 1978; Trapani et al., 2005), as in kitefin and cookie-cutter sharks (see page 21). In piranhas, this enables the establishment of an interlocked row of teeth (see page 64). However, it may have a wider adaptive value. Trapani et al. (2005) suggested that replacement of a single tooth from below could produce a localized weakness of the jaw, whereas replacement of the entire quadrant, coupled with a temporary cessation of feeding, which permits reformation of the bone of attachment, might circumvent this problem.

In many characins, the mandibular symphysis is knitted together by interdigitating processes from the subalveolar bone of the dentaries. The symphysis thus forms a hinge that allows the mandibular rami to be spread apart, but prevents twisting (Eastman, 1917; Gregory and Conrad, 1936; Vari, 1979).

African Characins

The Citharinidae are filter feeders and the Hepsetidae is a small family of fish-eating carnivores. The Distichodontidae include herbivores, predators on small prey, and carnivores, which grow up to 80 cm in length. Their dentition consists of one to three rows of bicuspid teeth on the premaxilla and dentary. The teeth have a fibrous attachment to the jaw bones in a pleurodont position or, in a few species, in an acrodont position. In several distichodontid genera, the intramandibular joint between the dentary and angular is mobile (Vari, 1979), as in many percimorph coral fishes (see page 46). The Citharinidae and Distichodontidae lack the hinge-like symphysis present in most characins (Vari, 1979).

The only African characins included here are the Alestidae.

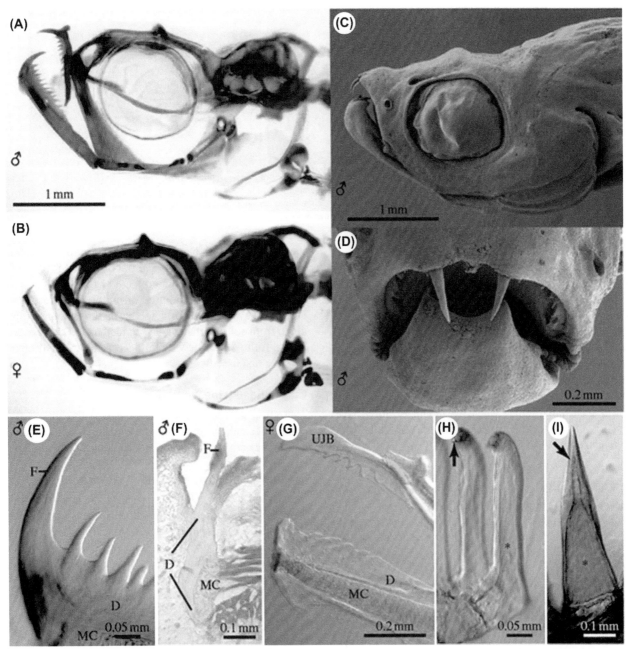

FIGURE 4.29 Dracula fish (*Danionella dracula*). (A) Male, paratype, 16.2 mm (BMNH, 2008.1.1.100–119), cleared and double stained, head skeleton, lateral view; (B) Female, paratype, 14.7 mm (BMNH, 2008.1.1.100–119), cleared and double stained, head skeleton, lateral view; (C) Male, paratype, 14.8 mm (BMNH, 2008.1.1.2–99), scanning electron micrograph, lateral view; (D) Same as (C), but close-up of mouth in anterior view; (E) Same as (A), but close-up of odontoid processes at anterior tip of lower jaw; (F) Male, 16.7 mm, histological section through canine-like odontoid process of the lower jaw, illustrating lack of pulp cavity and enameloid cusp; (G) Female, paratype, 15.4 mm (BMNH, 2008.1.1.100–119), cleared and double stained, close-up of jaws with serrate edges on jaw bones and rudimentary canine-like odontoid processes, lateral view; (H) Same as (B), but close-up of pharyngeal jaw teeth with pulp cavity (marked with *asterisk*) and enameloid cusp (marked with *arrow*); (I) *Hoplias malabaricus* (BMNH, 2005.7.5.778–863), 44 mm, cleared and double stained, close-up of individual premaxillary tooth with pulp cavity (marked with *asterisk*) and enameloid cusp (marked with *arrow*) for comparison with (E). *D*, dentary; *F*, fang-like odontoid process; *MC*, Meckel cartilage; *UJB*, upper jaw bone. *From Britz, R., Conway, K.W., Rüber, L., 2009. Spectacular morphological novelty in a miniature cyprinid fish,* Danionella dracula *n. sp. Proc. Biol. Sci. 276, 2179–2186. Courtesy Editors Proceedings of Biological Sciences.*

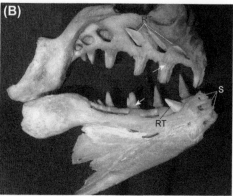

FIGURE 4.30 (A) Dentition of giant tigerfish (*Hydrocynus goliath*). *(Courtesy Wikipedia.)* (B) Dentition of African tigerfish (*Hydrocynus sp.*), medial view. The initially sharp, powerful and spaced teeth have undergone considerable wear. Their basal portions (*arrows*) are composed of plicidentine, so have ridged surfaces. Note the angulation of the replacing teeth (RT), which undergo considerable rotation during their eruption into a functional position. Flanges of bone at symphysis of lower jaw (S) form part of a hinge mechanism. Image width = 12 cm. *(Courtesy MoLS KCL. Catalogue no. V171.)*

Alestidae

The Alestidae includes omnivorous species, such as the **Congo tetra** (*Phenacogrammus interruptus*) and also large predatory species, such as the tigerfishes. The **tigerfishes** (*Hydrocynus* spp.) are voracious hunters, of which the **giant tigerfish** (*Hydrocynus goliath*: Fig. 4.30A) is the largest. There is a limited number of very large, spaced, dagger-like teeth that may show subsidiary points. The premaxilla carries six or seven teeth and the dentary five. The smallest teeth are at the back of the jaws (Fig. 4.30B). The upper and lower teeth interdigitate when the mouth is closed (Eastman, 1917). All teeth in each quadrant are replaced together (Trapani et al., 2005).

Neotropical Characins

This is a much larger and more diverse group than the African characins. They range in size from species of *Salminus* that grow to 100–130 cm in length to miniature forms that do not exceed 26 mm in length. Indeed, characiforms contribute about 40% of Neotropical miniature fishes (Toledo-Piza et al., 2014).

Guisande et al. (2012) identified two major lineages within the group. The members of **Lineage I** tend to inhabit slow-moving waters, such as lakes, swamps, and slow rivers and comprises the Serrasalmidae (carnivores, omnivores, and frugivores), the Anostomidae (herbivores), the Prochilodontidae, and the Curimatidae (all detritivores). The members of **Lineage II** tend to live in fast-moving waters, such as streams and creeks, and comprise the Characidae (omnivores, carnivores, and scale eaters), the Acestrorhynchidae (carnivores), Gastropelecidae (insectivores), and the Chalceidae (omnivores). The carnivorous Ctenoluciidae and Cynodontidae form a sister group to Lineage I.

The teeth of Neotropical characins are extremely varied. Guisande et al. (2012) identified 14 distinct tooth shapes. They observed in the characins as a whole a broad correlateion between six of their tooth forms and the diet (Fig. 4.31). Carnivores have sharp, pointed teeth; herbivores and alga eaters have flat or notched (bicuspid) teeth; frugivores have pointed teeth or flattened crushing teeth; omnivores have pointed or multicusped teeth, and detritivores have villiform teeth (in those that scraped algae from solid surfaces) or lack teeth (mud consumers). In the following, a few examples of Neotropical characins are discussed.

Curimatidae (Lineage I)

These fishes extract nutrients from mud. Although one generation of small, conical teeth develop and attach to the jaw bones, they are not replaced, so the adults are edentulous.

Chilodontidae, Prochilodontidae (Lineage I)

These fishes are also detritivores, but they scrape algae attached to submerged surfaces. Feeding is facilitated by the presence of distinct mobile lips, separated from the jaws by a groove. The dentition is unique because although the first generation of small conical teeth are attached to the jaw bones, numerous subsequent generations of complete small teeth remain within the soft tissues of the lips, moving outward with time (Castro and Vari, 2004). The tips of the functional (oldest) row pierce the surface of the lips and provide a rasping action as the lips move against the alga-covered substrate.

Serrasalmidae (Lineage I)

In broad terms, this family comprises fishes—loosely referred to as "pacus" and "piranhas"—with very different diets. The pacus are mainly herbivorous, whereas the piranhas are omnivorous or carnivorous. In both groups, teeth in each quadrant are replaced together (Roberts, 1967).

Pacus

Metynnis, Mylossoma, Mylesinus, Myleus, Colossoma, and *Piaractus* feed largely on vegetation, although they also consume insects and snails. The upper dentition of the **tambaqui**

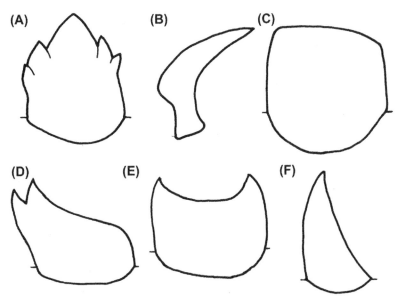

FIGURE 4.31 Diagram showing six different tooth types in Neotropical characiforms. (A) Multicispid; (B) Villiform; (C) Flat; (D) Crenulated; (E) Molariform; (F) Canine. *Modified from Guisande, C., Pelayo-Villamil, P., Vera, M., Hernández, A.M., Carvalho, M.R., Vari, R.P., Jiménez, L.F., Fernández, C., Martínez, P., Prieto-Piraquive, E., Granado-Lorencio, C., Duque, S.R., 2012. Ecological factors and diversification among Neotropical characiforms. Int. J. Ecol. Article ID: 610419, 20 pp.*

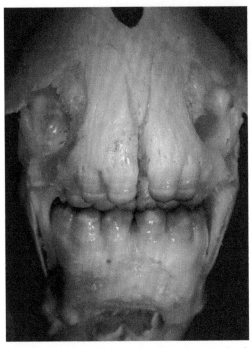

FIGURE 4.32 Anterior view of pacu (species unknown), showing massive rounded crushing teeth. Note the double row of multicusped molariform teeth in the upper jaw. *Courtesy Dr. R. Britz. © Trustees of the Natural History Museum, London.*

(*Colossoma macropomum*) consists of a double row of multicusped molariform teeth in the upper jaw, plus two rows on the roof of the mouth. The lower jaw carries a single row of similar teeth plus an inner pair of conical teeth at the front of the mouth. The upper and lower teeth of pacus together crush and break down seeds, nuts, and fruits (Fig. 4.32).

Piranhas

The popular reputation of piranhas as voracious predators hunting in shoals is much exaggerated, bearing in mind that they rarely exceed 20 cm in length and usually form small shoals of only about 10–100 fishes. Attacks on humans are rare and seem to occur when water levels are low or when food is scarce. The diet of piranhas varies considerably between species and according to food availability and season. *Pygocentrus* is closest to being an exclusive carnivore (piscivore), whereas *Serrasalmus* and *Pristobrycon* take a higher proportion of other foods, such as insects, crustaceans, and fins and scales of other fishes (Nico and Taphorn, 1988). Young individuals eat small invertebrates and fish fins make up a substantial part of their diet. The **wimple piranha** (*Catoprion mento*) is a specialized scale eater.

In most piranhas, the dentition consists of a limited number of relatively large, blade-like, multicusped teeth arranged in single rows and is unique among bony fishes because they are interlocked. Therefore, the row of teeth in each quadrant forms a single, serrated blade that shears past the opposing row and slices flesh from larger prey. This biting mechanism can be considered as even more efficient than that of sharks such as *Isistius*, in which rows of interlocked cutting teeth occur only in the lower jaw and the upper teeth are used for grasping (see page 21): the cutting action can only be completed by twisting movements of the head. *Catoprion* has double rows of teeth.

The **red-bellied piranha** (*Pygocentrus nattereri*) is an effective predator on other fishes. Its dental anatomy was described by Shellis and Berkovitz (1976). It has a constant dental formula throughout life; there are only six teeth on each premaxilla and seven on each dentary (Figs. 4.33 and 4.34), together with a few small blunt teeth on the ectopterygoid bones. In the upper jaw, the first five marginal teeth are tricuspid, whereas the sixth has four cusps. In the lower jaw, the most anterior tooth is tricuspid, whereas the second to sixth teeth have a prominent mesial main blade and a smaller distal round blade. The teeth are interlocked with their neighbors by the distal cusp fitting into a depression in the mesial aspect of the tooth behind, except that both mesial and distal cusps of the small third upper teeth engage with the adjacent teeth. The interlocking of the extremely sharp teeth forms a single cutting edge, as shown in Fig. 4.34.

The bases of the teeth of piranhas are fixed to the crest of the jaws by a fibrous attachment, but one that is so narrow (Fig. 4.9) that the joint is strong and almost inflexible. The tooth base is concave and rests on a saddle-shaped area of bone. This arrangement, together with the interlocking of the teeth, provides the tooth row with a firm, stable support.

Piranhas have a forward-jutting, deep-bodied lower jaw. The gape is large relative to total body length: a piranha 10-cm long has a 1-cm gape (Fig. 4.33). The high mechanical advantage for jaw closing indicates a powerful bite (Westneat, 2004). As the mouth is snapped shut, the tips of

FIGURE 4.33 Red-bellied piranha (*Pygocentrus nattereri*) with mouth open, showing the razor-sharp pointed teeth (six in the upper jaw, seven in the lower). Image width = 12 cm. *From Berkovitz, B.K.B., 2013. Nothing but the Tooth: A Dental Odyssey, Elsevier, London. Courtesy Elsevier, London.*

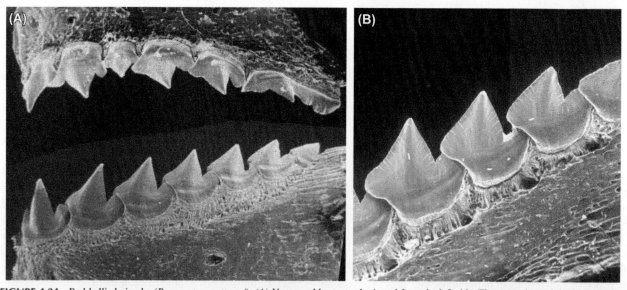

FIGURE 4.34 Red-bellied piranha (*Pygocentrus nattereri*). (A) Upper and lower teeth viewed from the left side. There are six teeth in the upper jaw and seven in the lower jaw. The enlarged upper sixth tooth is *arrowed*. Image width = 5 mm; (B) High-power view from Fig. 4.34A, showing interlocking of the back four teeth in the lower jaw. *From Shellis, R.P., Berkovitz, B.K.B., 1976. Observations on the dental anatomy of piranhas (Characidae) with special reference to tooth structure. J. Zool. Lond. 180, 69–84. Courtesy Editors Journal of Zoology, London.*

the lower teeth, which slope increasingly backward toward the rear of the jaw (Fig. 4.34), puncture the prey and the punctures are elongated into cuts by the razor-sharp edges of the teeth moving upward. The four posterior lower teeth slice like a pair of scissors against the large, sixth upper tooth, which is greatly expanded in length to form a long, serrated blade.

Characidae (Lineage II)

This family includes a wide variety of omnivorous and carnivorous species. The size covers a very wide range, from small **tetras** less than 20-mm long to the large carnivorous dorados (*Salminus* spp.), which can reach a length of 1 m.

The dentitions of carnivorous characids consist of sharp, triangular or conical teeth. This is illustrated in the **golden dorado** (*Salminus brasiliensis*) (Fig. 4.35A) and the **pike characin** (*Oligosarcus jenynsii*) (Fig. 4.35B). Both possess single rows of sharp, recurved teeth on the dentigerous bones, and the pike characin has one or two enlarged caniniform teeth (Mirande, 2010).

(A)

(B)

FIGURE 4.35 (A) Lower jaw of the golden dorado (*Salminus brasiliensis*). Scale = 5 mm; (B) Upper jaw of pike characin (*Oligosarcus jenynsii*). *From Mirande, J.M., 2010. Phylogeny of the family Characidae (Teleostei: characiformes): from characters to taxonomy. Neotrop. Ichthyol. 8, 385–568. Courtesy Editors Neotropical Ichthyology.*

In omnivorous characids, at least the anterior teeth are flattened and multicusped, with the cusps aligned mesio-distally and one central cusp being enlarged. *Astyanax* is an enormous genus, with about 130 species. The type species, the **Mexican tetra** (*Astyanax mexicanus*), has two rows of teeth on the premaxilla: four to five in the outer row and five to seven in the inner row. There are one or two teeth on the anterior part of each maxilla (Atukorala et al., 2013). On each dentary there are four large anterior ("rostral") teeth and four to six smaller posterior ("caudal") teeth. The upper teeth and anterior lower teeth have five cusps (median), whereas the posterior lower teeth have three cusps (median)[3]. The first two generations of teeth in *A. mexicanus* are single cusped and form outside the jaw bone. At the second tooth replacement, these are succeeded by multicusped teeth that form inside the jaw bone (Trapani et al., 2005). This species has a blind, cave-dwelling morph that differs from the wild type in tooth number and morphology and in the shape of the jaws (Atukorala et al., 2013). The cave morph has more mandibular teeth: median values of 11.5 anterior teeth and 7.5 posterior teeth. The anterior teeth also have more cusps than in the wild type (median 6.0), whereas the posterior teeth have a very similar number of cusps (median 3.0). With regard to the upper jaw, the only difference was a greater number of maxillary teeth in the cave-dwelling morph than in the wild type. The dentition of *Astyanax ajuricaba*, a recently discovered species, is very similar to that of the Mexican tetra; the premaxilla has two rows of five teeth, each with usually five cusps and the lower jaw has four large anterior teeth with four or five cusps and 7–10 small, unicuspid posterior teeth (Fig. 4.36).

Hemigrammus is another large genus of tetras. *Hemigrammus arua* has two rows of stout teeth on the premaxilla (Lima et al., 2009). The outer row generally has three closely set elongated teeth with three or four cusps, whereas the inner row generally has five teeth with five to seven cusps The anterior teeth are the largest. The number of teeth on the dentary can vary between 12 and 17. The anterior six or seven teeth are larger, elongated, and have five to seven cusps, whereas the posterior teeth are smaller and cylindrical with one to three cusps (Fig. 4.37).

The dentition of the **blue-bellied night wanderer** (*Cyanogaster noctivaga*), a miniature characid, generally resembles that of *Astyanax*, except that the maxillae are edentulous and on the premaxilla there are only four multicuspid teeth in the inner row and only two teeth in an outer row, lying between the second and third teeth of the inner row (Fig. 4.38A). There is also a plate with four to five longitudinal rows of conical teeth, on the proximal half of the fifth ceratobranchial (Mattox et al., 2013).

---

3. There seem to be errors of calculation in the averages reported by Atukorala et al. (2013). The medians provided here have been newly calculated from the raw data.

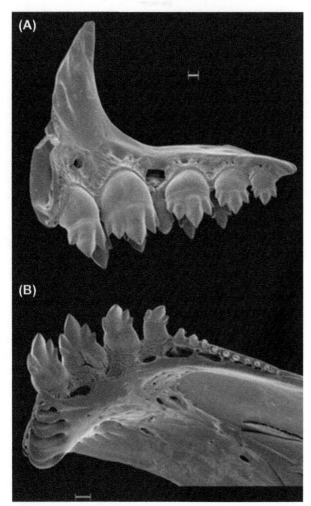

FIGURE 4.36 Scanning electron micrographs of teeth of *Astyanax aju-ricaba*. (A) Right premaxillary teeth in lingual view. Scale = 100 μm. (B) Right dentary teeth in lingual view. Scale = 200 μm. Channels in the bone behind the functional teeth lead down to the replacement teeth that develop intraosseously in cavities in the bone. *From Marinho, M.M.F., Lima, F.C.T., 2009.* Astyanax ajuricaba: *a new species from the Amazon basin in Brazil (Characiformes: Characidae). Neotrop. Ichthyol. 7, 169–174. Courtesy Editors Neotropical Ichthyology.*

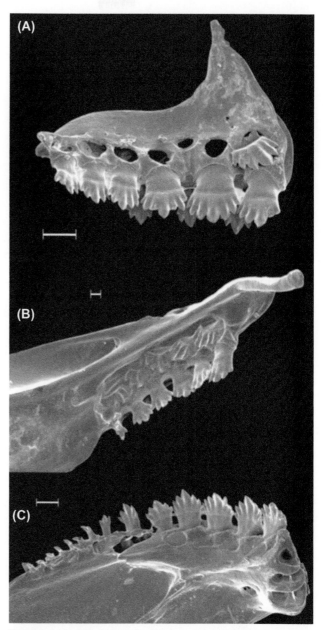

FIGURE 4.37 Dentition of *Hemigrammus arua*. Lingual views of (A) left premaxillary teeth, scale = 100 μm; (B) maxillary teeth, scale = 30 μm; (C) dentary teeth: channels in the bone behind the functional teeth lead down to the replacement teeth that develop intraosseously in a trench within the bone. Scale = 100 μm. *From Lima, F.C.T., Wosiacki, W.B., Ramos, C.S., 2009.* Hemigrammus arua, *a new species of characid (Characiformes: Characidae) from the lower Amazon, Brazil. Neotrop. Ichthyol. 7, 153–160. Courtesy Editors Neotropical Ichthyology.*

In another miniature fish genus, *Priocharax*, the dentition consists of single rows of small, conical teeth. There are 22–29 premaxillary teeth; 27–58 maxillary teeth; and 28–55 dentary teeth, according to species (Fig. 4.38B) (Toledo-Piza et al., 2014).

Scale-eating Characids

The Characidae include several known scale-eating fishes, including the **buck-toothed tetra** (*Exodon paradoxus*) and several species of *Roeboides*. As in other characiforms, the jaws are not protrusible, but these scale eaters possess teeth that project outward from the marginal dentigerous bones and allow scales to be scraped or plucked from the prey fishes. The earliest generations of

teeth have a normal orientation: protruding teeth appear later during development (Hahn et al., 2000; Novakowski et al., 2004). *Exodon*, in which scales are the major dietary item, displays some asymmetry of the jaws and a behavioral preference (less than 50%) for the direction of attack (Hata et al., 2011). *Roeboides* shows considerable variation between species in the extent of scale eating, and

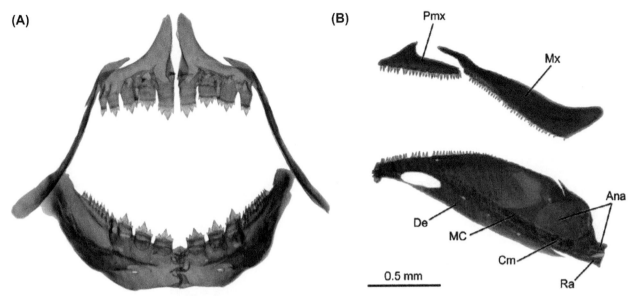

FIGURE 4.38 (A) Dentition of the blue-bellied night wanderer (*Cyanogaster noctivaga*). *(From Mattox, G.M.T., Britz, R., Toledo-Piza, M., Marinho, M.M.F., 2013. Cynogastor noctivago, a remarkable new genus and species of miniature fish from the Rio Grande, Amazon basin (Ostariophysi: Characidae). Ichthol. Explor. Freshwaters 2013;23:297–318. Courtesy Editors Ichthyological Exploration of Freshwaters.)* (B) Dentition of *Priocharax nanus*. *Ana*, anguloarticular, *Cm*, coronomeckelian, *De*, dentary, *MC*, Meckel's cartilage, *Mx*, maxilla, *Pmx*, premaxilla, *Ra*, retroarticular. *From Toledo-Piza, M., Mattox, G.M.T., Britz, R., 2014. Priocharax nanus, a new miniature characid from the rio Negro, Amazon basin (Ostariophysi: characiformes), with an updated list of miniature neotropical freshwater fishes. Neotrop. Ichthyol. 12, 229–246. Courtesy Editors Neotropical Ichthyology.*

this is correlated with the number and development of the protruding teeth (Novakowski et al., 2004). The proportion of scales in the diet also varies according to stage of growth, season, and habitat (Peterson and Winemiller, 1997; Novakowski et al., 2004).

Cynodontidae and Ctenoluciidae

According to the analysis of Guisande et al. (2012), these two families of primitive carnivorous fishes form a sister group to Lineage I. The **payara** (*Hydrolycus scomberoides*: Cynodontidae) is an aggressive freshwater fish with a fearsome reputation that can grow to a length of about 120 cm. They are also known as vampire fish because at the front of the lower jaw are two huge, recurved, fang-like teeth that may be up to 15-cm long (Fig. 4.39A). When the mouth is closed, these teeth lie in channels in the upper lip. Behind the main fangs are a number of widely spaced, medium-sized, pointed teeth that diminish in size toward the back of the jaws. These are in turn interspersed with numbers of small teeth. In the upper jaw, the first and last tooth on the premaxilla are enlarged and surrounded by a number of much smaller teeth. The teeth on the front half of the maxilla resemble those on the dentary, whereas the back half carries about 40 teeth of uniform size. This assembly of sharp teeth is used to impale smaller fishes, which are then swallowed.

In dentitions such as that of the payara, there is the problem of how such large teeth develop and erupt without compromising function. Part of the solution to this problem is seen in Fig. 4.39B,C. In a lateral-oblique view of the lower

jaw, it can be seen that the replacements for the large anterior fangs, and of the medium-sized posterior teeth, develop horizontally in the dentary, with the bases of the replacement teeth lying against those of the functional teeth. Replacement of these large teeth will require repositioning from the horizontal to the vertical axis, but the mechanism by which this is accomplished, and the speed with which it occurs, are unknown.

Siluriformes

Catfishes are characterized by their long, slender barbels that help them locate food. They are bottom dwelling animals and may have a sucker-like mouth. They are omnivorous and eat snails, worms, crustaceans, aquatic insects, and small fishes, as well as detritus. There are more than 3500 species. Although catfishes are unable to protrude their mouths like other fishes, their mouth can expand greatly, mainly by lowering the hyoid apparatus, and this generates enough suction to bring prey into the mouth, where it can be grasped between the upper and lower teeth (Gosline, 1973).

Catfishes typically have large numbers of teeth. They may be distributed in several rows of varying size in different species (Fig. 4.40).

Loricariidae

This is the largest catfish family, with more than 700 species, and the family that shows most variation and specialization of tooth form. A familiar characteristic of this group

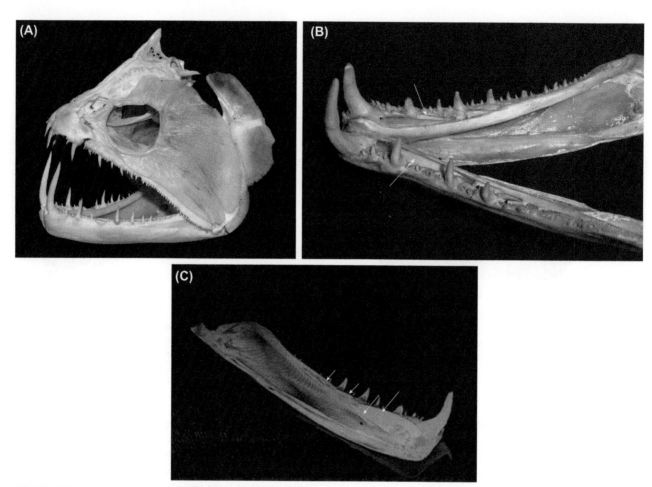

FIGURE 4.39 Payara (*Hydrolycus scomberoides*). (A) Lateral view of dentition. ©*MPcz, Shutterstock*; (B) Dorsolateral view of lower dentition. *Courtesy MoLSKCL. Catalog no. V229.* Image width=15 cm; (C) Lateral radiograph of dentary of specimen seen in Fig. 4.39B, showing recumbent replacement teeth. *Long arrows* indicate two or three replacements for fangs, *short arrows* indicate replacements for lateral teeth. Image width=23 cm. *Courtesy Drs. J. Brown, F. Ball and A. Samani.*

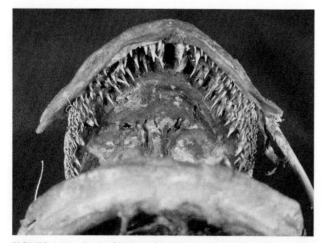

FIGURE 4.40 Teeth of Wels catfish (*Silurus glanis*), illustrating rows of teeth in the roof of the mouth. Image width=16 cm. *Courtesy RCSOMA/459.1.*

is the sucker mouth that allows the catfish to breathe and feed while attached to the substrate through suction. Unusually for catfishes, the premaxilla in loricariids is highly mobile, whereas the lower jaw has a mobile symphysis.

When the mouth is open the teeth point anteroventrally. During jaw closure, both upper and lower jaws rotate about their axes so that the teeth scrape the surface of the substrate and remove edible material. In some loricariids, hypomineralization of the dentine on the lingual aspect of the teeth makes them very flexible and reduces their tendency to fracture (Geerinckx et al., 2012).

Differences in diet between species are reflected in morphological variations in the teeth (Nelson et al., 1999; Lujan et al., 2011). Loricariid catfishes of the genus *Panaque* have spoon-shaped teeth adapted to scraping wood surfaces (Fig. 4.41). It has not, however, been established whether *Panaque* can extract nutrition from the wood itself. It is possible that it scrapes wooden surfaces for the sake of the associated algae and microfauna. Among species that live among, and feed on, coarse woody debris, the jaws and teeth vary according to the level of penetration of the wood, with levels classified as surface grazing, wood gouging, and macroinvertebrate probing. Thus, *Lamontichthys filamentosus* has a relatively weak lower jaw with numerous slender teeth and grazes algal films growing on wood surfaces (Fig. 4.42A).

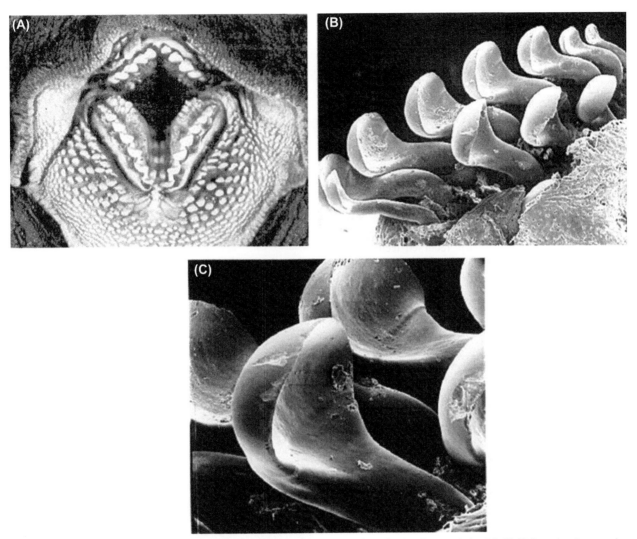

FIGURE 4.41 (A) Ventral view of the mouth of an undescribed species of *Panaque* from the *nigrolineatus* clade ×2.2; (B, C) Scanning electron micrographs of the dentary bone and specialized spoon-shaped teeth of an individual *Panaque cochliodon*. (B) ×20 and (C) ×35. *From Nelson J.A., Wubah D.A., Whitmer M.E., Johnson E.A., Stewart D.J., 1999. Wood-eating catfishes of the genus* Panaque: *gut microflora and cellulolytic enzyme activities. J. Fish. Biol. 54, 1069–1082. Courtesy Editors Journal of Fish Biology.*

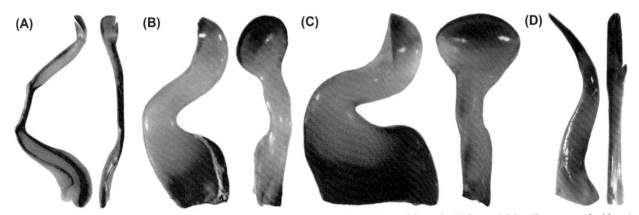

FIGURE 4.42 Individual teeth of different species of catfish. Alizarin red has stained the base of the teeth. (A) *Lamontichthys filamentosus* (detritivore); (B) *Panaquolus albomaculatus* (wood eater); (C) *Panaque bathyphilus* (wood eater); (D) *Spatuloricaria puganensis* (insectivore). *From Lujan, N.K., German, D.P., Winemiller, K.O., 2011. Do wood grazing fishes partition their niche? Morphological and isotopic evidence for trophic segregation in Neotropical Loricariidae. Funct. Ecol. 25, 1327–1338. Courtesy Editors Functional Ecology.*

The stronger jaws and stouter teeth of *Panaque albomaculatus* enable it to scrape away shallow layers of wood and associated microfauna (Fig. 4.42B), whereas *Panaque bathyphilus* has robust jaws and teeth that it uses to gouge deeper wood layers (Fig. 4.42C). *Spatuloricaria puganensis* does not obtain food by mechanically breaking down the structure of the wood; instead, it uses medial clusters of elongated teeth to probe and browse crevices for detritus and insect larvae (Fig. 4.42D).

Ariidae

Correlations have been made between diet and the morphology and number of palatal teeth in tropical ariid catfishes from Australia (Blaber et al., 1994). Fish-eating and polychaete-eating species have sharp, recurved teeth, but the number is smaller in the polychaete-eating group. Mollusk-eating species have a few large rounded teeth, adapted for crushing, located near the back of the mouth.

Pangasiidae

The **Mekong giant catfish** (*Pangasianodon gigas*) is the biggest freshwater catfish in the world, attaining a length of 2.5 m. Although the adult is toothless, juveniles have a marginal row of simple conical teeth on the upper and lower jaws, on the palate, and in the pharynx. There is only a single tooth replacement. The maximum number of functional teeth in a 9.5-cm-long specimen is 25. By the time the fish has reached a length of 17 cm, there are only four functional teeth present, with none at a length of 24 cm (Kakizawa and Meenakarn, 2003).

Protacanthopterygii

This subcohort contains moderately advanced teleosts, such as salmon and trout, pike, and barreleyes. The upper jaw is not protrusible and the tongue (glossohyal) is furnished with teeth.

Salmoniformes

In salmon and their relatives, the upper jaw retains the primitive condition in which the maxilla forms part of the margin of the mouth, carries teeth, and has restricted mobility, so is not protrusible. Mechanically, the jaws are adapted to biting and seizing prey, and the teeth are numerous, recurved, and sharp. These fishes possess a tongue bite apparatus, like osteoglossomorphs. Camp et al. (2009) found that in the **brook trout** (*Salvelinus fontinalis*), the apparatus exerts great force during raking, through powerful cranial muscles, and also maximizes the speed of prey processing.

The predatory dentition is exemplified by the **Atlantic salmon** (*Salmo salmar*) (Figs. 4.43 and 4.44). The marginal teeth are supplemented by numerous teeth on the palatines, vomers, and tongue. During spawning in freshwater, male salmon develop a hook-shaped nose/jaw and large fang-like teeth (Fig. 4.45).

FIGURE 4.43 Teeth of Atlantic salmon (*Salmo salar*). Image width = 13 cm. *Courtesy RCSOMA/452.1.*

FIGURE 4.44 Palatal teeth in of Atlantic salmon (*Salmo salar*). Image width = 6 cm. *Courtesy RCSOMA/A452.1.*

FIGURE 4.45 Male salmon in mating season, showing hooked jaws. Image width = 24 cm. © *Trustees of the Natural History Museum, London.*

The **brown trout** (*Salmo trutta*) has a similar dentition to that of salmon, but with two rows of teeth on the vomers instead of one row.

In **rainbow trout** [*Onchorhynchys mykiss* (*Salmo gairdneri*)] of body length 12–23 cm, the marginal teeth are arranged in single rows with 8 teeth on each premaxilla and 26–35 slightly smaller teeth on each maxilla (Fig. 4.46). On the roof of the mouth, the teeth form a double n-shaped arch, with the two vomerine rows closely apposed in the midline and the slightly smaller palatine rows inside the maxillary rows and parallel to them. Each row on the vomers consists of 19–25 teeth and on the palatines of 17–22 teeth. Each dentary bears 18–25 small, recurved teeth of similar size. The teeth on the tongue are larger and arranged in two parallel anteroposterior rows of four to six teeth, with sometimes a median tooth at the front of the tongue.

Esociformes

The **northern pike** (*E. lucius*) is a fearsome ambush predator. It has small teeth on the premaxilla and anterior part of the dentary, but large, narrow, pointed teeth toward the back of the dentary (Fig. 4.47). The roof of the mouth is furnished with a large number of curved, hinged teeth, pointing inward and backward, on the palatine and vomer bones (Fig. 4.48). Teeth also occur on the tongue (Fig. 4.47). Prey is initially grasped by the marginal teeth, guided backward by the hinged teeth in the roof of the mouth and swallowed head first.

Stomiati

This subcohort contains two orders: the **smelts** (Osmeriformes) and the Stomiatiformes, which includes more than 400 species of deep-sea fishes, eg, viperfishes, dragonfishes, and loosejaws.

Stomiatiformes

Viperfishes (*Chauliodus*), despite having few teeth, have one of the most specialized feeding apparatuses in the animal kingdom (Tchernavin, 1953). These small fishes, although no more than 15 cm long, have a fearsome appearance. On each premaxilla, there are four dagger-like teeth, of which the second is largest. These teeth have sharp, blade-like edges. There are several small sharp teeth on the maxilla, all directed backward (Fig. 4.49). There are six teeth on the dentary. Of these the first is by far the largest in the dentition, whereas the remaining five dentary teeth are small and dagger like. There are also three additional palatine teeth and

FIGURE 4.47 Lateral view of teeth of northern pike (*Esox lucius*). Image width = 18.5 cm. *Courtesy RCSOMA/ 454.2.*

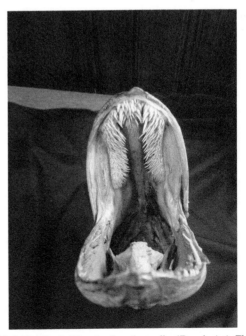

FIGURE 4.48 Roof of mouth of northern pike (*Esox lucius*). The teeth in the midline are hinged. Image width = 10 cm. *Courtesy of MoLSKCL. Catalog no. V116.*

FIGURE 4.46 Rainbow trout (*Oncorhynchus mykiss*) with mouth open. Lateral view, showing marginal teeth on premaxilla, maxilla, dentary, and tongue (basihyals). Image width = 9.5 cm.

paired upper pharyngeal teeth on the third and fourth gill arches. None of the teeth of the viperfishes are hinged, as they are in anglerfishes (Greven et al., 2009). Thus, for the jaw to close, the extremely long first lower tooth has to pass in a groove in the side of the skull so that its tip lies just above the eye. A functional tooth never falls out or becomes loose until the replacement tooth has erupted firmly into place in the jaw. The sharp edges on the very long teeth will lacerate prey (Greven et al., 2009), but it is unclear whether their main function is to impale prey or to trap it within the mouth.

Another specialization of the viper fish is its ability to swallow prey almost as big as itself, by opening its jaw extremely wide (Fig. 4.50). This involves two adaptations

FIGURE 4.49   Teeth of a Pacific viperfish (*Chauliodus macouni*). *Courtesy Dr. P. Yancey, Whitman College.*

FIGURE 4.50   Illustration of the large gape of a Pacific viperfish (*Chauliodus macouni*). *Courtesy Dr. P. Yancey, Whitman College.*

(Tchernavin, 1953), as shown in Fig. 4.51. First, specializations of the anterior part of the vertebral column allow the first 10 vertebrae to be raised almost vertically and to lift the cranium high. This rotates the suspensorium and brings the articular-quadrate joints (at rest posterior to the skull) forward. As the jaws move forward, they open and also expand at the articular-quadrate joints. These movements enable the hyoid apparatus to unfold so that the ceratohyals are directed backward instead of being folded inside the mandibular rami. This in turn increases the extent of protraction of the jaws. Because of the extremely wide gape, the seemingly oversized front teeth do not provide a hindrance to swallowing the prey, despite their being rigidly attached to the jaws.

The **black dragonfish** (*Idiacanthus atlanticus*), like anglerfishes, represents an extreme example of sexual dimorphism. They are deep-sea fishes living at depths of 2000 m. The females are long and slender, reaching a length of up to 40 cm. They possess numerous long, slender, sharp teeth, some of which are extremely long and fang-like, and they prey on other fishes (Fig. 4.52). By contrast, the males are no longer than 5 cm and have no teeth.

Eurypterygia

Aulopiformes

This order comprises more than 200 species, of which some are deep-sea predators. These include lizard fishes (Bathysauridae), so named because their heads and elongated mouths resemble those of lizards (Fig. 4.53). They have numerous sharp teeth dispersed over many bones in the mouth and gills (Fishelson et al., 2004). The **variegated lizardfish** (*Synodus variegatus*) has only a single row of teeth in the roof of the mouth (Fig. 4.54). The **greater lizardfish** (*Saurida tumbil*) is a shallow-water, tropical species. It has numerous teeth in rows on the maxilla and dentary. Two additional rows of teeth are situated on the internal and external pterygoid bones on the roof of the mouth. Numerous teeth are distributed over the hyoid and gill arches (Fig. 4.55).

Paracanthomorphacea

This division includes the cod family, the dories, and other families that prey on fish.

Zeiformes

The premaxillae of the **John Dory** (*Zeus faber*) are furnished with long dorsal processes, so the upper jaw is highly protrusible and enables highly efficient suction feeding. The John Dory stalks its prey of smaller fishes until close enough to capture them by suction: a procedure aided by its extremely compressed body, which presents a minimal profile to the

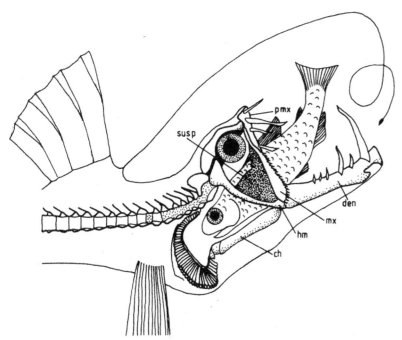

FIGURE 4.51 Diagram showing head of viperfish as it swallows a fish. Note the extreme upward bend of the anterior part of the spine, the anterior rotation of the suspensorium (susp: dark stippling) and the retraction of the ceratohyals (ch). *Den*, dentary; *hm*, hyomandibula; *mx*, maxilla; *pmx*, premaxilla. *Modified from Tchernavin, V.V., 1953. The Feeding Mechanisms of a Deep Sea Fish* Chauliodus Sloani *Schneider, British Museum (Natural History), London, pp. 1–101. Courtesy of the Trustees of the Natural History Museum, London.*

NORFANZ TAN0308

FIGURE 4.52 Teeth of black dragonfish (*Idiacanthus atlanticus*). *Courtesy NORFANZ NORFANZ Tan0308.*

FIGURE 4.53 Reef (variegated) lizardfish (*Synodus variegatus*). © Aliced/Dreamstime.

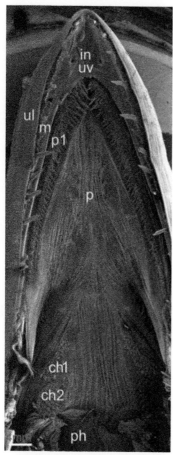

FIGURE 4.54 Upper jaw of reef (variegated) lizardfish (*Synodus variegatus*). Teeth present on maxilla (m) and on the palatine and ectopterygoid (p1). Strong cover of teeth is also visible on the pharyngobranchial tooth-plates (ch1 and ch2). There are also two internal pores (in) on the apical end of the jaw, possibly representing part of the nasal olfactory system. Scale = 4 mm. *From Fishelson, L., Golani, D., Galil, B., Goren, M., 2010. Comparison of taste bud form, number and distribution in the oropharyngeal cavity of lizardfishes (Aulopiformes, Syndontidae). Cybium 34, 269–277. Courtesy Editors Cybium.*

prey. Several rows of very small teeth are present on the premaxilla and dentary (Fig. 4.56). The jaws of John Dory are adapted to snapping shut very rapidly (Westneat, 2004).

Gadiformes

Gadidae

The upper and lower jaws of the **Atlantic cod** (*Gadus morhua*) have a row of large, recurved teeth plus a number of small recurved teeth. On the premaxilla the large teeth are on the labial margin, whereas on the dentary they are on the lingual margin (Fig. 4.57). The two sets of large teeth thus overlap during jaw closure. On the premaxilla there are about five irregular rows of small teeth lingual to the row of large teeth. On the dentary, there is only a single row of small teeth labial to the large teeth. Multiple rows of small teeth occur in the midline region of the premaxilla and also

on the vomers (Fig. 4.58). All of the oral teeth are ankylosed to the bone. The number of ankylosed teeth in a jaw increase with age (Holmbakken and Fosse, 1973).

The **whiting** (*Merlangius merlangus*) has a simpler dentition than the cod. The teeth on the jaws are less numerous, more equal in size and arranged more in single rows (Fig. 4.59).

Merlucciidae

Like the cod, **hakes** (*Merluccius*) possess rows of large teeth (on the lingual aspect of the dentary and the labial aspect of the upper jaw), parallel with rows of smaller teeth (Fig. 4.60). All of the teeth are conical and recurved and in the living animal look red because the tooth shaft is composed of vasodentine (Fig. 4.61). The large teeth are hinged, so that entry of prey is facilitated and escape prevented.

Euacanthomorphacea

Beryciformes

Fangtooths (*Anaplogaster*), of which there are only two species, inhabit the abyss at depths of 5000 m and feed on other fishes and squid. Although growing to a length of only about 7 cm, they have the longest teeth of any fish in proportion to body length. As their name suggests, these fishes have thin, pointed fang-like teeth (Fig. 4.62). The largest pair is adjacent to the midline of the lower jaw and to allow the mouth to close, they fit into spaces in the roof of the mouth. The young fishes feed on zooplankton and have small, villiform teeth.

Percomorphaceae

These perch-like fishes constitute the crown group of bony fishes. The percomorphs comprise about 40% of all living fishes, with more than 14,000 species in about 250 families. Members comprise many familiar types of fishes, such as basses, gobies, mackerels, perches, cichlids, and wrasses. They are divided into nine clades or series, of which we present representatives of the five larger series.

Gobiaria

This series is dominated by the gobies, but the order also contains nurse fishes, cardinal fishes, and sleepers.

Gobiiformes

The gobioids are bottom-living fishes that include both marine and freshwater representatives.

Gobiidae

Gobies are one of the largest families of fishes, with more than 2000 species. They are mainly small and are

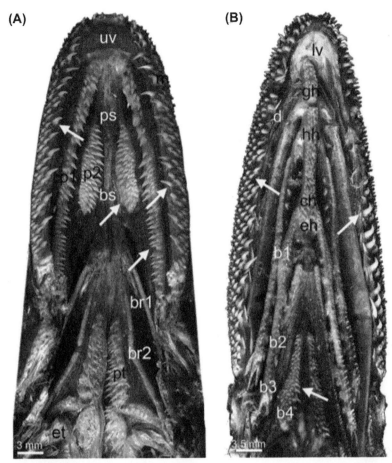

FIGURE 4.55 Dentition of greater lizardfish (*Saurida tumbil*). (A) Upper jaw. (B) Lower jaw. Note the wide distribution of teeth (*arrows*). In upper jaw, teeth are present on the palatine and ectopterygoid (p1) and on the pterygoid (p2) as well as on the premaxillae and maxillae. *br1, br2,* first and second branchial arches; *bs,* basisphenoid; *et,* epibranchial toothplate; *m,* maxilla; *ps,* parasphenoid; *pt,* pharyngobranchial toothplates; *uv,* breathing valve of the upper jaw. In lower jaw, *b1–4,* first–fourth branchial arches; *ch,* ceratohyal; *d,* dentary; *eh,* epihyal; *gh,* glossohyal (tongue); *hh,* hypohyal; *1v,* breathing valve of the lower jaw. *From Fishelson, L., Golani, D., Galil, B., Goren, M., 2010. Comparison of taste bud form, number and distribution in the oropharyngeal cavity of lizardfishes (Aulopiformes, Syndontidae). Cybium 34, 269–277. Courtesy Editors Cybium.*

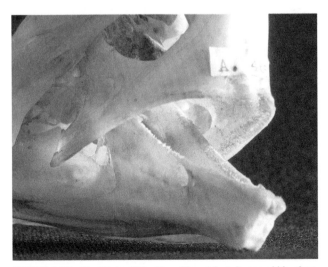

FIGURE 4.56 Dentition of John Dory (*Zeus faber*). Image width = 6 cm. *Courtesy RCSOMA/ A465.1.*

distinguished by a disc-shaped ventral sucker, formed from fused pelvic fins that allow them to adhere to rocks. They feed mainly on small invertebrates, sometimes mollusks, and a few consume algae. The diet of the **round goby** (*Neogobius melanostomus*; Benthophilinae) includes crayfishes, mussels, and aquatic insects. It has numerous sharp, pointed teeth on the premaxilla and dentary, the longest being at the front of the premaxilla (Fig. 4.63). On the pharyngeal tooth plates, changes in the shape of the teeth occur with age in association with changes in the diet. The plates of round gobies shorter than 50 mm bear narrow papilliform teeth suitable for eating soft-bodied prey. In gobies longer than 80 mm, the pharyngeal teeth are larger, rounded, and show wear consistent with a harder diet of mollusks.

The *Sicydiinae* are a small subfamily of freshwater gobies, usually found in fast-flowing freshwater streams. They are characterized by rounded ventral suction discs by which they

(A)

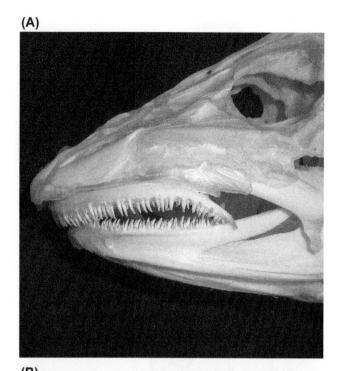

(B)

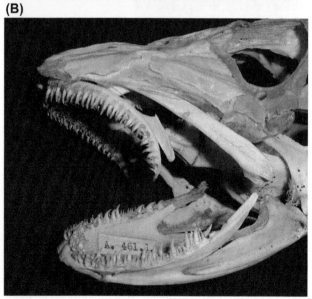

FIGURE 4.58 Ventral view of upper teeth of Atlantic cod (*Gadus morhua*). Image width=5 cm. *Courtesy RCSOMA/ 461.1.*

FIGURE 4.57 Teeth of Atlantic cod (*Gadus morhua*). (A) Jaws closed. Image width=7.5 cm. (© *Oxford University Museum of natural History. Catalog no. 17,419.*) (B) Jaws open. Image width=5.3 cm. *RCSOMA/461.1.*

cling to stones on the stream bed. The most detailed dental research has been undertaken on the **monk goby** (*Sicyopterus japonicus*) (Kakizawa et al., 1986): small, undistinguished-looking fishes that nevertheless have one of the most remarkable dentitions in the whole of the animal kingdom.

Like other members of the subfamily, monk gobies are small fishes that spawn in fast-flowing rivers. The eggs are carried down to the sea where they hatch. The larvae lack teeth and feed on plankton. After metamorphosis the

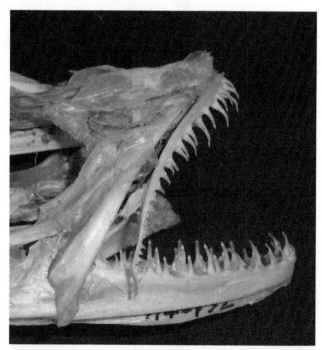

FIGURE 4.59 Jaws of a whiting (*Merlangius merlangus*). Image width=3.5 cm. *Courtesy RCSOMA/ 461.32.*

young monk gobies return from the sea and ascend their home rivers. As the gobies adjust to life in fresh water, they begin to feed on algae growing on stones on the river bed, and this requires an amazing metamorphosis of the

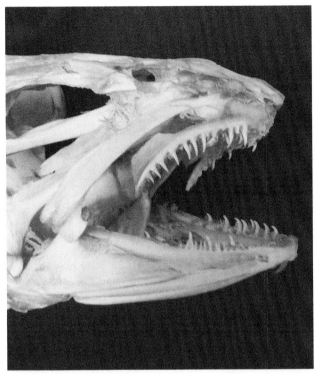

FIGURE 4.60   Jaws and dentition of European hake (*Merluccius merluccius*). *Courtesy RCSOMA/ 462.2.*

FIGURE 4.62   Teeth of fangtooth (*Anaplogastercarnuta* sp.). *Courtesy of NORFANZ, Mark Norman.*

FIGURE 4.61   Head of European hake (*Merluccius merluccius*), showing hinged teeth that have red shafts because they are composed of vasodentine. © *Richard Griffin/Dreamstime.*

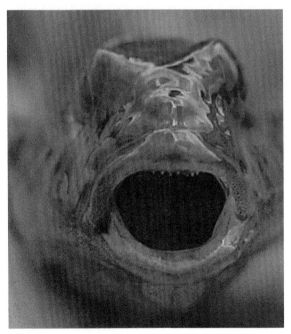

FIGURE 4.63   Mouth of round goby (*Neogobius melanostomus*), showing many teeth. *Courtesy Wikipedia.*

head region. First, the mouth is relocated from the very front of the head to a ventral position so that, when the goby attaches to a rock with its sucker, its mouth is in direct contact with the Aufwuchs. Second, the upper lip enlarges to form a second sucker at the front of the head. Third, the appearance of teeth inside the mouth enables the goby to scrape algae off the rocks. These dramatic changes in the morphology of rock-climbing gobies take place very rapidly, probably in just a few days. The rock-climbing monk goby scales waterfalls by alternately attaching to the rock via its mouth sucker and chest sucker, aided by

muscular contractions of its body. This allows the fishes to gain access to the water immediately behind the top edge of the waterfall, a habitat denied to other fishes.

In the rock-climbing monk gobies (eg, *S. japonicus*) there are major differences between the upper and lower dentitions (Mochizuki and Fukui, 1983; Moriyama et al., 2009, 2010). Although there are only a few, spaced, conical

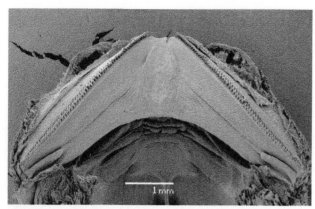

FIGURE 4.64 Rock-climbing monk goby (*Sicyopterus japonicas*): row of 50–60 erupted teeth on each side of the upper jaw. *From Moriyama, K., Watanabe, S., Iida, M., Fukui, S., Sahara, N., 2009. Morphological characteristics of upper jaw dentition in a gobiid fish* (Sicyopterus japonicus); *a micro-computed tomography study. J. Oral. Biosci. 51, 81–90. Courtesy Editors Journal of Oral Biosciences.*

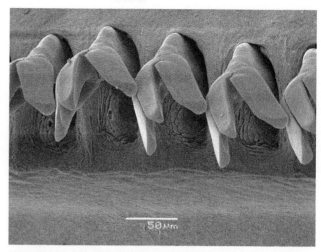

FIGURE 4.65 Rock-climbing monk goby (*Sicyopterus japonicas*): newly erupted upper teeth, each with three small cusps. *From From Moriyama, K., Watanabe, S., Iida, M., Fukui, S., Sahara, N., 2009. Morphological characteristics of upper jaw dentition in a gobiid fish* (Sicyopterus japonicus); *a micro-computed tomography study. J. Oral. Biosci. 51, 81–90. Courtesy Editors Journal of Oral Biosciences.*

teeth in the lower jaw, those in the upper jaw form a single, continuous row of up to 60 or more tiny teeth on each side (Fig. 4.64). Unlike the simple, conical lower teeth, the upper teeth when they first erupt have a more complex shape with three small cusps (Fig. 4.65). The teeth have a fibrous attachment to the bone of the jaw, giving them a degree of flexibility that is useful in helping the teeth act as scrapers.

The scraping action means that the teeth of rock gobies wear very rapidly. In compensation, tooth replacement is extremely rapid, almost certainly among the fastest in the animal kingdom. It is estimated that each tooth is replaced after only 9 days. At each tooth position in the upper jaw there is a phenomenal number of successional teeth: up to 25 sets in a 3-cm-long rock-climbing monk goby (Fig. 4.66) and 45 in a 10-cm fish, giving a total of about 2700 teeth in the goby's small head! The process of tooth succession is described in Chapter 10.

Pelagiaria

The Pelagiaria are fast-swimming pelagic fishes that feed on smaller fish. The premaxilla is not protrusible and there is only limited movement of the lower jaw during mouth opening. In other words, the jaws function as a pair of forceps to close on the prey, which is gripped, and prevented from escaping by rows of small teeth on the margins of the jaws. The dentition is homodont and consists of sharp, conical teeth.

Scombridae

The teeth of the **Atlantic mackerel** (*Scomber scombrus*) are pointed and slightly recurved (Fig. 4.67), whereas the **king mackerel** or **kingfish** (*S. cavalla*) has closely aligned

triangular blade-like teeth (Fig. 4.68). In the premaxilla, there may be more than 50 tooth loci, with tooth size increasing to about halfway along the series, before decreasing. In the dentary tooth loci may number up to about 40 and tooth size increases until about two-thirds of the way along the row. The **Atlantic bluefin tuna** (*Thunnus thynnus*) is one of the largest of the Pelagiaria and has major commercial importance. A voracious feeder on fish, such as anchovy, sand lance, herring, and sardine, its dentition consists of a single row of small conical teeth on the premaxilla and dentary (Fig. 4.69).

Trichiuridae

Silver scabbardfish and **black scabbardfish** (*Lepidopus caudatus* and *Aphanopus carbo*, respectively) have a single row of laterally flattened, pointed teeth with sharp edges, spaced along each jaw bone, with a few larger fang-like teeth on the premaxilla (Figs. 4.70 and 4.71). These fishes feed on crustaceans, cephalopods, and fishes.

Carangaria

This series contains fast-swimming, pelagic fishes, such as jacks, billfishes, and barracudas, and also benthic flatfishes.

Sphyraenidae

Barracudas are fierce predators and their dentition is adapted to grasping and cutting. There is an outer row of small, close-fitting flattened, triangular teeth at the margin of the premaxilla and a single row of much larger teeth with a similar shape at the margins of the dentary (Fig. 4.72). In addition, a number of large teeth are carried on the roof of

(A)

(B)

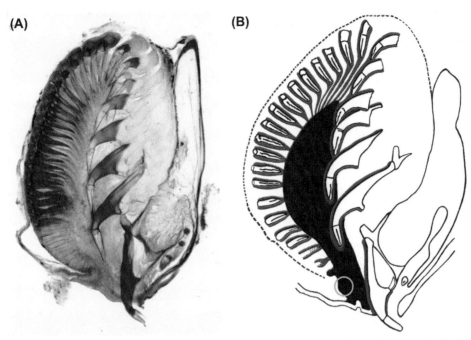

FIGURE 4.66 Section of the upper jaw (A) and accompanying line diagram (B) of a 3-cm specimen of rock-climbing monk goby (*Sicyopterus japonicus*), showing 26 generations of replacing teeth beneath each functional tooth. *From Moriyama, K., Watanabe, S., Iida, M., Sahara, N., 2010. Plate-like permanent dental laminae of upper jaw dentition in adult gobiid fish,* Sicyopterus japonicus. *Cell Tiss. Res. 340, 189–200. Courtesy Editors Cell Tissue Research.*

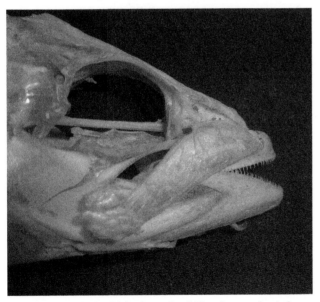

FIGURE 4.68 Teeth of king mackerel (*Scomberomorus cavalla*). Image width = 6.5 cm.

FIGURE 4.67 Teeth of Atlantic mackerel (*Scomber scombrus*). Image width = 3 cm. *Courtesy RCSOMA/ 471.1.*

the mouth. These fit into sockets in the lower jaw, allowing the jaws to close.

Although the premaxilla is not protrusible, rotation of the maxilla as the jaw opens causes the premaxilla to rotate into a more upright position, thus increasing the gape (Habegger et al., 2011). The mass-specific anterior bite force is below average; the success of the barracuda as a high-level predator seems to be because of its sharp teeth,

which require only a modest force to penetrate flesh, and the speed with which the jaws close on the prey (Westneat, 2004; Porter and Motta, 2004; Habegger et al., 2011).

Istiophoriformes

Xiphiidae and Istiophoridae

The billfishes, made up of the swordfishes (Xiphiidae) and sailfishes and marlins (Istiophoridae), include the fastest fish in the sea, reaching speeds of more than 80 km/h.

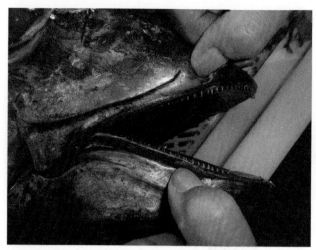

FIGURE 4.69 Dentition of Atlantic bluefin tuna (*Thunnus thynnus*). Image width = 12 cm.

FIGURE 4.70 Teeth of silver scabbardfish (*Lepidopus caudatus*). Image width = 3.5 cm.

FIGURE 4.71 Teeth of black scabbardfish (*Aphanopus carbo*). Image width = 14 cm. *Courtesy RCSOMA/ 470.1.*

FIGURE 4.72 Teeth of barracuda (*Sphyraena* sp.). Image width = 15 cm. *Courtesy RCSOMA/ 474.1.*

They are characterized by a sword- or spear-shaped projection from the upper jaw (Fig. 4.73) that, in the swordfish, starts to develop in juveniles about 100 mm in length. They are the apex predators among teleosts, feeding on squids, octopuses, and bony fishes. Together with sunfishes, billfishes are the largest teleosts. For example, the **blue marlin** (*Makaira nigricans*) reaches a length of 4.3 m and a mass of 910 kg. Billfishes possess a mass of heat-producing tissue in the head that warms the brain and eyes and presumably allows them to dive to considerable depths in search of prey.

The **swordfish** (*Xiphias gladius*) slashes with its sword to injure prey fishes, which it then consumes. It is uncertain whether sailfishes and marlins use the sword in the same way. Adult swordfishes have no teeth, but the larvae and juveniles possess fine, file-like teeth (Nakamura, 1951; Arata, 1954; Jones, 1958; Yasuda, 1978; Palco et al., 1981). At body length 9 mm, 18 sharp pointed teeth on the premaxilla interdigitate with 17 dentary teeth. During growth the number of teeth increases considerably, to about 140 on the maxilla and about 215 on the dentary. At a body length of 192 mm, there are great numbers of smaller, villiform teeth, together with several bands of villiform teeth on the inner margin of the main tooth row and some rudimentary teeth on the palate. However, all teeth are lost by a body length of 252 mm.

Adult sailfishes and marlins possess large numbers of small teeth in multiple rows on the jaws (Fig. 4.74).

Pleuronectiformes

This order contains the flatfishes, such as halibut, plaice, turbot, and sole. These are asymmetrical fishes, mostly benthic. During development, one eye migrates around the head so that in the adult both eyes lie on one side of the head. The eyes may be on the right side or the left side. There are more than 700 species of flatfishes in 11 families.

FIGURE 4.73 Swordfish (*Xiphias gladius*). Skeleton of adult, viewed from the front to show the sword projecting from skull. *Courtesy Wikipedia.*

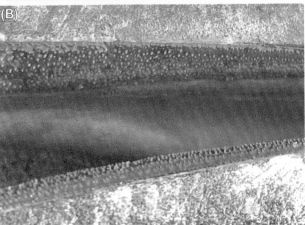

FIGURE 4.75 Jaws of Atlantic halibut (*Hippoglossus hippoglossus*). Image width = 21 cm. *Courtesy RCSOMA/ 477.2.*

FIGURE 4.74 (A) Teeth of sailfish (*Istiophorus*). (B) Higher power of Fig. 4.74A, showing large number of small teeth in jaws. © *Asther Lau Choon Siev/Dreamstime.*

Pleuronectidae

Atlantic halibut (*Hippoglossus hippoglossus*) is unusual in that only one eye is located on the dorsal surface. The mouth is very large and, viewed from the front, the jaws can be seen to be more or less symmetrical. Halibut feed on a wide variety of cephalopods, crustaceans, and fishes. The teeth on the premaxilla are disposed in two rows, with those in the outer row being generally larger and more curved than those in the inner row, except anteriorly where both rows contain large teeth (Fig. 4.75). On the dentary the teeth form a single row posteriorly and a double row of larger teeth at the front. The jaws of the **European plaice** (*P. platessa*) are asymmetrical, with the underneath side being smaller and flattened (Fig. 4.76). Plaice spend the day buried in the sand and emerge at night to feed on mollusks and crustaceans; these prey are crushed between the oral and pharyngeal teeth. The teeth form a single row in each jaw. Those on the

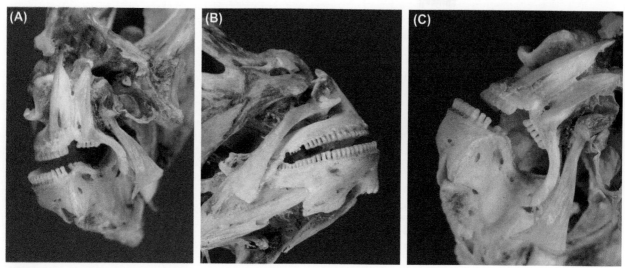

FIGURE 4.76 Dentition of European plaice (*Pleuronectes platessa*). (A) Frontal view showing asymmetry of skull: smaller left side faces downward, whereas larger right side faces upward in life. Image width=2.5 cm; (B) View of the dominant teeth on the right jaw. The teeth are large, slightly flattened labiolingually and closely set. Image width=2.8 cm; (C) Teeth on the left jaw of same specimen, showing that the teeth are much smaller and fewer in number than on the right. Image width=2.4 cm. *Courtesy RCSOMA/ 477.1.*

larger, upper side of the mouth are closely packed and blunt (Fig. 4.76A,B), whereas those on the smaller, lower side are fewer and smaller (Fig. 4.76A and C). The pharyngeal dentition also consists of strong crushing teeth.

Ovalentaria

This second largest series of perciform fishes contains diverse types of fishes (5300 species in at least seven orders): surf perches, damsel fishes, clown fishes, cichlids, silversides, flying fishes, garfishes, tooth carps, mullets, and blennies.

Cichliformes

Cichlidae

Cichlids are generally small, mainly freshwater fishes. In scientific importance they punch well above their weight for they are the most species-rich family of vertebrates. There are as many as 2000 species, accounting for about 7% of teleosts. The main groups of cichlids are found in East Africa (up to about 1300 species) and Central and South America (about 400 species), with smaller numbers of species in the Middle East and Asia.

The cichlids have exploited almost all food sources, and this is reflected in a wide range of adaptations of the teeth and jaws. The great adaptability of the oral dentition that makes this trophic radiation possible may be because, in cichlids, the role of food processing has been taken over by the pharyngeal jaws, leaving the oral jaws with the sole function of gathering food (Liem, 1973; Galis and Metz, 1998; Hulsey et al., 2006). In cichlids, the paired upper pharyngeal jaws, which are supported on the second, third, and fourth

pharyngobranchials, articulate by way of synovial joints with the neurocranium. The tooth plates of the lower pharyngeal jaw are united by a suture to form a single triangular plate in the midline and are also fused with the supporting fifth ceratobranchials (see Fig. 4.79). The upper and lower tooth plates have become mechanically uncoupled and so have acquired considerable freedom of movement (Galis and Drucker, 1996). In addition, the fourth external levator muscle that, in generalized percoid fishes, runs between the cranium and the fourth epibranchial and functions as a gill arch muscle instead inserts on the caudal wings of the lower pharyngeal tooth plate (Liem, 1973). This muscle, in concert with the other muscles connected to the lower tooth plate, protracts the lower pharyngeal jaw and brings it into occlusion with the upper pharyngeal jaw (Liem, 1973; Galis and Drucker, 1996). The mechanical stability of the pharyngeal tooth plates, combined with the freedom of movement of the lower pharyngeal jaw, increases the efficiency of food processing.

African Cichlids

The cichlids that have attracted most attention are those inhabiting the lakes of the Great Rift Valley of East and Central Africa. Each lake has a unique cichlid fauna. As the lakes are relatively young in geological terms, this distribution shows that cichlid speciation has been rapid and these populations have thus been of great interest for the study of adaptive radiation (eg, Fryer and Iles, 1972; Danley and Kocher, 2001; Salzburger and Meyer, 2004; Turner, 2007).

These radiations involve extreme diversification in diet and modes of feeding among cichlids. Methods of feeding include algae plucking and scraping, plant cropping,

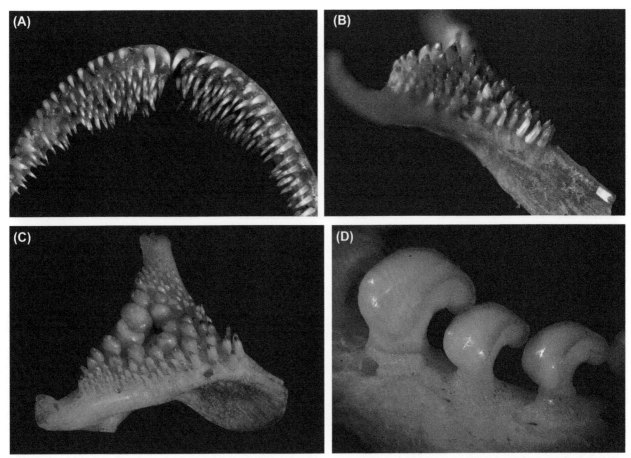

FIGURE 4.77 Variations in tooth form among African cichlids feeding on animal materials. (A) *Exochochromis anagenys* (fish eater), premaxillary teeth; (B) *Exochochromis anagenys*, lower pharyngeal teeth; (C) *Neolamprologus tretocephalus* (mollusk eater), lower pharyngeal teeth; (D) *Perissodus straeleni* (scale eater), dentary teeth. *Courtesy Dr. Ad. Konings.*

biting fins and removing scales, picking and sucking insects, eating small fish, and crushing mollusks (Fryer and Iles, 1972).

Adaptation to the various modes of feeding involves highly coordinated modifications of the jaws and of the oral and pharyngeal teeth. Cichlid teeth may be mono-, bi-, or tricuspid, but within each basic type there are numerous variations of shape, many of which are unique. The shape, arrangement, number, and distribution of teeth vary enormously according to the diet and method of feeding. Tooth size and form often vary within the same jaw. As in many other teleosts, the diet of cichlids can vary during the life cycle, and morphology will then vary with age. Because a particular feeding habit may have evolved in several different lineages, the dental adaptations often vary. In some cases an apparently specialized diet may be successfully exploited by fishes with relatively unspecialized dentitions.

To illustrate these points, several oral and pharyngeal dentitions exemplifying the variation in feeding methods are illustrated in Figs. 4.77 and 4.78. Fryer and Iles (1972) provide valuable descriptions of these and other dentitions.

Many cichlids feed on a range of foods derived from animals. The **three-spot torpedo** (*Exochochromis anagenys*) is a fish-eating predator. Its dentition comprises an anterior row of strong, sharp, recurved, monocuspid teeth, separated by a space from several rows of somewhat more slender teeth of similar form (Fig. 4.77A). These teeth seize and immobilize the prey after a side-on attack. The pharyngeal jaws are furnished with sharp conical teeth that are straight rather than recurved (Fig. 4.77B) and that are used to lacerate the flesh of the prey. Mollusk feeders, eg, the **five-barred cichlid** (*Neolamprologus tretocephalus*), crushes its prey between large, flat crowned pharyngeal teeth (Fig. 4.77C), which are fewer in number than in algae feeders. A few stout teeth may also be present on the jaws. Among the cichlids there exist several genera that have specialized in feeding on the scales of other species. The teeth of *Perissodus straeleni* have unique, strongly curved teeth (Fig. 4.77D). Takahashi et al. (2007) observed that *P. straeleni* uses its large hook-like teeth to scrape scales as it moves along the victim's body. In contrast, *Perissodus microlepis* has squat teeth with triangular tips. This fish grips a patch of scales and twists them off by rotating its body.

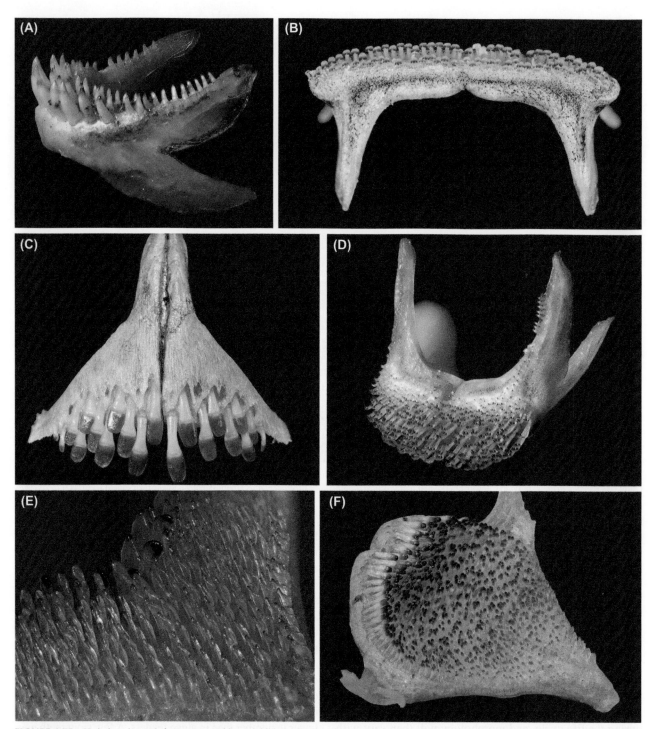

FIGURE 4.78 Variations in tooth form among African cichlids feeding on plant materials. (A) *Cynotilapia* sp. (plankton or Aufwuchs feeder), dentary teeth; (B) *Tropheus moorii* (algal cropper), dentary teeth; (C) *Eretmodus marksmithii* (Aufwuchs scraper), premaxillary teeth; (D) *Petrotilapia flaviventris* (Aufwuchs comber), dentary teeth; (E) *Protomelas taeniolatus* (alga feeder), lower pharyngeal teeth; (F) *Cyathopharynx foae* (detritus feeder), lower pharyngeal teeth. *Courtesy Dr. Ad. Konings.*

Cynotilapia sp. feeds on plankton or Aufwuchs. Its oral teeth are sharp and slightly recurved. At the front of the mouth there are three rows of teeth, of which the outer row is largest. Only the inner row of teeth extends to the back of the jaws (Fig. 4.78A). Fryer and Iles (1972) suggested that this genus has adopted plankton feeding relatively recently.

Its dentition is unspecialized and it snaps at prey, such as midge larvae, unlike other cichlid plankton eaters such as *Haplochromis cuanaeus* that use suction.

Many Great Lake cichlids feed on algae (Aufwuchs) that can be gathered by various methods. The mouth of the **blunthead cichlid** (*Tropheus moorii*) is wide, ventrally directed and transversely orientated. The jaws are furnished with an outer row of strong, bicuspid, spatulate teeth with recurved crowns, separated from an inner, irregular row of smaller teeth of similar morphology (Fig. 4.78B). The dentition of the **blue mbuna** (*Labeotropheus fuelleborni*) is similar, except that there is a single row of tricuspid teeth that are erect rather than recurved. Both species grip algae growing on rocks between their transverse rows of multicuspid teeth and detach it by pulling. *Eretmodus marksmithii* feeds by scraping algal deposits from rock surfaces. It has a relatively small number of spatulate, fingernail-shaped scraping teeth that are spaced apart (Fig. 4.78C). Replacement teeth develop labial to the functional row. It is clear that they develop and erupt synchronously, successive generations developing in alternate positions. They erupt for a considerable distance between the site of formation and the functional position (Fig. 4.78C). *Petrotilapia flaviventris* possesses numerous closely packed, elongated teeth with backwardly directed, spatulate tricuspid crowns (Fig. 4.78D). It uses these teeth as a comb, passing the array through the Aufwuchs to garner small particles of food such as diatoms or fragments of algae. In several feeders on algae, such as the **red empress** (*Protomelas taeniolatus*), *Tropheops*, and *Metriaclima*, the pharyngeal dentition consists of numerous backward-directed elongated, pointed teeth that are flattened laterally. Near the tip, the tooth is concave on the anterior aspect and the sharp tip is situated posteriorly (Fig. 4.78E). These teeth triturate the plant material passing through the pharyngeal jaws.

The food of the **featherfin cichlid** [*Cyathopharynx furcifer* (*foae*)] consists of loose detritus removed from rocks. The pharyngeal dentitions consist of a brush-like array of thin, closely packed teeth which are gently curved and blunt ended (Fig. 4.78F).

Fryer and Iles (1972) drew attention to several instances where dentition and diet were much less strongly correlated than in the examples just described. For example, *Haplochromis acidens* has the dentition of a piscivore, but consumes significant quantities of plant material and insects. The observations of Liem (1980) on the Lake Malawi cichlid *Petrotilapia tridentiger* show that even a high degree of specialization of the dentition can be misleading. This cichlid is morphologically specialized for scraping algae from rocks and has hinged teeth with a recurved, tricuspid spatulate tip well adapted to this function. However, it possesses a very wide repertoire of feeding methods: slow suction to feed on plankton and small invertebrates (three methods), fast suction to capture fishes, biting fins and scales of other fishes, scraping (two methods), and "manipulation" (by which the fish reduces prey without releasing it). While feeding by some of these methods, *Petrotilapia* can perform asymmetric movements of the jaws that are facilitated by a decoupling of the premaxilla and maxilla and other anatomical changes.

TABLE 4.2 Mean Oral and Pharyngeal Tooth Counts and Standard Length for Adult Malawi Cichlids

Genus/species	Pharyngeal	Oral	Standard length
Astatotilapia calliptera	178	176	73
Buccochromis heterotaenia	163	228	116
Chilotilapia euchilus	227	128	89
Copadichromis	235–410	129–137	85–92
Cynotilapia afra	314	67	75
Cyphotilapia moorii	154	223	131
Dimidiochromis	113–190	66–177	124–191
Docimodus evelynae	106	90	59
Fossorochromis rostratus	164	138	90
Labeotropheus	429–602	376–497	87–97
Labidochromis gigas	244	125	72
Melanochromis	308–312	230–375	72–73
Metriaclima	372–602	152–210	68–138
Nimbochromis	126–150	144–173	79–101
Otopharynx	198–229	141–163	89–91
Petrotilapia nigra	722	1170	96
Placidochromis milomo	226	179	97
Protomelas	140–169	88–208	85–149
Rhamphochromis esox	110	65	63
Stigmatochromis woodi	334	194	120
Taeniolethrinops praeorbitalis	271	300	91
Trematocranus placodon	88	196	141
Tropheops	296–478	253–349	72–170
Tyrannochromis	116–120	178–264	92–95

Data condensed to genus level where there are data for more than one species.
From Fraser, G.J., Hulsey, C.D., Bloomquist, R.F., Uyesugi, K., Manley, N.R., Streelman, J.T., 2009. An ancient gene network is co-opted for teeth on old and new jaws. PLoS Biol. 7, e1000031. Courtesy Editors Public Library of Science (PLoS One) Biology.

This versatility is also shown by other species that are omnivorous or adapted to feeding from surface deposits and, to a smaller extent, by species feeding on mollusks, arthropods, or plankton. Pursuit hunters or ambush hunters have a much more restricted range of feeding methods based on suction (Liem, 1980).

Table 4.2 displays mean tooth counts for the oral and pharyngeal teeth from 24 genera of adult cichlids (Fraser et al., 2009).

Neotropical Cichlids

The Neotropical cichlids tend to live in quiet waters, although some inhabit flowing water. Feeding specializations have not been investigated in the same detail as in the African cichlids, but it is known that they have adapted to eating a wide range of foodstuffs. Some are active predators of free-swimming fishes and invertebrates. This mode of feeding is associated with more protrusible jaws (Hulsey and García de León, 2005). In some genera, such as *Petenia*, jaw protrusibility can reach an extreme (Waltzek and Wainwright, 2003). Other Neotropical cichlids are more omnivorous, feeding on invertebrates and plant material. Food may be acquired from the mud of the river bottom, often by the process of **winnowing**, in which the fish burrows in the bottom sediment, takes in its prey along with sand and mud, and ejects the latter from the mouth and opercula before swallowing the food items. Neotropical cichlids also include plankton feeders and species that scrape Aufwuchs from rocks.

The oral dentition of Neotropical cichlids shows much less variation than in their African counterparts (Casciotta and Arratia, 1993). Most species have a row of unicuspid teeth on the premaxilla and dentary, with one to seven inner rows of smaller, unicusid teeth. In a few species, there is the same arrangement but with bicuspid teeth rather than unicuspid teeth. The **Poor man's tropheus** [*Hypsophrys* (*Neetroplus*) *nematopus*], which scrapes Aufwuchs from rocks, has an outer row of spatulate, incisor-like teeth with one to three inner rows of similarly shaped teeth.

The pharyngeal dentition shows more variation than the oral dentition (Casciotta and Arratia, 1993). The most prevalent tooth type is curved, with a sharp point, and may be uni- or bicuspid. The latter type has a small cusp on the concave aspect of the tooth. On the upper pharyngeal tooth plates the uni- or bicuspid teeth are curved posteriorly, whereas on the upper tooth plates they are curved anteriorly. The opposite orientation of upper and lower teeth suggests that anteroposterior movement of the pharyngeal jaws would shred prey very effectively. Rounded, crushing teeth, sometimes with a small apical cusp, occur in a variety of species. In the upper pharyngeal jaws, the three dentigerous plates may carry different types of tooth. In the lower jaws, two or three tooth types may occur on the same plate. It is noteworthy that crushing teeth are not confined to species specialized for mollusk eating, but form part of the pharyngeal dentition of many omnivorous cichlids (Casciotta and Arratia, 1993) and presumably have a role in crushing insects or crustaceans.

Plasticity of Cichlid Dentitions

Observations on the Lake Victoria cichlid **Alluaud's haplo** (*Astatoreochromis alluaudi*) and on the Neotropical *Herichthys* and *Amphilophus* (formerly *Cichlasoma*) have demonstrated considerable phenotypic plasticity in the pharyngeal teeth. Some populations of Alluaud's haplo consume hard-shelled mollusks and have heavily built, pharyngeal jaws with robust, dome-shaped teeth, whereas other populations do not seem to eat such tough mollusks and have much lighter pharyngeal jaws with fewer crushing teeth (Fryer and Iles, 1972). Experiments showed that the offspring of Alluaud's haplo raised on a soft diet instead of on snails did not adapt to eating snails and developed pharyngeal jaws that are lightly built, like those in wild, nondurophagous populations, and that have fewer crushing teeth (Fryer and Iles, 1972). Further laboratory experiments on this species (Huysseune, 1995, 2000) and on the Neotropical **Midas cichlid** [*Amphilophus* (*Cichlasoma*) *citrinellus*] (Meyer, 1990), showed that, in fishes reared on soft food, the pharyngeal dentition is dominated by slender, conical teeth and contain few crushing teeth, whereas fishes reared on hard foods acquire larger pharyngeal jaws occupied largely by round crushing teeth (Fig. 4.79). Moreover, in Alluaud's haplo reared on soft food the number of pharyngeal teeth increases with age, whereas in those reared on hard food the number remains constant or even decreases slightly, while the size of the teeth increases (Huysseune, 1995, 2000). Quantitative observations on wild-caught *Herichthys* (*Cichlasoma*) *minckleyi* (Trapani, 2004) were consistent with these results. In sum, the development of robust, crushing pharyngeal jaws in these cichlid species seems to be induced epigenetically, through repeated mechanical stresses associated with consumption of hard-shelled snails. This plasticity is only feasible in polyphyodont dentitions. A possible sequence of events that could translate mechanical stress into morphological change was outlined by Gunter and Meyer (2014), but unravelling this intriguing problem remains at an early stage.

The cichlids are notable for providing examples of interactions between behavior and morphology. Scale-eating cichlids show strong and consistent preferences for attacking the left or right side of the victim and this is correlated with morphological handedness of the jaws. Lee et al. (2012) observed that juvenile *Perissodus microlepis* showed strong preferences for removing scales from a particular flank and that this was not correlated with the degree of morphological handedness, which was not pronounced. They suggested that behavioral handedness is prior to, and the cause of, morphological handedness—an intriguing hypothesis that is worth further investigation.

The plasticity of the jaws and dentition have established cichlids as animals of choice in studying the mechanisms controlling mouth and jaw form, tooth number, and tooth shape (Albertson and Kocher, 2006; Fraser et al., 2008, 2009, 2013).

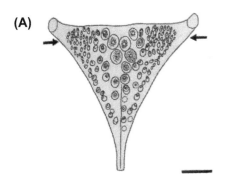

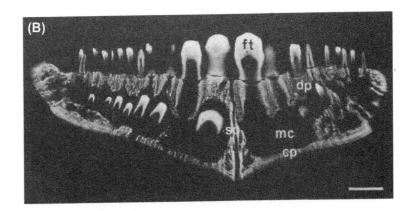

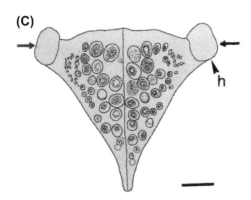

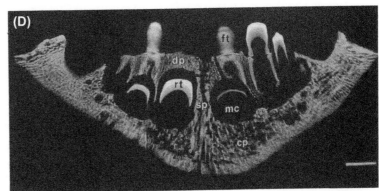

FIGURE 4.79 Alluaud's haplo (*Astatoreochromis alluaudi*). Dorsal view of the lower pharyngeal plate of a pond-raised soft food specimen (A) and a wild-caught durophagous specimen (C), both about 85-mm standard length. *Arrows* in (A) and (C) indicate the level of the transverse ground sections from which microradiographs are presented in (B) (soft food specimen) and (D) (durophagous specimen), respectively. *h*, horn, *ft*, functional tooth, *dp*, dentigerous plate, *sp*, sutural plate, *mc*, medullary cavity of jaw, *cp*, cortical plate, *rt*, replacement tooth. Scale in (A) and (C)=2 mm. Scale in (B) and (D)=1 mm. *From Huysseune A., 2000. Developmental plasticity in the dentition of a heterodont polyphyodont fish species. In: Teaford, M.F., Smith, M.M., Ferguson, M.J. (Eds.), Development, Function and Evolution of Teeth, Cambridge University Press, Cambridge, pp. 231–241. Courtesy Cambridge University Press.*

Beloniformes

This order includes the needlefishes (Belonidae), halfbeaks (Hemiramphidae), flyingfishes (Exocoetidae), sauries (Scomberesocidae), and ricefishes (Adrianichthyidae). Among beloniforms, jaw structure varies both ontogenetically and phylogenetically and seems to be related to diet. The greatest ontogenetic changes occur in needlefishes. As larvae they have short jaws of equal length, in juveniles the lower jaw elongates beyond the upper jaws, as in halfbeaks, and finally the upper jaw also elongates so that most adult needlefishes have jaws with nearly equal lengths. Such changes seem to be related to diet; juvenile needlefish feed primarily on plankton, whereas adults feed mainly on fish (Lovejoy, 2000).

Belonidae

Needlefishes have elongated bodies and the jaws are extended to form a long, slender beak that is furnished with large numbers of sharp, conical teeth (Fig. 4.80). The principal prey consists of small fishes that are caught by an

upward sweep of the beak. Cephalopods and crustaceans form smaller proportions of the diet.

Blenniiformes

Blennies are distributed over about 130 genera with approximately 850 species. Their diet can be based on vegetation; detritus; mollusks; coral; invertebrate worms; and the mucus, scales, and fin rays of other fish. Many species are omnivorous.

Blenniidae

The largest family, the **combtooth blennies**, feed mostly on invertebrates. The dentition of the **peacock combtooth blenny** [*Lipophrys canevae* (*Salaria parvo*)] is typical and consists of large recurved canines behind a single row of incisiform teeth in both jaws (Fig. 4.81A). In this position, the canines are unfavorably positioned for grasping prey and show little evidence of wear, although the caniniform tooth in the related *Lipophrys trigloides* can exhibit wear (Fig. 4.81B) (Kotrschal and Goldschmid, 1992). No

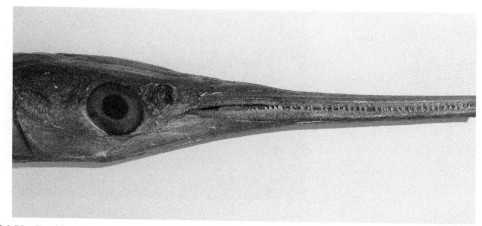

FIGURE 4.80   Dentition of needlefish (species unknown). Image width=20 cm. © *Trustees of the Natural History Museum, London.*

significant correlation was obtained between canine length and importance of animal prey in the diet, leading Kotrschal and Goldschmid (1992) to suggest that canines in comb-tooth blennies are predominantly used for predatory deterrence and agonistic interactions.

The number of teeth in blennies varies widely between taxa: **Caneva's blenny** (*Lipophrys canevae*) has 18–24 teeth in each jaw (Fig. 4.82A), and the **blue-dashed rock-skipper** (*Istiblennius periophthalmus*) has up to 150 teeth (Fig. 4.82B) (Fishelson and Delarea, 2004).

Sabre-tooth blennies are remarkable because their fang-like teeth are equipped with venom glands. Two genera (*Aspidontus* and *Plagiotremus*) mimic the color and behavior of cleaner wrasse, which pick parasites from the skin of larger fish when they present themselves at cleaning "stations." Fishes mistakenly present themselves to the sabre-tooth blenny expecting to be cleaned but must be unpleasantly surprised when the blenny takes a small bite from their flesh using the fang-like teeth in the lower jaw (Fig. 4.83). The enlarged caniniform teeth are thought to be used principally as defence weapons (Kotrschal and Gold-schmid, 1992).

Eupercaria

This is the largest and most diverse series of percomorph fishes, containing 6600 species divided into 13 or more orders. It contains grunts, snappers, angel fishes, drums, star gazers, wrasses, parrotfishes, sea breams, butterfly-fishes, anglerfishes, pufferfishes, sunfishes, sea basses, and gurnards.

Lutjanidae

The teeth in the upper jaw of the **red snapper** (*Lutjanus campechanus*) are mostly densely packed, fine or hair-like villiform teeth. However, there are also enlarged, anterior caniniform teeth (Fig. 4.84). The lower jaw, which projects slightly beyond the upper jaw, has larger villiform teeth. Adult red snapper feed on a variety of smaller fishes, crustaceans, and mollusks.

Sciaenidae

The drums or croakers typically feed on invertebrates and small fishes on the seabed. The **red drum** (*Scianops ocellatus*) and **black drum** (*Pogonias chromis*) have conspicuous pharyngeal teeth, with different morphologies related to differences in diet (Grubich, 2000). The red drum possesses large, pointed pharyngeal teeth for shredding shrimp, fish, and other soft-bodied prey. The black drum has molariform pharyngeal teeth adapted to transmitting the heavy forces needed to crush the hard-shelled, bivalve prey on which it subsists (Fig. 4.85). The upper pharyngeal jaw of the black drum is not retracted by the dorsal retractor muscle, as it is in other teleosts. This indicates that crushing hard-shelled marine bivalves requires a vice-like compression bite in contrast to the shearing forces that are applied to the weaker shelled prey of red drum.

The **corvina** (*Argyrosomus regius*) possesses conical teeth of varying size that may be distributed in more than a single row (Fig. 4.86). Enlarged teeth form an outer series in the upper jaw and an inner series on the lower jaw. Caniniform teeth may be present at the front of both jaws. Teeth are absent from the vomer and palatine bones.

Labriformes

This suborder consists of the Labridae (wrasses and parrotfishes) and Odacidae (weed whitings) and includes carnivores, herbivores, and omnivores.

Labridae

This is a large family with about 600 species of fishes that live in shallow water and occupy a range of habitats,

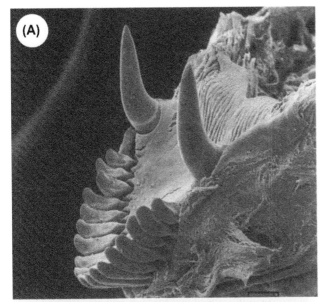

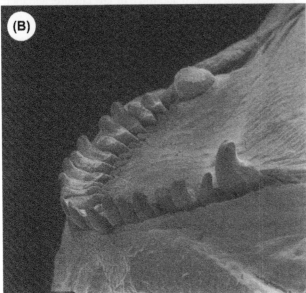

FIGURE 4.81 (A) Scanning electron micrograph (SEM) of lower jaw of the peacock combtooth blenny (*Lipophrys canevae*). Note the unworn caniniform tooth at the back of the tooth row. Scale = 1 mm; (B) SEM of lower jaw of a combtooth blenny *Lipophrys trigloides*. Note the wear on the caniniform tooth at the end of the tooth row. Scale = 1 mm. *From Kotrschal, K., Goldschmidt, A., 1992. Morphological evidence for the biological role of caniniform teeth in combtooth blennies (Blennidae, Teleostei). J. Fish. Biol. 41, 983–991. Courtesy Editors Journal Fish Biology.*

most notably tropical coral reefs. There are two groups: the **wrasses** and the **parrotfishes**. The wrasses are mostly carnivorous, although some small species feed on plankton or on mucus at the coral surface, whereas the parrotfishes include a large number of herbivores.

They are "pharyngognathous fishes," but their pharyngeal jaw apparatus has a unique feature. The fifth ceratobranchials are fused and support the fused lower pharyngeal tooth plates that form a single triangular structure, the caudal "wings" of

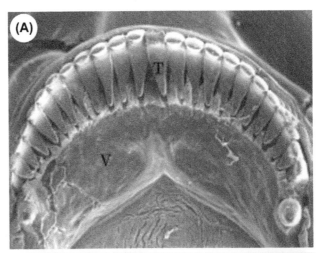

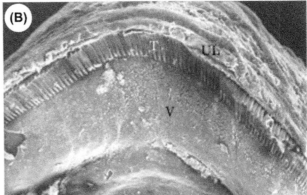

FIGURE 4.82 (A) Scanning electron micrograph (SEM) of lower jaw of Caneva's blenny (*Lipophrys canevae*). T, teeth, V, breathing valve. Image width = 2.7 mm; (B) SEM of upper jaw of blue-dashed rockskipper (*Istiblennius periophthalmus*). UL, upper lip, V, breathing valve. Image width = 3.2 mm. *From Fishelson, L., Delarea, Y., 2004. Taste buds on the lips and mouth of some blennid and gobiid fishes: comparative distribution and morphology. J. Fish. Biol. 65:651–665. Courtesy Editors Journal of Fish Biology.*

FIGURE 4.83 Fang of the bluestriped fangblenny (*Plagiotremus rhinorynchus*). *Courtesy Dr. A. Bshary.*

FIGURE 4.84 Red snapper (*Lutjanus campechanus*). Front view of mouth to show long pointed teeth. © *David Mailland/Dreamstime.*

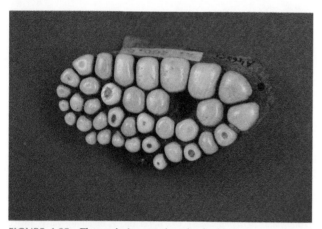

FIGURE 4.85 Flattened pharyngeal teeth of a black drum (*Pogonias chromis*). Image width=9 cm. *Courtesy RCSOMA/ 465.5.*

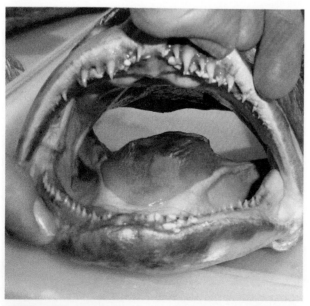

FIGURE 4.86 Dentition of corvina (*Argyrosomus regius*). Image width=6 cm.

which articulate with the cleithra (vertical elements of the pectoral girdle). These pharyngocleithral joints stabilize the jaw and provides a fulcrum that contributes to a powerful pharyngeal bite (Liem and Greenwood, 1981; Vandewalle et al., 2000). The pharyngeal jaws in various species are capable of shearing, crushing, and, in the Scaridae, grinding prey. However, the rigidity of the lower pharyngeal jaws restricts the pharyngeal opening and hence the size of the prey that can be ingested; this may account for the relative paucity of piscivory among labrids (Wainwright et al., 2012).

There is a corresponding diversity in the structures associated with feeding, and the functional morphology of feeding in labrids has attracted considerable research interest. Wainwright et al. (2004) concluded that adaptation to the wide ecological range involved independent variation of the components of the feeding system (oral jaws, hyoid apparatus, and pharyngeal jaws), thereby allowing the evolution of many different combinations of mouth size, upper jaw protrusibility, speed and force of jaw closure, and tooth morphology.

Wrasses

Carnivorous wrasses prey upon a wide variety of mollusks, crustaceans, and fishes. Prey is taken in a variety of ways and some conclusions regarding jaw mechanics and tooth morphology are possible, although much remains to be clarified (Clifford and Motta, 1998; Ferry-Graham et al., 2002; Wainwright et al., 2004). Suction is used in capturing prey in the water column, in detaching prey clinging to the substrate, or in winnowing, as used by Neotropical cichlids. Biting is used to remove pieces from certain types of prey and to capture elusive prey, and it is combined with suction in benthic feeding and winnowing. Mollusks and some crustaceans need to be crushed by application of considerable force and, in durophagous species, the posterior levator muscles that close the pharyngeal jaws are well developed. Both predators of active prey and durophagous species typically have several large procumbent anterior teeth (Fig. 4.87) and strong, but shorter, posterior teeth. The pharyngeal teeth of durophagous wrasses are hemispherical and may show signs of abrasion.

Cleaner wrasses (*Labroides* spp.) remove ectoparasites and additional nutrients, such as mucus, from the skin of other fishes, and they have limited protrusible jaws and anterior teeth that are smaller and less robust than those of predatory wrasses. The same features are found in predators of small crustaceans. Small wrasses, such as *Cirrhilabrus* spp., which feed upon zooplankton, have highly protrusible mouths and a reduced dentition.

The **Ballan wrasse** (*Labrus bergylta*) has strong conical oral teeth, adapted to seizing prey (Fig. 4.88). The teeth on the pharyngeal plates are rounded and adapted to crushing the prey of mollusks and crustaceans (Fig. 4.89). The **harlequin tuskfish** (*Choerodon fasciatus*) has a pair

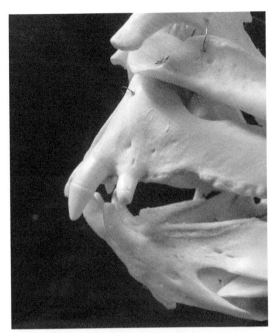

FIGURE 4.87 Jaws of blackspot tuskfish wrasse (*Choerodon schoen-leinii*), showing large, procumbent anterior teeth. Specimen size unknown (this species can reach 100 cm in total length). *Courtesy Prof D.R. Bellwood.*

FIGURE 4.88 Oral dentition of Ballan wrasse (*Labrus bergylta*). Image width = 5 cm. *Courtesy RCSOMA/ 468.1.*

FIGURE 4.89 Lower pharyngeal plate of Ballan wrasse (*Labrus ber-gylta*), furnished with rounded crushing teeth. Image width = 4.5 cm. *Courtesy Horniman Museum & Gardens and Dr. Paolo Viscardi. Catalog no. LDHRN-NH. 72.29.135.*

FIGURE 4.90 Harlequin tuskfish (*Choerodon fasciatus*), showing protruding, blue-pigmented anterior teeth. *Courtesy Dr. Richard Vevers, Australian Museum.*

Parrotfishes

Parrotfishes are distinguished from wrasses by many features of the feeding apparatus (Wainwright et al., 2004). They have small mouths and jaws that are capable of little protrusion but that have high mechanical advantage in opening and closing. The pharyngeal jaw apparatus (Gobalet, 1989) is uniquely specialized for grinding. The possession of a grinding pharyngeal mill has contributed to the success of parrotfishes as herbivores and coral grazers: both feeding activities require extensive breakdown of ingested food.

Streelman et al. (2002) identified two parrotfish lineages: a group mostly associated with seagrass and a group living on coral reefs. Members of the smaller "seagrass" group (20% of species) have a dentition that is not greatly different from that of wrasses: the teeth are arranged in diagonal rows and are not

of tusk-like anterior teeth on both premaxilla and dentary, which turn dark blue as the fish matures (Fig. 4.90). It uses these teeth to puncture and tear the exoskeletons of its invertebrate prey.

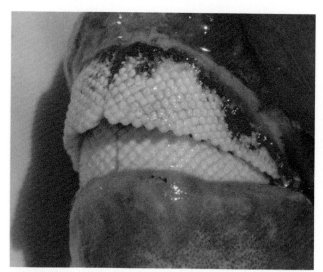

FIGURE 4.91 Mouth of green humphead parrotfish (*Bolbometopon muricatum*) showing the beak, which is partly covered by macerated algae. This species excavates the surface of coral. Note the alternating array of the constituent teeth and the resulting "castellated" or notched scraping edge of the beak. Width of specimen unknown (this species can reach 1.3 m in total length). *Courtesy Prof D.R. Bellwood.*

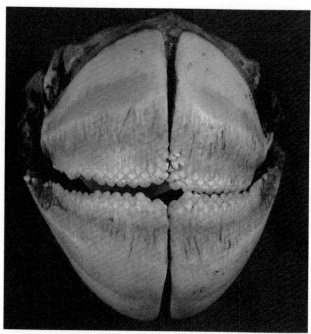

FIGURE 4.92 Prepared specimen of parrotfish beak (species unknown). The alternating arrangement of the component teeth is similar to that in *Bolbometopon* (Fig. 4.91). In this specimen, the shafts of the constituent teeth, cemented together by bone, are visible between the beak and the subalvolar jaw bones. *Courtesy MoLSKCL. Catalog no. V59.*

cemented together. Most of the members of this group browse on periphyton, seaweed, and seagrass. However, one Caribbean genus (*Sparisoma*) removes algae from coral reefs by scraping or excavating the surface. Members of the reef group possess strong beaks made up of teeth cemented together by bone. Parrotfish beaks resemble those of tetraodonts in shape, but they consist of superimposed rows of individual rounded teeth instead of stacks of elongated teeth (Figs. 4.91 and 4.92). It seems that rows of replacement teeth develop and move into position before their predecessors are shed. In the "reef" group of parrotfishes, the primitive mode of feeding is excavation of the coral surface, ie, removal of part of the substrate including coral organisms and algae. In this group, the beak is robust and covered with a thick layer of cement. At the cutting edge alternate teeth project to give a notched appearance (Figs. 4.91 and 4.92) (Bellwood and Choat, 1990). Two genera—*Scarus* and *Hippocarus*—scrape the surface without damaging it. These fishes have a less robust beak than the excavating species and do not have the notched cutting edge (Bellwood and Choat, 1990). In the clade consisting of these two genera plus *Chlorurus*, the intramandibular joint between the dentary and the articular is mobile, as in many coral reef species (see page 46) (Fig. 4.93), and it is thought that this mobility enlarges the gape and also allows the dental beak to adapt more closely to the coral surface during scraping or excavating (Wainwright et al., 2004). Members of the reef group are important in maintenance of the health of coral by preventing overgrowth of algae and sponges.

The specializations of the pharyngeal jaws of parrotfishes for grinding are described by Gobalet (1989), Bellwood (1994), and Carr et al. (2006). The upper pharyngeals are

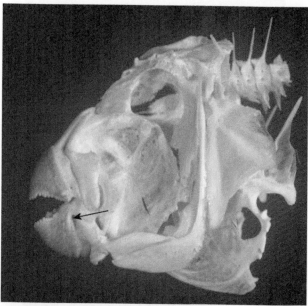

FIGURE 4.93 Skull of blunt-head parrotfish (*Chlorurus microrhinos*). The mobile intramandibular joint, between the dentary and articular bones, is indicated by an *arrow*. Width of specimen unknown (this species can reach 80 cm in total length). *Courtesy Prof. D.R. Bellwood.*

elongated anteroposteriorly (Fig. 4.94) and are connected by cruciate ligaments in the midline, so they do not move independently. The degree of movement of the upper pharyngeals relative to the neurocranium seems to be much less than in

FIGURE 4.94 Left upper pharyngeal tooth plate of parrotfish (species unknown). Occlusal view, anterior to left. Note the unworn, newly formed teeth at the anterior end, embedded in bone. At the posterior end, the teeth are worn, so that the enameloid forms the elevated margin of a central region of dentine. *Courtesy Megan Fisher and Australian Museum.*

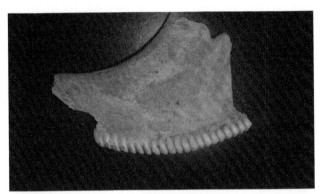

FIGURE 4.95 Side view of upper pharyngeal tooth plate of parrotfish (species unknown): anterior to left. Note the convex curved shape of the plate. The transition from newly formed teeth at the anterior end and the worn teeth at the posterior end is clearly seen. *Courtesy Megan Fisher and Australian Museum.*

FIGURE 4.96 Lower pharyngeal toothplate of parrotfish (species unknown): anterior to left. Note the presence of newly formed teeth at the posterior end and worn teeth at the anterior end. Image width = 5 cm. *Courtesy Horniman Museum & Gardens and Dr. Paolo Viscardi. Catalog no. LDHRN-NH. A3278.*

wrasses, at least during grinding, which is generated by anteroposterior motion of the lower pharyngeal against the almost static upper pharyngeals, and is powered by massive levator and adductor muscles.

The functional surfaces of the pharyngeal jaws carry longitudinal rows of incisiform teeth orientated transverse to the axis of the tooth plate (Figs. 4.94–4.96). The teeth are embedded in hard tissue variously described as bone or cement. In the upper pharyngeal plates, teeth form in the anterior region of the plate and become progressively more worn posteriorly (Figs. 4.94 and 4.96). At the posterior margin, the tooth plate is worn very thin. In the lower pharyngeals new teeth form posteriorly and wear anteriorly (Fig. 4.96). The teeth in both upper and lower plates are said to migrate away from the site of formation, but no work seems to have been done to establish the histological basis for the movement of the teeth relative to the plate. In the upper pharyngeals the teeth form up to three longitudinal rows (Fig. 4.95), whereas in the lower pharyngeals there are five or six rows of teeth that interdigitate (Fig. 4.96).

There are differences in the pharyngeal dentition between the groups of parrotfishes (Bellwood, 1994). In browsing species, the upper pharyngeals are relatively broad and carry three rows of teeth, often in a chevron arrangement. The breadth of the lower pharyngeals is greater than the width. Among reef species, the upper pharyngeals are more elongated than in browsers and the teeth are orientated transverse to the plate. *Sparisoma*, *Cetoscarus*, and *Bolbometopon* (excavators) possess three rows of teeth on each upper plate (although the outer row is much reduced, as in Fig. 4.94), whereas the tooth-bearing area of the lower pharyngeals is approximately square. In *Chlorurus* (excavators) and *Hipposcarus* and *Scarus* (scrapers), there is only one row of teeth occupying most of the width of the upper pharyngeal plate, together with an outer row of vestigial teeth. The tooth-bearing areas of the lower pharyngeals are elongated in reef species.

Spariformes

Sparidae

The breams and porgies are bottom-living fishes that feed mainly on invertebrates. The **black sea bream** (*Spondyliosoma cantharus*) eats seaweeds as well as invertebrates, and its dentition of small conical teeth is adapted to grasping (Figs. 4.3 and 4.97). The **common dentex** (*Dentex dentex*)

FIGURE 4.97 Teeth of black sea bream (*Spondyliosoma cantharus*). © *Dinoforlena/Dreamstime*

is an active predator of fishes, cephalads, and crustaceans, with a dentition consisting of strong, conical, sharp teeth of which the anterior teeth are largest (Fig. 4.98A). On the dentary several very small teeth lie inside the marginal teeth (Fig. 4.98B).

The **gilt-head (sea) bream** (*S. auratus*) has a diet of shellfish and a heterodont dentition adapted to grasping and crushing. The anterior teeth are stout and pointed, whereas the posterior teeth form a battery of flattened crushing teeth of various sizes (Fig. 4.99). The **sheepshead fish** (*Archosargus probatocephalus*) has a similar diet to *S. auratus* and a similar dentition, consisting of three to four pairs of anterior grasping teeth, used to pry shellfish from rocks, and three rows of posterior crushing teeth (Fig. 4.100A and B). The anterior teeth are remarkable in that they are not pointed but flattened and resemble in form the incisors of mammalian herbivores. These teeth develop deep within the jaw and erupt through the overlying bone (Fig. 4.100C). The height of the mature teeth shows that replacing incisor teeth have to erupt vertically for some distance. Once in place, the teeth become ankylosed by deposition of bone between the tooth and the surrounding jaw bone.

The herbivorous **Karanteen seabream** (*Crenidens crenidens*) possesses three rows of denticulate teeth (ie, flattened teeth with small cusps along the margins) (Fig. 4.101). The front row has 10 teeth, with four denticles at the tip and two on each side of the teeth the second row has teeth with four denticles and the third row has teeth with three denticles. Interspersed among these teeth are a few robust teeth (Fishelson et al., 2014).

Chaetodontiformes

This order consists of the butterflyfishes (Chaetodontidae) and ponyfishes (Leiognathidae).

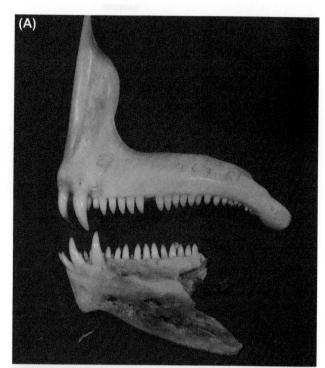

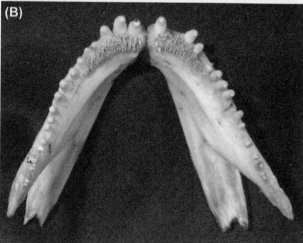

FIGURE 4.98 Common dentex (*Dentex dentex*). (A) Side view of upper and lower teeth. Image width=4 cm. RCSOMA/467.51; (B) Dorsal view of lower jaw showing numerous small teeth lying behind larger front teeth. Image width=6 cm. *Courtesy RCSOMA/ 467.5.*

Chaetodontidae

The butterflyfishes inhabit coral reefs. The Linnaean family name draws attention to the numerous bristle-like teeth that form clusters near the anterior margins of the jaw. Specializations of the feeding apparatus are associated with the method of feeding rather than with the diet (Motta, 1988). The **fourspot butterflyfish** (*Chaetodon quadrimaculatus*), which is omnivorous and browses on hard coral, has a wide mouth with numerous rows of teeth at the front of the mouth to form a transverse "brush" (Fig. 4.102A and B). In species that also excavate the coral surface the teeth are more

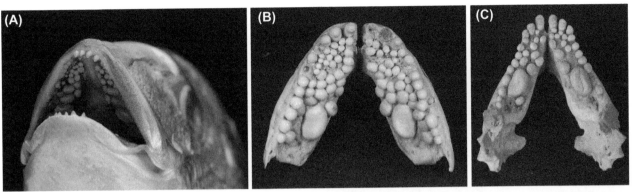

FIGURE 4.99   Gilt-head (sea) bream (*Sparus auratus*). (A) Mouth opening, showing pointed (grasping) marginal teeth and rounded (crushing) inner teeth. *(© Andrei Calangui/Dreamstime.)* (B) Premaxillae to show distribution of grasping and crushing teeth. Image width=5.7 cm; (C) Lower jaw to show similar distribution of grasping and crushing teeth. Image width=6 cm. *(Courtesy RCSOMA/ 466.3.)*

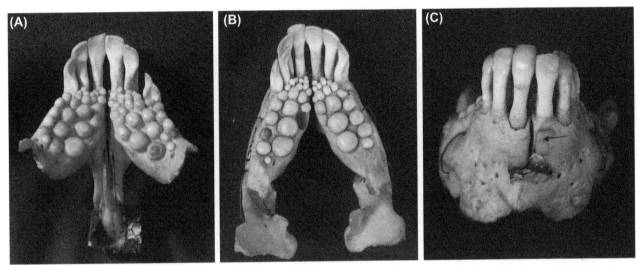

FIGURE 4.100   Sheepshead fish (*Archosargus probatocephalus*). (A) Teeth on the premaxillae. Image width=4 cm; (B) Teeth in the lower jaw. Image width=3.7 cm; (C) Front view of lower jaw, showing position of replacing incisiform tooth (*arrowed*) deep to its predecessor. Image width=3 cm. *Courtesy RCSOMA/ 467.1.*

robust. The **yellow longnose butterflyfish** (*Forcipiger flavissimus*) has a slender, elongated snout (Fig. 4.102C and D) that is used to probe crevices in the coral in search of prey, such as polychaetes and other larger invertebrate prey. The prey is grasped and torn using inwardly hooked teeth that form anterior clusters and also extend down the sides of the mouth. The **pebbled butterflyfish** (*Chaetodon multicinctus*) feeds on individual coral polyps and has forceps-like jaws furnished with relatively small anterior groups of teeth (Fig. 4.102E and F).

An individual butterflyfish tooth (Fig. 4.103) is long and slender, with an acutely pointed enameloid cap. The shaft dentine is composed at least partly of vasodentine. Motta (1984b) described the arrangement and structure of the dentition of the **millet-seed butterflyfish** (*Chaetodon miliaris*), which is an opportunistic feeder, taking zooplankton as well as benthic organisms. The teeth have a fibrous attachment that would allow them to flex when moved over the coral

surface. The dentary and premaxilla each carry multiple obliquely orientated rows of teeth arranged en echelon and in groups separated by tooth-free spaces.

Lophiiformes

Anglerfishes, of which there are about 320 species distributed in about 16 families, are found in a variety of habitats, from the continental shelf to the deeper regions of the oceans. These fishes lie in wait for their prey (fishes and various invertebrates), either on the seabed or floating in midwater. The feature that gives this group its name is the possession of a thin, fishing-rod–like process, the **illicium**, which is derived from the first dorsal fin ray and which extends forward over the mouth from the front of the head. At its tip is a lure, the **esca**, which is used to attract prey. In deep-sea anglerfishes the esca is luminous. The head is always large in relation to body length. The mouth is

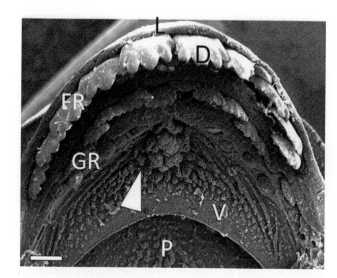

FIGURE 4.101 Teeth in the upper jaw of the Karanteen sea bream (*Crenidens crenidens*). Note the denticulate teeth (D) in front row (Fr). *GR*, tooth rows behind the front row. *P*, palate, *V*, oral valve. *Arrowhead* indicates group of taste buds. Scale = 1.2 mm. *From Fishelson, L., Golani, D., Diamant, A., 2014. SEM study of the oral cavity of members of the Kyphosidae and Girellidae (Pisces, Teleostei), with remarks on* Crenidens *(Sparidae), focusing on teeth and taste bud numbers and distribution. Zoology 117, 122–130. Courtesy Editors Zoology.*

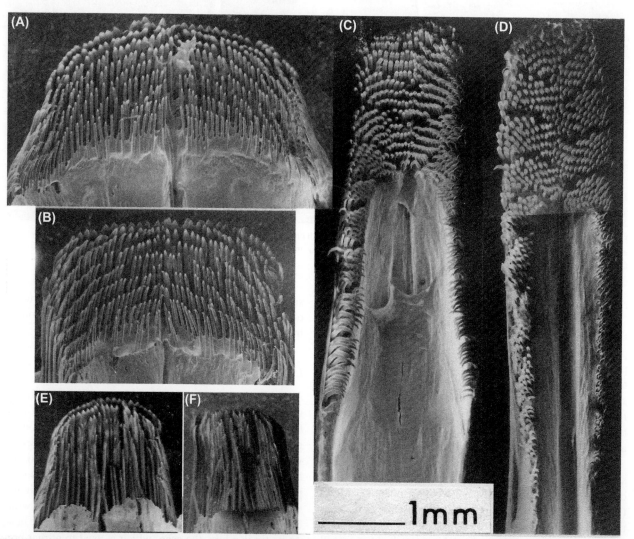

FIGURE 4.102 Dentitions of butterflyfishes. (A,C,E) Premaxillae; (B,D,F) Dentaries. (A, B) *Chaetodon quadrimaculatus*; (C, D) *Forcipiger flavissimus*; (E, F) *Chaetodon multicinctus. Modified from Motta, P.J., 1988. Functional morphology of the feeding apparatus of ten species of Pacific butterflyfishes (Perciformes: Chaetodontidae): an ecomorphological approach. Env. Biol. Fish. 22, 39–67. Courtesy Editors Environmental Biology of Fishes.*

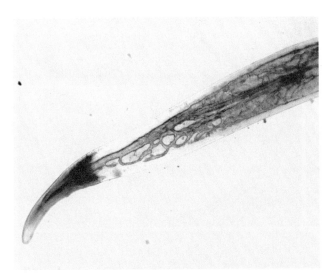

FIGURE 4.103 Isolated tooth of butterflyfish (species unknown). Curved, conical cap of enameloid at tip. Shaft dentine contains capillaries (vasodentine). Image width=2.1 mm. © *Museums at the Royal College of Surgeons, Tomes slide collection. Catalog no. 439.*

correspondingly wide, extending a long way round the head, and it is furnished with large, pointed teeth. Anglerfishes can capture prey of a wide range of sizes, and the large mouth enables them to ingest prey nearly as large as themselves.

Many species of anglerfish show extreme sexual dimorphism in which males are extremely diminutive and as small as 1/40th the size of females. After a certain stage of maturity the male digestive system may stop and the male must find a female or die. It achieves this by an acute olfactory system that senses pheromones released by the female. The male uses small hooked teeth at the front of its jaws to attach to the female. After this attachment, the male secretes enzymes that break down both the oral cavity of the male and the adjacent skin of the female. This allows the male and female tissues to merge at the level of the vascular systems and enables the male to receive nutrients from the female. Apart from its reproductive system, the remaining organs of the parasitic male degenerate, including its teeth. Female anglerfishes may host up to six males. There is much less information on the teeth of male anglerfishes than on those of the females. Bertelsen and Struhsaker (1977) described the presence of four teeth in the tip of the upper jaw and seven on the lower jaw of the young male *Thaumatichthys*.

Lophiidae

The **angler** or **monkfish** (*Lophius piscatorius*) inhabits coastal waters. It is flattened dorsoventrally and rests motionless on the seabed until prey comes within reach. It is well camouflaged by the patterns on the skin and by a fringe of lateral skin appendages that break up the fish's

outline. The enormous mouth (Fig. 4.104A) can take in very large prey, which is grasped with numerous sharply pointed teeth. On the premaxilla the teeth at the front are large and arranged in two rows, whereas further back lies a single row of smaller teeth. On the dentary, the teeth are arranged in two rows, with the inner row containing larger teeth (Fig. 4.104B). An additional row of teeth is present on the ectopterygoids, and small pointed teeth are also present on the pharyngeal jaws. All of these teeth are hinged (Fig. 4.104C and D) and allow passage of prey toward the pharynx, but they prevent its escape in the opposite direction.

Ceratioidei

This is the largest suborder of lophiiforms (162 species in 11 families). All live at depths of greater than 300 m and include both pelagic and benthic forms. They show a wide variety of shapes and sizes (Fig. 4.105) Many, such as the **ghostly sea devil** (*Haplophryne mollis*; Linophrynidae), are globose, with large heads (Fig. 4.106), whereas others are more elongated. In these deep-sea species, the esca is made visible in the darkness by the possession of symbiotic, bioluminescent bacteria. Pelagic forms spend most of their time floating while awaiting prey, maintaining position by using their pectoral fins. The ability to ingest large prey is highly advantageous in the deep-sea environment, where there is a low density of food organisms. The group shows extreme sexual dimorphism. The males are always tiny and are often parasitic, either temporarily or permanently, on the much larger females.

Thaumatichthyidae

This family comprises two genera of highly specialized deep-sea anglerfishes: *Lasiognathus* (Fig. 4.105H) and *Thaumatichthys* (Fig. 4.105I) and the family name, meaning "wonder fishes," certainly applies to the unique method of capturing prey (Bertelsen and Struhsacker, 1977). The upper jaw of female **Regan's strainer-mouth anglerfish** (*Lasiognathus saccostoma*) extends a long way in front of the lower jaw. The premaxillae are armed with long, curved, sharp teeth (Fig. 4.107) and are connected laterally to the maxillae by loose connective tissue. The ascending processes of the premaxillae articulate with projection of the ethmoid cartilage and the tip of each ascending process is connected to the maxilla and palatine of the opposite side by elastic ligaments (Fig. 4.108). In the resting position, the premaxillae are rotated about their long axes so that the teeth are directed outward (Fig. 4.108A and B) and are held there by tension in the premaxillary ligaments. Prey is attracted into the mouth by the esca, which lies behind the premaxillae. It has two to three bony hook-shaped denticles and is at the end of a very long illicium. The prey is captured

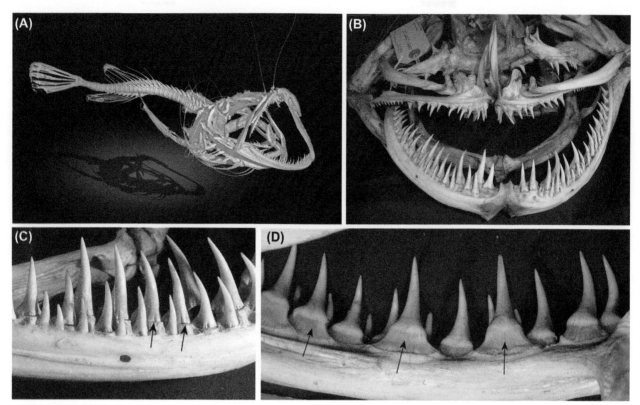

FIGURE 4.104   Angler (*Lophius piscatorius*). (A) Skeleton. *(Courtesy Wikipedia.)* (B) Jaws. Image width=20 cm. *(Courtesy RCSOMA/482.2.)* (C) Hinged teeth (labial view), showing discontinuity between pedicel and crown *(arrow)*. Image width=5 cm. (D) Hinged teeth (lingual view), showing fibrous attachment between crown and pedicel *(arrow)*. Image width=9 cm. *(Courtesy RCSOMA/ 482.2.)*

by rotation of the premaxillae, which brings the teeth downward and close off the mouth like the jaws of a trap (Figs. 4.107 and 4.108C). The action of the levator maxillae superioris muscles is believed to provide the force for closure of the jaws (Bertelsen and Struhsaker, 1977). Prey trapped by the teeth can be seized and transported toward the pharynx by the short lower jaw.

Tetraodontiformes

The Tetraodontiformes are represented by 10 families and about 360 species. Their name comes from the possession of a beak on both jaws, each formed by the fusion of separate, elongated teeth.

Tetraodontidae, Diodontidae

Comprising **pufferfishes** and **porcupine fishes**, respectively, these fishes are all slow swimmers with round, square, triangular or compressed bodies armored with platy or spiny scales or thick skin (Fig. 4.109). Their dentition has become converted into a hard beak suitable for crushing the hard-shelled invertebrates, such as mollusks and crustaceans, on which they feed. In pufferfishes (*Tetraodon* = "four teeth"), the upper and lower

beaks have visible midline sutures (Fig. 4.109), whereas in porcupine fishes (*Diodon* = "two teeth"), such sutures are absent.

The beak of pufferfishes consists of paired upper and lower elements (Figs. 4.109 and 4.110), each made up of a stack of elongated laminar teeth that consist almost entirely of enameloid and that are cemented together in a matrix of hard tissue (Britski et al., 1985) (Fig. 4.111). The individual teeth making up each element of the beak are members of the same tooth family, so that those at the cutting edge are older than those at the base. A cutting edge at the margin of the beak is maintained by differential wear of the tooth enameloid and of the surrounding bone. Tooth formation is intraosseous, so that as the teeth wear away at the cutting edge, the beak is replenished by addition of new teeth initiated and formed in a cavity at the base of the beak (Fig. 4.111). Soon after the formation of an individual tooth is complete, hard tissue is laid down around it, to incorporate it into the beak. The tissue cementing the individual teeth together has been described as "osteodentine" (Bristki et al., 1985), but it does not have the characteristic denteonal structure (Fig. 4.111) and is more likely to be bone of attachment, as seen in Figs. 4.11 and 4.14.

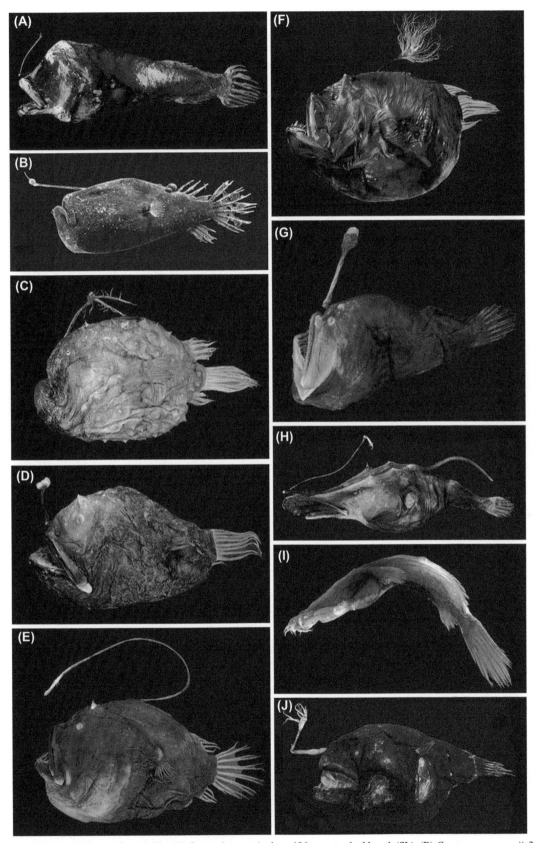

FIGURE 4.105 Variation of shapes of ceratioids. (A) *Centrophryne spinulosa*, 136-mm standard length (SL); (B) *Cryptopsaras couesii*, 34.5-mm SL; (C) *Himantolophus appelii*, 124-mm SL; (D) *Diceratias trilobus*, 86-mm SL; (E) *Bufoceratias wedli*, 96-mm SL; (F) *Bufoceratias shaoi*, 101-mm SL; (G) *Melanocetus eustalus*, 93-mm SL; (H) *Lasiognathus amphirhamphus*, 157-mm SL; (I) *Thaumatichthys binghami*, 83-mm SL; (J) *Chaenophryne quasiramifera*, 157-mm SL. *Courtesy of Wikipedia.*

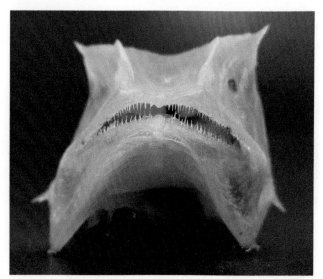

FIGURE 4.106 Anterior view of ghostly sea devil (*Haplophryne mollis*), showing globular form of fish and jaws furnished with numerous sharp teeth. *Courtesy Kerryn Parkinson, Australian Museum. AMS 1.21365-008.*

FIGURE 4.107 Jaws of female Regan's strainer-mouth angler fish (*Lasiognathus saccostoma*). Note the greater length of the upper jaw compared with the lower jaw. The mandible is furnished with a single row of recurved sharp teeth, whereas the upper jaw carries a single row of very long, curved teeth. The premaxillae are rotated in this specimen so that the fangs are held pointing outward, ready to capture prey. Image width = 12 cm. © *Trustees of the Natural History Museum, London*

During the ontogeny of the dentition of the parrotfish (*Monotrete*: Tetradontidae) (Fraser et al., 2012), the beak formed of stacked laminar teeth is preceded by a row of separate teeth connected by dentine. The transformation into a beak is discussed in Chapter 10.

Molidae

The **ocean sunfish** (*Mola mola*) is the largest teleost in the ocean. The average adult may weigh 1000 kg and the highest recorded weight is 2300 kg. The sunfish has an unusual disc-like shape that may measure 4-m dorsoventrally and 3 m in length (Fig. 4.112). It has a small mouth with a beak-like tooth plate in each jaw (Fig. 4.113). It also possesses curved, pointed pharyngeal teeth (Fig. 4.114). The main prey of ocean sunfish is jellyfish, which it must consume

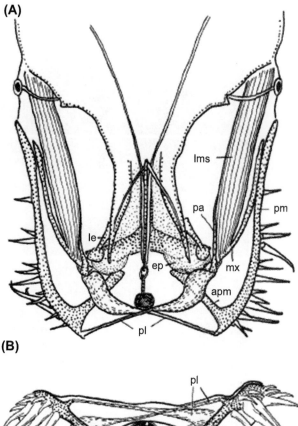

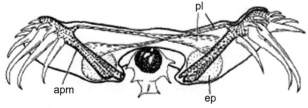

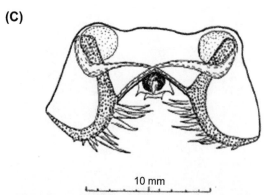

FIGURE 4.108 Jaws of female *Thaumatichthys binghami*. (A) Dorsal view, premaxillary fangs directed outward; (B) Anterior view of jaws in same position; (C) Anterior view of jaws, with premaxillary fangs directed ventrally. *apm*, articular process of premaxillary, *ep*, projection of ethmoid cartilage, *le*, lateralethmoid, *lms*, levator maxillae superioris, *mx*, maxilla, *pa*, palatine, *pl*, premaxillary ligaments, *pm*, premaxilla. *From the Galathea Report by Bertelsen, E., Struhsacker, P.J., 1977. The ceratioid fishes of the genus* Thaumatichthys. *Osteology, relationships, distribution and biology. Galathea Rep. 14, 7–40. Reproduced by kind permission of the Zoology Museum, Natural History Museum of Denmark.*

FIGURE 4.109 Anterior view of the stellate pufferfish (*Arothron stellatus*), showing beak with clear median suture which separates the beak into four quarters. © *Trustees of the Natural History Museum, London.*

FIGURE 4.112 Ocean sunfish (*Mola mola*). *Courtesy Wikipedia.*

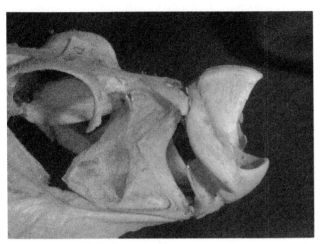

FIGURE 4.110 Skull of pufferfish (*Tetraodon* sp.) viewed from the side, showing the "beak." Image width = 10 cm. *Courtesy RCSOMA/ 480.1.*

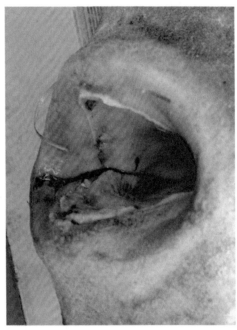

FIGURE 4.113 Mouth of ocean sunfish (*Mola mola*), showing the tooth "beak" in both upper and lower jaws. © *Kerryn Parkinson, Australian Museum.*

FIGURE 4.111 Longitudinal ground section of pufferfish beak (species unknown), showing the functional edge of the beak of a pufferfish. The bone enclosing the terminal tooth has worn away, exposing the enameloid of the tooth at the edge. Image width = 2.1 mm. © *Museums at the Royal College of Surgeons, Tomes slide collection. Catalog no. 488.*

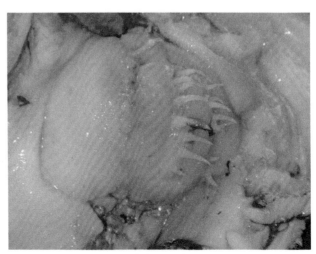

FIGURE 4.114 Hook-shaped pharyngeal teeth of *Mola mola*. © *Kerryn Parkinson, Australian Museum.*

in large quantities to maintain its bulk. This diet is supplemented by squid, crustaceans, eelgrass, fish larvae, and small fishes.

Balistidae

Triggerfishes and file fishes are slow-swimming fishes specialized for feeding on hard-shelled mollusks and crustaceans. The dentition of the **grey triggerfish** (*Balistes capriscus*) is typical and consists of eight large, strong, incisiform teeth in each jaw (Fig. 4.115).

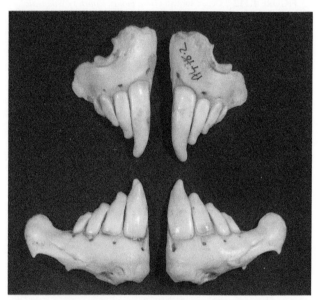

FIGURE 4.115 Anterior view of dentition of the grey triggerfish (*Balistes capriscus*). Image width=10 cm. *Courtesy RCSOMA/478.2.*

Acanthuriformes

Acanthuridae

Surgeon fishes are herbivores that eat algae growing on coral and rocks, although some species also feed on detritus (Fishelson and Delarea, 2014). They possess a row of labiolingually flattened, multicuspid teeth that are deeply embedded within the soft tissue of the jaws by a long shaft. The nature of the denticulations differs among the various species. Thus, in the **brown surgeonfish** (*Acanthurus nigrofuscus*), the teeth form a uniform row and each tooth bears one apical denticle and four others on each side (Fig. 4.116A). In the **bristletooth surgeonfish** (*Ctenochaetus striatus*), each tooth has about four cusps along one edge, and all the teeth on both jaws lean with their smooth edge toward the midline of the jaw (Fig. 4.116B) (Fishelson and Delarea, 2014). Clumps of algae are gripped between the teeth and detached by movements of the head. During this process, the algae may be gripped by soft tissue pads—the "retention plate"— that are located immediately behind the rows of teeth and that retain the separated algae within the mouth. **Unicorn fishes** (*Naso*) feed on zooplankton and have pointed teeth.

Pempheriformes

Polyprionidae

Atlantic wreckfish (*Polyprion americanus*) is a deepwater fish found in caves and sea wrecks; hence, the common name. They have hundreds of small conical teeth distributed in linear patches on both upper and lower jaws (Fig. 4.117) as well as teeth on the tongue.

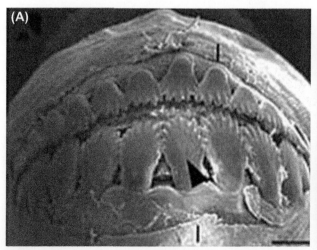
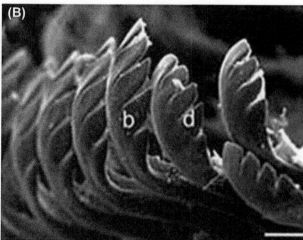

FIGURE 4.116 Teeth of surgeonfishes. Scanning electron micrographs. (A) Front teeth of the brown surgeonfish (*Acanthurus nigrofuscus*). Each tooth bears one apical denticle and four others on each side. Scale=1 mm; (B) The teeth of the bristletooth surgeonfish (*Ctenochaetus striatus*). Each tooth has about four cusps along one edge. Scale = 80 μm. *From Fishelson, L., Delarea, Y., 2014. Comparison of the oral cavity architecture in surgeonfishes (Acanthuridae, Teleostei), with emphasis on the taste buds and jaw "retention plates". Env. Biol. Fish. 97, 173–185. Courtesy Editors Environmental Biology of Fishes.*

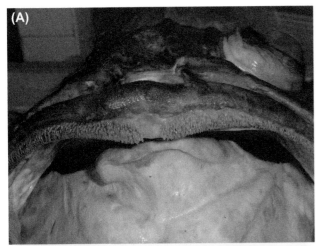

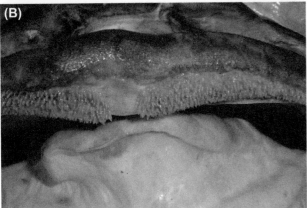

FIGURE 4.117 Dentition of Atlantic wreckfish (*Polyprion americanus*). (A) View of whole upper dentition. Image width = 11 cm; (B) Higher power view of teeth in Fig. 4.117A in upper jaw near the midline. Image width = 5.3 cm.

Centrarchiformes

Centrarchidae

The **pumpkinseed** (*Lepomis gibbosus*) is a freshwater fish of the sunfish family. The adult feeds primarily on mollusks that are sucked in and then crushed between the upper and lower pharyngeal jaws. This is reflected in the rounded crushing surfaces of the pharyngeal teeth (Fig. 4.118B). During snail crushing, the lower jaw is held relatively stationary and the upper jaw exerts the primary crushing force as it is pressed firmly against the snail shell. Upper jaw depression is caused by rotation of epibranchial 4 about the insertion site of the posterior oblique muscle.

Unlike the diet of the pumkinseed, the diet of the related **bluegill** (*Lepomis macrochirus*) consists of soft-bodied prey, and this is reflected in the morphology of the pharyngeal teeth, which are much more conical (Fig. 4.118A).

Pumpkinseeds are flexible in their diet and will eat alternative, softer prey if snails are unavailable. The effects of this change in diet on the feeding apparatus have been studied in the laboratory and by comparing pumpkinseeds inhabiting lakes in North America, which differ in the abundance of snails. In pumpkinseeds consuming a diet that includes a high proportion of snails, the mass of the muscles actively involved in moving the jaws, and the size of the three main bones involved, are significantly increased and the pharyngeal teeth are worn down (Wainwright et al., 1991). These morphological differences were shown to be because of phenotypic plasticity rather than natural selection, as the diet exerted the same effect on the feeding apparatus regardless of whether the test fishes originated from snail-rich or snail-poor lakes (Mittelbach et al., 1999). The plasticity demonstrated in the pumpkinseed is similar to that observed in the cichlid Alluaud's haplo (see page 86).

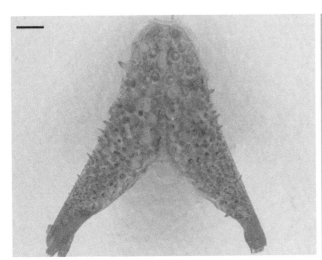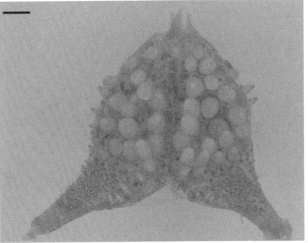

FIGURE 4.118 Images of alizarin red–stained lower pharyngeals of bluegill (*Lepomis macrochirus*) (113 mm in total length) (left), and pumpkinseed (*Lepomis gibbosus*) (111 mm in total length) (right). Scale = 1 mm. *Courtesy Dr. G. Andraso.*

Girellidae

This family includes **drummers**, **opaleyes**, and **blackfish**, mostly placed in the genus *Girella*. They are herbivores or omnivores. The marginal oral dentition comprises three to six rows of flattened, multicuspid teeth, of which the outer row is the largest (Fishelson et al., 2014). The jaw teeth of the **opaleye** (*G. nigricans*) are adapted to scraping. The crown is bent approximately at right angles to the shaft and the teeth have a hinged attachment to the pedicel, so that the teeth have a hoeing action (Norris and Prescott, 1959). Small multicuspid teeth are present on the vomers, whereas the pharyngeal teeth are recurved and may be mono- or multicuspid. In the **luderick** (*Girella tricuspidata*), there are about five rows of teeth, of which the front teeth are largest. As the name suggests the teeth have a tricuspid form, with a large central cusp (Fig. 4.119) (Fishelson et al., 2014). In young fish, the teeth are monocuspid.

Kyphosidae

The **sea chubs** are herbivorous fishes, feeding on Aufwuchs. The dentition consists of a single row of spatulate, closely set marginal teeth on the premaxilla and dentary (Fishelson et al., 2014). The crowns of these teeth form a large angle with the basal portion, which is horizontally orientated. Immediately within the marginal tooth row lie pads of soft tissue furnished with numerous small, conical teeth. Similar small, sharp teeth occur on the vomers and either side of the palate, on the tongue and also on the pharyngeal jaws. The marginal dentition seems to be adapted to gripping and tearing algae, whereas the sharp, pointed teeth within the mouth and pharynx presumably retain and shred ingested algae (Fishelson et al., 2014).

Perciformes

Serranidae

Groupers (Epinephelinae) are heavy-bodied carnivores that feed on fish and on invertebrates, such as octopuses and crustaceans. They have multiple rows of teeth along the margins of the jaws and in the roof of the mouth on the vomerine and palatine bones (Fig. 4.120). Some species

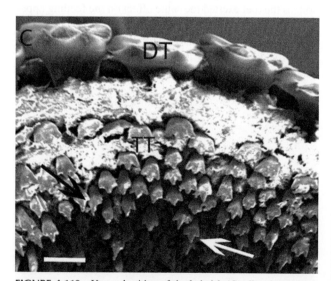

FIGURE 4.119 Upper dentition of the luderick (*Girella tricuspidata*), showing several rows of tricuspid teeth. Scanning electron micrograph. *DT*, outer row of denticulate teeth. *TT*, inner rows of tricuspid teeth. *Arrow* indicates teeth with sharp central denticles. Scale = 200 μm. *From Fishelson, L., Golani, D., Diamant, A., 2014. SEM study of the oral cavity of members of the Kyphosidae and Girellidae (Pisces, Teleostei), with remarks on* Crenidens *(Sparidae), focusing on teeth and taste bud numbers and distribution. Zoology 117, 122–130. Courtesy Editors Environmental Biology of Fishes.*

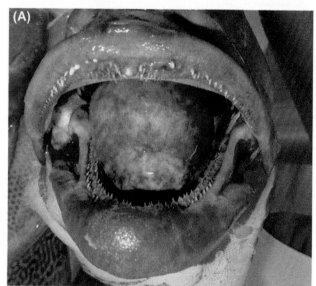

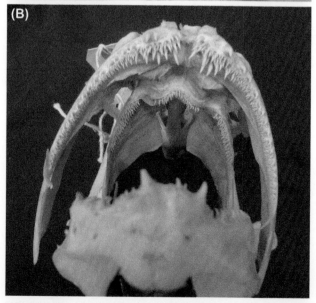

FIGURE 4.120 Dention of a grouper (*Epinephelus* sp.). (A) Fresh specimen viewed from front; (B) Ventral view of skull. Note the multiple rows of teeth in the upper jaws. *Courtesy RCSOMA/ 463.1.*

FIGURE 4.121   Dentition of ling cod (*Ophiodon elongatus*). *Courtesy Wikipedia.*

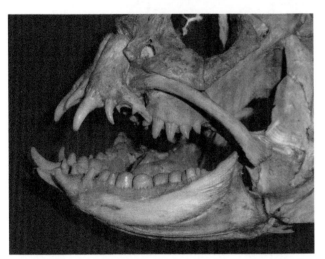

FIGURE 4.122   Anterolateral view of dentition of Atlantic wolffish (*Anarhichas lupus*). *Courtesy RCSOMA/ 473.14D.*

have enlarged canine-like teeth. The oral teeth are supplemented by crushing pharyngeal teeth.

Scorpaenidae

This family contains a variety of carnivorous fishes, such as **ling cod** (*Ophiodon elongatus*), scorpionfishes, and lumpsuckers. The jaws and pharyngeal plates of ling cod are armed with spaced, sharp, caniniform teeth. The premaxilla possesses an additional row of teeth (Fig. 4.121).

Anarhichadidae

The **Atlantic wolffish** (*Anarhichas lupus*) has a diet that includes shrimps and echinoderms and it has a similar heterodont dentition, with usually four anterior caniniform teeth in each jaw and two posterior rows of crushing teeth. Additional crushing teeth are present in the roof of the mouth (Figs. 4.122 and 4.123).

SARCOPTERYGII

Coelacanthimorpha (= Actinistia)

This subclass contains a single living order, Coelacanthiformes, in which there is one genus.

Coelacanthiformes

Coelacanths (*Latimeria*) were thought to be extinct until, in 1938, one was caught and identified in the Comoro archipelago. Populations have since been found at other sites in the Indian Ocean. Two species of coelacanth are now recognized: the **West Indian Ocean coelacanth** (*Latimeria chalumnae*) and the **Indonesian coelacanth** (*Latimeria menadoensis*).

Coelacanths prey on fish, and their dentition consists of numerous small, conical teeth on the upper and lower jaws, among which are a few much larger, fang-like teeth (Fig. 4.124). All teeth are ankylosed. Small conical teeth are also present on the pharyngeal arches (Nelson, 1969). Coelacanth teeth are covered by a thin layer of true enamel.

Dipnotetrapodomorpha

This class is divided into two subclasses. The Tetrapodomorpha comprises the amphibians, reptiles, and mammals, which are the subjects of the following chapters of this volume. The Dipnomorpha comprises a single order (Ceratodontiformes), with six living species of lungfishes in two suborders. All are omnivorous.

In juvenile lungfishes there is a single row of long, sharp teeth adapted to a diet of small worms and water fleas. However, in the adults the dentition is made up of paired tooth plates, each of which is a composite of many teeth fused together by bone (Smith, 1985, 2010; Kemp, 2002). In early development of the plates, rows of teeth attached to a common plate of bone are formed and become overgrown by

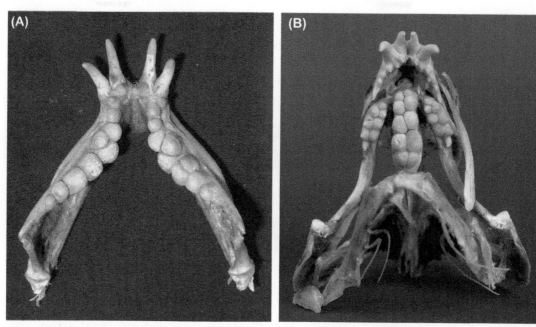

FIGURE 4.123 Teeth of Atlantic wolffish (*Anarhichas lupus*). (A) Lower jaw. Image width=6 cm; (B) Upper jaw. Image width=12 cm. *Courtesy RCSOMA/ 473.2.*

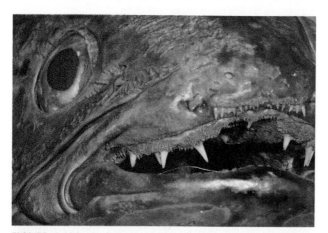

FIGURE 4.124 Coelacanth (*Latimeria chalumnae*): mouth opening, showing small, conical teeth, interspersed with larger fang-like teeth of similar morphology. © 2013 Danté Fenolio, Department of Conservation and Research–San Antonio Zoo. *Courtesy California Academy of Sciences.*

bone. Each plate consists of a characteristic number of tooth rows in a fan-shaped array, diverging toward the jaw margins. The plates subsequently grow by initiation of new tooth substance at the anterolateral margins of the tooth rows, followed by incorporation into the plate by growth of bone and also by outgrowth of dentine from the tooth (Fig. 4.125).

In contrast to all other bony fishes, the teeth of lungfishes are not shed or replaced, and the tooth plate continues to grow in thickness and area throughout life. Each tooth consists of a cone of orthodentine of which the outer surface is covered by true enamel, as in the coelacanths. Inside the cone of orthodentine there forms, in addition, a hypermineralized tissue unique to lungfishes—petrodentine—that has the mineral content and hardness of enamel or enameloid.

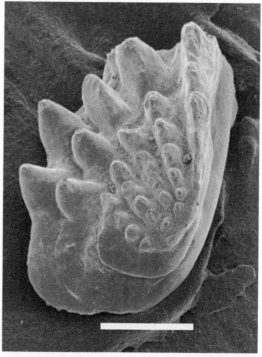

FIGURE 4.125 Developing tooth plate of an Australian lungfish (*Neoceratodus forsteri*), showing a number of small separate cusps that eventually fuse and are worn down to form the individual ridges of the tooth. Scanning electron micrograph. Scale=0.5 mm. *Courtesy Prof M.M. Smith.*

The tooth plates are used to break up the food, which includes both plant and animal foodstuffs, by crushing and shearing. Because of the core of petrodentine in each constituent tooth, the ridges on the plates are very resistant to wear and can break down even hard-shelled water snails.

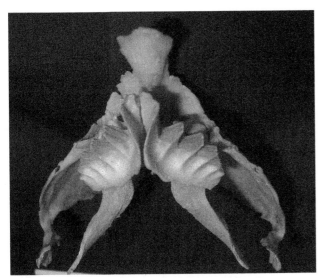

FIGURE **4.126** Australian lungfish (*Neoceratodus forsteri*). Tooth plates on upper jaw, each with six ridges. Image width = 13 cm. *Courtesy Professor M.M. Smith.*

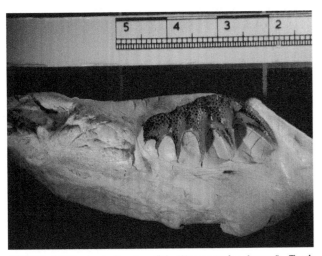

FIGURE **4.127** Australian lungfish (*Neoceratodus forsteri*). Tooth plate, showing six ridges. Image width = 15 cm. *Courtesy Professor M.M. Smith.*

Ceratodontiformes

Suborder Ceratodontoidei

In the **Australian lungfish** (*Neoceratodus forsteri*: Neoceratodontidae), the single tooth plate in each jaw quadrant has six ridges (Figs. 4.126 and 4.127).

Suborder Lepidosirenoidei

The **West African lungfish** (*Protopterus annectens*: Protopteridae), like the Australian lungfish, has a single pair of tooth plates in each jaw. However, these have only three ridges (Figs. 4.128 and 4.129).

The **South American lungfish** (*Lepidosiren paradoxa*: Lepidosirenidae), has teeth similar to those of *Protopterus*.

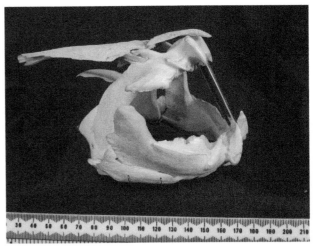

FIGURE **4.128** West African lungfish (*Protopterus annectens*). Right lateral view of dentition. Each tooth plate has three ridges. Image width = 15 cm. *Courtesy MoLSKCL. Catalog no. V60.*

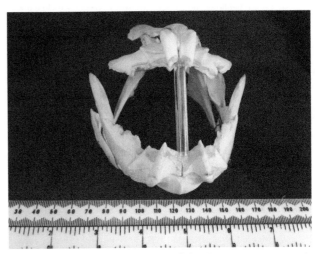

FIGURE **4.129** West African lungfish (*Protopterus annectens*). Dentition viewed from the front. Each tooth plate has three ridges. Image width = 14 cm. *Courtesy MoLSKCL and Dr. G Sales. Catalog no. V60.*

ONLINE RESOURCES

Fishbase: www.fishbase.org.

Suction feeding by *Petenia splendida*: http://www.youtube.com/watch?v=TbGILi8p5Y8&feature=rec-LGOUT-real_rn-HM.

Suction feeding by slingjaw wrasse (*Epibulus insidiator*): http://www.youtube.com/watch?v=pDU4CQWXaNY.

Moray eel feeding: https://www.youtube.com/watch?v=Rv2DkzOPBXw.

REFERENCES

Albertson, R.C., Kocher, T.D., 2006. Genetic and developmental basis of cichlid trophic diversity. Heredity 97, 211–221.

Alfaro, M.E., Brock, C.D., Banbury, B.L., Wainwright, P.C., 2009. Does evolutionary innovation in pharyngeal jaws lead to rapid lineage diversification in labrid fishes? BMC Evol. Biol. 9, 255–269.

Arata, G.F., 1954. A contribution to the life history of the swordfish, *Xiphias gladius* Linnaeus, from the south atlantic coast of the United States and the Gulf of Mexico. Bull. Mar. Sci. Gulf Caribb. 4, 183–243.

Atukorala, A.D.S., Hammer, C., Dufton, M., Franz-Odendaal, T.A., 2013. Adaptive evolution of the lower jaw dentition in Mexican tetra (*Astyanax mexicanus*). EvoDevo 4, 28.

Bellwood, D.R., 1994. A phylogenetic study of the parrotfishes family Scaridae (Pisces: Labroidei), with a revision of genera. Rec. Aust. Mus. 20 (Suppl.), 1–86.

Bellwood, D.R., Choat, J.H., 1990. A functional analysis of grazing in parrotfishes (family Scaridae): the ecological implications. Env. Biol. Fish. 28, 189–214.

Bemis, W.E., Giuliano, A., McGuire, B., 2005. Structure, attachment, replacement and growth of teeth in bluefish, *Pomatomus saltatrix* (Linnaeus, 1766), a teleost with deeply socketed teeth. Zoology 108, 317–327.

Berkovitz, B.K.B., 2013. Nothing but the Tooth: A Dental Odyssey. Elsevier, London.

Berkovitz, B.K.B., Shellis, R.P., 1978. A longitudinal study of tooth succession in piranhas (Characidae), with an analysis of the tooth replacement cycle. J. Zool. Lond. 184, 545–561.

Bertelsen, E., Struhsacker, P.J., 1977. The ceratioid fishes of the genus *Thaumatichthys*. Osteology, relationships, distribution and biology. Galathea Rep. 14, 7–40.

Betancur-R, R., Broughton, R.E., Wiley, E.O., Carpenter, K., López, J.A., Li, C., Holcroft, N.I., Arcila, D., Sanciangco, M., Cureton II, J.C., Zhang, F., Buser, T., Campbell, M.A., Ballesteros, J.A., Roa-Varon, A., Willis, S., Borden, W.C., Rowley, T., Reneau, P.C., Hough, D.J., Lu, G., Grande, T., Arratia, G., Ortí, G., 2013. The tree of life and a new classification of bony fishes. PLoS Curr. Tree Life Edition 1.

Betancur-R, R., Wiley, E., Bailly, N., Miya, M., Lecointre, G., Ortí, G., 2014. Phylogenetic Classification of Bony Fishes – Version 3. https://sites.google.com/site/guilleorti/home/classification.

Blaber, S.J.M., Brewer, D.T., Salini, J.P., 1994. Diet and dentition in tropical ariid catfishes from Australia. Env. Biol. Fish. 40, 159–174.

Bone, Q., Marshall, N.B., Blaxter, J.H.S., 1995. Biology of Fishes, second ed. Chapman and Hall, London.

Britski, H.A., Andreucci, R.D., Menezes, N.A., Carneiro, J.J., 1985. Coalescence of teeth in fishes. Rev. Bras. Zool. 2, 459–482.

Britz, R., Conway, K.W., Rüber, L., 2009. Spectacular morphological novelty in a miniature cyprinid fish, *Danionella dracula* n. sp. Proc. Biol. Sci. 276, 2179–2186.

Bruneel, B., Mathä, M., Paesen, R., Ameloot, M., Weninger, W.J., Huysseune, A., 2015. Imaging the zebrafish dentition: from traditional approaches to emerging technologies. Zebrafish 12, 1–10. http://dx.doi.org/10.1089/zeb.2014.0980.

Buddery, A., Kemp, A., Day, R.D., Tibbetts, I.R., 2009. Ultrastructure and the importance of wear in the dentition of the halfbeak (Pisces: Hemiramphidae) pharyngeal mill. J. Morphol. 270, 357–366.

Camp, A.L., Konow, N., Sanford, C.P.J., 2009. Functional morphology and biomechanics of the tongue-bite apparatus in salmonid and osteoglossomorph fishes. J. Anat. 214, 717–728.

Carr, A., Tibbetts, I.R., Kemp, A., Truss, R., Drennan, J., 2006. Inferring parrotfish (Teleostei: Scaridae) pharyngeal mill function from dental morphology, wear, and microstructure. J. Morphol. 267, 1147–1156.

Casciotta, J.R., Arratia, G., 1993. Jaws and teeth of american cichlids (Pisces: Labroidei). J. Morphol. 217, 1–36.

Castro, R.M.C., Vari, R.P., 2004. Detritivores of the South American fish family Prochilodontidae (Teleostei: Ostariophysi: characiformes): a phylogenetic and revisionary study. Smithson. Contrib. Zool. 622.

Claeson, K., Bemis, W.E., Hagadom, J.W., 2007. New interpretations of the skull of a primitive bony fish *Erpetoichthys calabaricus*. J. Morphol. 268, 1021–1039.

Clifton, K.B., Motta, P.J., 1998. Feeding morphology, diet, and ecomorphological relationships among five Caribbean labrids (Teleostei, Labridae). Copeia 953–966.

Danley, P.D., Kocher, T.D., 2001. Speciation in rapidly diverging systems: lessons from Lake Malawi. Mol. Ecol. 10, 1075–1086.

Day, S.W., Higham, T.E., Cheer, A.Y., Wainwright, P.C., 2005. Spatial and temporal patterns of water flow generated by suction-feeding bluegill sunfish *Lepomis macrochirus* resolved by particle image velocimetry. J. Exp. Biol. 208, 2661–2671.

Eastman, C.R., 1917. Dentition of *Hydrocyon* and its supposed fossil allies. Bull. Am. Mus. Nat. Hist. 37, 757–760.

Ferguson, A.R., Huber, D.R., Lajeunesse, M.J., Motta, P.J., 2015. Feeding performance of king mackerel, *Scomberomorus cavalla*. J. Exp. Zool. A Ecol. Genet. Physiol. 323, 399–413.

Ferry-Graham, L.A., Lauder, G.V., 2001. Aquatic prey capture in ray-finned fishes: a century of progress and new directions. J. Morphol. 248, 99–119.

Ferry-Graham, L.A., Wainwright, P.C., Westneat, M.W., Bellwood, D.R., 2002. Mechanisms of benthic prey capture in wrasses (Labridae). Mar. Biol. 141, 819–830.

Fink, W.L., 1981. Ontogeny and phylogeny of tooth attachment modes in actinopterygian fishes. J. Morphol. 167, 167–184.

Fishelson, L., Delarea, Y., 2004. Taste buds on the lips and mouth of some blennid and gobiid fishes: comparative distribution and morphology. J. Fish. Biol. 65, 651–665.

Fishelson, L., Delarea, Y., 2014. Comparison of the oral cavity architecture in surgeonfishes (Acanthuridae, Teleostei), with emphasis on the taste buds and jaw "retention plates". Env. Biol. Fish. 97, 173–185.

Fishelson, L., Golani, D., Diamant, A., 2014. SEM study of the oral cavity of members of the Kyphosidae and Girellidae (Pisces, Teleostei), with remarks on *Crenidens* (Sparidae), focusing on teeth and taste bud numbers and distribution. Zoology 117, 122–130.

Fishelson, L., Golani, D., Galil, B., Goren, M., 2010. Comparison of taste bud form, number and distribution in the oropharyngeal cavity of lizardfishes (Aulopiformes, Syndontidae). Cybium 34, 269–277.

Fraser, G.J., Bloomquist, R.F., Streelman, J.T., 2008. A periodic pattern generator for dental diversity. BMC Biol. 6, 32.

Fraser, G.J., Hulsey, C.D., Bloomquist, R.F., Uyesugi, K., Manley, N.R., Streelman, J.T., 2009. An ancient gene network is co-opted for teeth on old and new jaws. PLoS Biol. 7, e1000031.

Fraser, G.J., Britz, R., Hall, A., Johanson, Z., Smith, M.M., 2012. Replacing the first-generation dentition in pufferfish with a unique beak. PNAS 109, 8179–8184.

Fraser, G.J., Bloomquist, R.F., Streelman, J.T., 2013. Common developmental pathways link tooth shape to regeneration. Dev. Biol. 377, 399–414.

Fryer, G., Iles, T.D., 1972. The Cichlid Fishes of the Great Lakes of Africa: Their Biology and Evolution. Oliver and Boyd, Edinburgh.

Galis, F., Drucker, E.G., 1996. Pharyngeal biting mechanics in centrarchid and cichlid fishes: insights into a key evolutionary innovation. J. Evol. Biol. 9, 641–670.

Galis, F., Metz, J.A.J., 1998. Why are there so many cichlid species? Trends Ecol. Evol. 13, 1–2.

Geerinckx, T., Huysseune, A., Boone, M., Claeys, M., Couvreur, M., De Kegel, B., Mast, P., Van Hoorebeke, L., Verbeken, K., Adriaens, D., 2012. Soft dentin results in unique flexible teeth in scraping catfishes. Physiol. Biochem. Zool. 85, 481–490.

Gobalet, K.W., 1989. Morphology of the parrotfish pharyngeal jaw apparatus. Am. Zool. 29, 319–331.

Gosline, W.A., 1973. Considerations regarding the phylogeny of cypriniform fishes, with special reference to structures associated with feeding. Copeia 761–776.

Gregory, W.K., Conrad, G.M., 1936. The structure and development of the complex symphysial hinge joint in the mandible of *Hydrocyon lineatus* Bleeker, a characin fish. Proc. Zool. Soc. Lond. 106, 975–984.

Greven, H., Walker, Y., Zanger, K., 2009. On the structure of teeth in the viperfish *Chauliodus sloani* Bloch & Schneider, 1801 (Stomiidae). Bull. Fish. Biol. 11, 87–98.

Grubich, J.R., 2000. Crushing motor patterns in drum (Teleostei: Sciaenidae): functional novelties associated with molluscivory. J. Exp. Biol. 203, 3161–3176.

Guisande, C., Pelayo-Villamil, P., Vera, M., Hernández, A.M., Carvalho, M.R., Vari, R.P., Jiménez, L.F., Fernández, C., Martínez, P., Prieto-Piraquive, E., Granado-Lorencio, C., Duque, S.R., 2012. Ecological factors and diversification among Neotropical characiforms. Int. J. Ecol.:610419 20 pp.

Gunter, H.M., Meyer, A., 2014. Molecular investigation of mechanical strain-induced phenotypic plasticity in the ecologically important pharyngeal jaws of cichlid fish. J. Appl. Ichthyol. 30, 630–635.

Habegger, M.L., Motta, P.J., Huber, D.R., Deban, S.M., 2011. Feeding biomechanics in the great barracuda during ontogeny. J. Zool. Lond. 283, 63–72.

Hahn, N.S., Pavanelli, C.S., Okada, E.K., 2000. Dental development and ontogenetic diet shifts of *Roeboides paranensis* Pignalberi (Osteichthyes, Characinae) in pools of the upper rio Paraná floodplain (state of Paraná, Brazil). Rev. Bras. Biol. 60, 93–99.

Hata, H., Yasugi, M., Hori, M., 2011. Jaw Laterality and related handedness in the hunting behavior of a scale-eating characin, *Exodon paradoxus*. PLoS One 6 (12), e29349.

Hatooka, K., 1984. *Uropterybius nagoensis*, a new muraenid eel from Okinawa. Jpn. J. Ichthyol. 31, 20–22.

Holmbakken, N., Fosse, G., 1973. Tooth replacement in *Gadus callarias*. Z. Anat. Entwicklungsges 143, 65–79.

Holzman, R., Day, S.W., Mehta, R.S., Wainwright, P.C., 2008. Jaw protrusion enhances forces exerted on prey by suction feeding fishes. J. R. Soc. Interface 5, 1445–1457.

Hulsey, C.D., García de León, F.J., 2005. Cichlid jaw mechanics: linking morphology to feeding specialization. Funct. Ecol. 19, 487–494.

Hulsey, C.D., García de León, F.J., Rodiles-Hernández, R., 2006. Micro- and macroevolutionary decoupling of cichlid jaws: a test of Liem's key innovation hypothesis. Evolution 60, 2096–2109.

Huysseune, A., 1995. Phenotypic plasticity in the lower pharyngeal jaw dentition of *Astatoreochromis alluaudi* (Teleostei: Cichlidae). Arch. Oral. Biol. 40, 1005–1014.

Huysseune, A., van der heyden, C., Sire, J.-Y., 1998. Early development of the zebrafish (*Danio* rerio) pharyngeal dentition (Teleostei, Cyprinidae). Anat. Embryol. 198, 289–305.

Huysseune, A., 2000. Developmental plasticity in the dentition of a heterodont polyphyodont fish species. In: in Teaford, M.F., Smith, M.M., Ferguson, M.J. (Eds.), Development, Function and Evolution of Teeth. Cambridge University Press, Cambridge, pp. 231–241.

Jones, S., 1958. Notes on eggs, larvae and juveniles of fishes from Indian waters. Indian J. Fish. 5, 357–361.

Kakizawa, Y., Meenakarn, W., 2003. Histogenesis and disappearance of the teeth of the Mekong giant catfish, *Pangasianodon gigas* (Teleostei). J. Oral. Sci. 45, 13–221.

Kakizawa, Y., Kajiyama, N., Nagai, K., Kado, K., Fujita, M., Kashiwaya, Y., Imai, C., Hirama, A., Yorioka, M., 1986. The histological structure of the upper and lower jaw teeth in the gobiid fish, *Sicyopterus japonicus*. J. Nihon. Univ. Sch. Dent. 28, 175–187.

Kazancioğlu, E., Near, T.J., Hanel, R., Wainwright, P.C., 2009. Influence of sexual selection and feeding functional morphology on diversification rate of parrotfishes (Scaridae). Proc. R. Soc. Lond. B27, 3439–3446.

Kemp, A., 2002. Unique dentition of lungfish. Microsc. Res. Techn. 59, 435–448.

Konow, N., Bellwood, D.R., Wainwright, P.C., Kerr, A.M., 2008. Evolution of novel jaw joints promote trophic diversity in coral reef fishes. Biol. J. Linn. Soc. 93, 545–555.

Kotrschal, K., Goldschmidt, A., 1992. Morphological evidence for the biological role of caniniform teeth in combtooth blennies (Blennidae, Teleostei). J. Fish. Biol. 41, 983–991.

Lauder, G.V., 1980. Evolution of the feeding mechanism in primitive actinopterygian fishes: a functional analysis of *Polypterus, Lepisosteus,* and *Amia*. J. Morphol. 163, 283–317.

Lee, H.J., Kusche, H., Meyer, A., 2012. Handed foraging behavior in scale-eating cichlid fish: its potential role in shaping morphological asymmetry. PLoS One 7 (9), e44670. http://dx.doi.org/10.1371/journal.pone.0044670.

Liem, K.F., 1973. Evolutionary strategies and morphological innovations: cichlid pharyngeal jaws. Syst. Zool. 22, 425–441.

Liem, K.F., 1980. Adaptive significance of intra- and interspecific differences in the feeding repertoires of cichlid fishes. Am. Zool. 20, 295–314.

Liem, K.F., Greenwood, P.H., 1981. A functional approach to the phylogeny of the pharyngognath teleosts. Am. Zool. 21, 83–101.

Lima, F.C.T., Wosiacki, W.B., Ramos, C.S., 2009. *Hemigrammus arua*, a new species of characid (Characiformes: Characidae) from the lower Amazon, Brazil. Neotrop. Ichthyol. 7, 153–160.

Lovejoy, N.P., 2000. Reinterpreting recapitulation: systematics of needlefishes and their allies (Teleostei, Beloniformes). Evolution 54, 1349–1362.

Lujan, N.K., German, D.P., Winemiller, K.O., 2011. Do wood grazing fishes partition their niche? Morphological and isotopic evidence for trophic segregation in Neotropical Loricariidae. Funct. Ecol. 25, 1327–1338.

Mattox, G.M.T., Britz, R., Toledo-Piza, M., Marinho, M.M.F., 2013. *Cynogastor noctivago*, a remarkable new genus and species of miniature fish from the Rio Grande, Amazon basin (Ostariophysi: Characidae). Ichthol. Explor. Freshwaters 23, 297–318.

Marinho, M.M.F., Lima, F.C.T., 2009. *Astyanax ajuricaba*: a new species from the Amazon basin in Brazil (Characiformes: Characidae). Neotrop. Ichthyol. 7, 169–174.

Mehta, R.S., Wainwright, P.C., 2007. Raptorial jaws in the throat help moray eels swallow large prey. Nat. Lond. 449, 79–82.

Meyer, A., 1990. Ecological and evolutionary consequences of the trophic polymorphism in *Cichlasoma citrinellum* (Pisces: Cichlidae). Biol. J. Linn. Soc. 39, 279–299.

Miller, W.A., Radnor, C.J.P., 1973. Tooth replacement in bowfin (*Amia calva* – Holostei). J. Morphol. 140, 381–395.

Mirande, J.M., 2010. Phylogeny of the family Characidae (Teleostei: characiformes): from characters to taxonomy. Neotrop. Ichthyol. 8, 385–568.

Mittelbach, G.G., Osenberg, C.W., Wainwright, P.C., 1999. Variation in feeding morphology between pumpkinseed populations: phenotypic plasticity or evolution? Evol. Ecol. Res. 1, 111–128.

Mochizuki, K., Fukui, S., 1983. Development and replacement of upper jaw teeth in gobiid fish (*Sicyopterus japonica*). Jpn. J. Ichthyol. 30, 27–36.

Moriyama, K., Watanabe, S., Iida, M., Fukui, S., Sahara, N., 2009. Morphological characteristics of upper jaw dentition in a gobiid fish (*Sicyopterus japonicus*); a micro-computed tomography study. J. Oral. Biosci. 51, 81–90.

Moriyama, K., Watanabe, S., Iida, M., Sahara, N., 2010. Plate-like permanent dental laminae of upper jaw dentition in adult gobiid fish, *Sicyopterus japonicus*. Cell Tiss. Res. 340, 189–200.

Motta, P.J., 1984a. Mechanics and function of jaw protrusion in teleost fishes: a review. Copeia 1–18.

Motta, P.J., 1984b. Tooth attachment, replacement and growth in the butterflyfish, *Chaetodon miliaris* (Chaetodontidae, Perciformes). Can. J. Zool. 62, 183–189.

Motta, P.J., 1988. Functional morphology of the feeding apparatus of ten species of Pacific butterflyfishes (Perciformes: Chaetodontidae): an ecomorphological approach. Env. Biol. Fish. 22, 39–67.

Nakamura, H., Kamimura, T., Yabuta, Y., Suda, A., Ueyanagi, S., Kikawa, S., Honma, M., Yukinawa, M., Murikawa, S., 1951. Notes of the life-history of the sword-fish, *Xiphias gladius* Linnaeus. Jpn. J. Ichthyol. 1, 264–271.

Nelson, G.J., 1969. Gill arches and the phylogeny of fishes, with notes on the classification of vertebrates. Bull. Am. Mus. Nat. Hist. 141, 475–552.

Nelson, J.A., Wubah, D.A., Whitmer, M.E., Johnson, E.A., Stewart, D.J., 1999. Wood-eating catfishes of the genus *Panaque*: gut microflora and cellulolytic enzyme activities. J. Fish. Biol. 54, 1069–1082.

Nico, L.G., de Morales, M., 1994. Nutrient content of piranha (Characidae, Serrasalminae) prey items. Copeia 524–528.

Nico, L.G., Taphorn, D.C., 1988. Food habits of piranhas in the low Llanos of Venezuela. Biotropica 20, 311–321.

Norris, K.S., Prescott, J.H., 1959. Jaw structure and tooth replacement in the opaleye, *Girella nigricans* (Ayres), with notes on other species. Copeia 275–283.

Novakowski, G.C., Fugi, R., Hahn, N.S., 2004. Diet and dental development of three species of *Roeboides* (Characiformes: Characidae). Neotrop. Ichthyol. 2, 157–162.

Palco, B.J., Beardsley, G.L., Richards, W.J., 1981. Synopsis of the Biology of the Swordfish, *Xiphias gladius* Linnaeus. NOAA Tech Rep NMFS Circular 441. FAO Fisheries Synopsis No. 127.

Pasco-Viel, E., Charles, C., Chevret, P., Semon, M., Tafforeau, P., Viriot, L., Laudet, V., 2010. Evolutionary trends of the pharyngeal dentition in cypriniformes (Actinopterygii: Ostariophysi). PLoS One 5, e11293.

Pasco-Viel, E., Yang, L., Veran, M., Balter, V., Mayden, R.L., Laudet, V., Viriot, L., 2014. Stability versus diversity of the dentition during evolutionary radiation in cyprinine fish. Proc. R. Soc. B 281, 20132688.

Peterson, C.C., Winemiller, K.O., 1997. Ontogenic diet shifts and scale-eating in *Roeboides dayi*, a Neotropical characid. Env. Biol. Fish. 49, 111–118.

Porter, H.T., Motta, P.J., 2004. A comparison of strike and prey capture kinematics of three species of piscivorous fishes: Florida gar (*Lepisosteus platyrhincus*), redfin needlefish (*Strongylura notata*), and great barracuda (*Sphyraena barracuda*). Mar. Biol. 145, 989–1000.

Price, S.A., Wainwright, P.C., Bellwood, D.R., Kazancioglu, E., Collar, D.C., Near, T.J., 2010. Functional innovations and morphological diversification in parrotfish. Evolution 64, 3057–3068.

Roberts, T.R., 1967. Tooth formation and replacement in characoid fishes. Stanf. Ichthyol. Bull. 8, 231–247.

Salzburger, W., Meyer, A., 2004. The species flocks of East African cichlid fishes: recent advances in molecular phylogenetics and population genetics. Naturwiss 91, 277–290.

Sanderson, S.L., Cheer, A.Y., Goodrich, J.S., Graziano, J.D., Callan, W.T., 2001. Crossflow filtration in suspension-feeding fishes. Nature (London). 412, 439–441.

Sanford, C.P.J., Lauder, G.V., 1990. Kinematics of the tongue-bite apparatus in osteoglossomorph fishes. J. Exp. Biol. 154, 137–162.

Sazima, I., 1983. Scale-eating in characoids and other fishes. Env. Biol. Fish. 9, 87–101.

Schaeffer, B., Rosen, D.E., 1961. Major adaptive levels in the evolution of the actinopterygian feeding mechanism. Am. Zool. 1, 187–204.

Schwenk, K., Rubega, M., 2005. Diversity of vertebrate feeding systems. In: Starck, J.M., Wang, T. (Eds.), Physiological and Ecological Adaptations to Feeding in Vertebrates. Science Publishers, Enfield, New Hampshire, pp. 1–41.

Seehausen, O., 2006. African cichlid fish: a model system in adaptive radiation research. Proc. R. Soc. Lond. B273, 1987–1998.

Shellis, R.P., 1982. Comparative anatomy of tooth support. In: Berkovitz, B.K.B., Moxham, B., Newman, H.N. (Eds.), The Periodontal Ligament in Health and Disease. Pergamon, Oxford, pp. 3–24.

Shellis, R.P., Berkovitz, B.K.B., 1976. Observations on the dental anatomy of piranhas (Characidae) with special reference to tooth structure. J. Zool. Lond. 180, 69–84.

Shellis, R.P., Poole, D.F.G., 1978. The structure of the dental hard tissues of the coelacanthid fish *Latimeria chalumnae* Smith. Arch. Oral. Biol. 23, 1105–1113.

Sibbing, F.A., 1982. Pharyngeal mastication and food transport in the carp (*Cyprinus carpio* L.): a cineradiographic and electromyographic study. J. Morphol. 172, 223–258.

Smith, M.M., 1985. The pattern of histogenesis and growth of tooth plates in larval stages of extant lungfish. J. Anat. 140, 627–643.

Smith, M.M., Johanson, Z., 2010. The dipnoan dentition: a unique adaptation with a longstanding evolutionary record. In: Jørgenson, J.M., Joss, J. (Eds.), The Biology of Lungfishes. CRC Press, Boca Raton, pp. 219–244.

Stock, D.W., 2007. Zebrafish dentition in comparative context. J. Exp. Zool. Mol. Dev. Evol. 308B, 523–549.

Streelman, J.T., Alfaro, M., Westneat, M.W., Bellwood, D.R., Karl, S.A., 2002. Evolutionary history of the parrotfishes: biogeography, ecomorphology, and comparative diversity. Evolution 56, 961–971.

Streelman, J.T., Danley, P.D., 2003. The stages of vertebrate evolutionary radiation. Trends Ecol. Evol. 18, 126–131.

Soule, J.D., 1969. Tooth attachment by means of a periodontium in triggerfish (Balistidae). J. Morphol. 127, 1–5.

Takahashi, R., Moriwaki, T., Hori, M., 2007. Foraging behaviour and functional morphology of two scale-eating cichlids from Lake Tanganyika. J. Fish. Biol. 70, 1458–1469.

Tchernavin, V.V., 1953. The Feeding Mechanisms of a Deep Sea Fish *Chauliodus Sloani* Schneider. British Museum (Natural History), London, pp. 1–101.

Tibbetts, I.R., Carseldine, L., 2003. Anatomy of a hemiramphid pharyngeal mill with reference to *Arrhamphus sclerolepis krefftii* (Steindachner) (Teleostei: Hemiramphidae). J. Morphol. 255, 228–243.

Tibbetts, I.R., Carseldine, L., 2004. Anatomy of the pharyngeal jaw apparatus of *Zenarchopterus* (gill) (Teleostei: Beloniformes). J. Morphol. 262, 750–759.

Toledo-Piza, M., Mattox, G.M.T., Britz, R., 2014. *Priocharax nanus*, a new miniature characid from the rio Negro, Amazon basin (Ostariophysi: characiformes), with an updated list of miniature neotropical freshwater fishes. Neotrop. Ichthyol. 12, 229–246.

Tolentino-Pablico, G., Bailly, N., Froese, R., Elloran, C., 2007. Seaweeds preferred by herbivorous fishes. J. Appl. Phycol. http://dx.doi.org/10.1007/s10811-007-9290-4.

Trapani, J., 2001. Position of developing replacement teeth in teleosts. Copeia, 35–51.

Trapani, J., 2004. A morphometric analysis of polymorphism in the pharyngeal dentition of *Cichlasoma minckleyi* (Teleostei: Cichlidae). Arch. Oral. Biol. 49, 825–835.

Trapani, J., Yamamoto, Y., Stock, D.W., 2005. Ontogenetic transition from unicuspid to multicuspid oral dentition in a teleost fish: *Astyanax mexicanus*, the Mexican tetra (Ostariophysi: Characidae). Zool. J. Linn. Soc. 145, 523–538.

Turner, G.F., 2007. Adaptive radiation of cichlid fish. Curr. Biol. 17, R827–R831.

Vandewalle, P., Huysseune, A., Aerts, P., Verraes, W., 1994. The pharyngeal apparatus in teleost feeding. In: Bels, V.L., Chardon, M., Vandewalle, P. (Eds.), Biomechanics of Feeding in Vertebrates, Adv Comp Env Physiol vol. 18. Springer, Berlin, pp. 59–92.

Vandewalle, P., Parmentier, E., Chardon, M., 2000. The branchial basket in teleost feeding. Cybium 24, 319–342.

Vari, R.P., 1979. Anatomy, relationships and classification of the families Citharinidae and Distichodontidae (Pisces, Characoidea). Bull. Brit. Mus. Nat. Hist. Zool. 36, 261–344.

Wainwright, P.C., Osenberg, C.W., Mittelbach, G.G., 1991. Trophic polymorphism in the pumpkinseed sunfish (*Lepomis gibbosus* Linnaeus): effects of environment on ontogeny. Funct. Ecol. 5, 40–55.

Wainwright, P.C., Ferry-Graham, L.A., Waltzek, T.B., Carroll, A.M., Hulsey, C.D., Grubich, J.R., 2001. Evaluating the use of ram and suction during prey capture by cichlid fishes. J. Exp. Biol. 204, 3039–3051.

Wainwright, P.C., Bellwood, D.R., Westneat, M.W., Grubich, J.R., Hoey, A.S., 2004. A functional morphospace for the skull of labrid fishes: patterns of diversity in a complex biomechanical system. Biol. J. Linn. Soc. 82, 1–25.

Wainwright, P.C., Carroll, A.M., Collar, D.C., Day, S.W., Higham, T.E., Holzman, R.A., 2007. Suction feeding mechanics, performance, and diversity in fishes. Integr. Comp. Biol. 47, 96–106.

Wainwright, P.C., Smith, W.L., Price, S.A., Tang, K.L., Sparks, J.S., Ferry, L.A., Kuhn, K.L., Eytan, R.I., Near, T.J., 2012. The evolution of pharyngognathy: a phylogenetic and functional appraisal of the pharyngeal jaw key innovation in labroid fishes and beyond. Syst. Biol. 61, 1001–1027.

Wainwright, P.C., McGee, M.D., Longo, S.J., Hernandez, L.P., 2015. Origins, innovations, and diversification of suction feeding in vertebrates. Integr. Comp. Biol. 55, 1–12.

Waltzek, T.B., Wainwright, P.C., 2003. Functional morphology of extreme jaw protrusion in Neotropical cichlids. J. Morphol. 257, 96–106.

Westneat, M.W., 1991. Linkage biomechanics and evolution of the unique feeding mechanism of *Epibulus insidiator* (Labridae: Teleostei). J. Exp. Biol. 159, 165–184.

Westneat, M.W., 2004. Evolution of levers and linkages in the feeding mechanisms of fishes. Integr. Comp. Biol. 378–389.

Westneat, M.W., Wainwright, P.C., 1989. Feeding mechanism of *Epibulus insidiator*: evolution of a novel functional system. J. Morphol. 202, 129–150.

Yasuda, F., 1978. Embryonic and early larval stages of the swordfish. J. Tokyo Univ. Fish. 65, 91–97.

FURTHER READING

Ebeling, A.W., 1957. The dentition of Eastern Pacific mullets, with special reference to adaptation and taxonomy. Copeia 173–185.

Chapter 5

Amphibia

The subclass Amphibia contains about 7250 species of living amphibians, with about 150 new species added each year. In the classification used here (Frost et al., 2006), the subclass is divided into the following principal groups:

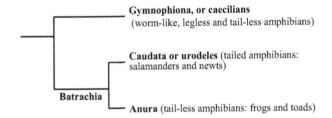

Gymnophiona, or caecilians
(worm-like, legless and tail-less amphibians)

Caudata or urodeles (tailed amphibians: salamanders and newts)

Batrachia

Anura (tail-less amphibians: frogs and toads)

Most amphibia have two stages during their life history: a larval stage that hatches from an egg and is usually aquatic and an adult stage that inhabits the land. Larvae undergo metamorphosis to reach the adult form, changing from a larva with gills to an adult, air-breathing form with lungs. Amphibians use their moist skin as a secondary respiratory surface. Exceptions to this general description include:

1. Caecilian species that give birth to live young (by vivipary or ovovivipary) and do not have an aquatic stage.
2. Plethodontid salamanders that are lungless and lay eggs on land, from which miniature adults hatch. They respire through their skin and through the tissues lining the mouth.
3. Urodeles that exhibit some degree of neoteny (pedomorphosis): the phenomenon of attaining reproductive maturity while still retaining the larval external morphology. Neotenous urodeles remain aquatic throughout life.

Virtually all amphibians are carnivorous and feed on small invertebrates, such as worms, insects, and crustaceans. The prey of most species is small relative to the size of the predator. Salamanders and frogs use vision to locate prey, whereas the burrowing caecilians primarily use olfaction. Caecilians possess numerous, relatively large teeth on the premaxillae, maxillae, vomers, palatines, and dentaries. They capture prey using a powerful bite. Among Batrachia, both tooth size and number are reduced. Salamanders

retain teeth on both upper and lower jaws and on the palate, whereas among anurans teeth occur on the dentary in only one species and the number of palatal teeth is reduced further; toads are completely edentulous. The small teeth of adult Caudata and Anura are mainly suited to grasping and preventing prey from escaping, although they probably also penetrate the integument of the prey and hence increase access for digestive enzymes. These terrestrial amphibians can capture prey by snapping with the jaws at passing organisms. Alternatively, they can use their protrusible, sticky-tipped tongue to capture prey. The tongue is rapidly protruded and the prey is pressed against the substrate and immobilized by the tip, then conveyed directly to the esophagus as the tongue retracts into the mouth. Among frogs and toads, the tongue is projected by muscular action, whereas in salamanders projection is achieved using the hyoid apparatus. Amphibians with protrusible tongues can switch between tongue prehension and jaw prehension according to the size of the prey, with the jaws being used for large prey (Monroy and Nishikawa, 2011). Aquatic species, including many fishes (Chapters 2–4), generally feed by suction.

The teeth of adult amphibians are pointed and recurved, with a cross section like that of a biconvex lens, and are covered with true enamel (see Chapter 11). They usually have two cusps, labial and lingual, of which the labial cusp tends to be the smaller. The cusps extend mesiodistally and thus form cutting edges (Greven, 1989), which presumably enhance the capacity of the tooth to penetrate the prey integument. Some metamorphosed amphibians, such as pipid anurans, some plethodontid salamanders, and several gymnophionian genera, have monocuspid teeth. Some anuran species—*Alytes obstetricans* (Alytidae), *Polypedates maculatus* (Rhacophoridae), *Agalychnis callydrias* and *Phyllomedusa bicolour* (Hylidae), and *Heterixalus madagascariensis* (Hyperoliidae)—possess tricuspid teeth, as does the urodele *Ambystoma mabeei* (Ambystomatidae) (Greven and Ritz, 2008).

Adult amphibian teeth are nearly always divided into a distal portion ("crown") and a basal pedicel, separated by a hypo- or unmineralized zone (Parsons and Williams, 1962). The pedicel is longer than the crown and is firmly ankylosed to the bone of the jaw. Marginal teeth are attached to

The Teeth of Non-Mammalian Vertebrates. http://dx.doi.org/10.1016/B978-0-12-802850-6.00005-9

the inside of the jaw bone (pleurodont) and palatal teeth to the crest of the bone (acrodont). The crown and the pedicel are developmentally a unit, as in teleosts (Chapter 9). The **dividing zone** separating the crown and pedicel consists of longitudinally orientated, unmineralized collagen fibers connecting the two mineralized parts of the joint (Greven et al., 1989; Wistuba et al., 2002; Davit-Béal et al., 2007) and sometimes contains clusters of mineral (Wistuba et al., 2002). Although the dividing zone may confer some flexibility on the tooth, it does not provide the degree of movement found in pedicellate teeth of teleosts, which have a wider unmineralized dividing zone, and true hinged teeth like those of teleosts do not occur among amphibians. It seems probable that the dividing zone is a zone of weakness that ensures that the tooth tip breaks off under excessive force, thus preventing damage to the jaw bone (Davit-Béal et al., 2007). In some species, such as the **greater siren** (*Siren lacertina*) and the **African clawed frog** (*Xenopus laevis*), the teeth lack a division into crown and pedicel (Davit-Béal et al., 2007; Greven and Laumeier, 1987). The **Sumaco horned treefrog** (*Hemiphractus proboscideus*), unusually, has teeth that are both monocuspid and unipartite (Shaw, 1989).

Larval Urodela and Gymnophiona also possess true teeth, but they are unicuspid, not bicuspid, and are tipped with enameloid, not enamel (see Chapter 11). Moreover, they lack the cutting edges typical of bicuspid teeth (Greven, 1989). In addition, monocuspid larval teeth are not divided into crown and pedicel. The larvae of Anura (tadpoles) lack true calcified teeth and instead possess horny tooth-like structures composed of keratin. Although structurally and developmentally distinct from true teeth, they will be referred to here as teeth, as no suitable alternative term exists.

Amphibian teeth are replaced continuously throughout life. Replacement teeth develop lingual to the functioning teeth.

GYMNOPHIONA

The Gymnophiona (or caecilians) comprise 200 species of limbless amphibians in 35 genera. Frost et al. (2006) divided them into three families, the Rhinatrematidae, Ichthyophiidae, and Caeciliidae, but Wilkinson et al. (2011) divided the Caeciliidae into seven families: Scolecomorphidae, Herpelidae, Caeciliidae, Typhlonectidae, Indotyphlidae, Siphonopedae, and Dermophiidae. Discovery of a new family, Chikiidae (Kamei et al., 2012), brings the total number to 10 families.

Apart from one aquatic species, these worm-like creatures (Fig. 5.1) burrow underground in moist soil and are the least known amphibians. They can reach lengths of 1.5 m. The term caecilian means "blind," but this is not strictly true, as they do have limited vision that allows some discrimination

between dark and light. Many species of caecilians exhibit extended parental care. Caecilians detect their prey of worms, termites, and other small invertebrates primarily through touch and olfaction. A pair of tentacles between the eyes and nostrils may be used as an additional olfactory sensing organ.

Most caecilian species lack an aquatic larval stage. One-quarter of species are oviparous, although the eggs are guarded by the females, whereas three-quarters are viviparous. They are the only order of amphibians that use internal insemination exclusively. Caecilians have a longer embryonic period than the Anura and Caudata and do not pass through a larval stage or a metamorphosis to the adult: the hatchlings are smaller versions of the adult.

Whereas most salamanders and frogs have weakly ossified skulls and use their protrusible tongues to catch prey, caecilians have a heavily ossified cranium, reflecting their fossorial lifestyle, and lack protrusible tongues. The maxilla is fused with the palatine to form the maxillopalatine bone, which is characteristic of all adult caecilians.

Compared with anurans and caudates, whose teeth are small, pointed, homodont, and form an inconspicuous component of the skull, the teeth of the burrowing caecilians are relatively larger, more heterodont, and function to restrain and deal with more active prey. A single row of teeth in the lower jaw fits between two rows in the upper jaw. In addition, there may be a specialized dentition in the developing young to obtain nutrients either from the reproductive tract or from the skin of the mother (Parker and Dunn, 1964; Wilkinson et al., 2013).

Caecilians use jaw prehension and their strong teeth to capture prey. Although they generally move slowly along their burrows, they are capable of a high-speed lunge (about 7 cm/s) to capture prey (Herrel and Measey, 2012). They have a powerful bite and have two sets of muscles for closing the jaw. The dentitions are well developed and consist of three main tooth rows. In the upper jaw there is

FIGURE 5.1 Narayan's caecilian (*Uraeotyphlus narayani*). *Courtesy Wikipedia.*

an outer premaxillary/maxillary row and an inner vomer/palatine row (Fig. 5.2). In the lower jaw there is a main row of teeth on the dentaries, but smaller teeth may be present on the coronoids and constitute an inner second row of "splenial" teeth.

The teeth are pedicellate. They may be monocuspid (eg, *Ichthyophis, Uraeotyphlus, Hypogeophis,* and *Geotrypetes*) or bicuspid (eg, *Dermophis, Gymnopis, Caecilia,* and *Typhlonectes*) and are often recurved (Wake and Wurst, 1979). The anterior teeth in the row tend to be larger than the posterior ones.

Tooth numbers for many species of caecilians have been reported by Taylor (1968, 1976, 1977). The number of teeth in the outer premaxillary/maxillary row and in the dentary row varies from 10 or less to 50, whereas the number of coronoid teeth may vary from 0 to 45. The number of teeth may increase with age by addition of teeth to the back of the

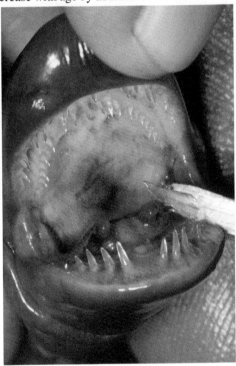

FIGURE 5.2 Upper jaw of a caecilian (*Caecilia* sp.) showing two rows of teeth in the upper jaw and a single row in the lower. © 2013 Dante Fenolio, Department of Conservation and Research, San Antonio Zoo. *Courtesy California Academy of Sciences.*

tooth row. The largest teeth are generally on the premaxilla, which may also possess a midline tooth.

Rhinatrematidae

These primitive caecilians are also known as the beaked caecilians as the mouth is positioned at the end of the snout rather than underneath it, giving a shape like the beak of a bird. The **two-coloured beaked caecilian** (*Epicrionops bicolor*) (Fig. 5.3) has about 22 teeth in the premaxillary/maxillary row and 28 in the vomeropalatine row, plus a midline tooth in both rows. There may be about 17 in the outer dentary row and slightly more in a second row on the coronoid bone.

The teeth of the **two-lined caecilian** (*Rhinatrema bivittatum*) (Fig. 5.4) are all monocuspid. The premaxillary/maxillary tooth row has a maximum number of 33 teeth, with the teeth on the maxilla being slightly larger. The inner vomeropalatine row has a maximum number of 37 smaller teeth. The maximum number of teeth on the dentary is 28 and these teeth have a similar size to those on the premaxillary/maxillary row. The teeth on the inner splenial row in the lower jaw (maximum number 31) are similar in size to the vomeropalatine teeth.

Ichthyophiidae

The **Banna caecilian** (*Ichthyophis bannanicus*) has a double row of teeth in both jaws (Fig. 5.5). The upper and outer premaxillary/maxillary row has approximately 20 teeth plus one midline tooth, and the upper and inner vomeropalatine row has 17 teeth on each side plus a midline tooth. The dentition in the lower jaw consists of two well-developed rows: a main row of about 20 teeth on the dentary and a shorter inner row of fewer coronoid teeth.

A ventral view of the dentition of the related **Ceylon caecilian** or **common yellow-banded caecilian** (*Ichthyophis glutinosus*) illustrates the two rows of teeth on the lower jaw (Fig. 5.6).

Scolecomorphidae

The dentition of **Kirk's caecilian** (*Scolecomorphus kirkii*) is illustrated in Fig. 5.7. There are three rows of

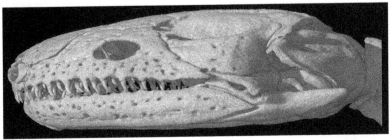

FIGURE 5.3 Dentition of the two-coloured beaked caecilian (*Epicrionops bicolor*). MicroCT image. Image width = 1.3 cm. *Courtesy Digimorph.org and Dr. J.A. Maisano.*

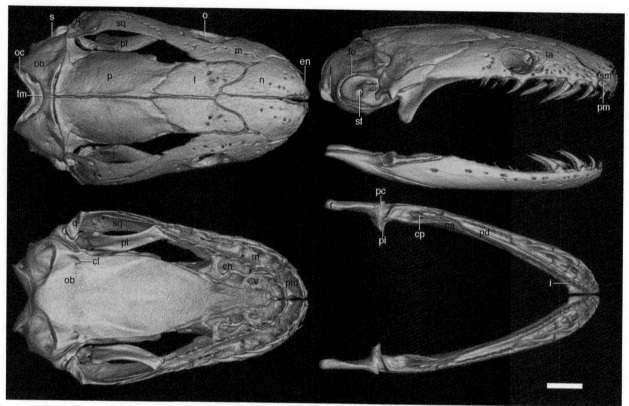

FIGURE 5.4   Dentition of *Rhinatrema bivittatum.* Scale = 1 mm. Abbreviations for relevant structures: *i*, inner mandibular tooth row; *m*, maxillopalatine; *pa*, pseudoarticular; *pd*, pseudodentary; *pm*, premaxilla; *q*, quadrate; *v*, vomer. *From Wilkinson, M., San Mauro, D., Sherratt, E., Gower, D.J., 2011. A nine-family classification of caecilians (Amphibia: gymnophiona). Zootaxa 2874, 41–64. Courtesy Editors Zootaxa.*

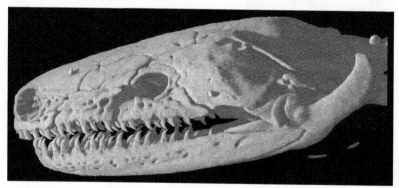

FIGURE 5.5   Dentition of the Banna caecilian (*Ichthyophis bannanicus*). Micro-CT image. Image width = 2 cm. *Courtesy Digimorph.org and Dr. J.A. Maisano.*

conical, monocuspid teeth that curve posteriorly. The teeth of each series are larger anteriorly, but only slightly so in the dentary series (Nussbaum, 1985). Tooth numbers of three species for both males and females are given in Table 5.1.

Herpelidae

The dentition of the **Cameroon caecilian** (*Herpele squalostoma*) is illustrated in Fig. 5.8.

Caeciliidae

This family contains two genera and more than 40 species.

The dentition of the **bearded caecilian** (*Caecilia tentaculata*) (Fig. 5.9) consists of uniformly monocuspid teeth and is characterized by the presence of a short, additional inner row of teeth at the front of the dentary. The premaxillary/maxillary row has a maximum number of 22 teeth, the posterior teeth being smaller. The inner vomeropalatine row has a maximum number of 20 teeth, which are smaller than the teeth in the outer row. The teeth on the dentary have a maximum number of 20 and are the same size as those in the premaxillary/maxillary row. The smaller inner teeth of the splenial row have a maximum number of four teeth (Maciel and Hoogmoed, 2009).

Oscaecilia koepekeorum has five premaxillary/maxillary teeth, seven vomeropalatine teeth, and five dentary

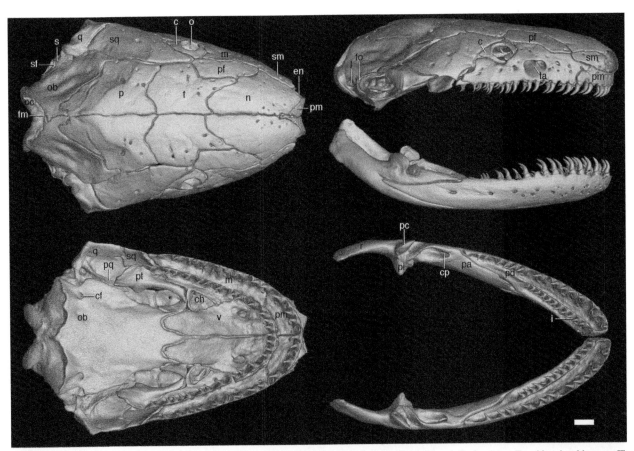

FIGURE 5.6 Dentition of the Ceylon caecilian or common yellow-banded caecilian (*Ichhyophis glutinosus*). Scale = 1 mm. For abbreviated key, see Fig. 5.4. *From Wilkinson, M., San Mauro, D., Sherratt, E., Gower, D.J., 2011. A nine-family classification of caecilians (Amphibia: gymnophiona). Zootaxa 2874, 41–64. Courtesy Editors Zootaxa.*

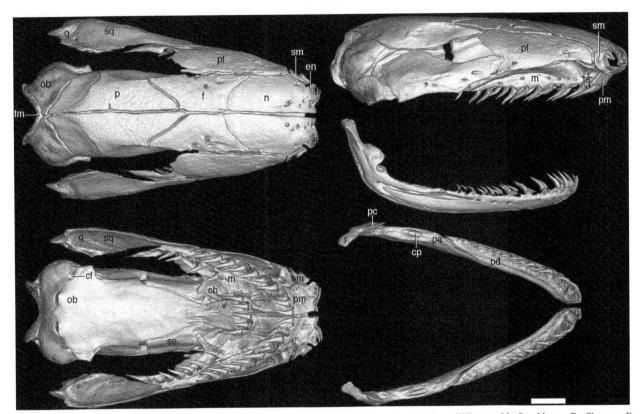

FIGURE 5.7 Dentition of *Scolecomorphus kirkii*. Scale = 1 mm. For abbreviated key, see Fig. 5.4. *From Wilkinson, M., San Mauro, D., Sherratt, E., Gower, D.J., 2011. A nine-family classification of caecilians (Amphibia: gymnophiona). Zootaxa 2874, 41–64. Courtesy Editors Zootaxa.*

teeth with one splenial tooth. In *Oscaecilia hypereumeces*, the tooth number is greater, with 8 premaxillary/maxillary teeth, 10, vomeropalatine teeth, 10 dentary teeth, and 3 splenial teeth (Wake, 1984). Tooth numbers for other Caeciliidae are listed in Table 5.2.

TABLE 5.1 Mean Tooth Numbers in Male and Female Specimens in Three Species of *Scolecomorphus* Adjusted to Body Size

Species	Sex	Premaxilla /Maxilla	Vomer and Palatine	Dentary
S. kirkii	Male	18.7	16.4	24.6
S. kirkii	Female	17.5	15.2	22.4
S. vittatus	Male	18.8	15.8	24.1
S. vittatus	Female	19.6	15.2	25.9
S. uluguruensis	Male	17.7	14.7	20.1
S. uluguruensis	Female	16.8	14.9	20.1

Data from Nussbaum, R.A., 1985. Systematics of Caecilians (Amphibia: Gymnophonia) of the Family Scolecomorphida. Occasional Papers of the Museum of Zoology. University of Michigan Number 713.

TABLE 5.2 Maximum Tooth Numbers on the Dentigerous Bones of Some Caeciliidae

Species	Premaxilla /Maxilla	Vomer and Palatine	Dentary	Splenial
Brasilotyphlus braziliensis	25	26[a,b]	14	
Brasilotyphlus guarantanus	25	27[a,b]	21	
Caecilia gracilis	20	19	19	6
Caecilia mertensi	21	23	24	4
Microcaecilia taylori	30	30[a]	23	
Microcaecilia unicolor	17	28	26	

[a]Bicuspid teeth. All other rows are monocuspid.
[b]Diastema between the vomerine and palatine tooth rows.
Data from Maciel, A.O., Hoogmoed, M.S., 2009. Taxonomy and distribution of gymnophiona of Brazilian Amazonia with a key to their identification. Zootaxa 2984, 1–53.

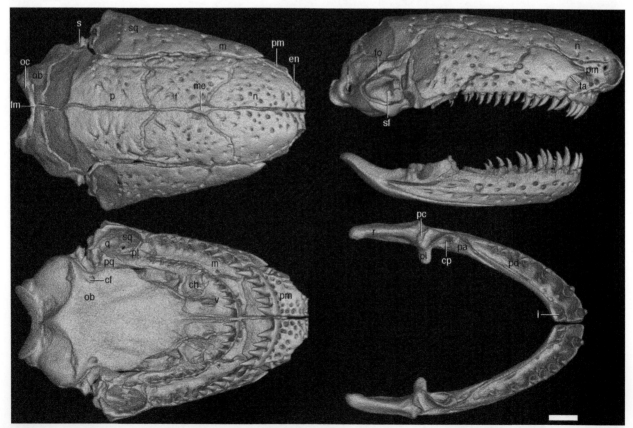

FIGURE 5.8 Dentition of *Herpele squalostoma*. Scale = 1 mm. For abbreviated key, see Fig. 5.4. *From Wilkinson, M., San Mauro, D., Sherratt, E., Gower, D.J., 2011. A nine-family classification of caecilians (Amphibia: gymnophiona). Zootaxa 2874, 41–64. Courtesy Editors Zootaxa.*

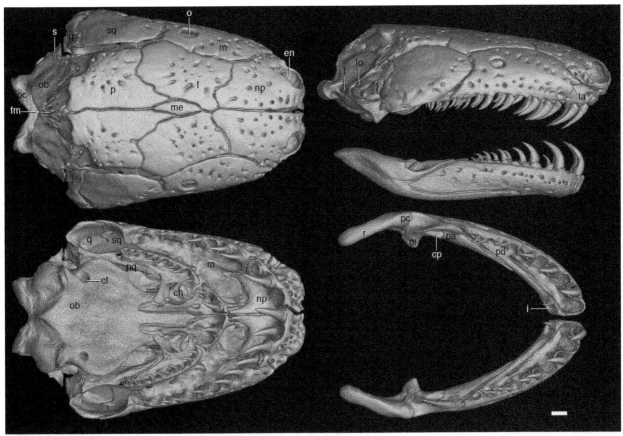

FIGURE 5.9 Dentition of *Caecilia tentaculata*. Scale = 1 mm. For abbreviated key, see Fig. 5.4. *From Wilkinson, M., San Mauro, D., Sherratt, E., Gower, D.J., 2011. A nine-family classification of caecilians (Amphibia: gymnophiona). Zootaxa 2874, 41–64. Courtesy Editors Zootaxa.*

Typhlonectidae

Members of this small group are viviparous and, unlike most other caecilians, are aquatic or semiaquatic and are sometimes referred to as **rubber eels**. Adult typhlonectids have four distinct rows of recurved, monocuspid teeth that may have blade-like lateral flanges. In most species the teeth are pointed. Teeth vary in size, curvature, and the development of lateral flanges within and between individuals. The tooth crowns are flexibly attached to their pedicels and can be displaced posteriorly, but they resist displacement in other directions (Bemis et al., 1983).

Typhlonectes natans has a formidable dentition (Fig. 5.10). The outer premaxillary/maxillary row bears approximately 23 teeth and the inner vomeropalatine row up to 28 teeth; there is a midline tooth in both rows. The dentary carries an outer row of about 21 teeth, whereas there is a short inner coronoid row of eight small teeth (Taylor, 1968; Wilkinson and Nussbaum, 1997).

The **Cayenne caecilian** (*Typhlonectes compressicauda*) (Fig. 5.11) has teeth with broadly dilated crowns (Wilkinson, 1991). It has a premaxillary tooth row with a maximum number of 52, with little size variation. The vomeropalatine row has a maximum number of 48

smaller teeth. There are up to 60 teeth on the dentary, which are larger than those in the premaxilla/maxillary row. There are up to 14 splenial teeth forming an inner row of small teeth in the lower jaw.

The teeth of **Kaup's caecilian** (*Potomotyphlus kaupii*) are distinctly narrower than those of *Typhlonectes*.

Tooth counts for some Typhlonectidae are shown in Table 5.3.

Atretochoana eiselti (length about 80 cm) is the largest lungless tetrapod known. Its unusual features include sealed choanae and complete loss of the pulmonary arteries and veins. It has a larger gape than other caecilians and enhanced cranial kinesis. The dividing zone of the teeth of *Atretochoana* seems to be exceptionally flexible. In one specimen, there were 9 (naso)premaxillary teeth, 18 maxillary teeth, 6 vomerine teeth, 14 palatal teeth, 29 dentary teeth, and 21 splenial teeth (Wilkinson and Nussbaum, 1997).

Indotyphlidae

Fig. 5.12 shows the dentition of **Battersby's caecilian** (*Indotyphlus battersbyi*). It possesses a short inner row of teeth at the front of the dentary and the presence of some

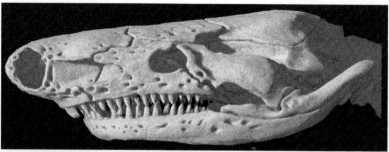

FIGURE 5.10   Dentition of a rubber eel (*Typhlonectes natans*). Micro-CT image. Image width=3 cm. *Courtesy Digimorph.org and Dr. J.A. Maisano.*

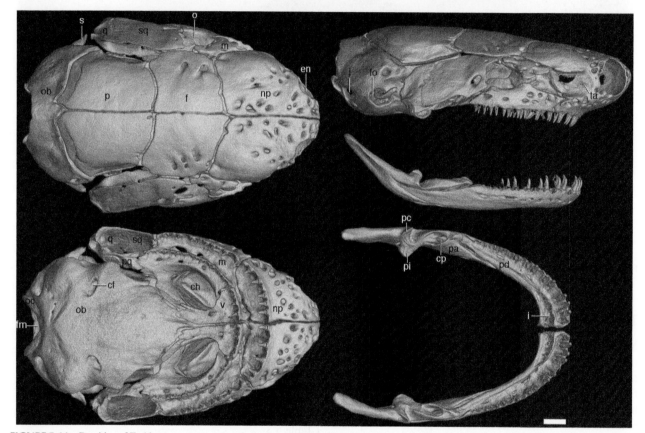

FIGURE 5.11   Dentition of *Typhlonectes compressicauda*. Scale=1 mm. For abbreviated key, see Fig. 5.4. *From Wilkinson, M., San Mauro, D., Sherratt, E., Gower, D.J., 2011. A nine-family classification of caecilians (Amphibia: gymnophiona). Zootaxa 2874, 41–64. Courtesy of Editors Zootaxa.*

bicuspid teeth. Tooth counts for the related *Indotyphlus maharashtraensis* give 19–25 premaxillary teeth, 22–26 vomeropalatine teeth, 19–22 dentary teeth, and 2–4 splenial teeth (Giri et al., 2004).

Siphonopedae

This family contains eight genera with more than 20 species.

The dentition of the **ringed caecilian** (*Siphonops annulatus*) comprises three rows of monocuspid teeth. The premaxillary/maxillary row has up to 43 teeth, the vomeropalatine row up to 47 teeth, and the dentary row up to 32 teeth (Fig. 5.13).

Microcaecilia iwokramae (*Caecilita iwokramae*) has a single (right) lung. It has two rows of teeth in the upper jaw: an outer premaxillary/maxillary row of about 11 teeth and an inner vomeropalatine row of 13 teeth. There is only a single row of eight teeth in the dentary (Fig. 5.14).

The recently discovered species *Microcaecilia dermatophaga* (Wilkinson et al., 2013) is so named because the young feed on the skin of the mother (see page 121). The teeth are pointed and gently curved. The premaxillary/maxillary teeth are large and monocuspid, but they are smaller posteriorly. The vomeropalatine teeth are much smaller, uniform in size, and bicuspid. The teeth on the dentary are the largest and are monocuspid.

Dermophiidae

The **Mexican burrowing caecilian** (*Dermophis mexicanus*) has about 20 teeth in the maxillary tooth row, 24 teeth in the vomeropalatine row, and 17 teeth in the dentary (Figs. 5.15 and 5.16).

TABLE 5.3 Maximum Tooth Numbers on the Dentigerous Bones of Some Typhlonectidae

Taxon	Premaxilla/ Maxilla	Vomer and Palatine	Dentary	Splenial
*Potomotyphlus kaupii**	60	53	64	12
*Nectocaecilia petersii**	38	34	30	6
Chthonerpeton#	19	15	16	3
Nectocaecilia#	16	27	13	3
Potomotyphlos#	27	27	27	6
Atretochoana#	27	20	29	21

*Data from Maciel, A.O., Hoogmoed, M.S., 2009. Taxonomy and distribution of gymnophiona of Brazilian Amazonia with a key to their identification. Zootaxa 2984, 1–53; #Data from Wilkinson and Nussbaum (1997).

Chikilidae

This family of caecilians was only discovered in 2012. An important dental feature in **Fuller's caecilian** (*Chikila fulleri*) classifying it as belonging to a new family is the presence of two rows of teeth in the lower jaw (Fig. 5.17).

DENTITIONS OF CAECILIAN LARVAE

In many viviparous species, the fetuses use their teeth to obtain nutrition by scraping off the oviduct epithelium (Parker, 1956; Wake and Dickie, 1998; Wilkinson et al., 2008). The scraping action may also stimulate the secretion of a nutrient substance that is then taken in by the growing fetus. There is even evidence that fetuses may imbibe liquid released from the cloaca of the mother. In some oviparous species the hatchlings use their teeth to peel off and digest the mother's nutritious skin (Kupfer et al., 2006; Wilkinson et al., 2008, 2013).

The teeth of fetal viviparous caecilians differ markedly from those of the adults, the latter of which are often bicuspid and arranged in single rows. In the fetal dentition:

1. The crowns are joined to the pedicel by a hinge joint.
2. The teeth often form several rows (polystichous), usually three to four, although there may be 12 rows in *Chthonerpeton petersi*. This results from the retention of replacement series on the dentigerous elements. The

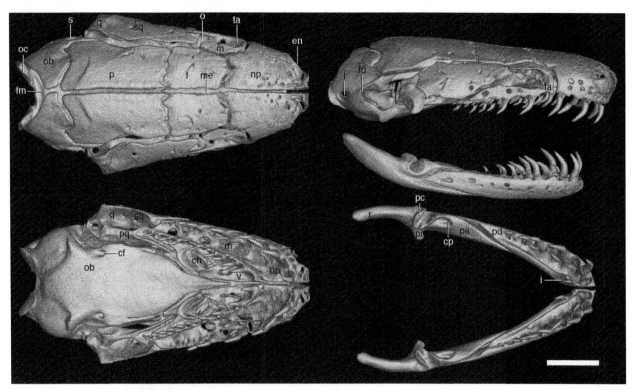

FIGURE 5.12 Dentition of Battersby's caecilian (*Indotyphlus battersbyi*). For abbreviated key, see Fig. 5.4. Scale=1 mm. *From Wilkinson, M., San Mauro, D., Sherratt, E., Gower, D.J., 2011. A nine-family classification of caecilians (Amphibia: gymnophiona). Zootaxa 2874, 41–64. Courtesy Editors Zootaxa.*

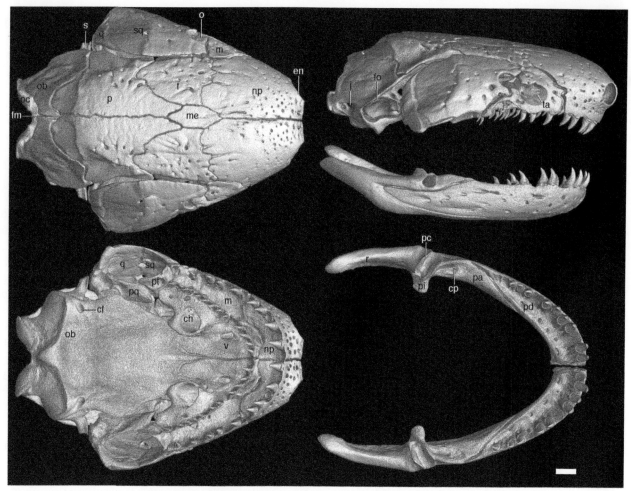

FIGURE 5.13 Dentition of the ringed caecilian, *Siphonops annulatus*. Scale = 1 mm. For abbreviated key, see Fig. 5.4. *From Wilkinson, M., San Mauro, D., Sherratt, E., Gower, D.J., 2011. A nine-family classification of caecilians (Amphibia: gymnophiona). Zootaxa 2874, 41–64. Courtesy Editors Zootaxa.*

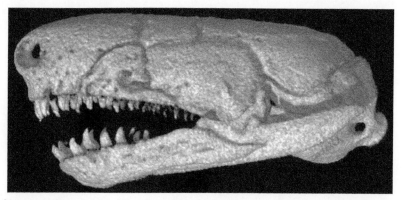

FIGURE 5.14 Dentition of *Microcaecilia iwokramae*. Micro-CT image. Image width = 10 mm. *Courtesy Digimorph.org and Dr. J.A. Maisano.*

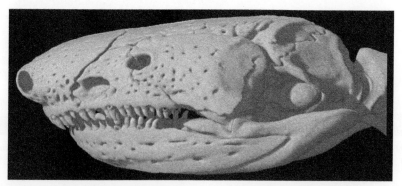

FIGURE 5.15 Dentition of the Mexican burrowing caecilian (*Dermophis mexicanus*) with jaws closed. Micro-CT image. Image width = 18 mm. *Courtesy Digimorph.org, University of Texas, and Dr. J.A. Maisano.*

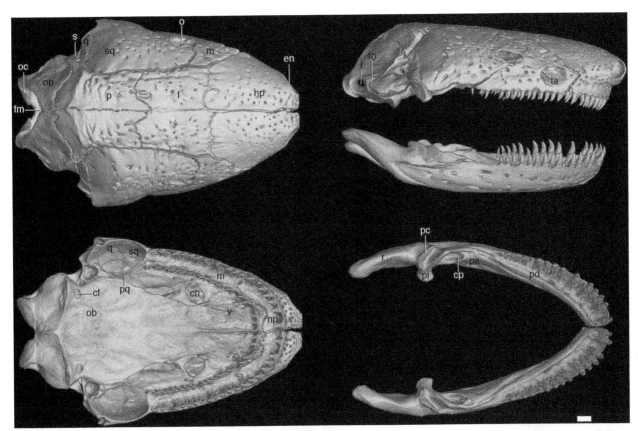

FIGURE 5.16 Dentition of the Mexican burrowing caecilian (*Dermophis mexicanus*). Marker bar = 1 mm. For abbreviated key, see Fig. 5.4. *From Wilkinson, M., San Mauro, D., Sherratt, E., Gower, D.J., 2011. A nine-family classification of caecilians (Amphibia: gymnophiona). Zootaxa 2874, 41–64. Courtesy Editors Zootaxa.*

supernumerary teeth are lost in subsequent development, so that adults have only one tooth row (monostichous) (Parker and Dunn, 1964; Wake, 1976).

3. The shape of the tooth crowns in fetuses, hatchlings, and juveniles differs from that of adults and varies from species to species and genus to genus. The crown may be spoon shaped, with or without a single terminal cusp, spatulate and multicuspid, or simple and recurved.

Although the adults of the **Kenyan caecilian** (*Boulengerula taitanus*) have pointed teeth with either one or two cusps (Fig. 5.18A and C), the dermatophagus young have very different teeth (Fig. 5.18C–F). The vomeropalatine teeth and the anteriormost three to four teeth of the premaxillary and dentary teeth are monocuspid, whereas the remaining teeth are multicusped and combine a pronounced blade-like labial cusp with a lingual cusp that has two or three subsidiary cusps (Fig. 5.18D), which may be short and blunt (Fig. 5.18E) or have more elongated, pointed processes resembling grappling hooks (Fig. 5.18F) (Kupfer et al., 2006).

In the **ringed caecilian** (*S. annulatus*) the adults have a single row of recurved, monocuspid teeth in the lower jaw (Fig. 5.13), whereas the durophagous nestlings have 22 spoon-shaped teeth on each side arranged alternately in three rows, each exhibiting multiple small, claw-like distal cusps (Fig. 5.19) (Wilkinson et al., 2008).

The upper jaw of near-hatching embryos of *Caecilia orientalis* possesses two rows of teeth: an outer premaxillary/maxillary row and an inner vomeropalatine row. The monocupid crowns are more rounded than in the adults. However, the teeth on the lower jaw differ markedly from the adult dentition. They form multiple rows, with five rows of teeth anteriorly and two to three rows more posteriorly. The crowns have a slightly bulbous ending on which lies four to six low spicules (Pérez et al., 2009). Similar features have been described for other caecilian embryos (Parker and Dunn, 1964; Wake, 1976, 1977a,b, 1980; Hraoui-Bloquet and Exbrayat, 1996).

CAUDATA (URODELA)

The tailed amphibians comprise the newts and salamanders, of which there are more than 650 living species. They are oviparous and their eggs are mainly fertilized internally, not externally like those of Anura. The larvae of urodeles differ from those of the Anura in that they are more similar in form to the adult. Embryological development is also longer and, when the larvae emerge, their outline is more elongated

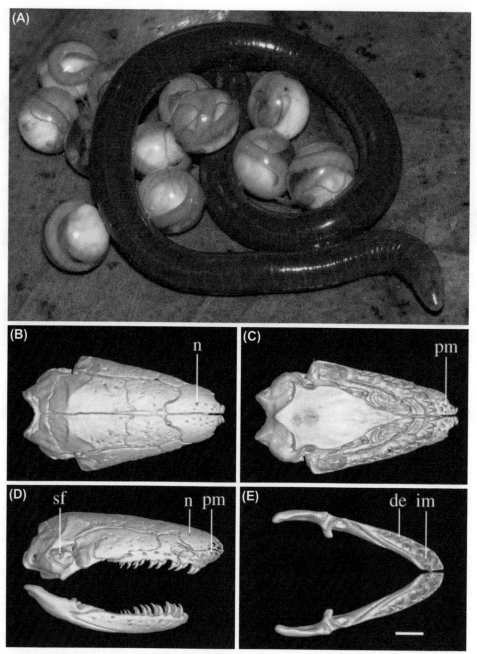

FIGURE 5.17 Dentition of Fuller's caecilian (*Chikila fulleri*). (A) *Chikila fulleri* in life, brooding egg clutch (in captivity). Scale = 10 mm. (B–E) Micro-CT images, showing cranium and mandibles; (B) Cranium in dorsal view; (C) Cranium in palatal view; (D) Cranium and mandible in right lateral view; (E) Mandibles in dorsal view. Scale (B, C) = 1 mm. *n*, nasal; *pm*, premaxilla; *sf*, stapedial foramen; *im*, inner mandibular (ie, "splenial") tooth row; *de*, dentary tooth row. *From Kamei, R.G., San Mauro, D., Gower, D.J., van Bocxlaer, I., Sherratt, E., Thomas, A., Babu, S., Bossuyt, F., Wilkinson, M., Biju, S.D., 2012. Discovery of a new family of amphibians from Northeast India with ancient links to Africa. Proc. R. Soc. Lond. B279, 2396–2399; Courtesy Editors Proceedings of the Royal Society B. Biological Sciences.*

and metamorphosis is less dramatic. Although most urodeles have aquatic larvae, some terrestrial forms do not, and a few species give birth to live young. Urodele larvae are more active than the tadpoles of Anura and feed on aquatic invertebrates rather than on plankton or algae. They possess true teeth on both the upper and lower jaws.

Neoteny is widespread among urodeles. The Amphiumidae, Proteidae, Cryptobranchidae, and Sirenidae retain, either partially or completely, larval features in the adult state. Thus, the adults retain their gills and spend their entire life underwater. Metamorphosis cannot normally be induced artificially in these families. In neotenous species of Ambystomatidae and Plethodontidae, however, metamorphosis can be induced by augmenting thyroid function. In other urodele families, neoteny can occur in response to environmental factors.

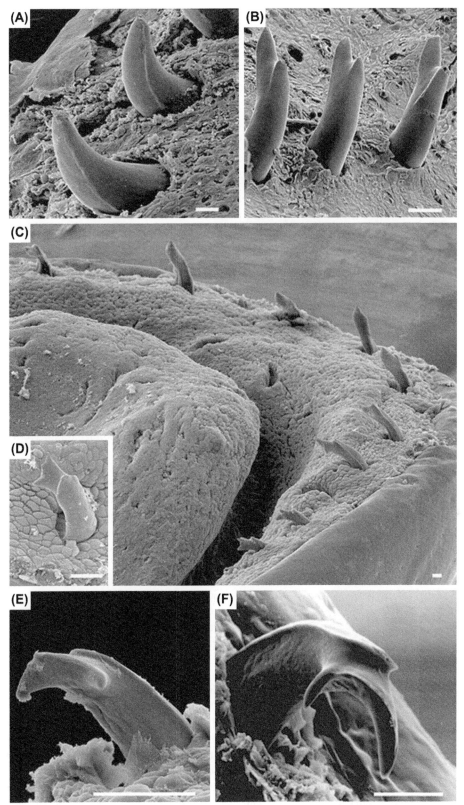

FIGURE 5.18 Dentition of adult and young *Boulengerula taitanus*. (A) Anterior view of two monocuspid, adult premaxillary teeth. (B) Labial view of three bicuspid, adult vomerine teeth. (C) Lateral view of a lower jaw of a young specimen (total length 69 mm), showing different dentary tooth crown morphologies. (D) Labial view of a posterior dentary tooth of young specimen in C. (E) Anterior premaxillary tooth of the same specimen. (F) Anterior premaxillary tooth resembling a grappling hook in a smaller specimen (total length 57 mm). Scales = 30 μm. *From Kupfer, A., Müller, H., Antoniazzi, M.M., Jared, C., Greven, H., Nussbaum, R.A., Wilkinson, M., 2006. Parental investment by skin feeding in caecilian amphibian. Nat. Lond. 440, 926–929. Courtesy Editors Nature.*

FIGURE 5.19 Top: Teeth in the lower jaw of a nestling (120 mm in total length) ringed caecilian (*Siphonops annulatus*), showing alternating dental tooth rows. Scale = 1.20 mm; bottom: detail of two dentary teeth. Scale = 120 μm. *From Wilkinson, M., Kupfer, A., Marque-Porto, R., Jeffkins, H., Antoniazzi, M.M., Jared, C., 2008. One hundred million years of skin shedding. Biol. Lett. 4, 358–361. Courtesy Editors Biology Letters.*

Ossification of the chondrocranium begins in the larvae just before feeding. The skull in fully developed salamander larvae is less extensively ossified than that in caecilian larvae, but far more so than in anuran larvae (Clemen and Greven, 2013). Salamander larvae are unique in that the entire bony palate is remodeled at metamorphosis. The

maxilla does not appear until late in larval development, with only the premaxilla forming the upper jaw. In some species, such as the mudpuppy (*Necturus*), the maxilla is absent in the adult.

The development of the dentition of urodeles from the larval stage to the adult is illustrated in Fig. 5.20, using the **smooth newt** (*Triturus vulgaris meridionalis*) as an example (Accordi and Mazzarani, 1992). In the larva, there is a single row of teeth on the premaxilla and multiple rows on the vomer and palatine bones. The maxilla develops late and similarly has a single row of teeth. A single row is present on the dentary and there are additional rows on the coronoid (splenial) (Fig. 5.20A). All these teeth are monocuspid and lack a dividing zone at the junction with the pedicel (ie, early larval nonpedicellate). They are replaced by monocuspid teeth with an incipient, hypomineralized dividing zone between the crown and pedicel (late larval subpedicellate teeth). During metamorphosis the patches of teeth on the palatines and coronoids are quickly lost and in the rest of the dentition the monocuspid teeth are progressively replaced by bicuspid teeth with the characteristic unmineralized dividing zone between the crown and pedicel (transformed teeth) (Fig. 5.20B–D). In adult *T. vulgaris*, the marginal teeth in both jaws have a modified bicuspid form (Fig. 5.20E) and the same is true of neotenous forms. However, in other species neotenous forms may retain teeth with larval characteristics.

The teeth of most salamanders are monostichous. However, in some species, such as the **ringed salamander** (*Ambystoma annulatum*), the premaxillae may carry up to five rows of teeth (Beneski and Larsen, 1989a). The teeth of adult salamanders are normally small, uniformly sized, and evenly spaced, which improves grip by increasing the roughness of the jaws rather than by piercing the integument of the prey, although some species have small recurved teeth. Plethodonts have a cluster of small teeth on the vomers, supported by the parasphenoid in the midline of the roof of the mouth. Tooth number tends to increase with age, with teeth being added at the back of the tooth row.

The bicuspid adult urodele tooth has labial and lingual cusps, of which the latter are usually more prominent. The three main forms of the cusps (Beneski and Larsen., 1989a) are conical, disc shaped (strongly developed cutting edges with a rounded apical form), or club shaped (blunt, without point or cutting edges). The morphology may be further complicated by ridging. Considerable variation in tooth morphology may be found between species of the same genus. Indeed, the dentition of **Mabee's salamander** (*Ambystoma mabeei*) is heterodont: on the premaxillary and maxillary teeth cusps are disc shaped, on the vomerine teeth the lingual cusp is pointed but the labial cusp is disc shaped, and on the dentary the teeth are tricuspid (Beneski and Larsen, 1989a).

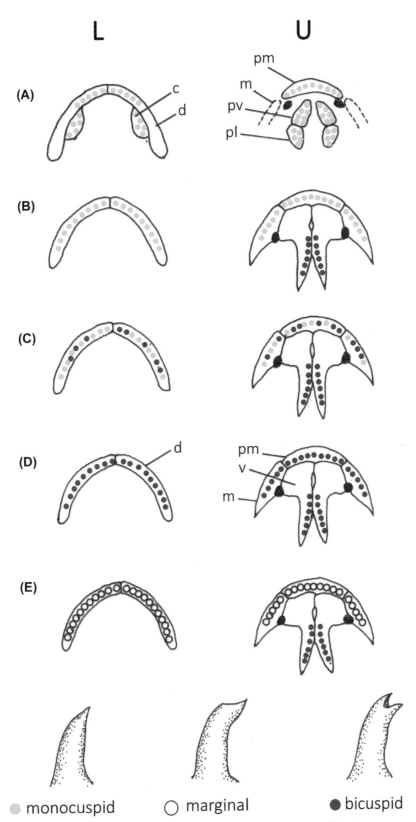

FIGURE 5.20   Dental development in the smooth newt (*Triturus vulgaris meridionalis*) from larva to adult. Arrangement and morphology of teeth: (A) in larva; (B) during metamorphosis; (C) 1 month after metamorphosis; (D) 3 months after metamorphosis. (E) In adult (and neotenous forms). *c*, coronoid; *d*, dentary; *L*, lower jaw; *U*, upper jaw; *m*, maxilla; *pl*, palatine; *pm*, premaxilla; *pv*, prevomer. *Redrawn after Accordi, F., Mazzarani, D., 1992. Tooth morphology in Triturus vulgaris meridionalis (Amphibia Urodela) during larval development and metamorphosis. Boll. Zool. 59, 371–376; Courtesy Editors Bolletino di Zoologia.*

Aquatic salamanders, including larval forms and terrestrial salamanders living in water during breeding, capture prey by suction feeding, sometimes with a lunge depending on the activity of the prey. Terrestrial salamanders capture small prey by using their tongues, which can often be extended for a considerable distance by the hyoid apparatus and not by muscular action as in anurans. The jaws play little part in this method of feeding as the prey is transferred directly to the esophagus by the tongue. With larger salamanders and larger prey, the jaws and teeth are involved in striking the prey. The presence of teeth on both the vomers and the dentaries provides increased grip of the jaws on prey compared to anurans.

There are more than 650 urodele species in nine families in two groups:

- **Cryptobranchoidei**: Hynobiidae (Asiatic salamanders), Cryptobranchidae (giant salamanders)
- **Diadectosalamandroidei**: Proteidae (mudpuppies), Sirenidae (sirens), Ambystomatidae (mole salamanders and Pacific giant salamanders), Salamandridae (newts and true salamanders), Rhyacotritonidae (torrent salamanders), Amphiumidae (amphiumas), Plethodontidae (lungless salamanders)

In most salamanders, the upper jaw is composed of a premaxilla (paired or unpaired) and paired maxillae bearing teeth. However, in the plethodontid **La Palma salamander** (*Bolitoglossa subpalmata*), teeth are absent from the upper jaw in young individuals, but present in adults. This may reflect the difference in diet between juveniles and adults. Juveniles use the well-developed tongue to transport small prey deep into the mouth, toward the vomerine teeth, whereas adults feed on larger prey, thus requiring the presence of maxillary teeth. Sirenids and most *Thorius* species (Plethodontidae) also lack teeth on the upper jaw.

Cryptobranchoidei

Hynobiidae

This group of Asiatic salamanders contains 64 species in nine genera.

The **Asian salamander** (*Ranodon sibiricus*) has a typical dentition consisting of single rows of teeth on the premaxillae, maxillae, vomers, and dentaries. More than nine tooth replacements are required to establish the adult dentition (Vassilieva and Smirnov, 2001) (see Chapter 10).

Tooth numbers have been described for the **spotted salamander** (*Hynobius naevis*). Numbers are comparable in both jaws and range from 55 to 78. Tooth numbers for vomerine teeth range from 35 to 61. Two genetically distinct groups have been described for this species (Tominaga et al., 2005). Among the distinguishing features was that

members of the group with a shorter body length also possessed a longer vomerine tooth series.

The **Shangchen stout salamander** (*Pachyhynobius shangchengenssis*) typically has a single row of bicuspid teeth on all the dentigerous bones. However, the dividing zone between crown and pedicel seems to be highly mineralized. Furthermore, there is evidence of sexual dimorphism. In the upper and lower jaw, the male possesses pedicellate teeth with a chisel- or spearhead-like crown, with the labial cusps largely reduced, whereas the main lingual cusps are flattened anteroposteriorly and exhibit sharp edges. In contrast, females have pedicellate, somewhat flattened teeth more variable in shape, but with small-bladed labial and large-bladed lingual cusps. Vomers of both sexes are similar, with typical bicuspid pedicellate teeth (Clemen and Greven, 2009).

Cryptobranchidae

This family contains two genera of giant salamander, both of which are neotenous and aquatic, living in streams and rivers. The **hellbender** (*Cryptobranchus allegeniensis*) is the third largest salamander and reaches a length of 30 cm. Its numerous teeth are small, bicuspid, and arranged in single rows (Figs. 5.21 and 5.22). The vomerine teeth are arranged parallel to, and behind, the premaxillary/maxillary tooth row. The premaxilla has about 16 teeth, the maxilla about 75 teeth, the vomer about 6 teeth, and the dentary about 65 teeth. This salamander already possesses bicuspid teeth in the late larval stage.

The **Chinese giant salamander** (*Andrias davidianus*) is the largest salamander and can reach a length of up to 180 cm. Similar to the hellbender, it has numerous small teeth arranged in single rows in the upper and lower jaws (Fig. 5.23), whereas the short, vomerine tooth row lies behind and parallel to the tooth row on the premaxilla (Fig. 5.24) (Greven and Clemen, 1979b).

Diadectosalamandroidei

Proteidae

The proteids are neotenous; they never lose their bright red gills during maturation and spend their entire lives underwater. Other distinguishing features of proteids are the absence of eyelids and of maxillary bones in the upper jaw. There are two living genera of Proteidae: mudpuppies (*Necturus*) and olms (*Proteus*).

There are five species of mudpuppies. The teeth in the upper jaw of the neotenous adult **common mudpuppy** (*Necturus maculosus*) are arranged in two V-shaped rows. Maxillae are absent and the outer anterior row of 15 teeth is attached to the premaxillae, whereas the inner row of 20 teeth is attached to the vomers and, posteriorly, to the palatopterygoid. The premaxillary teeth decrease progressively in size

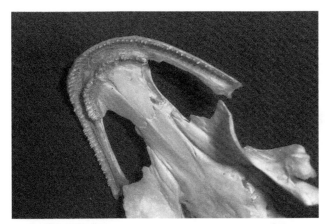

FIGURE 5.21   Upper jaw of hellbender (*Cryptobranchus alleganiensis*) showing outer and inner tooth rows. Note the presence of replacing teeth. Image width = 4 cm. *Courtesy Dr. C Underwood.*

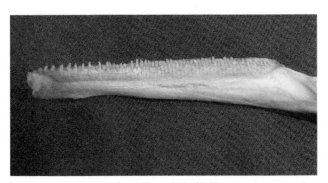

FIGURE 5.22   Inner surface of lower jaw of hellbender (*Cryptobranchus alleganiensis*). Note the numerous, developing, replacement teeth. Image width = 5 cm. *Courtesy Dr. C Underwood.*

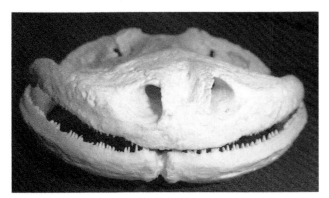

FIGURE 5.23   Model of dentition of Chinese giant salamander (*Andrias davidianus*), viewed from the front. Image width = 8 cm. *Courtesy UCL, Grant Museum of Zoology. Catalog no. LDUCZ-W330.*

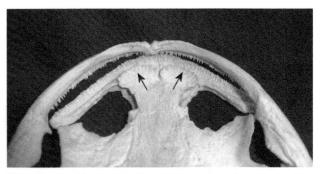

FIGURE 5.24   Model of dentition of Chinese giant salamander (*Andrias davidianus*), viewed from below. Note the vomerine tooth rows (*arrows*). Image width = 8 cm. *Courtesy UCL, Grant Museum of Zoology. Catalog no. LDUCZ-W330.*

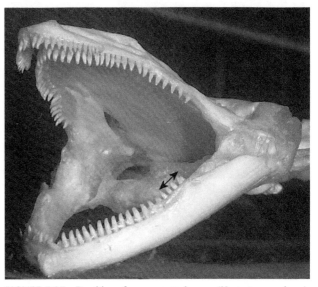

FIGURE 5.25   Dentition of common mudpuppy (*Necturus maculosus*). Note the splenial teeth forming an inner row at the back of the dentary row (*arrow*). Image width = 3.5 cm. *Courtesy MoLSKCL. Catalog no. W30.*

from front to back and are slightly smaller than teeth in the inner palatal row. In the lower jaw there is a single row of 14 teeth on the dentary, together with 6 posterior teeth on the coronoid bone medial to the dentary (Figs. 5.25 and 5.26). All the teeth are conical and unicuspid and possess a broad, partly mineralized dividing zone between the crown and the

pedicel (Greven and Clemen, 1979a). The monocuspid tooth morphology is a retained larval characteristic.

There is only a single species of **olm** or **proteus** (*Proteus anguinus*) and two subspecies. Uniquely for salamanders, the **olms** spends their whole lives underground in the water of caves and have eyes of reduced size. An olm's body is eel like and up to 40 cm in length. The limbs are small and thin, with the front legs having three digits (instead of the normal four) and the rear legs having two digits (instead of the usual five). Olms take about 14 years to reach sexual maturity.

The dentition of *Proteus* is illustrated in Fig. 5.27. On each dentary there are 21–25 teeth and on each coronoid there are 5–8 teeth (Dr. A Ivanovic, personal communication). In the upper jaw the number of teeth differs between the two subspecies. The **white olm** (*Proteus anguinus parkelj*) has 8–10 teeth on the premaxilla, 22–30 teeth on the vomer, and 6 teeth on the palatopterygoid. The **black olm** (*Proteus anguinus anguinus*) has 6–7 teeth on the premaxilla, 16–17

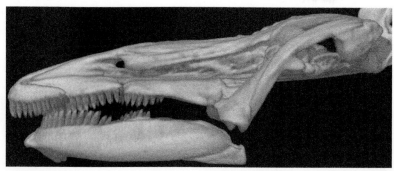

FIGURE 5.26 Dentition of common mudpuppy (*Necturus maculosus*). Note the splenial teeth at the back of the lower jaw. Image width=4.5 cm. *Courtesy Digimorph.org and Dr. J.A. Maisano.*

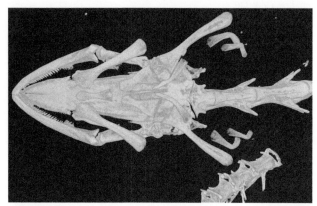

FIGURE 5.27 Skull of olm (*Proteus anguinus*). *D*, dentary; *S*, splenial; *PT*, palatopterygoid. Micro-CT image. *From Ivanovic, A., Aljancic, G., Arntzen, J.W., 2013. Skull shape differentiation of black and white olms (Proteus anguinus anguinus and Proteus a. parkelj): an exploratory analysis with micro-CT scanning. Contr. Zool. 82, 107–114. Courtesy Editors Contributions to Zoology.*

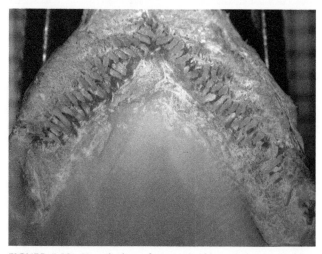

FIGURE 5.28 Ventral view of upper dentition of the greater siren (*Siren lacertina*). Teeth are only present on the vomer and palatine bones. Magnification ×8. *Courtesy Dr. M Wilkinson. © Trustees of the Natural History Museum, London.*

on the vomer, and 5–6 teeth on the palatpterygoid. There is also a difference in the orientation of the palatopterygoid teeth: in the white olm these are aligned with the arch of the vomerine teeth, whereas in the black olm they are oriented more transversely (Ivanovic et al., 2013).

Sirenidae

The sirens, the most ancient line of living urodeles, comprise two genera with about nine species. These obligate neotenous amphibians retain large external gills as adults and are fully aquatic and eel like, with very small forelimbs. They use suction to capture their prey, which includes worms, shrimps, and snails. Unusually for amphibians, the diet may also contain filamentous algae.

The Sirenidae lack teeth on the premaxilla and maxilla, which is not present in subadults. Fig. 5.28 shows the upper teeth of the **greater siren** (*Siren lacertina*). The vomerine and palatine teeth are monocuspid, lack ridges, and are arranged in a polystichous pattern. In the lower jaw teeth are absent from the dentary, but present on the splenials,

which are lingual to the dentary. The teeth are ankylosed horizontally (Figs. 5.29 and 5.30).

In the **lesser siren** (*Siren intermedia*), there may be up to 21 tooth loci on the vomer and 8 on the palate. Although the dental features are larval in character, the teeth of the **northern dwarf siren** (*Pseudobranchus striatus*) show evidence of a bipartite structure (Clemen and Greven, 1988).

Ambystomatidae

This family contains, in the classification of Frost et al. (2006), two genera: *Ambystoma* (mole salamanders) and *Dicamptodon* (Pacific giant salamanders).

The teeth of *Ambystoma* have been studied by Beneski and Larsen (1989a), particularly with reference to the premaxillary teeth. The teeth of all larvae initially have simple, conical, monocuspid crowns. In fully transformed individuals, the crowns are bicuspid. Between the larval and adult stages, teeth with incipient bicuspid crowns are often present at alternating loci between larval monocuspid teeth. Kerr (1960) reported that the changeover from one form to

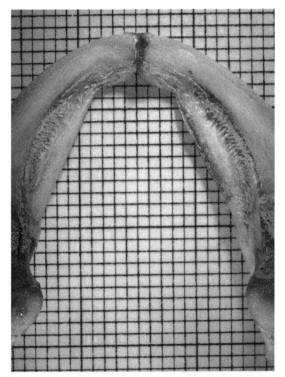

FIGURE 5.29 Lower jaw of the greater siren (*Siren lacertina*), showing teeth only present on the splenials toward the back of the jaw. Background scale = 1 mm. *Courtesy Dr. M Wilkinson.*

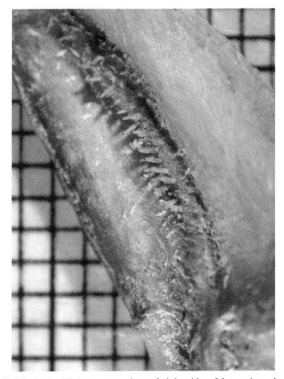

FIGURE 5.30 Higher power view of right side of lower jaw shown in Fig. 5.30, showing teeth on splenial somewhat horizontally aligned. Background scale = 1 mm. *Courtesy Dr. M Wilkinson. © Trustees of the Natural History Museum, London.*

the other is irregular and a bicuspid tooth may be replaced by a monocupid tooth. The shape of the adult crowns varies within the genus. Cusps tend to be conical in the subgenus *Rhyacosiredon*, predominantly disc shaped in the subgenus *Ambystoma*, and variously variously club shaped in *Linguaelapsus* (Beneski and Larsen, 1989a).

The **axolotl** or **Mexican salamander** (*Ambystoma mexicanum*) is neotenous. It is famous for its healing abilities and for its capacity to regenerate even whole limbs. It has, therefore, been used extensively as a model organism in studies of wound healing and tissue regeneration.

Axolotls are carnivorous and eat worms, insects, crustaceans, and small fish, which are swallowed whole. The dentition consists of numerous very small teeth on the dentigerous bones. At a week after hatching (stage 44), when its sexual organs are established, the maxilla has yet to develop. At this stage the upper dentition consists of an outer row of 12 teeth on the premaxilla and an inner row of 25 densely packed teeth on the vomer and palatine bones, whereas in the lower jaw there are 15 teeth forming an outer row on the dentary and 30 densely packed teeth on the splenial. With growth the main change is an increase in size. At a later stage of development, the maxilla appears and the tooth count increases. Thus, with a body length of 41 mm, the upper dentition (Fig. 5.31) has an outer row with about 12 teeth on the premaxilla and 8 teeth on the maxilla. The inner, polystichous row has about 25 teeth on the vomer and 18 teeth on the palatine. In the lower jaw at the same stage (Fig. 5.32) there are now about 30 on the dentary and 30 on the splenial.

In larvae, the teeth are monocuspid. With age, apart from increasing in number, they may also become bicuspid.

The related **spotted salamander** (*Ambystoma maculatum*) undergoes full metamorphosis and its teeth become bicuspid. There are about 60 teeth in each half of the jaw. In

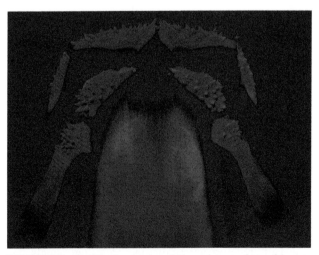

FIGURE 5.31 Dentition in upper jaw of 41-mm-long specimen of the axolotl (*Ambystoma mexicanum*). Image width = 9 mm. Alizarin red preparations scanned with fluorescent stereomicroscope. *Courtesy Dr. A. Pospisilova.*

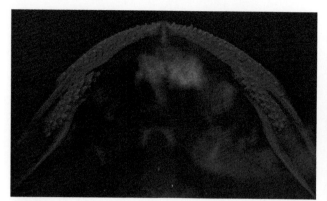

FIGURE 5.32 Dentition in lower jaw of 41-mm-long specimen of the axolotl (*Ambystoma mexicanum*). Image width=10mm. Alizarin red preparations scanned with fluorescent stereomicroscope. *Courtesy Dr. A. Pospisilova.*

addition, about 40 teeth are arranged in a sigmoid curve on each side of the palate (Fig. 5.33, bottom image).

Whereas the outer tooth row in the upper jaw of *Ambystoma* always follows the outer mouth shape, positioning of the inner tooth arcade differs between species, depending on the position of the palatal bones. Thus, in the axolotl, the inner tooth row is parallel with the outer (Fig. 5.33, middle image). In the spotted salamander, about 40 teeth form a sigmoid curve on each side of the palate, which is nearly perpendicular to the long axis of the skull (Fig. 5.33, bottom image). The vomerine teeth in the **northwestern salamander** (*Ambystoma gracile*) (Fig. 5.34) form an incomplete row that is less curved than in the **tiger salamander** (*Ambystoma tigrinum*) (Fig. 5.35).

Among the larvae of the tiger salamander, there occurs a cannibal morph that shows distinct differences in dental features from the noncannibals. In the normal (noncannibal) state, teeth are present in single rows on the maxilla, premaxilla, dentary, and coronoid. Polystichous rows of palatal teeth are present on the palatines and vomers (Fig. 5.36). At metamorphosis, the coronoid bones are resorbed, but the vomeropalatine teeth become monostichous and form a curved row inside the outer tooth row on the premaxilla/maxilla (Fig. 5.35) (Pederson, 1991). The teeth are typically sharply pointed, straight, monocuspid, and undivided, with no dividing zone being evident (Fig. 5.37A and D). The teeth of the cannibal morph are larger and recurved, and the vomerine and dentary teeth are significantly larger (Fig. 5.37B, C and E) (Pederson, 1991).

Dicamptodon is a genus of North American giant salamanders, comprising three species. The **Pacific giant salamander** (*Dicamptodon ensatus*) can grow up to 30cm in length. Aquatic larvae with filamentous external gills transform into four-legged salamanders that live on land and breathe air with lungs. Adults are sit-and-wait predators whose diet includes slugs, snails and other invertebrates as well as small vertebrates, such as other salamanders, lizards, and small rodents. However,

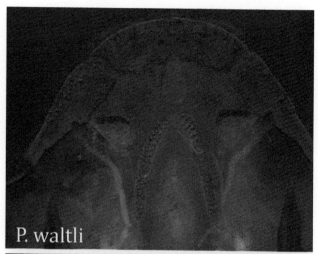

P. waltli

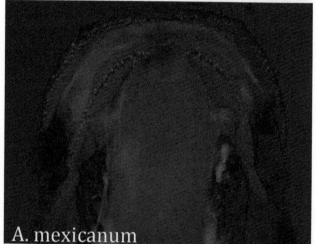

A. mexicanum

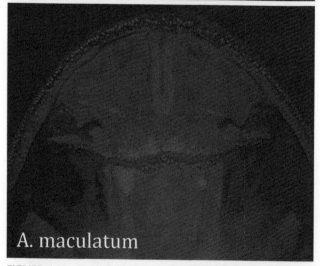

A. maculatum

FIGURE 5.33 Upper jaws of Iberian ribbed newt (*Pleuronectes waltl*) (top image), the axolotl (*Ambystoma mexicanum*) (middle image), and the spotted salamander (*Ambystoma maculatum*) (bottom image), comparing the shape of the inner dental row. Alizarin red preparations scanned with fluorescent stereomicroscope. *Courtesy Dr. A. Pospisilova and Dr. R. Cerny.*

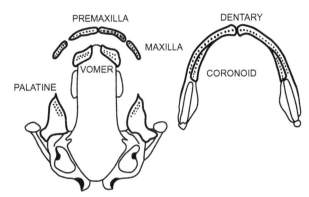

FIGURE 5.36 Line diagram of ventral view of the six dentigerous bones in the skull and mandible of the larval tiger salamander (*Ambystoma tigrinum*). *From Pederson, S.C., 1991. Dental morphology of the cannibal morph in the tiger salamander, Ambystoma tigrinum. Amphibia-Reptilia 12, 1–14; Courtesy Editors Amphibia-Reptilia.*

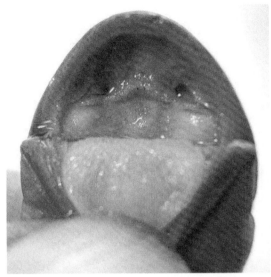

FIGURE 5.34 Roof of the mouth of the northwestern salamander (*Ambystoma gracile*), showing the vomerine teeth forming an incomplete row running in a near straight line between the internal nares. *Courtesy Dr. K Peterson.*

FIGURE 5.35 Roof of the mouth of the tiger salamander (*Ambystoma tigrinum*). The vomerine teeth form a more curved row than in *Ambystoma gracile. Courtesy Dr. K Peterson.*

they can remain as neotenous adults that retain their gills and continue to live in water where their diet would consist of aquatic invertebrates, fish, and other amphibians. Numerous small teeth are present on the dentigerous bones (Fig. 5.38), whereas an inner row of vomerine teeth parallels the curve of the maxillary tooth row and lies inside the internal nares (Fig. 5.39).

In the transformed Pacific giant salamander, there are, on average, about 89 bicuspid pedicellate teeth in the premaxillary/maxillary tooth row, whereas each vomerine tooth row has about 20 teeth. In larvae there are, on average, about 57

monocuspid, nonpedicellate teeth in the premaxillary/maxillary tooth row and 22 teeth in each vomerine tooth row. Tooth numbers for larvae in **Cope's giant salamander** (*Dicamptodon copei*) are lower, numbering, on average, 44 teeth in the premaxillary/maxillary tooth row and 19 teeth in each vomerine tooth row (Nussbaum, 1976). The tooth morphology of these two giant salamanders has been described by Beneski and Larsen (1989b), who showed that incipient bicuspid, subpedicellate teeth may be found during metamorphosis.

Salamandridae

All newts (subfamily Pleurodelinae) are salamanders, but not all salamanders are newts. Newts are specialized salamanders that do not have slippery skin, for it is rougher and drier. They are carnivorous and, on land, eat prey, such as worms, caterpillars, and insects, whereas in water they consume crustaceans, insects, and tadpoles. The prey of larval newts is mainly zooplankton: initially, small, slow-moving forms (eg, chydorids and daphniids), but later larger, more active forms (eg, copepods and chironomids). The ontogeny of the newt dentition is described on pages 126–127.

The teeth of adult newts are very small, numerous, and bicuspid and arranged mainly in two single rows in the upper jaw: one row on the premaxillae/maxillae, one on the vomers and one main row on the lower jaw on the dentary and sometimes on the coronoid/splenial (Figs. 5.40–5.44), although more than one row may be present at the front of the jaws (Fig. 5.40). The teeth are pedicellate (Fig. 5.43). Tooth number and the morphology of the vomerine tooth row can vary in different newt species (Fig. 5.44). In the marginal tooth rows, the lingual cusp is the larger, whereas in vomerine teeth, the two cusps are of more equal size.

During tooth ontogeny in the **Spanish ribbed newt** (*Pleurodeles waltl*) larvae initially have 23 teeth in the upper jaw, a row of 8 teeth on the premaxillae and two rows

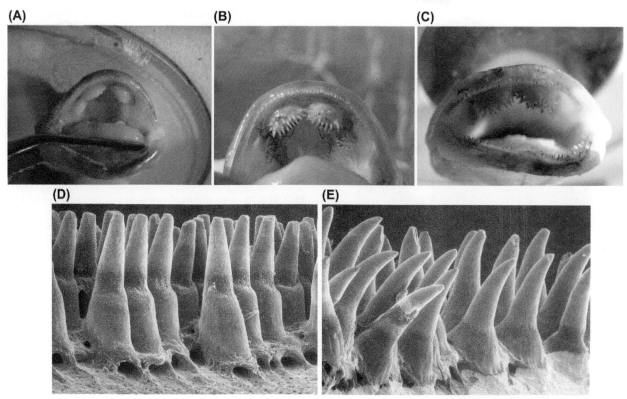

FIGURE 5.37 Teeth in larvae of tiger salamander (*Ambystoma tigrinum*). (A) Normal larva, upper jaw. Image width = 3.5 cm; (B) cannibal morph, upper jaw. Image width = 3 cm; (C) cannibal morph, dentition from the front. Image width = 3 cm; (D) scanning electron micrograph (SEM) of vomerine teeth of normal larva; (E) SEM of vomerine teeth of cannibal morph larva. *(A–C) Courtesy Dr. K McLean; (D) and (E) From Pederson, S.C., 1991. Dental morphology of the cannibal morph in the tiger salamander, Ambystoma tigrinum. Amphibia-Reptilia 12, 1–14. Courtesy Editors Amphibia-Reptilia.*

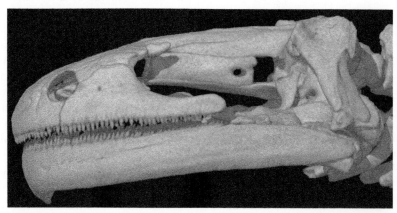

FIGURE 5.38 Lateral view of dentition of the Pacific giant salamander (*Dicamptodon ensatus*). Image width = 3 cm. *Courtesy Digimorph.org and Dr. J.A. Maisano.*

of 7 and 8 teeth on the vomers and palatines, respectively. A row of 25 mandibular teeth is present on the lower jaw with 16 splenial teeth. The adult dentition is stated to have about 14 premaxillary teeth, more than 30 maxillary teeth and 30 vomerine teeth in the upper jaw, and about 50 teeth on each dentary (Signoret, 1960) (see Fig. 5.34, top image).

Ontogeny of the dentition of the **Caucasian salamander** (*Mertensiella caucasica*) has been described by Vassilieva and Serbinova (2013). The functional dentition of

new hatchlings consists of one premaxillary tooth, two teeth on the dentary, two on the vomer and one palatine tooth, with more teeth at an earlier stage of development. Unusually, the first generation of teeth on the dentary and premaxilla showed evidence of a dividing zone between the crown and pedicel. Metamorphosis follows the same sequence as described for *Triturus* (see pages 126–127). In individuals of *Mertensiella caucasica* possessing all bicuspid teeth in single rows, the dentition consists of 22–23 tooth loci on

FIGURE 5.39 Ventral view of upper jaw of the Pacific giant salamander (*Dicamptodon ensatus*). *Courtesy Dr. K. Peterson.*

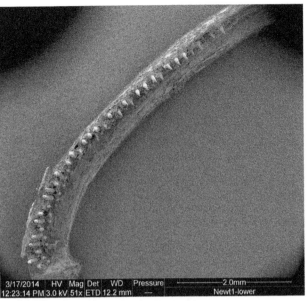

FIGURE 5.41 Lower dentition of a newt (species unknown). More than one row of teeth is present near the symphysis (lower left). Specimen provided by Mr M Cooke. Scanning electron micrograph prepared by Dr. G Vizcay, Centre for Ultrastructural Imaging, King's College London.

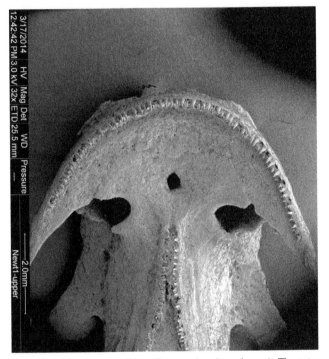

FIGURE 5.40 Upper dentition of a newt (species unknown). The outer row of bicuspid teeth is attached to the premaxillae and maxillae. There is an inner row of teeth on the vomers. Specimen provided by Mr M Cooke. Scanning electron micrograph prepared by Dr. G Vizcay, Centre for Ultrastructural Imaging, King's College London.

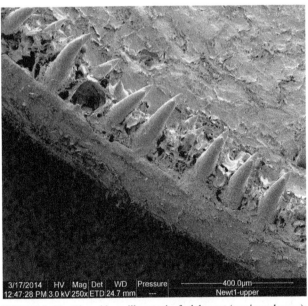

FIGURE 5.42 Bicuspid maxillary teeth of adult newt (species unknown). Specimen provided by Mr M Cooke. Scanning electron micrograph prepared by Dr. G Vizcay, Centre for Ultrastructural Imaging, King's College London.

the premaxilla, 26 loci on the maxilla, 56–57 loci on the dentary, and 58–60 loci on the vomer. Development of the dentition in the **Asian clawed salamander** (*Onychodactylus japonicus*) is similar to that in *Mertensiella* (Vassilieva et al., 2013b). The fact that the first generation of teeth on the dentary and premaxilla in *Onychodoactylus* and *Mertensiella* are subpedicellate suggested to Vassilieva et al. (2013b) that nonpedicellate teeth did not develop on these bones.

Amphiumidae

Amphiumas are eel-like salamanders with vestigial legs. They lack eyelids and a tongue. They have sharp prominent marginal teeth on the upper jaw and dentaries, as well as a second row of teeth on the vomers aligned parallel to those on the maxillae (Fig. 5.45).

The **two-toed amphiuma** (*Amphiuma means*) has 20 teeth on each side of the upper jaw (4 of which are on

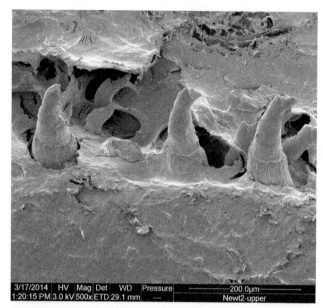

FIGURE 5.43 High-power view of pedicellate teeth of adult newt (species unknown). *C*, crown; *P*, pedicel; *D*, dividing zone. Specimen provided by Mr M Cooke. Scanning electron micrograph prepared by Dr. G Vizcay, Centre for Ultrastructural Imaging, King's College London.

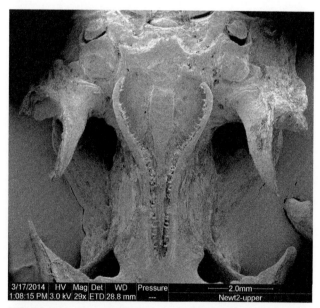

FIGURE 5.44 Vomerine tooth rows on roof of mouth of newt (species unknown). Specimen provided by Mr M Cooke. Scanning electron micrograph prepared by Dr. G Vizcay, Centre for Ultrastructural Imaging, King's College London.

the premaxilla), 16 teeth on each side of the lower jaw, and 14–15 on each vomer. The **three-toed amphiuma** (*Amphiuma tridactylum*) possesses more teeth on each half of the jaws: 4 teeth on the premaxilla, 31–32 teeth on the maxilla, 24 on the lower jaw, and 26–28 teeth on the vomer (Rose, 1968).

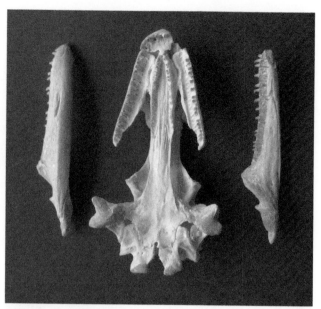

FIGURE 5.45 Dentition of *Amphiuma* sp. *Courtesy Horniman Museum and Gardens and Dr. Paolo Viscardi. Catalog no. LDHRN-NH.88.337.303.*

Plethodontidae

There are four subfamilies and nearly 400 species of Plethodontidae or lungless salamanders, which respire through the skin and oral mucosa. The name of the family (pleth = many, dont = teeth) refers to the large clusters of teeth on the posterior portions of the vomers (Figs. 5.46 and 5.47), which are supported by the parasphenoid, and which are important in feeding. However, the very small **pygmy salamander** (*Desmognathus wrightii*) may lack teeth. Plethodontids use their projectile tongues to catch and then press the prey against the vomerine tooth patches. A ventral view of the dentition of the **northern dusky salamander** (*Desmognathus fuscus*) reveals that, in addition to the teeth on the premaxilla, maxilla, and dentary, there is a single row of teeth on the vomers, together with the multiple vomerine teeth (Figs. 5.46 and 5.47).

Tooth ontogeny has been studied in three species of Costa Rican plethodontids that lack a larval stage. In these direct developers the first-generation teeth are bicuspid with a dividing zone (and equivalent to metamorphosed teeth), and no larval tooth generation is recapitulated. The species therefore hatch in a postmetamorphic developmental stage. Maximum tooth number for the **La Palma salamander** (*B. subpalmata*) is 5 on the premaxilla, 31 on the maxilla, 44 on the dentary, and more than 50 for the vomerine tooth patches. For the **common worm salamander** (*Oedipina uniformis*), the maximum number of vomerine teeth is more than 90 (Ehmcke and Clemen, 2000).

Male plethodontids develop enlarged conical teeth at the front of the upper jaw (and sometimes lower jaw) during the reproductive season only. This local change is under the control of androgens, but clearly does not affect all teeth

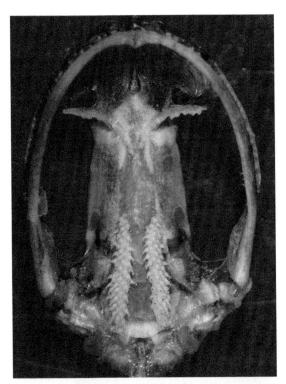

FIGURE 5.46   Ventral view of the upper jaw of the northern dusky salamander (*Desmognathus fuscus*). There are single rows of teeth on the premaxilla, maxilla, and vomer and multiple rows on the parasphenoids. *Courtesy Dr. M Wilkinson. © Trustees of the Natural History Museum, London.*

FIGURE 5.47   High-power view of parasphenoids of the upper jaw of the northern dusky salamander (*Desmognathus fuscus*), as in Fig. 5.46, showing large clusters of teeth. *Courtesy Dr. M Wilkinson. © Trustees of the Natural History Museum, London.*

(Davit-Béal et al., 2007). These enlarged teeth seem to be used by the male to inflict small wounds on the dorsum of the female. Secretions from the chin glands of the male may then enter the circulatory system of the female and stimulate her to mate. Although the maxillary teeth of both males and females, and the premaxillary teeth of females, have the usual bicuspid morphology (Fig. 5.48A and B), the enlarged anterior teeth superficially look monocuspid (Fig. 5.48C). However, they often retain a vestigial labial cusp (Fig. 5.48C and D) and are covered with enamel rather than enameloid, so they should be looked upon as a variant of the normal bicuspid adult tooth rather than a reversion to the larval monocuspid tooth type (Ehmcke and Clemen, 2000; Ehmcke et al., 2004).

The **arboreal salamander** (*Aneides lugubris*) has numerous small, sharp teeth (Fig. 5.49) and the normally bicuspid teeth in the premaxilla of males change to larger, monocuspid teeth during the breeding season.

In the **splayfoot salamander** (*Chiropterotriton priscus*), the row of prominent teeth on the maxilla overlap those of the dentary (Fig. 5.50). The anterior vomerine teeth are aligned transversely, whereas the teeth on the posterior vomers, supported by the parasphenoid, are arranged in longitudinal rows and increase in number posteriorly. Teeth are added laterally and shed medially.

The **Barton Springs salamander** (*Eurycea sosorum*) has numerous small, pointed teeth (Fig. 5.51) and feeds primarily on small aquatic crustaceans, but it can supplement its diet with other items, such as earthworms.

In the **two-lined salamander** (*Eurycea bislineata*), maxillary and dentary teeth are larger but less numerous in males than in females. Males average 24 maxillary teeth and 30 dentary teeth, compared with 27 maxillary teeth and 35 dentary teeth in females. There are five premaxillary teeth in males, and they are longer and wider than the eight premaxillary teeth in females. During the breeding season, males show changes in the morphology of the premaxillary teeth, which become monocuspid instead of bicuspid. When the breeding season ends, males lose teeth, especially in the premaxilla, and several of the new replacement teeth revert to the bicuspid morphology (Stewart, 1958).

The **Sierra Juarez hidden salamander** (*Thorius adelos*) has numerous maxillary teeth (up to 26), which are larger than those of any other *Thorius* species (Rovito et al., 2013). There are also numerous teeth on the roof of the mouth (Fig. 5.52).

ANURA

This order contains the frogs and toads. They lack a tail, which is related to their possession of long hind legs used for hopping and jumping. The Anura is by far the largest group of amphibians, with about 6400 species divided into more than 440 genera. Traditionally, the Anura were divided into three suborders: Archaeobatrachia ("primitive

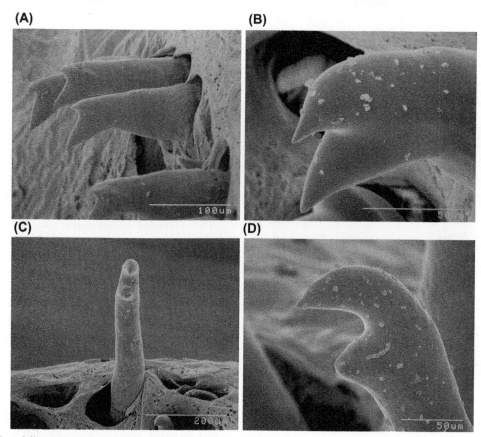

FIGURE 5.48   Sexual dimorphism in the dentition of the crater salamander (*Bolitoglossa marmorea*) as revealed by scanning electron microscopy. (A) Female, bicuspid maxillary tooth; (B) male, bicuspid maxillary tooth; (C) male, monocuspid premaxillary tooth (low power); (D) male, monocuspid premaxillary tooth (high power). The tooth is very prominent and bent caudally, whereas the lingual tip is small and blunt. Blades are lacking. *From Ehmcke, J., Wistuba, J., Clemen, G., 2004. Gender-dependent dimorphic teeth in four species of Mesoamerican plethodontid salamanders (Urodela, Amphibia). Ann. Anat. 186, 223–230; Courtesy Editors Annals Anatomy.*

FIGURE 5.49   Dentition of the arboreal salamander (*Aneides lugubris*). *Courtesy Val Johnson.*

frogs"), Mesobatrachia ("transitional frogs"), and Neobatrachia ("advanced frogs"). However, only the last group is monophyletic (Frost et al., 2006), and in this chapter families previously assigned to Archaeobatrachia and Mesobatrachia are designated non-Neobatrachia.

TEETH OF LARVAL ANURANS

Anurans hatch as **tadpoles** that have either internal or external gills. They consume a wide variety of foods, such as detritus, plankton, algae, and insect larvae (eg, mosquitoes), mainly by scraping periphyton from submerged stones and leaves, or by filter feeding. Differences in diet are to some extent correlated with the considerable variation in the mouths of tadpoles.

The mouth opening of tadpoles is surrounded by a complex known as the oral disc (Figs. 5.53 and 5.54), which may be positioned ventrally (especially in suctorial feeders), anteroventrally, terminally (in some carnivores), or upturned (in some surface feeders) (Wasserug, 1976). The oral disc is divided into anterior and posterior labia. The periphery of the oral disc contains marginal papillae, the innermost of which are termed submarginal papillae (Fig. 5.54). The papillae have chemoreceptive and tactile functions and are also used to control water flow around the oral disc. The labia carry two types of structure that are hardened by keratinization and are used in gathering food. At the margins of the mouth are a pair of upper and lower **jaw sheaths**, which together form a beak-like structure supported by

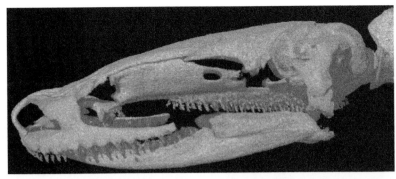

FIGURE 5.50   Dentition of splayfoot salamander (*Chiropterotriton priscus*), Image width=1 cm. *Courtesy Digimorph.org and Dr. J.A. Maisano.*

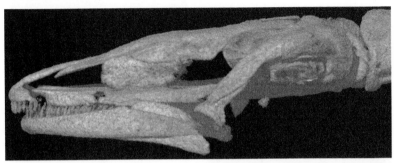

FIGURE 5.51   Dentition of Barton Springs salamander (*Eurycea sosorum*). Image width=6 mm. *Courtesy Digimorph.org and Dr. J.A. Maisano.*

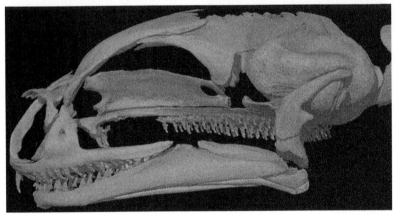

FIGURE 5.52   Dentition of the Sierra Juarez hidden salamander (*Thorius adelos*). Image width=4 mm. *Courtesy Digimorph.org and Dr. J.A. Maisano.*

supra-rostral and infra-rostral cartilages. Outside the jaw sheaths, the oral disc carries several rows of **labial teeth** on transversely arranged tooth ridges.

The jaw sheaths consist of a thick, superficial layer of keratinized epidermal cells that is continuously replenished as the superficial layers are worn away (Altig, 2007; Alibardi, 2010). The edges of the jaw sheaths may be smooth or serrated, with 30–80 serrations. The outline of the anterior (upper) sheath typically forms a smooth arc, whereas that of the posterior (lower) sheath is generally V-shaped.

Labial teeth form within the epidermis and migrate toward the surface (Altig, 2007). At each tooth site there is thus a functional tooth at the surface and one or more replacement teeth still buried within the epidermis (Fig. 5.55). Each tooth

has three parts, which are not always clearly distinguishable: a distal head; a body; and a basal, hollow sheath. The head can have a variety of shapes and curvatures and most have serrated tips, with up to 18 points depending on species. Most tadpoles have a single row of teeth per tooth ridge (monoserial), but there are some species of tadpole with biserial and even tri- or multiserial rows. The most common arrangement is for two anterior tooth rows (A-1 and A-2) and three posterior tooth rows (P-1, P-2, and P-3), although the number of rows varies from zero into double figures. Thus, one species of *Boophis* tadpole has 10 anterior and 4 posterior rows, a species of *Heleophryne* has 4 anterior and 15 posterior rows, and the sand-eating tadpole of *Boophis picturatus*, which subsists on organic matter caught between sand particles,

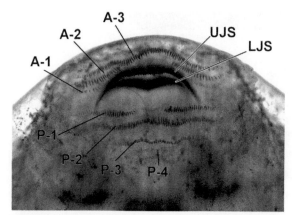

FIGURE 5.53 Whole mount preparation of tadpole illustrating oral disc. *UJS*, upper jaw sheath; *LJS*, lower jaw sheath; A-1, A-2, A-3, anterior tooth rows; P-1, P-2, P-3, P-4, posterior tooth rows. *Courtesy RCSOMA.*

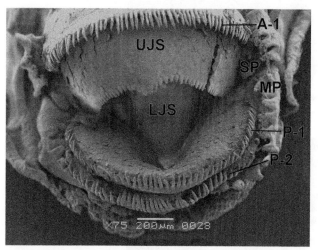

FIGURE 5.54 Scanning electron microscopy of oral disc of tadpole of the Indian skipper frog (*Euphlyctis cyanophlyctis*) at stage 25. A-1, anterior tooth row; P-1, first posterior tooth row; P-2, second posterior tooth row; *UJS*, upper jaw sheath; *LJS*, lower jaw sheath; *MP*, marginal papillae; *SP*, submarginal papillae. *From Lalremsanga, H.T., Hooroo, R.N.K., Lalronunga, S., 2013. Oral morphology of the tadpoles of Euphlyctis cyanophlyctis (Schneider, 1799) with notes on feeding behaviour. Sci. Technol. J. 1, 58–69; Courtesy Editors Science and Technology.*

FIGURE 5.55 Section of tadpole jaw showing keratinized teeth with successional columns of replacing keratinocytes which keratinize near the surface (*arrows*).

has no teeth (Altig and McDiamid, 1999; Grosjean et al., 2011). Filter-feeding tadpoles, such as those of pipids and rhinophrynids, also lack keratinized mouthparts. When teeth are present, they generally decrease in size towards the edges of the tooth row and the tooth row may be continuous or have medial gaps.

The huge differences in the size, shape, number, and spacing of labial teeth in anuran tadpoles are important features in identifying and classifying species (Trueb, 1973; Khan and Mufti, 1994; Altig, 2007). The **labial tooth row formula** (LTRF) has been established to notate the organization of the tooth rows. Tooth rows on the anterior labium are numbered distal to proximal from the mouth. Thus, A-1 denotes the first anterior and most distal row and A-2 a second row nearer the mouth. Tooth rows on the

posterior labium are numbered from proximal to distal from the mouth as P-1, P-2, and so on (see Figs. 5.53 and 5.54). Medial gaps in any tooth row are indicated in parentheses. Thus, an LTRF of 4/3 indicates four rows of teeth in the anterior labium (A-1 to A-4) and three rows in the posterior labium (P-1 to P-3), whereas a formula of 4(2-3)/3 would indicate that rows A-2 and A-3 both have medial gaps.

During feeding (Wasserug and Yamashita, 2001), the oral disc is applied to the surface of the substrate and held in place by the gripping action of the teeth. Periphyton is scraped away by the action of the beak. The beak and its supporting cartilages show considerable flexibility, which helps the beak to adapt to the contours of the substrate. At the end of a feeding episode, the tooth rows detach from the substrate, from the outside in, and are also used to abrade the surface. The combined actions of the beak and teeth produce a suspension of food particles that is ingested by the tadpole.

A principal tenet of biology is that structure is related to function, so oral discs showing different morphological characteristics should be associated with different feeding mechanisms. However, it has been difficult to ascribe specific functions to the circumoral appendages, such as the rows of labial teeth and the surrounding papillae. For example, there is considerable overlap in tooth form between tadpoles with different diets and habitats (Vera Candioti and Altig, 2010), suggesting that variation in tooth shape has little effect on food-gathering efficiency. In addition, adaptations of feeding behavior can enable tadpoles of a single species to make use of a variety of food types. For example, *Leptodactylus lanyrinthicus*, besides gathering food by scraping or by filter feeding, also eats eggs and hatchlings of other anurans (De Sousa et al., 2014).

Figs. 5.56–5.60 illustrate the oral discs of the tadpoles of five species of anurans (Khan and Mufti, 1994). The absence of the majority of features on the oral disc of the **ornate narrow-mouthed frog** (*Microhyla ornata*) (Fig. 5.56) reflects the fact that this species is a filter feeder that does not require

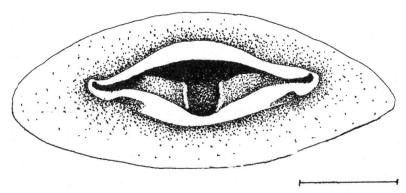

FIGURE 5.56 Oral disc of the ornate narrow-mouthed frog (*Microhyla ornata*). Scale = 1 mm. *From Khan, M.S., Mufti, S.A., 1994. Oral disc morphology of amphibian tadpole and its functional correlates. Pak. J. Zool. 26, 25–30; Courtesy Editors Pakistan Journal of Zoology.*

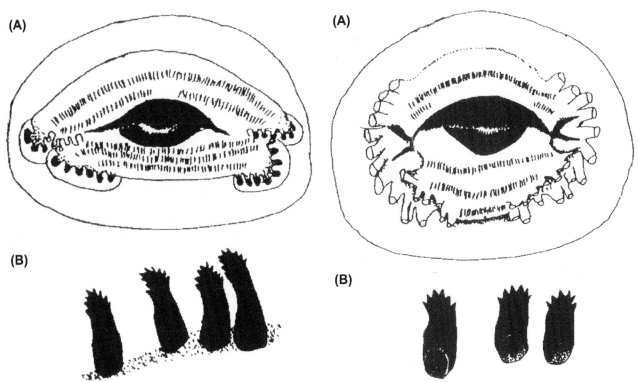

FIGURE 5.57 (A) Oral disc of the marbled toad (*Bufo stomaticus*). (B) Enlargement showing part of a tooth row. Scale = 0.25 mm. *From Khan, M.S., Mufti, S.A., 1994. Oral disc morphology of amphibian tadpole and its functional correlates. Pak. J. Zool. 26, 25–30; Courtesy Editors Pakistan Journal of Zoology.*

FIGURE 5.58 (A) Oral disc of long-legged cricket frog (*Fejervarya syhadrensis*). (B) Enlargement showing part of a tooth row. Scale = 0.25 mm. *From Khan, M.S., Mufti, S.A., 1994. Oral disc morphology of amphibian tadpole and its functional correlates. Pak. J. Zool. 26, 25–30; Courtesy Editors Pakistan Journal of Zoology.*

teeth to obtain food. Tadpoles of the **marbled toad** (*Bufo stomaticus*) have flattened teeth with lateral cusps: an adaptation to scratching and rasping the surfaces of submerged vegetation (Fig. 5.57). On the oral disc of the **long-legged cricket frog** (*Fejervarya syhadrensis*), row A-2 is interrupted by a large medial space, with few teeth at the margins. Tooth row P-3 is also short and the teeth have finer serrations (Fig. 5.58). The **Indian skipper frog** (*Euphlyctis cyanophlyctis*) has a small number of tooth rows (1/2) and the thin, blunt-tipped teeth curve toward the mouth (Figs. 5.54 and 5.59). It is thought that the teeth rake detritus toward the mouth where

it is cut up by the strong broad beak. The **Indian bullfrog** (*Hoplobatrachus tigerinus*) is carnivorous. The oral disc is positioned anteriorly and lacks palps and papillae, so that the tadpole can see its prey. The tadpole also has a high number of biserial tooth rows (5/5) and the teeth are sharp and pointed to help grip its prey (Fig. 5.60).

To assess the possible role of teeth, Venesky et al. (2010) assessed the impact of variation in tooth number on the feeding kinematics of tadpoles of the **southern leopard frog** (*Lithobates sphencephalus*). They found a significant positive relationship between the duration of the

(A)

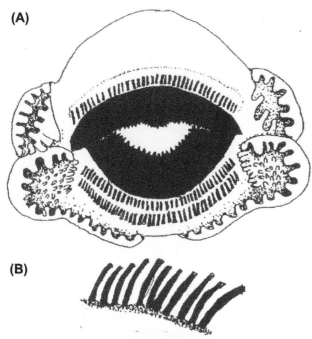

(B)

FIGURE 5.59 (A) Oral disc of Indian skipper frog (*Euphlyctis cyanophlyctis*). Scale = 1 mm. (B) Enlargement showing part of a tooth row. *From Khan, M.S., Mufti, S.A., 1994. Oral disc morphology of amphibian tadpole and its functional correlates. Pak. J. Zool. 26, 25–30; Courtesy Editors Pakistan Journal of Zoology.*

closing phase of the gape cycle and the number of labial teeth present. Tadpoles with fewer teeth were in contact with the food substrate for a shorter duration and thus foraged less effectively than tadpoles with an intact dentition.

The tadpole of the **vampire tree frog** (*Rhacophorus vampyrus*), discovered in southern Vietnam, has mouthparts that distinguish it from all other frogs (Rowley et al., 2010; Vassilieva et al., 2013a). The upper jaw sheath has a few huge, widely spaced, hook-shaped serrations that face backward into the buccal cavity. There is no lower jaw sheath, but two large, forward-facing, keratinized black hooks (Figs. 5.61 and 5.62) are present on the reduced lower labium. The female of this species deposits her eggs in small water pools in tree trunks, and because these tadpoles grow up in rather cramped surroundings, with little food, female vampire tree frogs return to their offspring to deposit unfertilized eggs for them to feed on. The "fangs" of the tadpoles are assumed to be an adaptation to the unusual nature of its diet in a confined space.

DENTITIONS OF ADULT ANURANS

Frogs and toads are carnivores that eat mainly small invertebrates, such as caterpillars, earthworms, beetles, and spiders, although large anurans may eat small vertebrates and some are even cannibalistic. Some anurans are active foragers, especially large carnivorous toads that hunt for

(A)

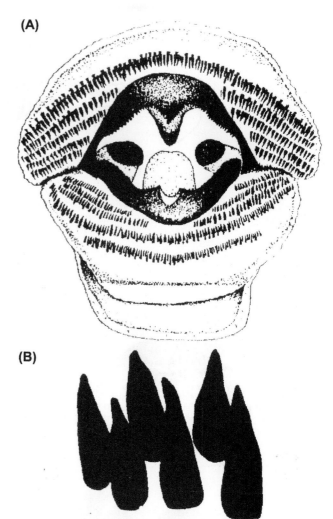

(B)

FIGURE 5.60 (A) Oral disc of Indian bullfrog (*Hoplobatrachus tigerinus*). Scale = 1 mm. (B) Enlargement showing part of tooth row. *From Khan, M.S., Mufti, S.A., 1994. Oral disc morphology of amphibian tadpole and its functional correlates. Pak. J. Zool. 26, 25–30; Courtesy Editors Pakistan Journal of Zoology.*

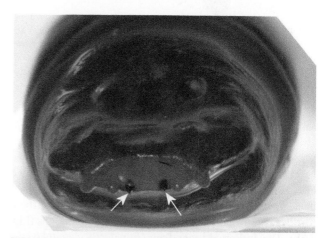

FIGURE 5.61 Anterior view of head of tadpole of vampire tree frog (*Rhacophorus vampyrus*). Keratinized "fangs" in lower jaw are *arrowed*. *Courtesy Jodi J L Rowley/Australian Museum.*

prey at night. However, probably most anurans are ambush predators that may snap at passing prey but that usually flip their sticky tongue outward and downward to immobilize prey. The tongue is attached anteriorly and free posteriorly, and it rotates as it is projected outward, so that the

sticky dorsal surface is then on the under surface and is the first to contact the prey. The tongue is usually propelled by rapid downward movement of the lower jaw combined with activity of muscles on the floor of the mouth, and the momentum imparted to the tongue causes it to extend. The distance to which the tongue can be extended varies, but in neobatrachians can be considerable. For example, in the **marine toad** (*Bufo marinus*), the tongue can be protruded 2.3–5.9 cm and prey capture takes only about 0.3 s (Nishikawa and Gans, 1996). The tongue transfers the prey directly to the esophagus. In most species, the tongue can only be projected in the direction in which the head is pointing, but some species can aim the tongue independently to some extent (Monroy and Nishikawa, 2011).

The role of anuran teeth in food capture is to grip prey and prevent its escape. Teeth are completely absent from the lower jaw in all but one species of frog, namely, the **Guenther's marsupial frog** (*Gastrotheca guentheri*). The teeth in the upper jaw of frogs always have a simple conical shape and function simply to help prevent the prey from escaping. The teeth are arranged as a single row of small teeth in the upper jaw on the premaxilla and maxilla (Fig. 5.63A–C), and a small number of additional teeth are usually present on the vomers

FIGURE 5.62 Lateral view of head of tadpole of vampire tree frog (*Rhacophorus vampyrus*). Keratinized "fangs" in lower jaw are *arrowed*. *Courtesy Jodi J L Rowley/Australian Museum.*

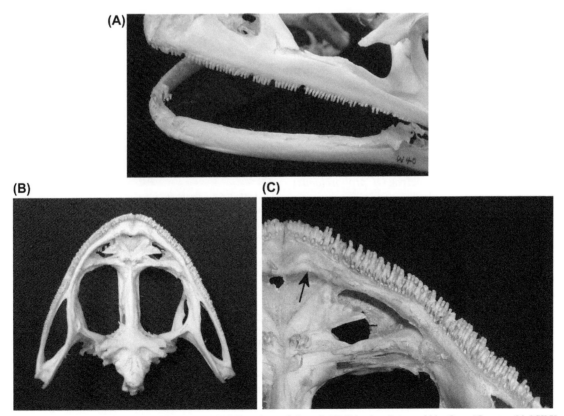

FIGURE 5.63 Dentition of the American bullfrog [*Rana (Lithobates) catesbeianus*]. (A) Side view. Image width=6 cm. *(Courtesy MoLSKCL. Catalog no. W40.).* (B) Ventral view. Image width=5 cm. *(Courtesy MoLSKCL. Catalog no. W46.).* (C) Ventral view of B at higher power. Note the unerupted developing replacement teeth palatal to the functional teeth and the vomerine teeth (*arrow*). Image width=4 cm. *(Courtesy MoLSKCL. Catalog no. W46.)*

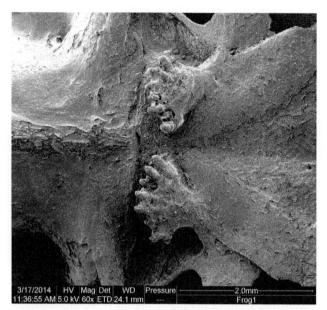

FIGURE 5.64 Teeth on vomers in a frog (species unknown). Specimen provided by Mr M Cooke. Scanning electron micrograph prepared by Dr. G Vizcay, Centre for Ultrastructural Imaging, King's College London.

(Figs. 5.63B, C and 5.64). The number of teeth tends to increase with age. The rim of the edentulous lower jaw fits closely behind the upper teeth and limits the chances of the prey escaping. Teeth are totally absent in the true toads (Bufonidae).

The teeth are usually bicuspid (Fig. 5.65), although some species have monocuspid teeth [eg, **African clawed frog** (*X. laevis*)] and others tricuspid teeth [eg, **red-eyed tree frog** (*Agalychnis callydrias*)] (Greven and Ritz, 2008). Some species lack the dividing zone between tooth and pedicel [eg, **phantasmal poison frog** (*Epipedobates tricolor*) and **strawberry poison dart frog** (*Epipedobates anthoni*)], whereas **Cranwell's horned frog** (*Ceratophrys cranwelli*) and **Sumaco horned treefrog** (*H. proboscideus*) have undivided monocuspid teeth.

In addition to the small teeth in the upper jaw, some species also possess a pair of bony projections (odontoids) near the midline in the lower jaw (Fabrezi and Emerson, 2003). In some species, odontoids are insignificant bumps on the bones of the jaw, but in others they are much larger fang-like structures that are used in biting (Mendelson and Pramuk, 1998; Shaw and Ellis, 1989). Odontoids have evolved independently within the families Ranidae, Hylidae, Myobatrachidae, and Leptodactylidae and also seem to have evolved on three to five occasions within the Ranidae. Comparisons of morphology, behavior, and diet among fanged frog taxa reveal that the fangs may be the result of either sexual or natural selection. Thus, odontoids may be used by males to obtain dominance either by display or by fighting (Fabrezi and Emerson, 2003). Alternatively, they could be used in acquiring food or could provide a defence against predators (Setiadi et al., 2011).

(A)

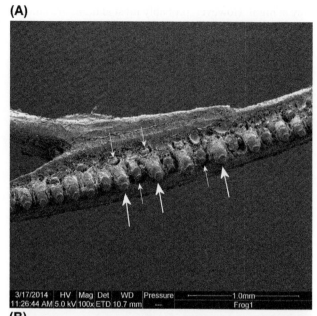

(B)

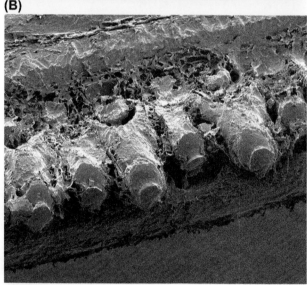

FIGURE 5.65 Teeth on a maxilla of a frog (species unknown). (A) Low-power view showing bicuspid functional teeth (*large arrows*) and the replacement teeth on the palatal side (*small arrows*). (B) High-power view of Fig. 5.65A showing bicuspid teeth. Specimen provided by Mr M Cooke. Scanning electron micrograph prepared by Dr. G Vizcay, Centre for Ultrastructural Imaging, King's College London.

Non-Neobatrachian Anurans

Pipidae

This is a primitive group of anurans that lack a tongue and are exclusively aquatic. Some have teeth but others do not. The highly aquatic **African clawed frog** (*X. laevis*) differs from other anuran taxa in that true teeth begin to develop on the upper jaw in tadpoles during the last stages of larval life, although teeth do not erupt until the end of metamorphosis. The adult has numerous small teeth that form a row along the premaxilla and maxilla (Fig. 5.66). The teeth are nonpedicellate and the number increases with age. In newly metamorphosed frogs, there are about 20 teeth on each side (of which the first 6 lie on the premaxilla). The number increases to about 30 by 18 months and to 60 in adults at least 3 years of age (Shaw, 1979). *Xenopus* sucks in food and water and expels the water before the mouth is completely closed.

(A)

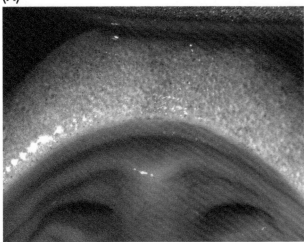

(B)

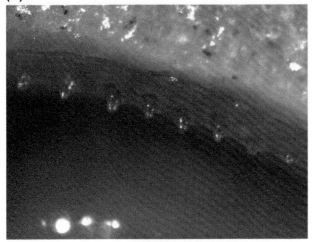

FIGURE 5.66 *Xenopus laevis.* (A) Dentition. × 5.66; (B) Higher power view of teeth in upper jaw. *Courtesy Dr. M. Gaete.*

Bombinatoridae

This family contains the fire-bellied toads that, as their name suggests, are brightly colored: a warning sign of their toxicity. When a fire-bellied toad is disturbed, toxin is secreted, mainly through the skin of their hind legs. Although not true toads, they have warty skin. The **Oriental fire-bellied toad** (*Bombina orientalis*) has the typical single row of small teeth on the premaxilla and maxilla of the upper jaw (Fig. 5.67). Their diet consists mainly of insects; worms; and small, aquatic arthropods.

Neobatrachia

This group contains more than 90% of extant species of frogs and includes about 25 families containing more than 5000 species.

Hemiphractidae

In this family, the females carry their eggs on their backs. In the **Sumaco horned treefrog** (*H. proboscideus*), the teeth are strongly recurved, monocuspid, but possessing two tubercles on the mesial and distal margins. In addition, odontoid fangs of bone are present on the dentaries. The pointed, recurved teeth and odontoid pegs are used to subdue relatively large vertebrate prey.

Amphignathodontidae

Guenther's marsupial frog (*G. guentheri*), so-called as the female possesses a dorsal brood pouch, is found in South and Central America. It has been placed by some in a separate family, Amphignathodontidae. Guenther's marsupial frog is unique among all frogs because it has a row of teeth on the lower jaw (Figs. 5.68 and 5.69) (Duellman and Trueb, 1986). In the holotype, there are 7 teeth on each premaxilla, 46 teeth on the right maxilla and 47 teeth on the left maxilla, 43 teeth on the right dentary and 44 teeth on the left dentary, and 8 teeth on the left vomer and 6 teeth on the right vomer. Current understanding is that all

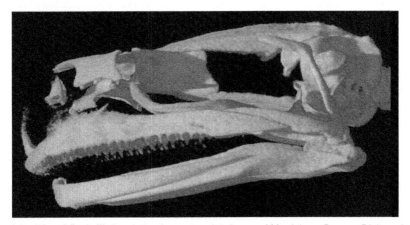

FIGURE 5.67 Dentition of the Oriental fire-bellied toad (*Bombina orientalis*). Image width = 1.1 cm. *Courtesy Digimorph.org and Dr. J.A. Maisano.*

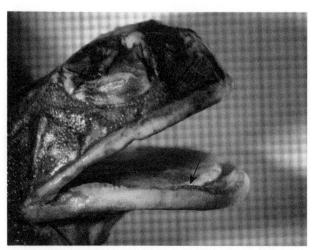

frogs lost the teeth in the lower jaw more than 200 million years ago. However, in this one instance, it is assumed that the row of teeth in the lower jaw has reevolved within the past 20 million years (Wiens, 2011).

Brachycephalidae

The genus *Eleutherodactylus* is enormous and currently contains 605 species. It is likely that further genetic investigation of the genus will result in its being divided among several genera (Frost et al., 2006). Differences in the number of maxillary teeth may help define the species of frog, and their numbers are known for many species of the rain frogs in this genus (Goin, 1959). For example, in the Jamaican species of *Eleutherodactylus*, the average number of maxillary teeth in *Eleutherodactylus lynni* is 89, in *Eleutherodactylus*

cavernicola 75, in *Eleutherodactylus grabhami* 73, in *Eleutherodactylus candalli* and *Eleutherodactylus unicolor* 72, in *Eleutherodactylus gossei* 59, in *Eleutherodactylus andreus* 52, and in *Eleutherodactylus alticola* 50.

Hylidae

Tooth counts have been made for the maxilla in more than 50 species of hylid frogs, the true tree frogs. The lowest average number was recorded in the **cricket frog** (*Acris gryllus blanchardi*) (28) and the highest average number in the **rusty tree frog** (*Hyla maxima*) (115). The **spotted chorus frog** (*Pseudacris clarki*) is sexually dimorphic, with males having larger and fewer teeth (average 41.7) than females (average 44.2) (Goin, 1958). In the genus *Acris*, the largest frogs do not always have the most teeth. The largest in the group (*Acris gryllus blanchardi*) has the smallest number (average 28), whereas the smallest (*Acris gryllus crepitans*) has the largest number (average 39).

In *H. proboscideus*, the teeth are, unusually, both monocuspid and unipartite. In addition, odontoids are found on the dentary, angular, and palatine bones (Shaw, 1989).

There is evidence that adult males of *Pseudacris clarki* have fewer but larger maxillary teeth than females (an average of just less than 42 versus 44).

Some hylids have poisonous skin, eg, dendrobatids. Two species—*Corythomantis greeningi* and *Aparasphenodon brunoi*—can inject the venom into other animals by using bony skull spines that penetrate the skin.

Ceratophryidae

The **horned frogs** (*Ceratophrys*) are voracious feeders and can eat prey, such as rats, lizards, and even other frogs. Their teeth are larger and more curved than those of most other frogs (Fig. 5.70). *Ceratophrys* possesses large, fang-like odontoids that seem to have evolved by natural selection,

(A)   (B)

(C)   (D)

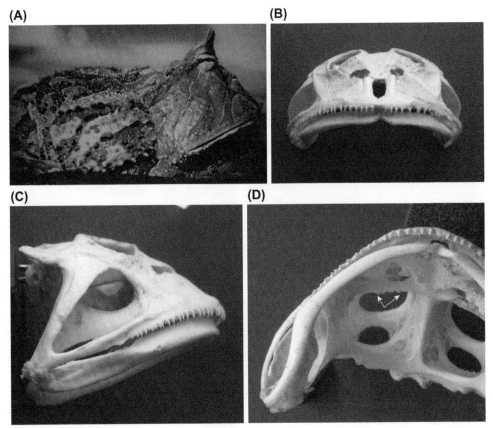

FIGURE 5.71 Surinam horned frog (*Ceratophrys cornuta*). (A) Frog, showing width of mouth (*Courtesy Wikipedia.*). (B) Front view, showing the teeth and the wide mouth. Image width=7 cm. (C) Side view of the teeth. Image width=7 cm. (D) Ventral view of the roof of the mouth, showing a small number of palatine teeth aligned transversely in upper jaw (*arrow*). Image width=5 cm (B–D). *(Courtesy Horniman Museum and Gardens and Paolo Viscardi. Catalog no. LDHRN-NH.34.70.)*

rather than by sexual selection (Fabrezi and Emerson, 2003). The fangs of males and females are of equal size, and the odontoids form part of a suite of adaptations, also including a heavily ossified skull and monocuspid, nonpedicellate teeth, to eating large prey.

Tooth counts on the maxilla of 13 species of the southern frogs (*Leptodactylus*) vary from an average of 45 in *Leptodactylus melanonotus* to 72 in *Leptodactylus dengleri* (Goin, 1959).

The **Surinam horned frog** (*Ceratophrys cornuta*) has a very large mouth, said to be 1.5 times wider than the length of the body (Fig. 5.71A–C). A few transversely aligned teeth are present on the palate (Fig. 5.71D). It has been reported there are no vomerine teeth.

Dendrobatidae

The **poison dart frogs**, of which there are more than 175 species, secrete poison from the skin. The toxicity of the skin is variable but the poisons from four species of the genus *Phyllobates* are among the deadliest of poisons and have long been used by indigenous peoples of Central and South America to coat their poisonous arrows. Some species lack true teeth in the upper jaw. The **golden poison frog** (*Phyllobates terribilis*) is one such species. However, there is an extra bone plate in the lower jaw with small projections that gives the appearance of teeth.

Dicroglossidae

The **Indian five-fingered pond frog** (*Euphlyctis hexadactylus*) has the typical frog dentition, with numerous, small, bicuspid teeth in a single row along the upper jaw (Fig. 5.72A) and oblique rows of about 10 backward-sloping teeth on the vomers (Fig. 5.72B).

Limnonectes species, such as the **big-headed Luzon Island fanged frog** (*Limnonectes macrocephalus*) (Fig. 5.73) and the **fanged river frog** (*Limnonectes macrodon*) (Fig. 5.74A) possess well-developed odontoid fangs in the lower jaw. In these species, the fangs seem to be the result of sexual selection. They are relatively larger in males, for which they are used when engaged in combat, than in females, who do not engage in combat. In addition, large prey does not form a large proportion of the diet (Barej et al., 2014). The **Indian bullfrog** (*H. tigerinus*) has less prominent odontoid fangs (Fig. 5.74B) and two prominent oblique rows of vomerine teeth.

(A)

(B)

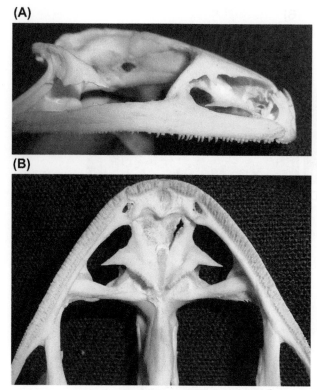

(A)

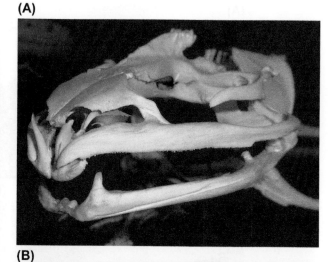

(B)

FIGURE 5.72 Dentition of the Indian five-fingered pond frog (*Euphlyctis hexadactylus*): (A) Side view. Image width=6.5cm. (B) Ventral view, showing the two oblique rows of palatine teeth. Image width=6cm. *Courtesy UCL, Grant Museum of Zoology. Catalog no. LDUCZ-5361.*

FIGURE 5.73 Odontoid pegs in a big-headed Luzon Island fanged frog (*Limnonectes macrocephalus*). ©*Rafe Brown, University of Kansas.*

FIGURE 5.74 (A) Oblique view of the skull of the fanged river frog (*Limnonectes macrodon*). Note the large odontoid peg (OP) and the small teeth in the upper jaw. Image width=8.5cm. (B) Side view of the skull of the Indian bullfrog (*Hoplobatrachus tigerinus*). Note the small odontoid peg teeth in the upper jaw. Image width=6cm. © *Trustees of the Natural History Museum, London.*

Ranidae

This family represents the true frogs and, together with the Leptodactylidae (southern or tropical frogs), contains the largest number (well exceeding 100) of genera. The teeth are bicuspid, with a small cusp located on the labial slope of the conical tooth tip. Teeth are continuously replaced. The replacement teeth are initiated low down on the palatal (lingual) side

of the functional teeth (Fig. 5.63C). There is no more than one functional and one replacement tooth at each tooth position.

The dentition of the **European common frog** (*Rana temporaria*) has the typical anuran features. There is a single row of about 40 small teeth on each side of the upper jaw, with about 8 teeth on the premaxilla and about 30 teeth on the maxilla (Fig. 5.75). There are four to five teeth on each vomer.

The **American bullfrog** [*Rana* (*Lithobates*) *catesbeiana*] is one of the larger frogs. There are single rows of about 10–15 teeth on each premaxilla and 60–70 teeth on each maxilla (Fig. 5.63A–C). All teeth are pedicellate and progressively diminish in size posteriorly. There may be 5–10 teeth on each vomer (Fig. 5.63C).

FIGURE 5.75 Dentition of European common frog (*Rana temporaria*).

FIGURE 5.76 Dentition of the edible frog (*Pelophylax* kl. *esculentus*). *Courtesy MoLSKCL. Catalog no. W114.*

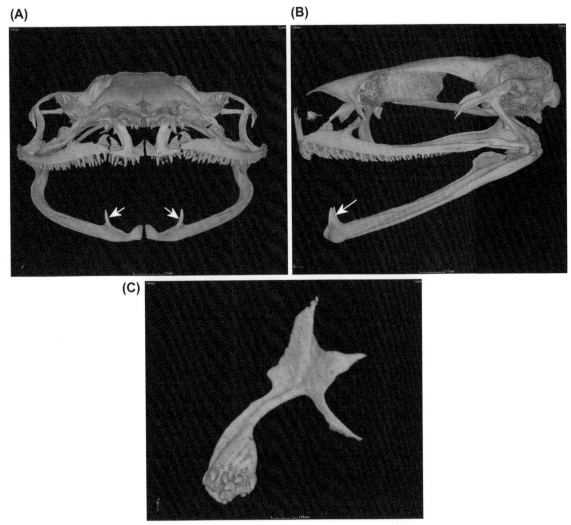

FIGURE 5.77 Dentition of West African torrent-frog (*Odontobatrachus natator*). (A) Front view. Scale = 2 mm. (B) Lateral view. Scale = 1.5 mm. (C) Teeth on the vomer. In (A) and (B), note the prominent odontoid peg at the front of each dentary. Scale = 0.5 mm. *(A) and (B) from Barej, M.F., Schmitz, A., Günther, R., Loader, S.P., Mahlow, K., Rödel, M.O., 2014. The first endemic West African vertebrate family-a new anuran family highlighting the uniqueness of the Upper Guinean biodiversity hotspot. Front. Zool. 11, 8. Courtesy Editors Frontiers of Zoology. (C) Courtesy Dr. M F Barej.*

The **northern leopard frog** [*Rana* (*Lithobates*) *pipiens*] has 10–12 teeth on the premaxilla and up to about 80 teeth on the maxilla. The largest teeth are in the anterior one-third of the series, and then they diminish in size posteriorly. There are four to five small teeth forming transverse rows on the vomers.

The hind legs of the **edible frog** (*Pelophylax esculentus*), a hybrid between the **pool frog** (*Pelophylax lessonae*) and the **marsh frog** (*Pelophylax ridibundus*), are listed on restaurant menus, particularly in France. This frog has approximately 50 teeth on the upper jaw (Fig. 5.76).

Petropedetidae or Odontobatrachidae

The **West African torrent-frog** was assigned by Frost et al. (2006) to *Petropedetes natator*. More recently, however, Barej et al. (2014) reassigned it to a new species (*Odontobatrachus natator*) and a new family (Odontobatrachidae). The dentition of this frog is unusual in that it has exceptionally long, pointed, recurved maxillary teeth together with two, prominent tusk-like odontoid pegs (Fig. 5.77A and B). A few teeth are also present on the vomers (Fig. 5.77C).

ONLINE RESOURCES

Altig R, McDiarmid RW, Nichols KA, Ustach PC: Tadpoles of the United States and Canada: A tutorial and key. http://www.pwrc.usgs.gov/tadpole/default.htm.

REFERENCES

Accordi, F., Mazzarani, D., 1992. Tooth morphology in *Triturus vulgaris meridionalis* (Amphibia Urodela) during larval development and metamorphosis. Boll. Zool. 59, 371–376.

Alibardi, L., 2010. Cornification of the beak of *Rana dalmatina* tadpoles suggests the presence of basic keratin-associated proteins. Zool. Stud. 49, 51–63.

Altig, R., 2007. A primer for the morphology of anuran tadpoles. Herpetol. Conserv. Biol. 2, 71–74.

Altig, R., McDiamid, R., 1999. Body plan: development and morphology. In: McDiamid, R., Altig, R. (Eds.), Tadpoles: The Biology of Anuran Larvae. University of Chicago Press, Chicago, pp. 24–51.

Barej, M.F., Schmitz, A., Günther, R., Loader, S.P., Mahlow, K., Rödel, M.O., 2014. The first endemic West African vertebrate family-a new anuran family highlighting the uniqueness of the Upper Guinean biodiversity hotspot. Front. Zool. 11, 8.

Bemis, W.E., Schwenk, K., Wake, M.H., 1983. Morphology and function of the feeding apparatus of *Dermophis mexicanus* (Amphibia: gymnophiona). Zool. J. Linn. Soc. 77, 75–96.

Beneski, J.T., Larsen, J.H., 1989a. Interspecific, ontogenetic, and life-history variation in the tooth morphology of mole salamanders (Amphibia, Urodela, Ambystomatidae). J. Morphol. 199, 53–69.

Beneski, J.T., Larsen, J.H., 1989b. Ontogenetic alterations in the gross tooth morphology of *Dicamptodon* and *Rhyacotriton* (Amphibia, Urodela, Dicamptodontidae). J. Morphol. 199, 165–174.

Clemen, G., Greven, H., 1988. Morphological studies on the mouth cavity of Urodela. IX. Teeth of the palate and the splenials in *Siren* and *Pseudobranchus* (Sirenidae, Amphibia). J. Zool. Syst. Evol. Res. 26, 135–143.

Clemen, G., Greven, H., 2009. Sex dimorphic dentition and notes on the skull and hyobranchium in the hynobiid *Pachynobius shangchengensis* Fei, Qu & Wu, 1983 (Urodela: Amphibia). Vert. Zool. 59, 61–69.

Clemen, G., Greven, H., 2013. Remodelling of the palate: an additional tool to classify larval salamandrids through metamorphosis. Vert. Zool. 63, 207–216.

Davit-Béal, T., Chiska, H., Delgado, S., Sire, J.-Y., 2007. Amphibian teeth: current knowledge, unanswered questions, and some directions for future research. Biol. Rev. 82, 49–81.

De Sousa, V.T.T., Nomura, F., Venesky, M.D., Rossa-Feres, D.C., Pezzuti, T.L., Andrade, G.V., Wassersug, R.J., 2014. Flexible feeding kinematics of a tropical carnivorous anuran tadpole. J. Zool. Lond. 293, 204–210.

Duellman, W.E., Trueb, L., 1986. Biology of Amphibians. John Hopkins University Press, Baltimore.

Ehmcke, J., Clemen, G., 2000. Teeth and their sex-dependent dimorphic shape in three species of Costa Rican plethodontid salamanders (Amphibia: Urodela). Ann. Anat. 182, 403–414.

Ehmcke, J., Wistuba, J., Clemen, G., 2004. Gender-dependent dimorphic teeth in four species of Mesoamerican plethodontid salamanders (Urodela, Amphibia). Ann. Anat. 186, 223–230.

Fabrezi, M., Emerson, S.B., 2003. Parallelism and convergence in anuran fangs. J. Zool. Lond. 260, 41–51.

Frost, D.R., Grant, T., Fainovich, J.N., Bain, R.H., Haas, A., Haddad, C.F.B., De Sá, R.O., Channing, A., Wilkinson, M., Donnelan, S.C., Raxworthy, C.J., Campbell, J.A., Blotto, B.L., Moler, P., Drewes, R.C., Nussbaum, R.A., Lynch, J.D., Green, D.M., Wheeler, W.C., 2006. The amphibian tree of life. Bull. Am. Mus. Nat. Hist. 297, 1–371.

Giri, V., Gower, D.J., Wilkinson, M., 2004. A new species of *Indotyphlus* Taylor (Amphibia: gymnophiona: Caeciliidae) from the Western Ghats, India. Zootaxa 739, 1–19.

Goin, C.J., 1958. Notes on the maxillary dentition of some hylid frogs. Herpetologica 14, 117–121.

Goin, C.J., 1959. Notes on the maxillary dentition dentition of some frogs of the genera *Eleutherodactylus* and *Leptodactylus*. Herpetologica 15, 134–136.

Greven, H., 1989. Teeth of extant amphibians: morphology and some implications. Progr. Zool. 35, 451–455.

Greven, H., Clemen, G., 1979a. Morphological studies on the mouth cavity of urodeles. IV. The teeth of the upper jaw and the palate in *Necturus maculosis* (Rafinesque) (Proteidae: Amphibia). Arch. Histol. Jpn. 42, 445–457.

Greven, H., Clemen, G., 1979b. Morphological studies on the mouth cavity of urodeles VI. The teeth of the upper jaw and the palate in *Andrias davidianus* (Blanchard) and *A. japonicus* (Temminck) (Cryptobranchidae: Amphibia). Amphibia-Reptilia 1, 49–59.

Greven, H., Laumeier, I., 1987. A comparative SEM study on the teeth of 10 anuran species. Anat. Anz. 164, 103–116.

Greven, H., Ritz, A., 2008. Tricuspid teeth in anurans (Amphibia). Acta Biol. Benrodis 15, 67–74.

Greven, H., Hillmann, G., Themann, H., 1989. Ultrastructure of the dividing zone in teeth of *Salamandra salamandra* (L.) (Amphibia, Urodela). Progr. Zool. 35, 477–479.

Grosjean, S., Randrianiaina, R.D., Strauss, A., Vences, M., 2011. Sand-eating tadpoles in Madagascar: morphology and ecology of the unique larvae of the treefrog *Boophis picturatus*. Salamandra 47, 63–76.

Herrel, A., Measey, G.J., 2012. Feeding underground: kinematics of feeding in caecilians. J. Exp. Zool. 317A, 533–539.

Hraoui-Bloquet, S., Exbrayat, J.-M., 1996. Les dents de *Typhlonectes compressicaudus* (Amphibia: gymnophiona) au cours du développement. Ann. Sci. Nat. Zool. Paris 17, 11–23.

Ivanovic, A., Aljancic, G., Arntzen, J.W., 2013. Skull shape differentiation of black and white olms (*Proteus anguinus anguinus* and *Proteus a. parkelj*): an exploratory analysis with micro-CT scanning. Contr. Zool. 82, 107–114.

Kamei, R.G., San Mauro, D., Gower, D.J., van Bocxlaer, I., Sherratt, E., Thomas, A., Babu, S., Bossuyt, F., Wilkinson, M., Biju, S.D., 2012. Discovery of a new family of amphibians from Northeast India with ancient links to Africa. Proc. R. Soc. Lond. B279, 2396–2399.

Kerr, T., 1960. Development and structure of some actinopterygian and urodele teeth. Proc. Zool. Soc. Lond. 133, 401–424.

Khan, M.S., Mufti, S.A., 1994. Oral disc morphology of amphibian tadpole and its functional correlates. Pak. J. Zool. 26, 25–30.

Kupfer, A., Müller, H., Antoniazzi, M.M., Jared, C., Greven, H., Nussbaum, R.A., Wilkinson, M., 2006. Parental investment by skin feeding in caecilian amphibian. Nat. Lond. 440, 926–929.

Lalremsanga, H.T., Hooroo, R.N.K., Lalronunga, S., 2013. Oral morphology of the tadpoles of *Euphlyctis cyanophlyctis* (Schneider, 1799) with notes on feeding behaviour. Sci. Technol. J. 1, 58–69.

Maciel, A.O., Hoogmoed, M.S., 2009. Taxonomy and distribution of gymnophiona of Brazilian Amazonia with a key to their identification. Zootaxa 2984, 1–53.

Mendelson, J.R., Pramuk, J.B., 1998. Neopalatine odontoids in *Bufo alvarius* (Anura: Bufonidae). J. Herpetol. 32, 586–588.

Monroy, J.A., Nishikawa, K.C., 2011. Prey capture in frogs: alternative strategies, biomechanical trade-offs and hierarchical decision making. J. Exp. Zool. 315, 61–71.

Nishikawa, K.C., Gans, C., 1996. Mechanism of tongue protrusion and narial closure in the marine toad *Bufo marinus*. J. Exp. Biol. 199, 2511–2529.

Nussbaum, R.A., 1976. Geographic variation and systematic of salamanders of the genus *Dicamptodon* Strauch (Ambystomatidae). Misc. Publ. Mus. Zool. Univ. Mich. 149, 1–104.

Nussbaum, R.A., 1985. Systematics of Caecilians (Amphibia: Gymnophonia) of the Family Scolecomorphida. Occasional Papers of the Museum of Zoology. University of Michigan Number 713.

Parker, H.W., 1956. Viviparous caecilians and amphibian phylogeny. Nature 178, 250–253.

Parker, H.W., Dunn, E.R., 1964. Dentitional metamorphosis in the amphibia. Copeia 75–86.

Parsons, T.S., Williams, E.E., 1962. The teeth of amphibia and their relation to amphibian phylogeny. J. Morphol. 110, 375–389.

Pederson, S.C., 1991. Dental morphology of the cannibal morph in the tiger salamander, *Ambystoma tigrinum*. Amphibia-Reptilia 12, 1–14.

Pérez, O.D., Lai, N.B., Buckley, D., del Pino, E.M., Wake, M.H., 2009. The morphology of prehatching embryos of *Caecilia orientalis* (Amphibia: gymnophiona: Caeciliidae). J. Morphol. 270, 1492–1502.

Rose, F.L., 1968. Ontogenetic changes in the tooth number of *Amphiuma tridactylum*. Herpetologica 29, 182–183.

Rovito, S.M., Parra-Olea, G., Hanken, J., Bonett, R.M., Wake, D.B., 2013. Adaptive radiation in miniature: the minute salamanders of the Mexican highlands (Amphibia: Plethodontidae: *Thorius*). Biol. J. Linn. Soc. 109, 622–643.

Rowley, J.J.L., Le, T.T.D., Tran, T.A.D., Stuart, B.L., Hoang, D.L., 2010. A new tree frog of the genus *Rhacophorus* (Anura: Rhacophoridae) from southern Vietnam. Zootaxa 272745–272755. See also: http://australianmuseum.net.au/media/vampire-flying-frog-discovery#sthash.f9REjzQT.dpuf.

Setiadi, M.I., McGuire, J.A., Brown, R.M., Zubairi, M., Iskandar, D.T., Andayani, N., Supriatna, J., Evans, B.J., 2011. Adaptive radiation and ecological opportunity in Sulawesi and Philippine fanged frog (*Limnonectes*) communities. Am. Nat. 178, 221–240.

Shaw, J.P., 1979. The time scale of tooth development and replacement in *Xenopus laevis* (Daudin). J. Anat. 129, 323–342.

Shaw, J.P., 1989. Observations on the polyphyodont dentition of *Hemiphractus proboscideus* (Anura: Hylidae). J. Zool. Lond. 217, 499–510.

Shaw, J.P., Ellis, S.A., 1989. A scanning electron microscopic study of the odontoids and teeth in *Hemiphractus proboscideus* (Anura: Hylidae). J. Zool. Lond. 219, 533–543.

Signoret, J., 1960. Céphalogénèse chez le triton *Pleurodeles waltlii* Michah, après traitement de la gastrula par le chlorure de lithium. Mém. Soc. Zool. 32, 1–117.

Stewart, M.M., 1958. Seasonal variation in the teeth of the two-lined salamander. Copeia 190–196.

Taylor, E.H., 1968. The Caecilians of the World: A Taxonomic Review. University of Kansas Press, Lawrence, KS.

Taylor, E.H., 1976. A Caecilian Miscellany, vol. 50. Science Bulletin University Kansas, pp. 187–231.

Taylor, E.H., 1977. The Comparative Anatomy of Caecilian Mandibles and Their Teeth, vol. 51. Science Bulletin, University of Kansas, pp. 261–282.

Tominaga, A., Matsui, M., Nishikawa, K., Tanabe, S., Sato, S., 2005. Morphological discrimination of two genetic groups of a Japanese salamander, *Hynobius naevius* (Amphibia Caudata). Zool. Sci. 22, 1229–1244.

Trueb, L., 1973. Bones, frogs and evolution. In: Vial, J.L. (Ed.), Evolutionary Biology. University of Missouri Press, Columbia, pp. 65–132.

Vassilieva, Serbinova, 2013. Bony skeleton in the Caucasian salamander *Mertensiella caucasica* (Urodela: Salamndridae) and embryonization effect. Russ. J. Herpetol. 20, 85–96.

Vassilieva, A.B., Smirnov, S.V., 2001. Development and morphology of the dentition in the Asian salamander, *Ranodon sibiricus* (Urodela: Hynobiidae). Russ. J. Herpetol. 8, 105–116.

Vassilieva, A.B., Galoyan, E.A., Poyarkov, N.A., 2013a. *Rhacophorus vampyrus* (Anura: Rhacophoridae) reproductive Biology: a new type of oophagous tadpole in Asian treefrogs. J. Herpetol. 47, 607–614.

Vassilieva, A.B., Poyakov, N.A., Iizuka, K., 2013b. Pecularities of bony skeleton development in Asian clawed salamanders (*Onychodactylus*; Hypnobiidae). Biol. Bull. 40, 1–11.

Venesky, M.D., Wasserug, R.J., Parris, M.J., 2010. The impact of variation in labial tooth number on the feeding kinematics of tadpoles of the Southern leopard frog (*Lithobates sphenocephalus*). Copeia 481–486.

Vera Candioti, M.F., Altig, R., 2010. A survey of shape variation in keratinized labial teeth of anuran larvae as related to phylogeny and ecology. Biol. J. Linn. Soc. 101, 609–625.

Wake, M.H., 1976. The development and replacement of teeth in viviparous caecilians. J. Morphol. 148, 33–64.

Wake, M.H., 1977a. The reproductive biology of caecilians: an evolutionary perspective. In: Guttman, S., Taylor, D. (Eds.), Reproductive Biology of the Amphibians. Plenum Press, New York, pp. 73–102.

Wake, M.H., 1977b. Fetal maintenanance and its evolutionary significance in the Amphibia-Gymnophiona. J. Herpetol. 11, 379–386.

Wake, M.H., 1980. Fetal tooth development and adult replacement in *Dermophis mexicanus* (Amphibia: gymnophiona): fields versus clones. J. Morphol. 166, 203–216.

Wake, M.H., 1984. A new caecilian from Peru (Amphibia: Gymnophonia). Bonn. Zool. Beit. 35, 1–3.

Wake, M.H., Dickie, R., 1998. Oviduct structure and function and reproduction modes in amphibians. J. Exp. Zool. 282, 477–506.

Wake, M.H., Wurst, G.Z., 1979. Tooth crown morphology in caecilians (Amphibia: Gymnophonia). J. Morphol. 159, 331–342.

Wasserug, R.J., 1976. Oral morphology of anuran larvae: terminology and general description. Occ. Pap. Mus. Nat. Hist. Univ. Kans. 48, 1–23.

Wasserug, R.J., Yamashita, M., 2001. Plasticity and constraints on feeding kinematics in anuran larvae. Comp. Biochem. Physiol. A 131, 183–195.

Wiens, J.J., 2011. Re-evolution of lost mandibular teeth in frogs after more than 200 million years, and re-evaluating Dollo's Law. Evolution 65, 1283–1296.

Wilkinson, M., 1991. Adult tooth crown morphology in the Typhlonectidae (Amphibia: gymnophiona) – a reinterpretation of variation and its significance. J. Zool. Syst. Evol. Res. 29, 304–311.

Wilkinson, M., Nussbaum, R.A., 1997. Comparative morphology and evolution of the lungless caecilian *Atretochoana eiselti* (Taylor) (Amphibia: gymnophiona: Typhlonectidae). Biol. J. Linnn Soc. 62, 39–109.

Wilkinson, M., Kupfer, A., Marque-Porto, R., Jeffkins, H., Antoniazzi, M.M., Jared, C., 2008. One hundred million years of skin shedding. Biol. Lett. 4, 358–361.

Wilkinson, M., San Mauro, D., Sherratt, E., Gower, D.J., 2011. A nine-family classification of caecilians (Amphibia: gymnophiona). Zootaxa 2874, 41–64.

Wilkinson, M., Sherratt, E., Starace, F., Gower, D.J., 2013. A new species of skin-feeding caecilian and the first report of reproductive mode in *Microcaecilia* (Amphibia: gymnophiona: Siphonopidae). PLoS One 8, e57756.

Wistuba, J., Greven, H., Clemen, G., 2002. Development of larval and transformed teeth in *Ambystoma mexicanum* (Urodela, Amphibia): an ultrastructural study. Tiss. Cell 34, 14–27.

FURTHER READING

De Vree, F., Gans, C., 1994. Feeding in tetrapods. In: Bels, V.L., Chardon, M., Vandewalle, P. (Eds.), Biomechanics of Feeding in Vertebrates, Advances in Comparative and Environmental Physiology 18. Springer, Berlin, pp. 93–118.

Smirnov, S.V., Vassilieva, A.B., 1995. Anuran dentition: development and evolution. Russ. J. Herpetol. 2, 120–128.

Chapter 6

Reptiles 1: Tuatara and Lizards

REPTILES: GENERAL

Reptiles are cold blooded (ectothermic), have a dry scaly skin and, apart from snakes and limb-reduced lizards, have four limbs ending in claws. Fertilization occurs internally. The eggs differ from those of amphibians in having four extraembryonic membranes and amniotic fluid surrounded by a shell to protect the egg from desiccation when laid on land. Some lizards, eg, the viviparous lizard *Zootoca* (*Lacerta*) *vivipara*, give birth to live young. There is therefore no water-dependent larval stage and a small but fully formed animal emerges from the large, yolky, egg.

Living reptiles are divided into three main groups, as in this classification by Benton (2015):

Infraclass Neodiapsida
Order Testudinata (turtles, tortoises),
Infraclass Lepidosauromorpha
Order Rhynchocephalia (tuatara): 1 species
Order Squamata (lizards and snakes): 8 infraorders
Infraclass Archosauromorpha
Order Crocodylia (crocodiles, alligators, gharials): 3 infraorders

Extant Testudinata (or chelonians) lack teeth; instead, their jaws are sheathed by hard, keratinized beaks. The form of the beak shows adaptation for different functions. For example, the beaks of herbivorous species have serrated edges suitable for cutting tough plant material, whereas those of carnivorous forms have knife-sharp ridges used for slicing through crabs and mollusks. However, these reptiles will not be considered further in this book.

The lepidosauromorphs are reptiles with overlapping scales. The Rhynchocephalia shared a common ancestor with Squamata about 250 million years ago and seem to have been initially the more diverse of the two groups, but they were eventually replaced by lizards across most of the planet and are now reduced to a single living member, the New Zealand **tuatara** (*Sphenodon punctatus*). The snakes split from the lizards during the Jurassic period. Of the 10,000 species of living reptiles, about 6000 are lizards and about 3500 are snakes. The Squamata is not only the largest order of reptiles, but also the third largest order of vertebrates, after birds and perciform bony fishes. A major phylogeny was published by Pyron et al. (2013).

The crocodylians and birds are the only surviving members of the Archosauromorpha, which separated from the Lepidosauromorpha in the Late Permian. The archosaurs also included the dinosaurs, represented today by the birds. The extant crocodylians possess several features that distinguish them from the Lepidosauromorpha, including a four-chambered heart and a type of diaphragm separating the thorax from the abdomen. With regard to the mouth, crocodylians, like mammals, possess a secondary palate that separates the breathing and feeding pathways, whereas in all other reptiles the nares open directly into the oral cavity, as they do in amphibians. The mode of tooth attachment in crocodiles is also different from that in all other reptiles.

In this chapter, we discuss the tuatara and the lizards, whereas in the following chapters we describe the dentitions of snakes and crocodylians. First, however, it is appropriate to describe some general aspects of the reptilian skull and dentition.

REPTILIAN SKULL

Among living reptiles, there is considerable variation in the skull, especially in the posterior region. The quadrate loses its connection with the hyomandibula and the latter becomes incorporated into the evolving auditory system as an ear ossicle, the **columella** (homologous with the **stapes** of mammals). In the earliest reptiles, the temporal region was covered by a complete shield of dermal bones: the **anaspid** condition. In the living crocodylians and squamates, the skull has one or two openings known as **temporal fenestrae** in the postorbital part of the skull. Only the tuatara has two complete fenestrae (**diapsid** condition). Among crocodylians, especially the caimans, the upper fenestra tends to be reduced and covered over. The skulls of chelonians, crocodylians, and the tuatara are solid, rigid structures, but among lizards and snakes there is considerable reduction of the dermal skull bones in the postorbital region.

The Teeth of Non-Mammalian Vertebrates. http://dx.doi.org/10.1016/B978-0-12-802850-6.00006-0

A feature of the squamate skull that has attracted much attention is **cranial kinesis**: movement at the joints between different elements of the skull (Bellairs, 1969; Metzger, 2002; Evans, 2008). In both lizards and snakes, the pterygoids can slide anteroposteriorly against the sides of the braincase. A widespread form of kinesis, also found among both lizards and snakes, is **streptostyly**, in which the quadrate is able to rotate at its articulation with the squamosal, so that the distal portion can swing backward and forward. This capacity stems from the reduction of the contacts between the quadrate and squamosal dorsally, and the quadrate and pterygoid ventrally. In the lizard skull, movement may also be possible at certain joints between elements of the cranium. Among snakes, the core of the cranium is consolidated, so that these intracranial movements are not possible. However, there is often considerable mobility of the palatal bones, suspensorium, snout (prokinesis), and elements of the upper and lower jaws. The details of cranial kinesis, and its possible roles in feeding, are discussed later in relation to lizards (see pages 161–162) and snakes (see page 202).

The lower jaw remains a composite structure, with the jaw joint formed between the articular and the quadrate. In the tuatara and some lizards, but above all in snakes, the mandibular symphysis is flexible and allows some independent movement of the two rami. The jaw joint of the tuatara and the agamids allows some anteroposterior (propalinal) movement of the lower jaw relative to the upper jaw but in nearly all reptiles, as in other non-mammal vertebrates, the jaw joint allows only a simple hinge action. Lateral movement of the upper and lower jaws past each other, as occurs during chewing in many mammals, does not occur among living reptiles.

The premaxillae of squamates form as separate centers of ossification and fuse to form a single bone (premaxilla or intermaxillary) (Evans, 2008). Tooth numbers given in this chapter therefore relate to the unified premaxilla.

TEETH OF REPTILES

The teeth form single rows on the premaxillae, maxillae, and dentaries. Crocodylians lack teeth on other bones. A very detailed account of the distribution of teeth in the palate of squamates was published by Mahler and Kearney (2006). Many lizards have teeth on the pterygoids. Teeth on the palatines are much rarer and generally, although not always, coexist with pterygoid teeth. Mahler and Kearney observed teeth on the vomers of only one species (*Ophisaurus apoda*). The tuatara has teeth on the palatines and occasionally in the vomers, whereas snakes have teeth on both palatines and pterygoids.

The teeth of Lepidosauromorpha are **ankylosed**, in either a **pleurodont** position (on the inner slope of the bone) or **acrodont** position (on the crest of the bone). Those of Crocodylia are **thecodont**, ie, embedded in sockets in the jaw bones, to which they are attached by a periodontal ligament.

The teeth of most reptiles are continuously replaced throughout life. This allows for growth, replacement of damaged teeth, and a change in tooth morphology associated with a change in diet. In certain reptiles, especially those with acrodont dentitions, tooth replacement may be partially suppressed (eg, Agamidae) or completely suppressed (eg, Chamaeleonidae). In most squamates only one or two developing replacement teeth are associated with each functional tooth, but more may be associated with the poison fangs of snakes and the teeth of the **Komodo dragon** (*Varanus komodoensis*). Tooth replacement in reptiles in considered in detail in Chapter 10.

Among squamates, there are two main modes of replacement of pleurodont teeth. Most species follow the **iguanid** mode, in which the replacement tooth is initiated at the base of its predecessor, on its inner (lingual or palatal) aspect (eg, Figs. 6.16, 6.33A, 6.40, and 6.81B). As it grows, the replacement tooth moves into the pulp cavity of the functional tooth through a resorption pit, the size of which is proportional to that of the replacement tooth. As pleurodont teeth are attached over a large area of bone, most of the functional tooth and its bone of attachment can be resorbed without weakening its attachment, until eventually only its tip is left. Once this tip finally shed the erupting replacement tooth can move into position. An alternative—**varanid**—mode of tooth replacement is typically found in varanid and helodermatid lizards (Rieppel, 1978). This mode of replacement is associated with wide spacing of the teeth, because the replacement tooth is not initiated on the inner aspect of its predecessor, but immediately distal to it. Resorption is local and the replacing tooth does not cause a resorption pit. The functional tooth may be lost at a later stage and replacement may be more rapid than in the iguanid mode. The replacement tooth erupts into the space from behind. In some lizards, eg, anguids, the mode of replacement is intermediate between the iguanid and varanid modes (Rieppel, 1978). In crocodylians, replacement teeth develop within the same socket as the functional teeth. This contrasts with the socketed teeth of mammals, where the replacement teeth form in separate crypts.

EGG TEETH

One of the adaptations that enabled reptiles to live on land without the need to return to water to breed was the evolution of eggs with impermeable, hard, or leathery shells that protected their developing young from desiccation. To hatch, the young must be able to pierce the tough shell, and this is accomplished with the aid of a specialized structure at the front of the beak or upper jaw, which is lost soon after hatching. In turtles and crocodiles (as well as all birds) there is a horny, tough, thickening of skin at the tip of the snout known as the **caruncle**. However, squamate reptiles develop an extraoral **egg tooth** in this position. This is a true tooth, with a core of dentine covered by a thin enamel layer, attached to the premaxilla (Fig. 6.1). In snakes, apart from pythons and primitive forms such as *Ilysia* and *Xenopeltis*, the egg tooth is usually the only tooth to develop on the premaxilla. Interestingly, monotreme mammals have both a caruncle and an egg tooth.

Lizards

Gekkota and Dibamidae are unique in having, at the time of hatching, two blunt egg teeth (Ananieva and Orlov, 2013) (Figs. 6.2 and 6.3A and B). In other lizards, although two egg teeth initially develop, one on each side, only the right one enlarges and migrates to the midline (Ananieva and Orlov, 2013; Berkovitz, 2013) (Figs. 6.3C and D, 6.4 and 6.5). It is

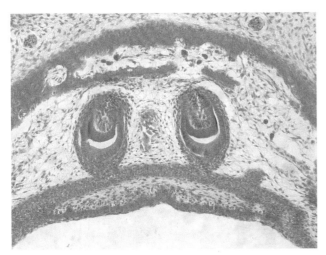

FIGURE 6.1 Demineralized section showing developing egg teeth on either side of the midline in a gecko (*Paroedura picta*) Hematoxylin and eosin. Image width = 1.8 mm. *Courtesy Dr. O. Zahradníček.*

A newly-hatched gecko shows the tiny twin "egg teeth" which are used to slit open the egg.

FIGURE 6.2 Two egg teeth on snout of hatchling gecko. *Courtesy of Jennifer Boeke. jbscresties.com*

prominent and larger than the oral teeth by the time of hatching and is lost within 2–3 days of hatching. The egg tooth in squamates is curved, with sharp margins, and projects forward from the premaxilla. It develops early in oviparous lizards at the front of the premaxilla (eg, *Lacerta*). In viviparous lizards, eg, the **slow worm** (*Anguis*), an egg tooth is still present but it is smaller and lost 5–7 days after birth (Ananieva and Orlov, 2013).

Snakes

In the majority of snakes, the egg tooth occupies a median position throughout its development, with no evidence that it originated on one side or the other (Smith et al., 1953). Additional rudimentary toothlets develop in relation to each premaxilla, but they are usually resorbed before birth; these toothlets perhaps represent teeth once present in the premaxillae (A.S. Tucker, personal communication). However, in snakes such as pythons (*Python* spp.) and **corn snake** (*Elaphe gluttata*), two closely positioned tooth germs start to develop precociously in the midline, close to the oral surface. During development they fuse to form a single egg tooth in the midline by the time of hatching (Fig. 6.6). Some egg teeth in snakes may show macroscopic signs of this fusion.

DIET AND FEEDING

Because juvenile reptiles must be able to feed for themselves, erupted teeth are present when they are born or emerge from the egg. Snakes, crocodylians, and most lizards are carnivorous. However, many lizards are omnivorous and a small proportion are even herbivorous. The teeth in most carnivorous species are adapted for grasping and for intraoral transport of prey, and they are conical, pointed, and often recurved. The fangs of snakes can be very long and in venomous forms are often modified so that venom is transported from the poison gland, through the fang to the tip, where it is injected into the prey. Most reptile dentitions are homodont, but some show variations in tooth form along the tooth row. For example, some carnivorous reptiles specialize in eating mollusks or crustaceans and their posterior teeth are often low and rounded with flattened crushing surfaces. The teeth of many lizards do not have the simple conical form, but they may be compressed, sometimes with mesial and distal cutting edges, and may bear cusps or serrations. Such teeth are often associated with feeding on plant material. Some of these variations are illustrated in Fig. 6.7 (Zahradníček et al., 2014).

The crocodylians and agamid lizards have blunt, muscular tongues with restricted mobility that play little part in processing food. Most lizards have tongues that are more or less protrusible and some, notably iguanians, can extend the tongue to capture small prey by adhesion to the tip, which is made sticky

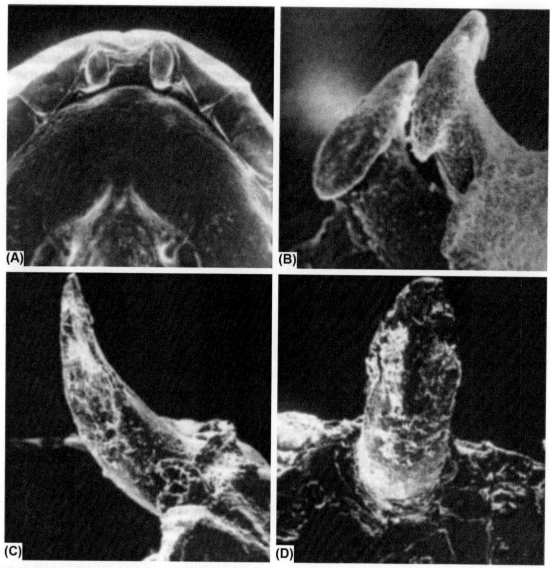

FIGURE 6.3 A,B: Egg teeth of the wonder gecko (*Teratoscincus scincus*). (A) Ventral view ×30. (B) Lateral view ×50. (C, D) Egg teeth of the agamid lizard *Physignathus cocincinus*. (C) Lateral view ×200. (D) Ventral view ×200. *From Ananieva, N.B., Orlov, N.L., 2013. Egg teeth of squamate reptiles and their phylogenetic significance. Biol. Bull. 40, 600–605. Courtesy Editors Biology Bulletin.*

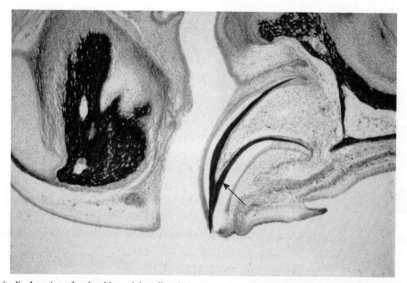

FIGURE 6.4 Median longitudinal section of prehatching rainbow lizard (*Agama agama*). Note the dentine core of the egg tooth (*arrow*). Demineralized section stained with Masson trichrome. Image width=1.25 mm. *Courtesy Dr. J.S. Cooper. From Berkovitz, B.K.B., 2013. Nothing but the Tooth, Elsevier, London.*

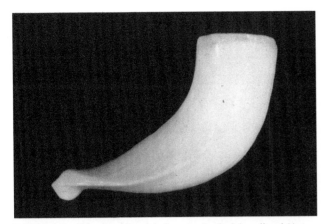

FIGURE 6.5 Side view of model of egg tooth of rainbow lizard (*Agama agama*). Note the curvature and the sharp lateral ridge. Image width = 1 mm. *Courtesy Dr. J.S. Cooper. From Berkovitz, B.K.B., 2013. Nothing but the Tooth, Elsevier, London.*

FIGURE 6.6 Egg tooth in midline of upper jaw of newly hatched python. *Courtesy Wikipedia.*

by mucous secretions. This method of capturing prey is developed to an extreme in chameleons, in which the tongue can be projected a distance equal to the length of the animal. Many lizards and all snakes use their tongue as an adjunct to the sense of olfaction. By protruding the tongue, the tip can pick up particles from the air, and these particles can be transferred to the **vomeronasal** or **Jacobson's organ** to provide chemosensory information. This activity is most highly developed in snakes and varanid lizards, whose long, forked tongues can be protruded well in advance of the mouth and retracted into a sheath.

RHYNCHOCEPHALIA

A diversity of tooth shapes and tooth arrangement in fossil rhynchocephalians suggests that they had a range of diets, including herbivory. Today, the lone survivor is carnivorous and feeds on a wide variety of prey, including beetles, wetas, spiders, snails, lizards, and young seabirds.

The acrodont teeth of *Sphenodon* are firmly fixed to the jaw bones so that there is no clear boundary between the teeth and the jaw bone. There is a second row of upper teeth on the palatine bone, parallel to the row on the maxilla. The teeth on the lower jaw bite between the two upper rows of teeth (Robinson, 1976; Jones et al., 2012).

Four stages of dental development can be differentiated (Harrison, 1901a,b; Jones et al., 2012):

1. a series of embryonic teeth that are resorbed without ever functioning: a phenomenon seen in many reptile species;
2. a hatchling dentition comprising three premaxillary teeth, approximately 10 maxillary teeth (alternately larger and smaller) and approximately 16 teeth on each dentary;
3. a series of additional teeth that are added distal to the maxillary, palatine and dentary tooth rows;
4. one or more larger, successional teeth that appear after hatching and replace the anterior part of the hatchling dentition.

The dentition of an adult tuatara contains members of the second, third, and fourth series. Interpretation of the dentition may be difficult because of the acrodont ankylosis and because hatchling teeth retained in the adult may show considerable wear (Robinson, 1976; Jones et al., 2012).

In adults, each beak-like premaxilla bears a large chisel-shaped tooth that succeeds three small hatchling teeth. This tooth initially bears two cusps, but they are worn down until the tooth resembles a mammalian incisor (Figs. 6.8–6.10).

The maxilla bears a row of 15–20 mainly conical teeth of varying size (Figs. 6.8–6.10). The anterior 2–4 teeth are moderately enlarged. These teeth are followed by a large successional caniniform tooth. There then follows the remains of a few small teeth retained from the hatchling stage. Finally, there is an additional series of larger teeth that are added on sequentially at the back of the tooth row. Commencing a little behind the maxillary caniniform tooth and lying parallel and medial to the maxillary teeth is a shorter row of about 12 teeth on each palatine (Figs. 6.9 and 6.10), the first of which is enlarged and caniniform. Vestigial teeth may also be present on each vomer.

Each dentary bears a single row of 17–20 single-cusped teeth. One of the anterior teeth is enlarged and caniniform (Figs. 6.8–6.10) and fits into a notch between the premaxilla and the maxilla. Immediately behind this tooth are some small, retained hatchling teeth. Behind

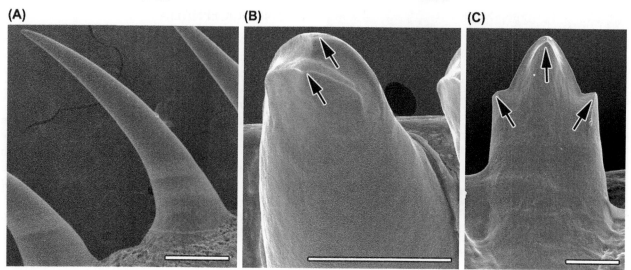

FIGURE 6.7 Tooth crown variety in reptiles (scanning electron micrographs of adult teeth). (A) Pointed conical tooth of a Burmese python (*Python molurus*). (B) Teeth with two crests in a pictus gecko (*Poroedura pictus*). (C) Tricuspid tooth in an Allison's (blue-headed) anole (*Anolis allisoni*). Scale (A)=0.5mm. Scale (B and C)=100μm. *From Zahradníček, O., Buchtová, M., Dosedelová, H., Tucker, A.S., 2014. The development of complex tooth shape in reptiles. Front. Physiol. 5, 1–7.*

FIGURE 6.8 Lateral view of dentition of tuatara (*Sphenodon punctatus*). Image width=6cm. *Courtesy MoLSKCL. Catalog no. X12.*

these hatchling teeth, the teeth are larger and are added on one at a time to extend the tooth row posteriorly. Only the larger, anterior teeth are replaced, but not the posterior teeth.

Small prey can be captured by flicking out the large sticky tongue. Larger prey may be immobilized by head movements and the use of the large anterior teeth. When the lower jaw is raised, the teeth of the dentary pass into the space between the maxillary and palatal rows of teeth in the upper jaw. This three-point contact generates a shearing

force that efficiently breaks down the exoskeleton of invertebrate prey. The shearing action is further increased by propalinal (back and forth) movement of the lower jaw by about one tooth position: a motion that depends on mobility at the mandibular symphysis (Gorniak et al., 1982; Jones et al., 2012). Clearly, such a system of breaking up food would be rendered inefficient if there was continual replacement of the teeth.

Because of tooth wear and the lack of tooth replacement, the smaller hatchling teeth retained in older specimens may contain little true dental tissue, but they are composed mainly of bone that is confluent at the tooth bases. Thus, the diet of older animals exhibiting considerable wear may include softer prey items, such as worms and slugs, that are easier to break down.

LIZARDS

There are more than 6000 species of lizards, belonging to five major groups (Pyron et al., 2013): Gekkota, Scincoidea, Lacertoidea (including Amphisbaenia), Anguimorpha, and Iguania. In the phylogeny of Pyron et al. (2013), the family Dibamidae, which contains two genera of insectivorous, limbless, burrowing lizards, is placed as the sister group to all other squamates.

Lizards do not have a fixed dental formula and the number of teeth usually increases with age (eg, Montanucci, 1968). Many lizards undergo a period of hibernation during which growth and tooth replacement may cease. A midline tooth is often present on the premaxilla and replaces the egg tooth.

The teeth of most lizards have sharp points and are usually recurved, but may be compressed laterally, possess

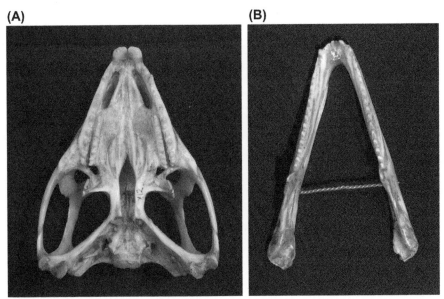

FIGURE 6.9   Dentition of tuatara (*Sphenodon punctatus*). (A) Ventral view of upper dentition, showing two rows of teeth on the maxillae and palatine bones. Image width=7cm. (B) Occlusal view of lower dentition. Image width=5cm. *Courtesy MoLSKCL. Catalog no. X12.*

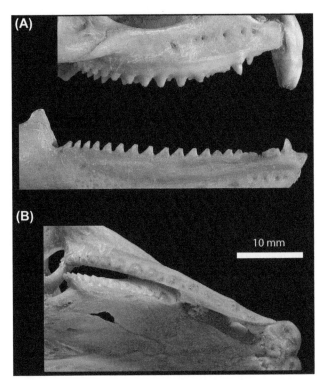

FIGURE 6.10   Detailed views of dentition of the tuatara (*Sphenodon punctatus*). (A) Lateral view. (B) Ventral view of the right half of the upper dentition. *Courtesy P. Stokes, South Australian Museum, for photography, Dr. M. N. Hutchinson for access to specimens and Dr. MEH Jones.*

cutting edges, and have tricuspid or serrated crowns (see Fig. 6.7). In many species, the dentition is homodont, but heterodonty occurs in several species. The variation along the tooth row could be reflected as a difference in size; a simple change from pointed to multicusped or serrated

crowns; but may, as in molluscivorous species, involve a transition between grasping anterior teeth and crushing posterior teeth. A change in tooth shape with age may be correlated with a shift in diet during growth.

Most lizards feed on invertebrates. About 90% of species probably include less than 10% of plant material in their diet (Cooper and Vitt, 2002) and, alternatively, only a few taxa, such as the monitors, are carnivores taking prey similar in size to themselves. The main food source consists of insects and other arthropods. Lizards use two main methods of capturing prey, using their tongue or their jaws (Table 6.1) (Metzger, 2002). The Iguania capture prey by adhesion to the sticky mucus coating the tip of their protrusible fleshy tongues, often after the prey has been immobilized by pressing it against the substrate by the tongue. This mode of prey capture is typical of ambush or "sit-and-wait" predators and is most highly developed in the Chamaeleonidae. More active, "wide-foraging" lizards in other lizard families use their jaws to seize prey. Skinks and girdled lizards use both methods of capturing prey.

Once captured, prey may be transported to the esophagus and then swallowed whole. When the tongue is used to capture prey, it is usually passed directly back to the esophagus as the tongue retracts. Most lizards use the tongue and hyoid apparatus to move prey back through the mouth (**hyolingual transport**) (Table 6.1) (Bels et al., 1994). Other species, especially those processing larger prey morsels, eg, monitor lizards, use **inertial feeding**, in which the food object is moved backward by first releasing it and then thrusting the head forward and seizing the object again in a position nearer the pharynx (De Vree and Gans, 1994). For larger prey, some reduction may be

TABLE 6.1 Distribution of Cranial Kinesis, Prey Capture Method, and Mode of Intraoral Transport Among Squamata

Infraorder	Family	Streptostyly	Mesokinesis	Prehension Mechanism	Intraoral Transport
Rhynchocephalia					
	Sphenodontidae	No	No	Tongue	Hyolingual
Squamata					
Iguania	Iguanidae	Variable	No	Tongue	Hyolingual
	Agamidae	Variable	No	Tongue	Hyolingual
	Chamaeleonidae	?	No	Tongue	Hyolingual
Gekkota	Gekkonidae	Yes	Yes	Jaws	Hyolingual
	Pygopodidae	Variable	?/Yes	Jaws	Hyolingual
Scincoidea	Scincidae	?	?/No	Tongue+jaws	Hyolingual
	Cordylidae	?	?	Tongue+jaws	Hyolingual
	Xantusiidae	?	?	Jaws	?
Lacertoidea	Lacertidae	?	?	Jaws	Hyolingual
	Teiidae	?/Yes	?	Jaws	Hyolingual+inertial
Anguimorpha	Anguidae	?/Yes	?/Yes	Jaws	Hyolingual
	Xenosauridae	?	?	Jaws	?
	Helodermatidae	No	No	Jaws	Hyolingual
	Lanthanotidae	?	?	?	?
	Varanidae	Yes	Yes	Jaws	Inertial
Serpentes		Yes	No	Jaws	Other

Modified from Metzger, K., 2002. Cranial kinesis in Lepidosauria: skulls in motion. In Aerts, P., D'Août, K., Herrel, A., van Damme, R. (Eds.), Topics in Functional and Ecological Vertebrate Morphology, Shaker Publishing, Aachen, pp. 15–46.

necessary. Prey may be dismembered outside the mouth, eg, by being torn while wedged in crevices, by being shaken vigorously, or by being dragged against the substrate, and the fragments are then ingested (De Vree and Gans, 1994). Alternatively, prey may be repeatedly bitten within the mouth during transit toward the esophagus (Bels et al., 1994). Such intraoral processing is essential in lizards such as *Dracaena* that feed on snails and similar hard-shelled prey so that they can separate the soft body from the protective shell. Reilly et al. (2001) described a process termed **palatal crushing** in which the prey is pressed against the palate by the tongue. This activity seems to play a more important part in processing of prey among Iguania, many of which possess teeth on the pterygoids (Montanucci, 1968; Mahler and Kearney, 2006), than in other lizards.

Several families of lizards have omnivorous members that include a sizable proportion of plant material in their diet. The families with the highest proportion (>30%) of omnivorous species are among the Iguania, the Iguanidae, Corytophanidae, Agamidae, and Tropiduridae and among the Scincoidea,

the Gerrhosauridae and Xantusiidae. The Lacertidae and Scincidae also have numerous omnivorous members (Cooper and Vitt, 2002). Herbivorous lizards (diet including >90% plant material) probably account for only about 1% of lizard species. The family with the largest number of herbivores is the Iguanidae, but herbivores are also found among the Agamidae (*Uromastyx* spp.), Tropiduridae, Teiidae, Scincidae, Gerrhosauridae, and Cordylidae (Cooper and Vitt, 2002).

The plant material consumed by omnivorous lizards tends to be fruits, shoots, buds, and flowers, ie, those parts that can be gathered comparatively easily, by plucking, and that have a relatively high nutrient content. Because these parts tend to be seasonal, it is likely that the amount of plant material taken by omnivores varies through the year. The diet of herbivores contains higher proportions of leaves, the plant food source that is most abundant and also available throughout the year. Grass, the staple food of many large mammalian herbivores, but which is also very difficult to break down because of the high content of cellulose fibers and the presence of phytoliths, does not figure in the diet of herbivorous lizards.

Several hypotheses have been proposed to account for the adoption of herbivory among lizards (Sokol, 1967; Pough, 1973; Cooper and Vitt, 2002). Limited availability of animal prey could be a factor and could be manifested in insular or arid habitats. However, although there is a correlation of herbivory with insularity, the importance of prey availability is not clear. Herbivorous lizards tend to be larger than carnivorous lizards. Pough (1973) found that in the Iguanidae, Agamidae, Scincidae, Gerrhosauridae, and Cordylidae, lizards weighing 50–100 g are mostly carnivorous, whereas lizards larger than 300 g are mostly herbivorous. The tendency for lizards to consume more plant material as body size increases is probably explained by the increasing energy requirements of the animal, combined with increasing difficulty of capturing prey and hence a greater energetic cost of predation (Pough, 1973). It is thus energetically more efficient for large lizards to browse on plants, even though, on a weight basis, the energy extracted from plant material is about one-half that from animal material. Large carnivorous lizards do exist in families, eg, Varanidae, which can increase the aerobic metabolic rate more than most lizards can.

In most lizards the jaw joint acts as a simple hinge, so the lower tooth row sweeps past the upper tooth row as the jaw closes on the food. Among carnivorous lizards, the main actions of the teeth on the prey are puncturing, as the sharp teeth penetrate the cuticle, and tearing, by shear forces generated by the relative movement of the two tooth rows. Among mammalian herbivores, occlusion between the opposing tooth rows, combined with lateral movement which grinds the food, is considered to be essential for breaking down plant material. Lizard dentitions provide neither of these actions, although the lower teeth of *Uromastyx* can slide back and forth to some extent (Reilly et al., 2001). Instead, herbivorous lizards process plant material in essentially the same way as carnivorous species break up their prey. However, the efficiency of the dentition is enhanced by modification of tooth shape. Rather than being conical, the teeth of herbivores tend to be compressed laterally and to possess mesial and distal cutting edges, which may be multicusped or serrated. These modifications convert the rows of teeth into serrated edges that improve the chopping action of the dentition, although the efficiency of food reduction is probably much lower than in mammalian omnivores or herbivores.

The operation of the jaws and dentition during feeding could potentially be modified by cranial kinesis. **Streptostyly** is present in many lizards; ie, the quadrate is capable of rotation about a transverse axis at the joint with the squamosal, and in some species lateral rotation may also be possible. In addition, lizard skulls may show movement between various skull bones. **Mesokinesis** is a hinge-like rotation about the transverse frontal-parietal suture that can result in the raising or lowering of the snout. Mesokinesis

may be coupled with streptostyly by way of the pterygoids through a pterygoid/quadrate articulation: hence, streptostylic rotation of the quadrate can cause protraction or retraction of the pterygoids and elevation or depression of the muzzle. This type of kinesis seems to be present in Gekkonidae, Pygopodidae, and Varanidae (Metzger, 2002), although Herrel et al. (2007) considered it to be present in *Cordylus* and absent from varanids. In **metakinesis**, the axis of rotation is located at the back of the skull and involves the supraoccipital bone sliding beneath the parietal bone, as well as movement between the paroccipital processes and suspensorium, and sliding or rotatory action between the basisphenoid and pterygoids. **Hypokinesis** occurs at sutures between the bony elements of the palate.

It was once thought that cranial kinesis in one form or another was widespread or even universal among lizards, but there is actually much uncertainty on this point, particularly as to the occurrence of metakinesis (Table 6.1). Most studies have concentrated on larger species, and there is a lack of adequate data on many taxa. Diagnoses of kinesis by manipulation of skulls have not always been substantiated by experimental observations on feeding animals (Metzger, 2002; Herrel et al., 2007). Detailed understanding has also been hampered by lack of measurement techniques that do not interfere with the physiology of feeding.

Many hypotheses about the role of kinesis have been proposed (De Vree and Gans, 1994; Metzger, 2002; Evans, 2008). Although some hypotheses have been contradicted by experimental evidence on feeding lizards, others remain conjectural for want of evidence, and considerable additional work in this area is clearly needed. Possible functions for streptostyly include assisting cropping of plant material by herbivores, allowing propalinal (back and forth) motion of the lower jaw to enhance shearing, or promoting accurate occlusion in agamids. Mesokinesis, whether coupled with streptostyly or not, potentially allows the anterior component of the skull, the muzzle, to be raised and lowered. The suggestion that mesokinesis thereby increases the gape is not supported experimentally. It is possible that possession of a mesokinetic skull makes for more precise control during prehension. Coupled mesokinesis and streptostyly seems to favor faster jaw closure (Herrel et al., 2007). Mesokinetic skulls may also increase contact between the teeth and the prey (Patchell and Shine, 1986b) or may improve the efficiency of recurved teeth in predatory varanids (Rieppel, 1979). As Metzger (2002) remarked, several potentially valid hypotheses about the role of kinesis are not mutually exclusive, and it is probable that the function of a particular form of kinesis varies between species. It is likely that a major benefit of mobility between different elements of the skull is an enhanced capacity to resist the stresses associated with prey capture (De Vree and Gans, 1994; Evans, 2008). These stresses are unpredictable and are therefore better countered by the presence of several soft tissue interfaces that can absorb stress than by reinforcement of the skull (Evans, 2008).

Lizards possess salivary glands, producing mixed mucous and serous secretions, that lubricate the lining of the mouth and also the food to facilitate intraoral transport and swallowing. In addition, some lizards, as well as many snakes, produce venomous serous secretions that can facilitate capture of prey or can be used in defense. Recent studies suggest that the potential to produce venom appeared once among squamates, in the common ancestor of a clade—the **Toxicofera**—consisting of the iguanian and anguimorph lizards together with the snakes (Fry et al., 2006, 2012). It is thought that venom constituents evolved by duplication of genes for a variety of normal bodily proteins, often with bioactive or bioregulatory functions. Later mutations in the copies of the genes expressed by serous salivary glands enhanced the toxicological properties of some of these proteins (Fry et al., 2012). Although their oral secretions contain potentially toxic substances, the overwhelming majority of toxicoferan lizards are nonvenomous; successful prey capture does not depend on the pharmacological effects of the oral fluids on the prey. Use of oral secretions as venoms requires a capability to produce and store the toxic components in quantity and a system for transferring the venom into the prey. There are only a handful of lizards that possess these capabilities, namely, the helodermatids and some varanids (see pages 182–183 and 188).

GEKKOTA

The Gekkota includes six families of limbed geckos and the limbless Pygopodidae.

Geckos have several distinctive features. They are well known for the special adaptations of their toes that allow them to cling to most surfaces. They are unique among lizards for their use of chirping sounds to communicate with each other.

Some geckos may consume plant material, but usually in small quantities, and only about 4% of species are considered omnivorous or herbivorous.

Geckos typically have homodont dentitions consisting of numerous pleurodont teeth. Most species have 20–40 teeth on each dentary. Each maxilla has slightly fewer teeth than the dentary, whereas the premaxilla has 9–13 teeth. Thus, there may be well over 100 teeth in the whole dentition (Edmund, 1969). In the eublepharids *Goniurosaurus* and *Aeluroscalabotes* and the gekkonid *Cyrtodactylus louisiadensis*, there are 50 or more teeth in both dentary and maxilla (Nikitina and Ananieva, 2009). Table 6.2 provides tooth numbers on the three dentigerous bones for two families of geckos: Eublepharidae and Gekkonidae (Nikitina and Ananieva, 2009). Tooth numbers increase with age. For example, the number of maxillary teeth in the

white-spotted wall gecko (*Tarentola annularis*: Phyllodactylidae) increase in the 4 years after hatching by addition of approximately 15 teeth to the original mean number of 62 (Edmund, 1969).

Tooth size generally decreases from anterior to posterior along the tooth row but, in at least one species, the beautifully colored **wonder gecko** (*Teratoscincus*: Sphaerodactylidae) (Fig. 6.11), the largest teeth are in the middle of the row.

The teeth of geckos are straight and conical or slightly flattened laterally. In most species, the tooth crown has two centrally placed and lingually tilted cusps or ridges (Fig. 6.7), separated by a concave sulcus. The labial cusp is the more prominent, the opposite to the bicuspid teeth of amphibians. The deepest portion of the sulcus is closer to the labial cusp rather than to the lingual cusp. In some gekkonids the lingual cusp is reduced or absent, whereas other geckos have more than two cusps.

In geckos that display coupled mesokinesis and streptostyly, the bite force is lower than in species with nonkinetic skulls and the time required for intraoral transport of prey is much longer (Herrel et al., 2007). Geckos are capable of closing their jaws faster than species with nonkinetic skulls and might thus have an advantage in capturing fast-moving prey (Herrel et al., 2007).

Carphodactylidae

The **Northern leaf-tailed gecko** [*Saltuarius* (*Phyllurus*) *cornutus*] is named after its large, leaf-shaped tail. Instead of the adhesive toe disks of many geckos, this arboreal gecko has clawed toes that assist in climbing over rough surfaces. There are more than 60 teeth in each jaw quadrant (Fig. 6.12).

Phyllodactylidae

The **turnip-tailed gecko** (*Thecadactylus solimoensis*) has a typical homodont dentition composed of many cone-shaped teeth (Fig. 6.13).

Gekkonidae

This is the largest family of geckos and contains about 1000 species in more than 50 genera. They do not have teeth on the palate (Mahler and Kearney, 2006).

The **tokay gecko** (*Gekko gecko*) (Fig. 6.14A and B) and **Smith's green-eyed gecko** (*Gekko smithii*) (Fig. 6.15) demonstrate the typical gecko dentition, with each half of the dentition made up of about 100 sharp teeth. In the Tokay gecko, the **white-line gecko** (*Gekko vittatus*), and the **day geckos** (*Phelsuma*), the lingual cusp on the teeth is absent or only faintly defined.

Fig. 6.16 shows the typical relationship of the replacing teeth to the functional teeth in the **ocelot gecko**

TABLE 6.2 Tooth Numbers in a Variety of Geckos From the Families Gekkonidae and Eublepharidae

Family	Species	Dentary	Maxilla	Premaxilla
Gekkonidae	*Alsophylax pipiens*	20–21 (21)	16–19 (17)	9–10
	Crossobamon eversmanni	26–33 (29)	23–29 (26)	11
	Cosymbotus platiurus	36	31	11
	Cyrtopodion caspius	25–33 (30)	24–28 (26)	9–10
	Cyrtopodion fedtschenkoi	27–33 (30)	23–32 (28)	10–11
	Cyrtopodion louisadensis	58–62 (61)	54–59 (56)	13
	Gekko gekko	31–39 (34)	29–37 (33)	9–11
	Gekko vittatus	37	36	
	Teratoscincus bedriagai	31–32 (31)	24	10
	Hemidactylus frenatus	30–33 (32)	27–30 (29)	10–11
	Mediodactylus korschyi	24/	21	10
	Mediodactylus russowi	23–27 (25)	19–24 (22)	10
Sphaerodactylidae	*Teratoscincus keyserlingii*	28–34 (31)	24–29	9
	Teratoscincus microlepis	26	25	11
	Teratoscincus prezewalskii	27–31 (29)	29	10
	Teratoscincus roborowskii	32	30	9
	Teratoscincus scincus rustamovi	27–31 (30)	26–31 (30)	10–13
	Teratoscincus scincus scincus	21–33 (27)	21–27(23)	9–10
Eublepharidae	*Aeluscalabotes felinus*	53	44	19
	Coleonyx mitratus	35–42 (39)	28–32 (30)	12
	Eublepharis macularius	31–47 (37)	30–47	11–13
	Eublepharis turkmenicus	37–38 (37)	35	9
	Goniosaurus kuroiwae	50–52 (51)	46–50 (48)	13
	Goniosaurus murphyi	51	43	13
	Hemitheconyx caudicinctus	34–45 (37)	26–29 (28)	12–13

Mean numbers in parenthesis.
From Nikitina, N.G., Ananieva, N.B., 2009.Characteristics of dentition in gekkonid lizards of the genus Teratoscincus and other Gekkota (Sauria, Reptilia). Biol. Bull. 36, 237–242. Courtesy Editors Biology Bulletin.

FIGURE 6.11 Wonder gecko (*Teratoscincus scincus*). *Courtesy Wikipedia.*

(*Paroedura picta*). The replacing teeth are seen at varying stages of development by the lingual side of the predecessors, which show varying degrees of resorption at the bases of the teeth.

The genus *Uroplatus* has six species. The **giant leaf-tailed gecko** (*Uroplatus fimbriatus*) (Fig. 6.17) may have more marginal teeth than any other living amniote. There may be nearly 170 teeth in the upper jaw and 150 teeth in the lower jaw (Wellborn, 1933; Bauer and Russell, 1989), in contrast with other geckos, which have total counts of about 50–80 in the upper jaw and about 46–121 in the lower jaw (Edmund, 1969; Bauer and Russell, 1989). Such high tooth counts are not related to size alone; they must be associated in some way with diet, food handling,

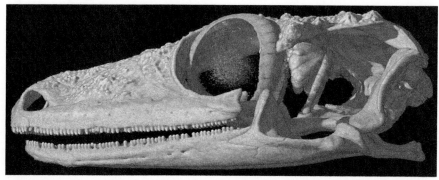

FIGURE 6.12   Northern leaf-tailed gecko [*Saltuarius (Phyllurus) cornutus*]. CT image of head, showing dentition. Image width=4.4 cm. *Courtesy Digimorph.org and Dr. J.A. Maisano.*

FIGURE 6.13   Teeth of the turnip-tailed gecko (*Thecadactylus solimoensis*). *Courtesy Dr. Morley Read. Shutterstock.com.*

(A)

(B)

FIGURE 6.14   Tokay gecko (*Gekko gekko*). (A) Live animal. *(Courtesy Reptiles4all.)* (B) Dentition of tokay gecko. Image width=6 cm. *(Courtesy Professor S.E. Evans.)*

and foraging strategies (Bauer and Russell, 1989). They may be functionally associated with the extreme rostral depression found in *Uroplatus* (Wellborn, 1933). A cave-dwelling species of the Asian gecko genus *Cyrtodactylus* from Thailand was described as having 11 teeth in the premaxilla, 45 teeth in the maxilla, and 51 teeth in the dentary (Bauer et al., 2002).

Eublepharidae

This family contains about 30 species in six genera. **Kuroiwa's eyelid gecko** (*Goniurosaurus kuroiwae*) is unique in having four cusps per tooth (Fig. 6.18). The angle of the most labial cusp approaches 90°; that of the most lingual cusp is much more rounded in profile, and the intervening cusps show intermediate angulations. The outer ridges of each cusp are approximately parallel. The subrectangular tooth crown accommodates the first three cusps, whereas the fourth cusp is on the lingual face of the tooth. The cusps on both the upper and lower teeth are lingually directed.

Tooth numbers for some Gekkonidae and Eublepharidae (Nikitina and Ananieva, 2009), and for four species of **banded geckos** of the genus *Coleonyx* (Kluge, 1962) are listed in Table 6.2. Average tooth numbers are premaxilla nearly 13 (range 11–14), maxilla 28 (maximum 37), and dentary 34 (maximum 44). Tooth numbers increase with age.

FIGURE 6.15   Dentition of Smith's green-eyed gecko (*Gekko smithii*). Image width = 5 cm. *Courtesy of the Hunterian, University of Glasgow. Catalog no. GLAM:Z5986.*

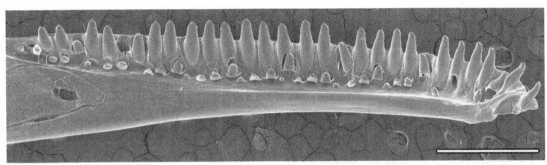

FIGURE 6.16   Scanning electron micrograph of the lower jaw of the ocelot gecko (*Paroedura picta*). Note the replacing teeth developing at the base of the functional teeth, which show varying degrees of resorption. Scale = 1 mm. *From Zahradniček, O., Buchtová, M., Dosedelová, H., Tucker, A.S., 2014. The development of complex tooth shape in reptiles. Front. Physiol. 5, 1–7. Courtesy Editors Frontiers of Physiology.*

Diplodactylidae

In five species of *Rhacodactylus*, a genus of medium-to-large geckos, total tooth number increases ontogenetically, at least until sexual maturity, and varies between species (Bauer and Russell, 1988). The mean adult tooth number varies from 105 for the **gargoyle gecko** (*Rhacodactylus auricularis*) to 183 for **Sarasin's giant gecko** (*Rhacodactylus sarasinorum*).

The **crested gecko** (*Correlophus ciliatus*) (Fig. 6.19A) has small, tightly packed, bicuspid and isodont teeth seemingly adapted to crushing hard-bodied insects. The teeth of **Leach's giant gecko** (*Rhacodactylus leachianus*) and the **rough-snouted giant gecko** (*Rhacodactylus trachyrhynchus*) are less numerous, pointed, predominantly monocuspid, and more recurved (Fig. 6.19B and C). The most specialized dentition is that of the gargoyle gecko (*Rhacodactylus auriculatus*), which has the fewest, most recurved and elongated teeth (Fig. 6.19D). This dentition seems to be adapted to subduing larger, soft-bodied prey. However, it is not always easy to correlate form of teeth to diet as little is known about diet in the wild and most geckos take a variety of prey types (Bauer and Russell, 1988).

Pygopodidae

This family contains gekkotans that have reduced or absent legs and resemble snakes in having long slender bodies. Like snakes they have no eyelids but, unlike snakes, they have external ears, fleshy unforked tongues, lack venom glands, and cannot constrict their prey. A list of tooth numbers for a range of pygopodid genera is given in Table 6.3 (Kluge, 1976; Patchell and Shine, 1986a).

The dentitions have been compared in three genera (Patchell and Shine, 1986a). *Pygopus* and *Delma* resemble limbed gekkotan relatives in that they possess peg-like, ankylosed teeth used for crushing invertebrate prey. However, **Burton's legless lizard** (*Lialis burtonis*) (Fig. 6.20) has a very different diet, consisting entirely of lizards and snakes. The skull is elongate and highly mobile, with mesokinetic and hypokinetic joints, coupled to a streptostylic jaw joint. Consequently, the snout can be depressed from the horizontal, resulting in improved contact between the upper teeth and the prey (Patchell and Shine, 1986b). There are about 100 small, slender, pointed, recurved teeth on each side. However, the most unusual feature is the presence of hinged, rather than solid, ankylosed teeth. Such hinged teeth fold backward when force is applied

(A)

(B)

FIGURE 6.17 Giant leaf-tailed gecko (*Uroplatus fimbriatus*). (A) Live animal. *(Courtesy Wikipedia.)* (B) Micro-CT scan of skull of *U. fimbriatus*. Image width = 4.5 cm. *(Courtesy Professor A.M. Bauer.)*

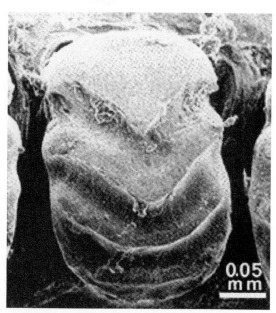

FIGURE 6.18 Kuroiwa's ground gecko (*Goniurosaurus kuroiwae*): scanning electron micrograph of tooth showing presence of four cusps. *From Sumida, S.S., Murphy, R.W., 1987. Form and function of the tooth crown structure in gekkonid lizards (Reptilia, Squamata, Gekkonidae). Can. J. Zool. 65, 2886–2892. Courtesy Editors Canadian Journal of Zoology.*

to the leading edge and snap back into the upright position once the prey has passed by. All these specializations allow *Lialis* to deal with tough, large prey (Patchell and Shine, 1986b).

An extreme dentition, further demonstrating the wide range of dentitions in the Pygopodidae, is that of the **sedgelands worm lizard** (*Aprasia repens*). The diet consists chiefly of ants, and this lizard has only two teeth at the front of the dentary (Fig. 6.21A and B). In this, it resembles blind snakes that subsist on a similar diet and that also have a very reduced dentition (see pages 207–209).

SCINCOIDEA

This is the largest infraorder of lizards, containing approximately 2000 species. It comprises four families: Scincidae (skinks), Cordylidae (girdled or spiny tail lizards), Gerrhosauridae (plated lizards), and Xantusiidae (night lizards).

Scincidae

Skinks make up one of the most numerous and diverse of squamate families, containing approximately 1300 species. Skinks tend to resemble lacertids, but many lack an obvious neck and their legs are relatively small. Indeed, many have reduced legs or none at all (eg, *Typhlosaurus*). Because skinks many dig and burrow, the head is often blunt ended. They are mainly insectivorous, but a small proportion (15%) are considered omnivorous or herbivorous. Marginal teeth form single rows on the premaxilla, maxilla, and dentary in all genera. The paired premaxillae bear small numbers of teeth. Palatal teeth, confined to the pterygoid bone, occur in only a limited number of genera of skinks, such as *Eumeces* (Fig. 6.22) and *Corucia*, and expression is variable. The palatal teeth are small and generally few in number and simple in form (Mahler and Kearney, 2006). The number of maxillary teeth increases significantly with head length in a large number of species and the number of dentary teeth is usually strongly correlated with the number of maxillary teeth (Greer and Chong, 2007).

The dentitions of six skinks are illustrated in Fig. 6.23. The teeth of most skinks have a cylindrical shaft with a rather blunt conical tip, a shape that is described as "peg-shaped." Typically, only the tip of the tooth protrudes from the oral mucosa. The teeth of the smaller, insectivorous skinks are more slender and more sharply pointed (Fig. 6.23C). Durophagous species (eg, *Hemisphaeriodon gerrardii*) possess a large molariform tooth toward the

(A)

(B)

(C)

(D)

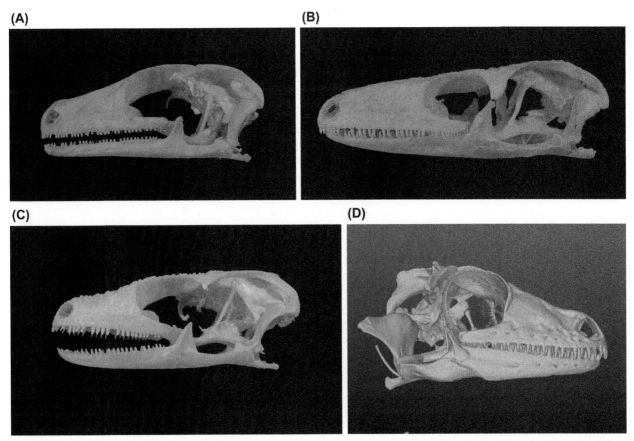

FIGURE 6.19 Dentitions of various species of *Rhacodactylus* (micro-CT images). (A) Crested gekko [*Correlophus* (*Rhacodactylus*) *ciliatus*]. Image width=4 cm. (B) Leach's giant gecko (*Rhacodactylus leachianus*). Image width=7 cm. (C) Rough-snouted giant gecko (*Rhacodactylus trachrhynchus*). Image width=4 cm. (D) Gargoyle gecko (*Rhacodactylus auricularis*). Note the caniniform teeth. Image width=3.2 cm. *From Bauer, A.M., Russell, A.P., 1988. Dentitional diversity in* Rhacodactylus *(reptilia: Gekkonidae). Mem. Qld. Mus. 29, 311–321. Courtesy Editors Memoirs of the Queensland Museum.*

back of each tooth row. *Coeranoscincus* has characteristic recurved teeth that are believed to be specializations for preying on earthworms (Fig. 6.23B). The primarily vegetarian species of the *Egernia cunninghami* species group have many small, closely packed teeth, with laterally compressed crowns that together form a cutting edge along the upper and lower tooth rows (Fig. 6.23D) (Hutchinson, 1993).

The dentitions of the **five-lined skink** [*Plestiodon* (*Eumeces*) *fasciatus*] and of the **water skink** (*Amphiglossus spendidus*), both consisting of the typical peg-shaped teeth, are illustrated in Figs. 6.24 and 6.25. **Schneider's or Berber's skink** (*Eumeces schneideri*) has small conical teeth of approximately equal size along the marginal tooth rows, whereas the related **Algerian orange-tailed skink** (*Eumeces algeriensis*) is reported to have more flattened teeth.

The **bobtail** or **shingleback blue-tongued skink** (*Tiliqua rugosa*) has a broad, stumpy tail. The teeth are not the same size along the tooth row, as is usual among skinks: the teeth toward the back of the maxilla and dentary are enlarged (Figs. 6.26 and 6.27). The diet of this skink consists largely

of plant material, together with a variety of invertebrates, including snails.

The **pink-tongued skink** [*Hemisphaeriodon* (*Cyclodomorphus*) *gerrardii*] is durophagous, feeding largely on slugs and snails, and it shows the most specialized dentition of any skink. The majority of teeth are blunt and cylindrical and there is an enormously enlarged molariform tooth toward the back of both upper and lower tooth rows (Fig. 6.23F). These enlarged teeth do not occlude; instead, the upper tooth closes in front of the lower and produces a shearing action rather than a direct crushing effect on the food.

The **Solomon Islands skink** (*Corucia zebrata*) is herbivorous, feeding on the leaves, flowers, fruits, and growing shoots of several different species of plants. There are approximately 20 teeth on the dentary and slightly less on the maxilla. The tooth crowns are compressed laterally and have mesial and distal cutting edges, with a main central cusp and traces of two lateral cusps (Figs. 6.28 and 6.29).

Tooth numbers are known for the **ground skink** (*Scincella lateralis*) (Townsend et al., 1999). Juveniles possess mean tooth numbers of 9.2 (premaxilla), 37.1 (maxilla), and 43.0 (dentary), whereas adult males possess 9.6 (premaxilla), 41.6 (maxilla), and 49.2 (dentary). There are no differences in tooth numbers between adult males and adult females. In this species, there is an obvious increase in tooth numbers with age. However, in the Canarian scincid (*Chalcides*), there is no such increase. Fifteen days after birth, there are 15 tooth loci in the two lower jaw quadrants, but at 6 years of age, there are between 14 and 18. For the upper jaw, there are 9 teeth on the premaxilla (the odd number owing to the presence of a midline tooth replacing the original egg tooth) and 14 on the maxilla, with no change with age (Delgado et al., 2003).

Xantusiidae

The **night lizards** are so named because they were once thought to be nocturnal, but they are now known to be diurnal. These small lizards inhabit rock crevices or damp logs and have flat heads and bodies. They may spend their entire life under the same cover, feeding on insects and sometimes plants. There are three genera with more

TABLE 6.3 Maximum Number of Teeth in Various Species of the Pygopodidae

Sub-family	Species	Premaxilla	Maxilla	Dentary
Lialiinae	*Lialis burtonis*	10–11	35–39	60–69
	Lialis jicari	8–9	35–39	50–59
Pygopodinae	*Aclys concinna*	12–13	25–29	30–39
	Aprasia aurita	2–3	0–4	0–9
	Aprasia repens	4–5	0–4	0–9
	Aprasia parapulcella	4–5	0–4	0–9
	Aprasia pseudopulcella	4–5	0–4	0–9
	Aprasia pulcella	0–1	0–4	0–9
	Aprasia striolata	4–5	0–4	0–9
	Delma australis	12–13	15–19	30–39
	Delma fraseri	12–13 (13)	20–24 (17)	20–29 (25)
	Delma impar	12–13	15–19	20–29
	Delma inornata	12–13	20–24	20–29
	Delma molleri	12–13	15–19	20–29
	Delma nasuta	12–13	25–29	30–39
	Delma tincta	12–13	15–19	30–39
	Ophidiocephalus taeniatus	2–3	10–14	10–19
	Paradelma orientalis	10–11	15–19	20–29
	Pletholax gracilis	6–7	5–9	10–19
	Pygopus lepidopodus	12–13	15–19	20–29
	Pygopus nigriceps	10–11 (11)	10–14 (13)	20–29 (18)

Data From Kluge, A.G., 1976. Phylogenetic relationships in the lizard family Pygopodidae: an evaluation of theory, methods and data. Misc. Publ. Mus. Zool. Univ. Mich. No 152, 1–80, Supplemented with data for two species (in bold) from Patchell, F.C., Shine, R., 1986a. Hinged teeth for hard-bodied prey: a case of convergent evolution between snakes and legless lizards. J. Zool. Lond. 208, 269–275.

FIGURE 6.20   Dentition of Burton's legless lizard (*Lialis burtonis*). Image width = 3.7 cm. *Courtesy Digimorph.org and Dr. J.A. Maisano.*

than 30 species. Xantusiids have homodont dentitions, with 7–8 teeth on the premaxillae, 9–15 on each maxilla, and 12–18 on each dentary. No palatal teeth are present in any species. In most species the teeth are simple and conical: the dentition of the **yellow-spotted night lizard** (*Lepidophyma flavimaculatum*) (Fig. 6.30) is typical. *Xantusia* is insectivorous and Savage (1963) considered

that *Cricosaura* and *Lepidophyma* were also primarily insectivores. However, the diet of **Smith's tropical night lizard** (*Lepidophyma smithii*) has been found to consist largely of figs (Cooper and Vitt, 2002). The **island night**

(A)

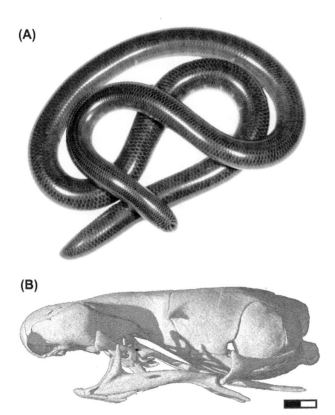

(A)

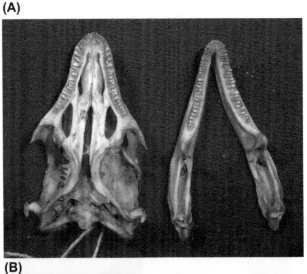

(B)

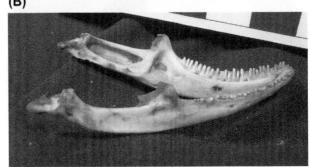

FIGURE 6.21 Sedgelands worm lizard (*Aprasia repens*). (A) Body form. (B) CT scan of skull, showing a greatly reduced dentition consisting of just two teeth in the dentary. Scale = 1 mm. *Courtesy Prof A.M. Bauer.*

FIGURE 6.22 Dentition of Schneider's or Berber's skink (*Eumeces schneideri*). (A) Ventral view. The perygoid teeth are too small to be seen. Image width = 7 cm. (B) Lingual view of mandible, showing replacing teeth. Scale = 1 cm divisions. *Courtesy RCSOMA/A411.2.*

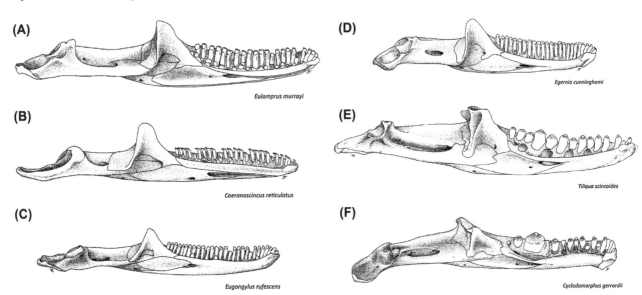

FIGURE 6.23 Variation in mandibular dentition of skinks. (A) *Eulamprus murrayi*; (B) *Coeranoscincus reticulatus*; (C) *Eugongylus rufescens*; (D) *Egernia cunninghami*; (E) *Tiliqua scincoides*; (F) *Hemisphaeriodon* (*Cyclodomorphus*) *gerrardii*. Original illustrations by Jennie Thurmer. *Courtesy Dr. M. Hutchinson and the South Australian Museum.*

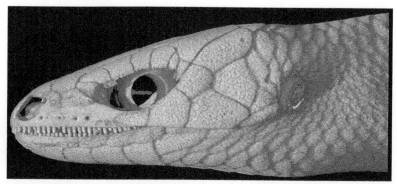

FIGURE 6.24 Dentition of five-lined skink [*Plestiodon (Eumeces) fasciatus*]. Image width=2.2 cm. *Courtesy Digimorph.org and Dr. J.A. Maisano.*

FIGURE 6.25 Dentition of water skink (*Amphiglossus spendidus*). Image width=2.2 cm. *Courtesy Digimorph.org and Dr. J.A. Maisano.*

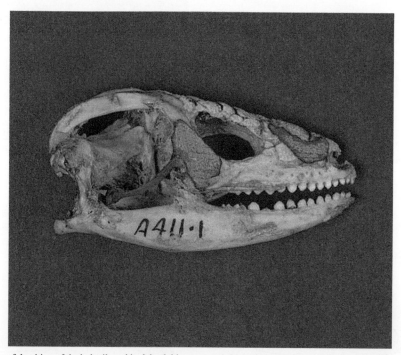

FIGURE 6.26 Lateral view of dentition of the bobtail or shingleback blue-tongued skink (*Tiliqua rugosa*). Image width=7 cm. *Courtesy RCSOMA/411.1.*

FIGURE 6.27   Lingual view of dentition of the bobtail or shingleback blue-tongued skink (*Tiliqua rugosa*). Image width = 8 cm. *Courtesy RCSOMA/ 411.1.*

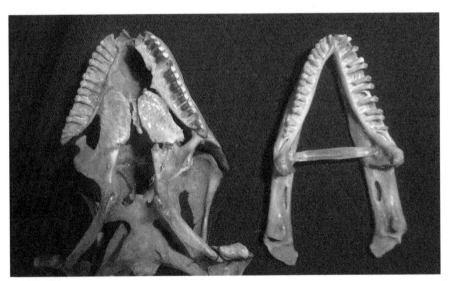

FIGURE 6.28   Dentition of the Solomon Islands skink (*Corucia zebrata*). Image width = 9 cm. *Courtesy Professor S.E. Evans.*

lizard [*Xantusia* (*Klauberina*) *riversiana*] includes a large proportion of vegetation in its diet and takes all parts of plants, from stems to fruits (Savage, 1963; Cooper and Vitt, 2002). The dentition is correlated with this diet, in that the teeth are larger than in other xantusiids and are tricuspid with mesial and distal cutting edges (Savage, 1963).

Cordylidae

About 30% of **girdled** or **spiny-tail lizards** are herbivorous or omnivorous (Cooper and Vitt, 2002). Cordylids have teeth with a typical morphology for lizards: generally bicuspid, frequently compressed, and incurved. Depending on species, the premaxilla carries 7 teeth, the maxilla 16–25

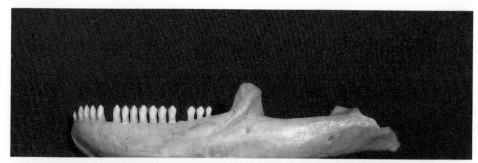

FIGURE 6.29   Lateral view of lower dentition of the Solomon Islands skink (*Corucia zebrata*). Image width = 8 cm. *Courtesy Professor S.E. Evans.*

FIGURE 6.30   Dentition of the yellow-spotted night lizard (*Lepidophyma flavimaculatum*). Image width = 3.3 cm. *Courtesy Digimorph.org and Dr. J.A. Maisano.*

teeth, and the dentary 18–26 teeth (Edmund, 1969). The **East African spiny-tailed lizard** (*Cordylus tropidosternon*), like geckos, displays coupled mesokinesis and streptostyly and shares with them a faster jaw closing speed, a relatively low bite force and prolonged intraoral prey transport (Herrel et al., 2007).

LACERTOIDEA

This group of lizards consists of two clades. One clade comprises the two families Teiidae (whiptails and tegus) and Gymnophthalmidae (speckled lizards). The other clade comprises the family Lacertidae together with the six families making up the Amphisbaenia (worm lizards).

Teiidae

Teiids can be recognized by their long tails, the large rectangular ventral scales that form distinct transverse rows, and their fork-like tongues. There are 10 genera containing more than 230 species, all living in the Americas. Many of the smaller, whiptail species consist of only females, which reproduce by parthenogenesis. Teiids are sometimes referred to as macroteiids, to distinguish them from the Gymnophthalmidae, or microteiids.

The teeth are supported by a pleurodont ankylosis, but the bases of the teeth are surrounded by a plate of bone, giving the false impression of a partial socket. This type of attachment has been called **subpleurodont** or **subthecodont**. At first sight the teiid dentition may seem homodont, but detailed inspection often shows that the number of cusps on the teeth can vary from one to three in different parts of the same jaw. Although the majority of dentitions show sharp cusps for dealing with insect prey, the teeth of *Dracaena* are molariform and specialized for crushing hard-bodied prey. Table 6.4 details modal tooth counts for macroteiid genera for one-half of each jaw (Presh, 1974).

There are three species of *Teius*, and the numbers of teeth in specimens of each are shown in Table 6.5 (Brizuela and Albino, 2009). The dentitions are heterodont in that the number of cusps on the teeth varies from one to three. Unusually, the cusps may be oriented labiolingually or mesiodistally (ie, respectively transverse or longitudinal to the tooth row) (Brizuela and Albino, 2009) (Fig. 6.31). In the **four-toed tegu** (*Teius teyou*), the anterior teeth are monocuspid and are followed by bicuspid teeth. Among the latter, teeth with mesiodistally oriented cusps are situated in front of teeth with labiolingually oriented cusps. Tricuspid teeth, with cusps aligned either mesiodistally or labiolingually, were found to be present in 27% of specimens.

TABLE 6.4 Modal Tooth Numbers for One Side of the Jaws in Various Genera of Macroteiids

Genus	Premaxilla	Maxilla	Dentary
Ameiva	9	19	22
Callopistes	4	14	16
Cnemidophorus	7	18	21
Crocodilurus	4	14	15
Dicridon	3	15	15
Dracaena	4	10	12
Kentropyx	7	18	12
Tupinambis	6	15	14

Derived from a large number of specimens representing a variable number of different species within each genus.
Modified from Presch, W., 1974. A survey of the dentition of the macroteiid lizards (Teiidae: Lacertilia). Herpetologica 30, 344–349. Courtesy Editors Herpetologica.

TABLE 6.5 Tooth Counts for Half the Jaw of the Three Species of *Teius*

Species	Premaxilla	Maxilla	Dentary	Pterygoid
Teius teyou	6	13–16	14–18	2–3
Teius oculatus	6	11–15	13–15	1
Teius suquiensis	6	13–14	14–17	1–4

Data from Brizuela, S., Albino, A.M., 2009. The dentition of the neotropical lizard genus *Teius Merrem* 1820 (Squamata Teiidae). Trop. Zool. 22, 183–193. Courtesy Editors Tropical Zoology.

Where both types are present, teeth with mesiodistally orientated cusps lie anteriorly (Brizuela and Albino, 2009). The teeth interdigitate with those of the opposite jaw and show little signs of wear, indicating that there is little intra-oral processing of food (ie, shearing). *Teius* feeds mainly on insects, which only need to be punctured and crushed.

Young **golden tegu** (*Tupinambis teguixin*) (Fig. 6.32) are insectivorous and have a dentition adapted to puncturing the chitinous exoskeleton (Figs. 6.33A and B). The premaxillae carry tricuspid teeth with cusps of the same size. The posterior regions of the maxillae and dentaries are occupied by tricuspid teeth with small mesial and distal cusps. The anterior teeth on the maxillae and dentaries are monocuspid, sharply pointed, and recurved. During growth, the number of teeth is said to increase and, with continuing tooth succession, the posterior tricuspid teeth are replaced by blunt crushing teeth (Presch, 1974): changes correlated with an increased variety of prey. In the adult dentition (Fig. 6.34), the anterior teeth are conical with mesial and distal cutting edges, each furnished with a low cusp, so are suited to grasping and puncturing prey, whereas the most posterior teeth are enlarged and robust, with flattened surfaces which are used to crush snails. The premaxilla carries about 13 small, recurved teeth. On the maxillae, tooth size increases to a maximum at the 3rd and 11th tooth positions. In the lower jaw, which carries up to 15 teeth, tooth size increases to a maximum between the third and about the sixth positions, then decreases somewhat to a constant size in the posterior part of the tooth row (Presch, 1974; Dalrymple, 1979; Estes and Williams, 1984). The positions of the largest teeth coincide with elevations of the crest of the jaw bone, so that the tooth rows have undulating profiles. This seems to be an adaptation that prevents hard prey objects from sliding forward as crushing pressure is applied.

The **Argentinian red tegu** or **red tegu** (*Tupinambis rufescens*), is one of the largest teiid lizards. It is omnivorous: as well as insects, rodents, birds and fish, its diet

(A) **(B)**

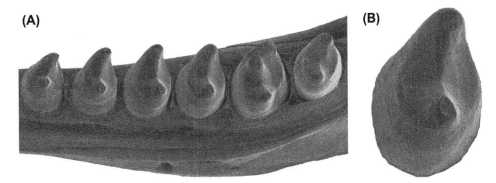

FIGURE 6.31 Teeth of four-toed tegu (*Teius teyou*). (A) Dorsal view of posterior end of the right dentary. The teeth are bicuspid with the intercuspal ridge aligned transversely, except for the second tooth from the right, which is tricuspid. Image width=9 cm. (B) High-power view of 13th tooth along the dentary with the intercuspal ridge aligned transversely. Image width=1 mm. *From Brizuela, S., Albino, A.M., 2009. The dentition of the neotropical lizard genus* Teius *Merrem 1820 (Squamata Teiidae). Trop. Zool. 22, 183–193. Courtesy Editors Tropical Zoology.*

FIGURE 6.32 Young golden tegu (*Tupinambis teguixin*), with partial view of dentition. *Courtesy Ryan M. Bolton.*

FIGURE 6.34 Dentition of adult golden tegu (*Tupinambis teguixin*), showing heterodonty. Image width=7.3 cm. *Courtesy Horniman Museum & Gardens and Dr. Paolo Viscardi. Catalog no. LDHRN-NH 36.59.*

(A)

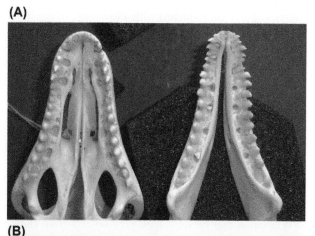

(B)

FIGURE 6.33 Dentition of young golden tegu (*Tupinambis teguixin*). (A) Occlusal view. Image width=8 cm. (B) Lateral view. Image width=5 cm. *Courtesy Horniman Museum & Gardens and Dr. Paolo Viscardi. Catalog no. LDHRN-NH 36.101.*

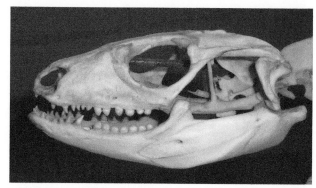

FIGURE 6.35 Dentition of red tegu (*Tupinambis rufescens*). Actual image length=12 cm. © *Trustees of the Natural History Museum, London.*

includes a high proportion of fruit and other vegetable material: over 30%, according to Cooper and Vitt (2002). Its dentition resembles that of the golden tegu (Fig. 6.35).

The **Caiman lizard** (*Dracaena guianensis*) has a very specialized dentition that allows it to deal with a diet of snails, even as a hatchling, in which blunt teeth are already present at the back of the jaws. The premaxillary teeth and the first three teeth on the maxilla are conical and blunt and of similar size. The remaining teeth on the maxilla (numbering about 10) are large and rounded to form a crushing platform (Figs. 6.36 and 6.37). The lower dentition is similar to the upper. Adult Caiman lizards have fewer teeth than other large teiids, with 14 teeth in the maxilla and 12 teeth in the dentary (compared with, eg, the adult golden tegu with 21 and 15 teeth, respectively). The teeth show a very firm, subpleurodont attachment.

In Caiman lizards, a delay between loss of a tooth and eruption of its successor would reduce the efficiency of crushing. This seems to account for the fact that some replacement teeth erupt lingually while their predecessors are still

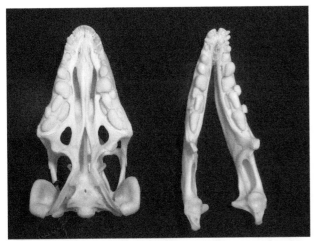

FIGURE 6.36 Dentition of northern caiman lizard (*Dracaena guianensis*): lateral view. Actual image length=10cm. *Courtesy Professor S.E. Evans.*

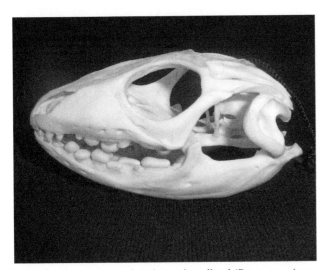

FIGURE 6.37 Dentition of northern caiman lizard (*Dracaena guianensis*): occlusal views. Note the trace of a double row of teeth toward the back of the jaws. Actual image length=9.5cm. *Courtesy Professor S.E. Evans.*

functioning so that a double row of teeth may sometimes be encountered at the back of the tooth row (Figs. 6.36 and 6.37).

Dracaena picks up snails with its front teeth and lifts the head, so that the snail rolls to the region of the back teeth, where it is crushed by repeated bites (Dalrymple, 1979). Shell fragments are cleared from the mouth by the tongue. A comparison of the cranial anatomy of *Dracaena* with that of *Tupinambis*, which is omnivorous, suggests that the bite force would be larger in *Dracaena*, on the basis of the size of the jaw-closing muscles and the mechanical advantage of the lower jaw (Dalrymple, 1979). Schaerlaeken et al. (2012) confirmed this for juveniles, but not for adults. However, their study sample included few adults and further work seems to be required.

In *Crocodilurus* and *Callopistes*, as in young *Tupinambis*, the dentition consists of conical, recurved pointed teeth that are best adapted to grasping and holding large, living prey.

FIGURE 6.38 Dentition of European green lizard (*Lacerta viridis*). Image width=5cm. *Courtesy MoLSKCL. Catalog no. X25.*

Crocodilurus and *Callopistes* have further reduced the number of triconodont teeth, and the majority of teeth are simple, sharply conical structures. In addition, the second, third or fourth maxillary teeth are enlarged into "canine teeth" and perhaps serve to hold, or aid, in tearing food (Presch, 1974).

Pterygoid teeth are present in most genera of Teiidae, except *Tupinambis*. The teeth are small, conical, and few in number. Palatine and vomerine teeth have not been reported (Mahler and Kearney, 2006).

Lacertidae

Typically, the dentition in this family is pleurodont and homodont. The teeth are generally tricuspid, with a main central cusp and small mesial and distal cusps. Teeth are commonly found on the pterygoids but are absent from the palatines. At hatching, many species have a similar dental formula of 2 premaxillary teeth, 13 maxillary teeth, and 15 dentary teeth. However, with subsequent growth, different species may develop different dental formulae, perhaps reflecting different growth patterns and different types of prey. Slight changes in tooth morphology also occur with age.

The dentition of the adult **European green lizard** (*Lacerta viridis*) is illustrated in Fig. 6.38. There are seven to nine conical teeth on the premaxilla, plus a midline tooth remarkable for its bilateral symmetry. Each maxilla bears up to 20 teeth that are larger than those on the premaxilla. Each maxillary tooth has a small mesial cusp that becomes more pronounced on the posterior teeth. The profile of the tooth row has a gentle wave form with peaks at the tooth positions 4–5 and 13–14, which is due to variations in the crest of the jaw bone and variations in tooth size. On each dentary there are up to 27 teeth that increase in size from the front to the back of the tooth row, although there is a peak in tooth level and size near the front. As in the maxilla, the last few teeth in the dentary are small. There is a diastema in the midline where the dentaries are joined by a flexible symphysis. A few of the teeth at the front of the dentary have a simple conical shape but, as in the maxilla, the reminder of the teeth have an additional smaller mesial cusp (Cooper, 1963).

There may be up to 10 peg-shaped pterygoid teeth visible on the palate. However, careful examination of dry specimens has revealed a series of smaller, medially situated rows hidden beneath the mucosa. This implies that, although a series of pterygoid teeth continue to develop throughout life in an anterior and lateral position, only the largest series is functional and the earlier series are retained but hidden beneath the oral mucosa (Cooper, 1963).

The teeth of young green lizards are tricuspid, with a main central cusp and smaller mesial and distal cusps (Fig. 6.39). With age and repeated tooth replacement, there is a gradual increase in tooth size and the posterior cusp is gradually lost, resulting in the bicuspid adult tooth form. The model in Fig. 6.40 shows the pleurodont tooth in the dentary, together with the presence of underlying replacing teeth at various stages of development.

In the **ocellated lizard** (*Lacerta lepida*) (Fig. 6.41), the number of teeth increases during growth from neonate to adult: from about 5 to about 10 teeth on the premaxilla, from about 12 to about 22 teeth on each maxilla, and from about 15–28 teeth on each dentary (Mateo and Lopez-Jurada, 1997). An uneven row of teeth runs anteroposteriorly near the middle of the pterygoid, while medial and posteromedial to this row, smaller teeth are scattered in subparallel rows, extending all the way to the medial edge of the pterygoid (Fig. 6.41).

With increasing age, there is more variation in the size of the teeth and larger, recurved, canine-like teeth appear in the anterior region of the maxilla. As in *L. viridis*, there is also a change in tooth form: initially it is tricuspid but the smaller lateral cusps are gradually lost. This change in tooth morphology with age may be related to a change in the nature of the

diet. Juveniles have a relatively diverse diet including soft-bodied insects, but older individuals eat harder and tougher insects, as well as a high proportion of plant material, such as shoots, flowers, and fruits. The simplified, conical adult teeth may perforate the exoskeleton of the prey more effectively than the tricuspid juvenile teeth and would be adapted to plucking parts of plants.

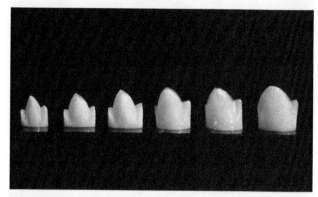

FIGURE 6.39 Models of European green lizard teeth, showing the gradual change in morphology from a tooth with three cusps in a young specimen (left) to one with two cusps in an adult tooth (right), following a number of tooth replacements. *Courtesy Dr. J.S. Cooper.*

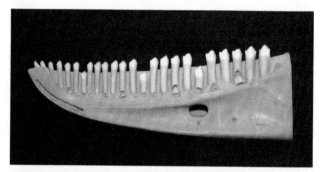

FIGURE 6.40 Model of the dentary of European green lizard (*Lacerta viridis*), showing pleurodont functional teeth and replacing teeth at different stages of development. *Courtesy Dr. J.S. Cooper.*

(A) **(B)**

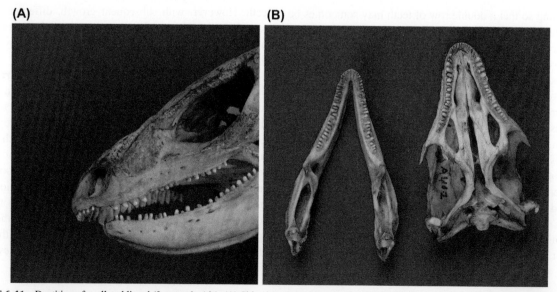

FIGURE 6.41 Dentition of ocellated lizard (*Lacerta lepida*). (A) Side view. Actual image size = 5 cm. (B) Occlusal view. Image width = 8 cm. *Courtesy RCS. Catalog no. A410.2.*

There are several living species of **Canarian wall lizards** (*Gallotia*), including some giant forms. Like the ocellated lizard, members of the *Gallotia* are omnivorous and the diet includes significant quantities of plant material. The dentition shows some degree of heterodonty: the first four to six teeth are monocuspid and the remainder are laterally compressed, with two or more cusps depending on species, eg, generally two in the **Atlantic lizard** (*Gallotica atlantica*) and four to six in the **Tenerife lizard** (*Gallotia galloti*) (Fig. 6.42).

AMPHISBAENIA

There are more than 180 species of amphisbaenians, also known as worm lizards, in six families: Rhineuridae, Bipedidae, Blanidae, Cadeidae, Trogonophidae, and Amphisbaenidae. The Rhineuridae are considered to be the sister group to all the remaining Amphisbaenia (Pyron et al., 2013). The Amphisbaenidae (170 species in 11 genera) contains the overwhelming number of worm lizards. The position of this group among reptiles has been controversial, but the molecular evidence indicates that they are related to the Lacertidae (Pyron et al., 2013).

Amphisbaenians resemble snakes in that they usually lack limbs and have long bodies but, unlike snakes, the right lung is reduced instead of the left. Some species lay eggs, but others retain the eggs within their bodies and give birth to live young. Amphisbaenians are burrowers, mostly less than 15 cm in length, and live in tunnels of their own construction for virtually their entire lives. They are found in tropical and subtropical regions and require moist soils that permit the construction and maintenance of tunnel systems. In accord with this lifestyle, they lack external ears and have only rudimentary eyes. The word amphisbaena means to "move in both ways," as the animals can move backward and forward with equal ease. The body scales of the skin are arranged in rings so that, superficially, amphisbaeinians look like large earthworms (Fig. 6.43). The skin is loosely attached to the body and they move using an accordion-like motion.

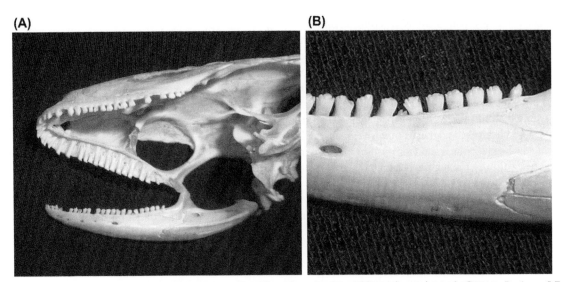

FIGURE 6.42 *Gallotia* sp. (A) Dentition. (B) Higher power view of lower jaw, showing multicusped posterior teeth. *Courtesy Professor S.E. Evans.*

FIGURE 6.43 Iberian worm lizard (*Blanus cinereus*). *Courtesy Wikipedia.*

The main advantage of a fixed tunnel system would seem to be the increased efficiency in the search for prey (Gans, 1968). As is to be expected, in fossorial worm lizards the skull is stout and compact. The nostrils are depressed into the snout and are provided with internal chambers, thus preventing the penetration of soil particles during tunneling. The different methods and mechanical demands of burrowing are reflected in differences in their unique cranial morphology (Zangerl, 1944; Vanzolini, 1951; Gans, 1968, 1978). The snout may be rounded or tapered, with a ridge on the top. Four distinct skull types that correlate with the movements of the head during tunnel excavations and constructions can be identified:

- **Shovel-snouted skull** (eg, *Rhineura*: Rhineuridae). The snout is flattened dorsoventrally and there is a sharp craniofacial angle. This skull morphology is associated with digging carried out by forcing the head forward and slightly downward, and then lifting it dorsally to pack the soil onto the top of the tunnel. The sides of the tunnel are smoothed with the pectoral musculature;
- **Spade-snouted skull** (Trogonophidae). The snout is flattened dorsoventrally and there is a strong craniofacial angle. The burrowing method is however distinct from that seen in *Rhineura*. Trogonophids use the sharp sides of their head (called lateral canthi) to shave off soil from the front of the tunnel in an oscillatory motion. Soil is pushed and packed using the sides of the head and the body;
- **Keel-snouted skull** (*Anops*: Amphisbaenidae). The skull is compressed laterally, and these amphisbaenians dig by ramming the head forwards and then pushing and packing the soil rearwards by forcing the head alternately left and right;
- **Round-snouted (bullet-headed) skull** (the majority of amphisbaenians). The skull is used as a battering ram, followed by pushing the head in different directions to pack soil.

Amphisbaenians are carnivores and find their prey by sound and scent. They have a varied diet, including small arthropods, worms, insects, and small vertebrates that they encounter while tunneling. Some species will come to the surface at night and can capture prey and drag it down into their tunnels. Although often considered generalist and opportunistic feeders, the few studies on this topic indicate that they may be more discriminating and capable of selecting particular prey items (Martin et al., 2013). Amphisbaenians have strong, heavy jaws and a single fused premaxilla. The teeth are sharp, stout, recurved, and can rapidly crush and cut up any prey consumed. Their dentition is specialized anteriorly where a single tooth in the middle of the upper jaw fits between a pair of teeth at the front of the lower jaw to form sharp forceps to capture or cut up prey. The rest of the teeth in the lower jaw fit inside those of the upper. The teeth also alternate so that they interlock when the jaw is closed.

There is less variation in tooth numbers in amphisbaenians than in most other infraorders of lizards (Zangerl, 1944: Gans, 1960). In general terms, there are up to seven teeth on the premaxilla, including a strong, median tooth flanked by a variable number of smaller teeth (that may be reduced to one tooth or none). The mode for the number of teeth on each dentary is seven (range six to nine), with usually five teeth on each maxilla. Teeth are absent from the palate (Vanzolini, 1951). Whereas the teeth of most amphisbaenians are pleurodont, those of trogonophids are acrodont (Gans, 1960).

Rhineuridae

The **Florida worm lizard** (*Rhineura floridana*) is the only member of the genus *Rhineura*. It has a shovel-snouted skull (Fig. 6.44), and the stout teeth forming its dentition consist of one premaxillary tooth, four teeth on each maxilla, and six teeth on each dentary. Of the four maxillary

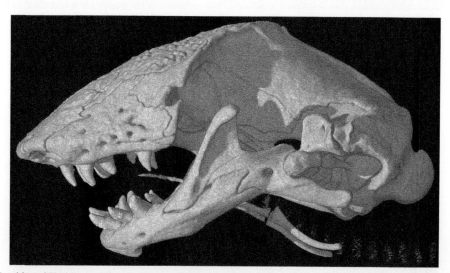

FIGURE 6.44 Dentition of Florida worm lizard (*Rhineura floridana*). Image width = 1.5 cm. *Courtesy of Digimorph.org and Dr. J.A. Maisano.*

teeth, the second is the largest, followed by the first. The last maxillary tooth is the smallest. The first tooth on the dentary is the largest in the lower jaw.

Bipedidae

These amphisbaenians are distinguished by the possession of limbs. The **Mexican mole lizard** or **five-toed worm lizard** (*Bipes biporus*) possesses strong, paddle-like forelimbs (Fig. 6.45), whereas the hind legs have disappeared, leaving behind only vestigial bones that are visible in radiographs. It has a round-snouted skull. There are seven premaxillary teeth, decreasing in size behind the large median tooth. The three teeth on each maxilla and the six teeth on each dentary are similar in size (Fig. 6.46). It seems to be a generalist predator that takes mostly ants, but also some soft- and hard-bodied invertebrates (Kearney, 2003).

FIGURE 6.45 Mexican mole lizard (*Bipes biporus*), with strong, paddle-like forelegs. *Courtesy Wikipedia.*

Trogonophidae

These worm lizards have acrodont teeth. In the **checkerboard worm lizard** (*Trogonophis wiegmanni*), there are three to five premaxillary teeth, four teeth on each maxilla, and eight teeth on each dentary: all are pointed and recurved (Gans, 1960). The long median premaxillary tooth faces backward and is flanked by smaller, more rounded teeth. Of the four maxillary teeth, the second is by far the largest, with the third and fourth being small and of equal size and the long bases of the teeth are in contact. The first four dentary teeth are nearly twice the size of the posterior four dentary teeth. This species eats a variety of invertebrates, which may include snails and the teeth are more rounded than in other worm lizards (Fig. 6.47).

In other species of Trogonophidae, there is one less tooth present on the premaxilla as well as a reduction in the number of teeth on the maxilla and dentary that, as in *Trogonophis*, are joined at their bases. Thus, the **short worm lizard** (*Pachycalamus brevis*) has six teeth on the dentary, **Zarudnyi's worm lizard** (*Diplometopon zarudnyi*) has three teeth on the maxilla and six on the dentary, and *Agamodon anguliceps* has two teeth on the maxilla and five on the dentary (Gans, 1960).

Amphisbaenidae

The skull and dentitions of the Amphisbaenidae have been described by Zangerl (1944). **King's worm lizard** (*Anops kingi*) has a keel-snouted skull. The dentition comprises seven premaxillary teeth, three teeth on each maxilla, and seven on each dentary (Fig. 6.48). The teeth are conical and recurved and are of varying size, with the largest maxillary tooth being near the middle of the tooth row.

The **speckled worm lizard** (*Amphisbaena fuliginosa*) (Fig. 6.49) has a round-snout skull and the dentition consists

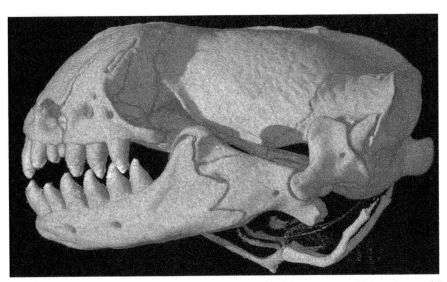

FIGURE 6.46 Dentition of Mexican mole lizard (*Bipes biporus*). Image width = 8.5 mm. *Courtesy Digimorph.org and Dr. J.A. Maisano.*

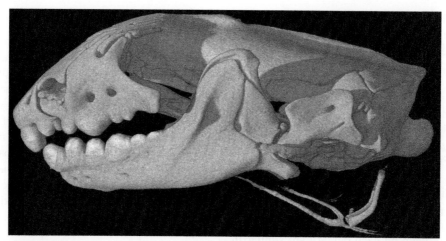

FIGURE 6.47   Dentition of checkerboard worm lizard (*Trogonophis wiegmanni*). Image width = 1.5 cm. *Courtesy Digimorph.org and Dr. J.A. Maisano.*

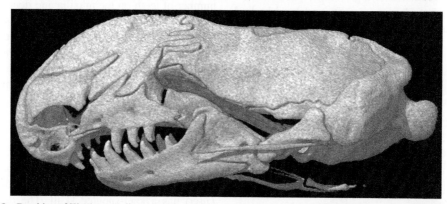

FIGURE 6.48   Dentition of King's worm lizard (*Anops kingii*). Image width = 1.1 cm. *Courtesy Digimorph.org and Dr. J.A. Maisano.*

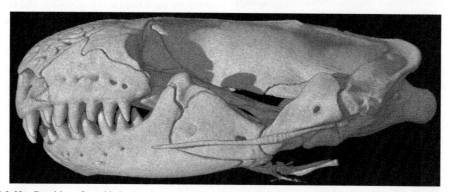

FIGURE 6.49   Dentition of speckled worm lizard (*Amphisbaena fuliginosa*). *Courtesy Digimorph.org and Dr. J.A. Maisano.*

of seven premaxillary teeth, five teeth on each maxilla, and seven to nine on each dentary. The largest tooth on the maxilla is the second from the front, followed by the fourth and fifth teeth. The first one or two teeth on the dentary are small, whereas the remainder are larger, caniniform, and approximately equal in diameter and height. The teeth on the dentary are oriented obliquely forward from their bases and terminate in recurved points.

The dentition of the **red worm lizard** (*Amphisbaena alba*) (Fig. 6.50) is similar to that of the speckled worm lizard (Gans, 1957). Fig. 6.51 shows the "forceps" formed by the median upper tooth and the and the two front lower teeth. There is evidence that in this species tooth replacement becomes infrequent or stops altogether beyond a certain body size (Gans, 1957).

Snails form a large proportion of the diet of **Ridley's worm lizard** (*Amphisbaena ridleyi*) and it is the only amphisbaenid with a dentition adapted to durophagy (Pregill, 1984) (Fig. 6.52). There are seven premaxillary teeth, with the large median tooth being flanked by two rows of three smaller, slightly recurved teeth that diminish in size posteriorly. On each maxilla there are five teeth that become

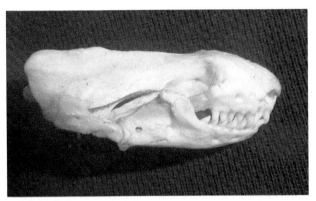

FIGURE 6.50 Dentition of the red worm lizard (*Amphisbaena alba*). Image width=4 cm. *Courtesy UCL, Grant Museum of Zoology. Catalog no. LDUCZ-X311.*

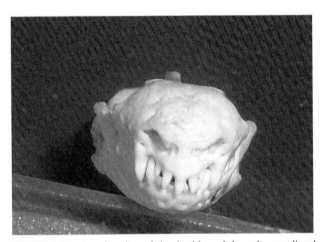

FIGURE 6.51 Anterior view of the dentition of the red worm lizard (*Amphisbaena alba*), showing the enlarged midline tooth on the premaxilla fitting between the two most anterior teeth on the dentary. Image width=3.2 cm. *Courtesy Courtesy UCL, Grant Museum of Zoology. Catalog no. LDUCZ-X311.*

FIGURE 6.52 Dentition of Ridley's worm lizard (*Amphisbaena ridleyi*). Image width=1.1 cm. *From Pregill, G., 1984. Durophagous feeding adaptations in an amphisbaenid. J. Herpetol. 18, 186–191. Courtesy Editors Journal of Herpetology.*

less curved posteriorly. The first and fourth teeth are the largest. There are six teeth on each dentary (occasionally seven), with a similar gradation of shape as on the maxilla. The second and last teeth are largest, and the last tooth is separated from the tooth in front by a diastema. When the jaws close, the large second lower tooth fits in front of the first maxillary tooth, over the small premaxillary teeth, whereas the fourth upper maxillary tooth fits into the diastema between the fifth and sixth dentary teeth. These interlocking teeth, closed by powerful jaw muscles, crush the shells of the snails that comprise most of the diet. The dentition does not change during growth.

ANGUIMORPHA

This infraorder contains seven families: Xenosauridae, Helodermatidae, Anniellidae, Anguidae, Shinisauridae, Lanthanotidae, and Varanidae. The Shinisauridae, Lanthanotidae, and Varanidae are often grouped together as Varanoidea. Recent studies indicate that the Anguimorpha, Iguania, and snakes are descended from a common ancestor that possessed venom, so together are designated Toxicofera. Venom glands are derived from the serous, protein-secreting portion of salivary glands that are located along the margins of the oral cavity and secrete into the region of the tooth bases (Weinstein et al., 2009; Fry et al., 2010). In the representatives of the anguids and xenosaurids that have been studied so far (Fry et al., 2006, 2010, 2012), the maxillary labial gland is absent and the mandibular glands are mixed, mucus-secreting and serous protein-secreting organs. Each gland is organized into several compartments, each consisting of a mucus-secreting region dorsal to serous tissue; secretions flow along separate ducts to the base of each tooth. The role of venom in prey capture by anguids and xenosaurids has not been clarified (Weinstein et al., 2009; Fry et al., 2010). In varanoids and helodermatids, however, the venom system is more highly developed and seems to be important in feeding or defense (see below).

Table 6.6 provides counts of teeth in the upper jaw for several anguimorph species (Bhullar, 2011).

Xenosauridae

This family of lizards is distinguished by possessing knob-like scalation and flattening of the head and body, associated with living in cracks within cliff faces. There are eight living species, although a ninth has been described (Woolrich-Pina and Smith, 2012). Xenosauridae are considered to be the sister group to a clade containing Helodermatidae+(Anniellidae+Anguidae), although other molecular analyses lead to slightly different conclusions (Pyron et al., 2013).

The dentition of the **flathead, knob-scaled lizard** (*Xenosaurus grandis*) consists of 9 premaxillary teeth and 18 maxillary teeth (Fig. 6.53). The teeth are pointed, slightly recurved, and of similar size. Teeth are absent from the pterygoids.

Helodermatidae

This family traditionally comprises only one genus and two species: **beaded lizard** (*Heloderma horridum*: four subspecies) and the smaller **Gila monster** (*Heloderma suspectum*: two subspecies). However, a reassessment of beaded lizards concluded that the four subspecies should be elevated to full species status (Reiserer et al., 2013). *Heloderma* consumes the eggs of birds and reptiles and attacks a variety of prey, including small vertebrates, earthworms, and insects. The teeth of *Heloderma* are sharp and pointed, with bases formed of plicidentine (Kearney and Rieppel, 2006). Teeth are replaced in the varanid mode, in which the replacing teeth develop distal to their predecessors and move mesially. The functional teeth are consequently spaced apart.

Both species are venomous. In the venom glands, which are located in the lower jaw, the serous, venom-producing compartments are separate from the mucus-producing compartments and each venom compartment is encapsulated and has a well-defined central lumen. Some of the posterior compartments are fused and the total number is reduced to six. It was once thought that there was only a single venom duct in *H. horridum* and several in *H. suspectum* (Weinstein et al., 2009), but studies by Fry et al. (2010) suggest that there are six ducts in both species. The teeth of *Heloderma* possess longitudinal grooves. In the absence of musculature for venom injection, the prey is grasped for some time and secretion from the glands is stimulated by repeated biting movements of the jaws. The venom migrates along the grooves in the teeth by capillarity and enters the prey. The venom in both species

TABLE 6.6 Numbers of Upper Teeth in Various Anguimorphs

Family	Species	Premaxilla	Maxilla	Pterygoid
Xenosauridae	*Xenosaurus agrenon*	9	16	0
	Xenosaurus newmanorum	9	18	0
	Xenosaurus platyceps	9	18	0
	Xenosaurus rectocollaris	9	18	0
	Xenosaurus rackhami	9	16	0
Helodermatidae	*Heloderma suspectum*	9	7	Large row
Anguidae	*Ophisaurus ventralis*	9	16	Large row
Shinisauridae	*Shinisaurus crocodilurus*	7	12	Large row
Lanthanotidae	*Lanthanotus borneensis*	≥9	11	Large row
Varanidae	*Varanus exanthematicus*	9	9	Large row
	Varanus niloticus	9	11	0

Data from Bhullar, B.A.S., 2011. The power and utility of morphological characters in systematics: a fully resolved phylogeny of *Xenosaurus* and its fossil relatives (Squamata: Anguimorpha). Bull. Comp. Zool. 160, 65–181. Courtesy of Editors Bulletin Comparative Zoology.

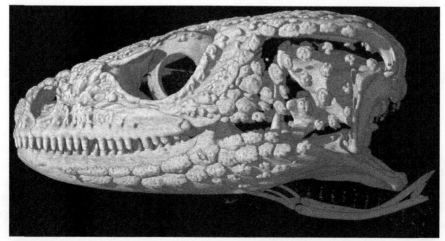

FIGURE 6.53   Dentition of the knob-scaled lizard (*Xenosaurus grandis*). Image width=3.8 cm. *Courtesy Digimorph.org and Dr. J.A. Maisano.*

contains a large number of bioactive molecules, such as hyaluronidase, phospholipase A, serotonin, and kallikreins, that can have a wide variety of hemotoxic or neurotoxic effects (Fry et al., 2012). In view of the rather inefficient venom delivery system, it has often been thought that the main role of the venom is for defense. However, bites by these lizards can produce marked clinical effects in humans, so it cannot be excluded that they play a useful role in predation.

The **Rio Fuerte beaded lizard** (*Heloderma horridum exasperatum*) has a short, broad, rounded snout and a relatively flat, short tail that stores fat. The premaxilla is a single bone holding, on average, 9 teeth (range 5–11), whereas each maxilla carries 6 or 7 teeth and each dentary usually has 9 teeth. The teeth are large, monocuspid, curved, and sharp (Fig. 6.54), with prominent grooves on the mesial and distal aspects. The teeth in the middle of the row are the largest

FIGURE 6.54 Dentition of Rio Fuerta beaded lizard (*Heloderma horridum*). *Courtesy Dr. Mark O'Shea.*

(Bogert and Martin Del Campo, 1956). Alternate groups of teeth tend to be replaced at the same time so that gaps occur within the dentition (eg, a group of the first, fourth, and seventh teeth are replaced together, followed by the second, fifth, and eighth). Conical and curved teeth are also consistently present on the pterygoid (up to eight) and palatine bones (one to a few). Unlike the marginal teeth, the teeth on the pterygoid and palatal bones lack grooves (Mahler and Kearney, 2006).

When feeding, small eggs may be taken into the mouth and pierced and broken by the anterior teeth. Then, the food is positioned further back in the mouth where it is crushed without further contact between the teeth (Herrel et al., 1997). Large eggs may be pierced outside the mouth and the contents lapped up by the tongue.

The dentition of the **Gila monster** (*H. suspectum*) is similar to that of *H.horridum* except that it has more maxillary teeth (eight to nine rather than seven) and often one less premaxillary tooth (eight rather than nine) (Bogert and Martin Del Campo, 1956) (Figs. 6.55–6.57). The sharp, lance-shaped teeth possess longitudinal grooves, but those on the distal aspect are not well developed. Teeth are only variably present on the pterygoid bones, numbering up to a maximum of four, but are not present on the palatines (Bogert and Martin Del Campo, 1956).

Anguidae

Palatal teeth occur in many anguids, either on both the pterygoid and palatine bones or only on the pterygoids. The **European legless lizard** [*Pseudopus* (*Ophisaurus*) *apodus*] is the only anguid that might have teeth on the pterygoid, palatine, and vomer bones (Mahler and Kearney, 2006).

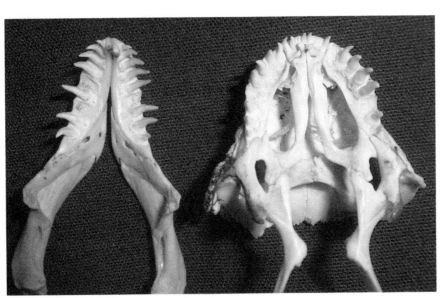

FIGURE 6.55 Dentition of Gila monster (*Heloderma suspectum*). Left: lower dentition; right: upper dentition. Image width=7 cm. *Courtesy UCL, Grant Museum of Zoology. Catalog no. LDUCZ-X94.*

The majority of young anguid lizards show a mode of tooth replacement between the varanid and the iguanid modes (Rieppel, 1978, 1979). The replacement teeth do not arise directly lingual to the predecessor tooth but somewhat distolingually and move into resorption pits toward the final stages of replacement. This intermediate form of tooth replacement was studied in two species of **Canarian skink**: *Chalcides sexlineatus* and *Chalcides viridanus*. Although present in young specimens, in adults this method of tooth replacement progressively changed to the iguanid method, indicating the importance of growth in determining the relationship between the functional and replacing tooth (Delgado et al., 2003). Because replacement occurs in alternate tooth positions at similar times, the teeth seem spaced.

The **slow worm** (*Anguis fragilis*) is a legless lizard that can grow to a length of 50 cm and superficially resembles a snake. However, many features, such as the presence of eyelids and ear openings, distinguish the slow worm from snakes.

Anguis is ovoviviparous. At birth, there are 9 premaxillary teeth, including a median tooth, 10 on each maxilla and 11 on each dentary. By the second year of life, there are still 9 teeth on the premaxillae, but only 9 on each maxilla and 10 on each dentary. This reduction of tooth numbers with age is extremely rare among lizards and may be the result of a single, slightly enlarged tooth replacing two others.

The adult slow worm has relatively large, well-spaced, sharp teeth that are recurved, with the tips facing inward and backward (Cooper, 1966) (Figs. 6.58 and 6.59). The premaxillary teeth are small. The first maxillary tooth is larger than the premaxillary teeth and the size increases to a maximum in the fifth or sixth position, decreasing again posteriorly. With successive replacements the teeth become larger, with the diameter increasing relatively more than the height. However, the tooth crowns never become flattened as in *Varanus niloticus*. Tooth replacement is highly coordinated. Alternate teeth are replaced almost simultaneously along the whole tooth row so that only one-half the teeth may seem to be in function. After the functional tooth is shed, the replacing tooth moves into position while still only about one-third of its final size. It subsequently increases in size, maintaining a backward inclination. Eruption, including growth to full height, reorientation to the final inclination and complete ankylosis, is rapid, taking only 3 days (Cooper, 1966).

VARANIDAE AND THEIR RELATIVES

There is considerable molecular and other evidence that the Shinisauridae, Lanthanotidae, and Varanidae form a clade (Pyron et al., 2013). Owing to their characteristic (varanid) mode of tooth replacement, where the replacing teeth develop behind their predecessors and move forward, the functional teeth in Lanthanotidae and Varanidae are spaced

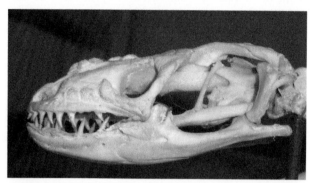

FIGURE 6.56 Dentition of the Gila monster (*Heloderma suspectum*). Lateral view of skull. Actual image length = 10 cm. *Courtesy © Trustees of the Natural History Museum, London.*

FIGURE 6.57 CT image of the dentition of the Gila monster (*Heloderma suspectum*). Image width = 7.2 cm. *Courtesy Digimorph.org and Dr. J.A. Maisano.*

apart. The basal region of the teeth of these two families consists of plicidentine rather than orthodentine (Kearney and Rieppel, 2006), and the folded structure is reflected as longitudinal grooving on the surface of the tooth base. They also share similar venom gland structure. As in Helodermatidae, the serous (venom-producing) compartments are separate from the mucus-producing compartments. Venom is stored within the glands but a muscular system to deliver it to the teeth is absent. There is evidence that, even in the absence of a pressurized delivery system, venom may be important in prey capture among varanids, at least in the **Komodo dragon** (*V. komodoensis*) (Fry et al., 2009). Although potentially bioactive molecules have been identified in the secretions of the venom/salivary glands of

Shinisauridae and Lanthanotidae, there seems to be no evidence that these lizards actually use venom in feeding.

Shinisauridae

This family contains a single species, the **Chinese crocodile lizard** (*Shinisaurus crocodilurus*). This semiaquatic lizard is so named because its tail resembles that of a crocodile. The dentition consists of fairly large, slightly curved, sharp teeth sited to handle its aquatic prey (Fig. 6.60). In the upper jaw on each side there are 3–4 premaxillary teeth and 13–14 maxillary teeth (McDowell and Bogert, 1954; Conrad, 2004). The premaxillary teeth are smaller than the maxillary teeth, and a midline tooth is variably present. The last three to four maxillary teeth are smallest. The bases of the teeth are expanded, but there are no striations indicative of the presence of plicidentine. There are between 8 and 10 pterygoid teeth that are smaller than the margin teeth, but more strongly recurved.

The teeth on the dentary number 14–15. They are similar in shape and size to the maxillary teeth, with the last three to four teeth decreasing in size.

Lanthanotidae

This family contains a single living species, the **earless monitor lizard** (*Lanthanotus borneensis*), restricted to the island of Borneo. It is a semiaquatic, burrowing species with reduced eyes and limbs. Little is known about its diet,

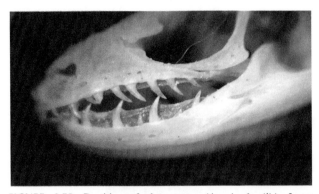

FIGURE 6.58 Dentition of slow worm (*Anguis fragilis*). Image width = 1 cm. *Courtesy Horniman Museum & Gardens and Paolo Viscardi. Catalog no. LDHRN-NH.36.62.*

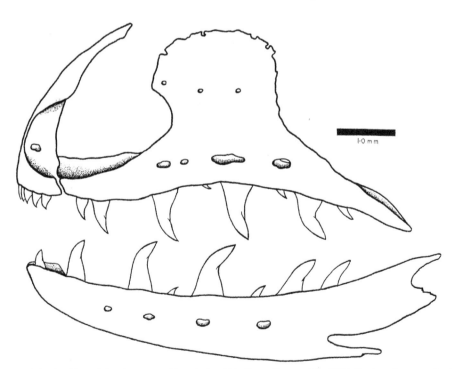

FIGURE 6.59 Diagram of the dentition of the slow worm (*Anguis fragilis*). *From Cooper, J.S., 1966. Tooth replacement in the slow worm (Anguis fragilis). J. Zool. Lond. 150, 235–248. Courtesy Editors Journal of Zoology, London.*

but its pointed, recurved teeth suggest a diet of soft-bodied invertebrates. The dental formula on each side of the jaws is 4 premaxillary teeth, 11–12 maxillary teeth, and 12 dentary teeth (McDowell and Bogert, 1954) (Fig. 6.61). The teeth are more compressed and more sharply recurved than those of *Shinisaurus*. The base is striated because it is composed of plicidentine. In addition, more teeth are present on the pterygoid (11–13 compared to 9–10) and one or two teeth are also present on the palatines (McDowell and Bogert, 1954).

Varanidae

This family contains nearly 80 species of **monitor lizards** that all belong to one genus, *Varanus*, and *Varanus* shows the widest size range of any vertebrate genus. The **Komodo dragon** (*V. komodoensis*), the largest of all lizards, can weigh more than 70 kg and reach a length of 3 m, whereas the smallest monitor (*Varanus brevicauda*) weighs only about 17 g and has a total length of 20 cm. Varanids have long necks and powerful claws. They have a long, forked tongue like that of snakes that, together with a large Jacobson's organ, plays a major role in detecting prey by olfaction. Monitor lizards may be venomous but, although a case for a role of venom in prey capture has been made for the Komodo dragon, its role in feeding or defense among other varanids is uncertain.

Varanid skulls are mesokinetic, possibly amphikinetic, and the snout can be raised or lowered relative to the cranium. Teeth are absent from the palate (Mahler and Kearney, 2006). Tooth numbers for many species of *Varanus* are given by Mertens (1942).

Three species [*Varanus bitatawa*, *Varanus mabitang*, and **Gray's monitor** (*Varanus olivaceus*)] are arboreal and subsist largely on fruit. The diet of Gray's monitor includes about 55% of fruit (Cooper and Vitt, 2002). It possesses blunt posterior teeth that are presumably used to crush fruit.

Most monitors are opportunistic predators and scavengers that generally eat anything that they can subdue. Depending on size, the diet can include mammals, birds, reptiles, fishes, arthropods, eggs, crustaceans, and mollusks. When the diet does not include a high proportion of shelled prey, the teeth

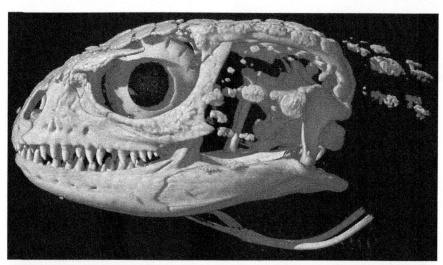

FIGURE 6.60 Dentition of the Chinese crocodile lizard (*Shinisaurus crocodilurus*). Image width = 3.7 cm. *Courtesy Digimorph.org and Dr. J.A. Maisano.*

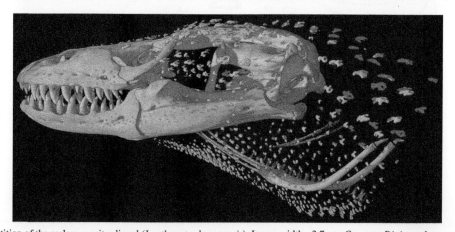

FIGURE 6.61 Dentition of the earless monitor lizard (*Lanthanotus borneensis*). Image width = 3.7 cm. *Courtesy Digimorph.org and Dr. J.A. Maisano.*

are uniformly recurved, pointed, and laterally compressed, so they are sabre like with mesial and distal cutting edges that can be serrated. Tooth attachment is subpleurodont and anky-losed at an angle that minimizes stress when the teeth are loaded vertically (Rieppel, 1979). Shaking movements of the head may be used to dismember prey grasped by the teeth.

The **Bengal** or **common Indian monitor** (*Varanus bengalensis*) usually has 8–9 premaxillary teeth, 10–11 teeth on each maxilla, and 11–12 on each dentary. It has a typical varanid dentition, with sharp, pointed, spaced teeth (Fig. 6.62) and preys mainly on arthropods and mollusks, but it also captures small vertebrate animals and birds. A cineradio-graphic study of feeding by *V. bengalensis* indicates that the upper jaw is elevated at the mesokinetic joint as the mouth opens and depressed as it closes. The motion of the upper jaw drives the recurved teeth into the prey. The holding effect of the teeth is thus augmented without requiring longer teeth and allows relatively large prey to be captured (Rieppel, 1979).

The **water monitor** (*Varanus salvator*) (Fig. 6.63) is a large species and reaches a length of about 1.5 m. There are 8–9 premaxillary teeth, 12–13 teeth on each maxilla, and 11–13 teeth on each dentary. The water monitor has sharp,

pointed recurved teeth and eats a wide range of mainly aquatic prey, including fish, frogs and crabs, but also birds, rodents, and snakes.

The **Komodo dragon** (*V. komodoensis*) is the largest of the monitor lizards, inhabiting the islands of Indonesia, including Komodo and Flores. It has about 60 large teeth in total in the jaws, with 7 premaxillary teeth, 13 teeth on each maxilla, and 12 teeth on each dentary. The teeth are sharp and strongly recurved, with serrated margins (Fig. 6.64), and they can reach lengths of up to 2.5 cm. Only the tips of the teeth are exposed as they are sheathed by gum tissue. Whereas in other monitor lizards, as with most reptiles, there are only one or two replacement teeth at each tooth position, there may be four to five developing teeth near each functional tooth in the Komodo dragon, suggesting that tooth replacement occurs comparatively rapidly (Fig. 6.65).

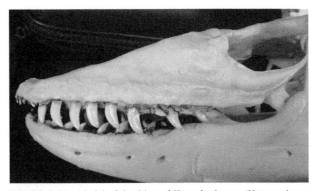

FIGURE 6.64 Model of dentition of Komodo dragon (*Varanus komodoensis*). Image width=23 cm. *Courtesy Professor S.E. Evans.*

FIGURE 6.62 Dentition of young Bengal or common Indian monitor lizard (*Varanus bengalensis*). Image width=20 cm. © *Trustees of the Natural History Museum, London.*

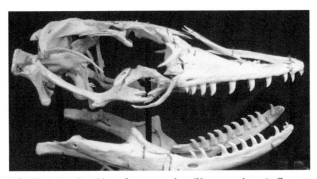

FIGURE 6.63 Dentition of water monitor (*Varanus salvator*). *Courtesy MoLSKCL. Catalog no. X39.*

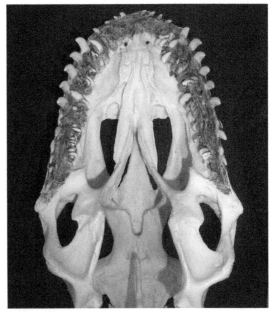

FIGURE 6.65 Model of upper jaw of Komodo dragon (*Varanus komodoensis*), showing the series of replacing teeth. Image width=12 cm. *Courtesy Professor S.E. Evans.*

FIGURE 6.66 Dentition of a juvenile African savannah monitor (*Varanus exanthematicus*) with sharp pointed teeth. *Courtesy Professor S.E. Evans.*

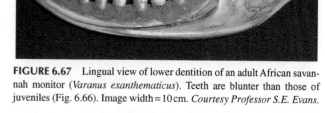

FIGURE 6.67 Lingual view of lower dentition of an adult African savannah monitor (*Varanus exanthematicus*). Teeth are blunter than those of juveniles (Fig. 6.66). Image width = 10 cm. *Courtesy Professor S.E. Evans.*

In the absence of competition, the Komodo dragon is an aggressive and fearless hunter and the adults prey on large mammals, such as deer. Fry et al. (2009) presented evidence that the dragon uses venom in the capture of such prey. This would, of course, curtail the period when the lizard is at risk of injury by struggling prey. It was suggested that venom enters the victim along grooves on the tooth surface and would penetrate deeply into the lacerations produced by the sabre-like teeth.

The skull of the Komodo dragon seems to be adapted to minimizing stresses set up by pulling forces on the jaws rather than to exerting a large bite force (Moreno et al., 2008; Fry et al., 2009). This is consistent with the characteristic method used by the lizard to deflesh prey larger than itself (D'Amore and Blumenschine, 2009). Holding its head at an angle to the body of the prey, the lizard closes the teeth on one side of its mouth on the prey and then withdraws its snout in an arc-like trajectory. This action drags the teeth through the flesh, distal side first. The process is repeated until a morsel of flesh is detached. The morsel is then swallowed by inertial feeding. The Komodo dragon does not crush bones when reducing prey.

In monitor lizards that include a large proportion of shelled prey in their diet as adults, the posterior teeth of adults may be wider and blunter than those in juveniles, an adaptation to durophagy.

The **African savannah monitor** (*Varanus exanthematicus*) is smaller than the water monitor. The dentition of juveniles is adapted to a diet mainly of insects and consists of sharp, recurved teeth (Fig. 6.66). There are 9 premaxillary teeth, 9 teeth on each maxilla, and 10–11 teeth on each dentary. However, with increasing size and tooth replacement, the teeth toward the back of the jaw become much blunter, allowing for a wider variety of prey, including snails (Fig. 6.67). The **white-throated monitor** (*Varanus albigularis*), previously included in *V. exanthematicus*, also possesses blunt crushing posterior teeth. It has a varied diet including arthropods, mollusks, eggs, birds, and snakes.

FIGURE 6.68 Lateral view of dentition of adult Nile monitor (*Varanus niloticus*), showing rounded teeth at the back of the jaws. Image width = 15 cm. *Courtesy Catal RCSOMA/ 448.2.*

The **Nile monitor** (*Varanus niloticus*) has 9 premaxillary teeth, 10–11 teeth on each maxilla, and 11–12 teeth on each dentary. The dentition in this species undergoes a change in morphology with age. All the teeth are initially laterally compressed, recurved, and pointed, with mesial and distal serrated cutting edges. With age, although the premaxillary teeth retain the same morphology, successive replacements result in the posterior teeth becoming rounder, shorter, and wider (Figs. 6.68 and 6.69). Simultaneously, the tooth row becomes shorter relative to the length of the lower jaw, whereas the jaw becomes deep, massive, and bowed ventrally, so that a gap between upper and lower dentitions develops posteriorly (Rieppel and Labhardt, 1979). The overall effect is to increase the crushing force exerted by the posterior teeth on shelled prey.

IGUANIA

This suborder consists of the Agamidae, Chamaeleonidae, and several families containing iguanas and their relatives.

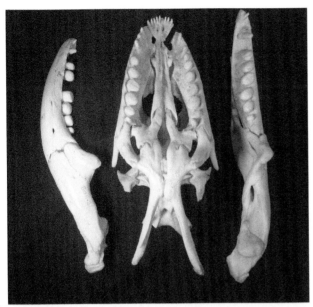

FIGURE 6.69 Occlusal views of the upper and lower dentitions of Nile monitor (*Varanus niloticus*), showing the conical grasping teeth at the front of the jaws and rounded crushing teeth at the back. Image width = 16 cm. *Courtesy RCSOMA/ A408.2.*

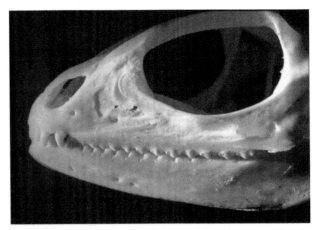

FIGURE 6.70 Dentition of the rainbow agama (*Agama agama*). Note the tricuspid form of the posterior teeth. *Courtesy Dr. D.F.G. Poole.*

The iguanians include many omnivorous and herbivorous species as well as carnivores. Because of the iguanid mode of tooth replacement, the teeth are close-set in the jaws, not spaced out as in varanoids. The teeth of agamids and chameleons are mainly or wholly acrodont and those of the remaining taxa are pleurodont. On this account, the suborder is sometimes divided into Acrodonta and Pleurodonta. Although included in the Toxicofera, the iguanians retain labial glands in both upper and lower jaws, and the glands have a simple structure with mixed serous and mucus regions. Although much remains to be discovered about venom in this group, it is likely that the iguanians diverged early in the evolution of the Toxicofera and that they remain in an "incipient" stage of venom development (Fry et al., 2012).

Acrodonta

Agamidae

This family has six subfamilies and more than 300 species. About one-half of agamids are primarily insectivorous and one-half omnivorous or herbivorous. A row of marginal teeth is present on the premaxilla, maxilla, and dentary. No other teeth are present on any other bone. Typically, their dentition is heterodont and is made up of teeth with two different tooth shapes. A few anterior teeth in both jaws are pointed, conical, and pleurodont. In the upper jaw these are found both on the premaxilla and anterior part of the maxilla. Further back in the maxillae the teeth are broader, compressed laterally, and acrodont.

The adult dentition of the **rainbow agama** (*Agama agama*) displays the typical heterodonty. The premaxilla bears a small median tooth in the position of the egg tooth, which is replaced soon after hatching. At the front of each maxilla in the adult are two caniniform, recurved teeth, of which the second is slightly larger. At the front of each dentary are two similar pleurodont caniniform teeth. Behind these anterior teeth in both jaws is a row of laterally flattened, triangular teeth that increase in size posteriorly (Fig. 6.70). The upper acrodont teeth are tricuspid, with a large central cusp and small mesial and distal cusps. There may be up to 19 posterior teeth in each jaw (Cooper et al., 1970). The first two or three small anterior acrodont teeth are displaced by the enlarging pleurodont teeth at the front of the jaws. The teeth in males are larger than those in females.

The buccal surfaces of the lower teeth slide against the palatal surfaces of the upper teeth to produce an efficient slicing and crushing action. Small prey may be ingested using the tongue. Larger prey can be grasped by the anterior teeth and then crushed by the posterior teeth. The anterior teeth are replaced at least three times, so they increase in size with age. The posterior acrodont teeth are not replaced, so become worn down, although maintaining a sharp edge. Eventually the function of these teeth is taken over by a crest of bone of the jaws.

In the hatchling rainbow agama, nine teeth are present in each jaw, together with an egg tooth in the midline of the premaxilla. This number increases by the addition of acrodont teeth at the posterior end of the tooth row.

In other agamids, the number of teeth is similar to that of *Agama agama*, with between 15 and 20 in each jaw quadrant, as illustrated by the **blue-headed tree agama** (*Agama agricollis*) (Fig. 6.71), and some teeth may show a more definite tricuspid form (Cooper et al., 1970).

It has been suggested that **Hardwicke's spiny-tailed lizard** should be renamed *Saara hardwickii* instead of *Uromastyx hardwickii*, but Pyron et al. (2013) found that

FIGURE 6.71 Dentition of the blue-headed tree agama (*Agama agricollis*). Actual width=3.7 cm. *Courtesy Horniman Museum & Gardens and Dr. Paolo Viscardi. Catalog no. LDHRN-NH 0167.*

the subfamily Uromasticinae is monophyletic and that this redesignation is unnecessary. *Uromastyx hardwickii* is herbivorous. All the teeth are acrodont and none are replaced (although limited replacement may occur in one or two anterior teeth in related species). Unlike other agamids, which have a single tooth on the premaxilla, Hardwicke's spiny-tailed lizard has two teeth on each side of the premaxilla, with the first tooth being the larger. In the hatchling, there are 9 teeth on the maxilla and 10 teeth on the dentary. Tooth number increases with age by addition at the back of each tooth row so that old specimens may have evidence of about 20 teeth in each upper and lower jaw (Cooper and Poole, 1973). The teeth are predominantly single cusped, but with traces of two low additional cusps at the margins. The teeth increase in size posteriorly and are stout with a lingual ridge (Fig. 6.72)

In *Uromastyx*, as in *Sphenodon*, the lower jaw can move forward and backward (Reilly et al., 2001). This propalinal movement, which is made possible by motion between the quadrate and the articular rather than through streptostyly, increases the effectiveness of shearing on plant material. As the teeth in the lower jaw occlude inside the upper, characteristic wear facets are produced on the buccal aspect of the lower teeth and on the palatal aspect of the upper teeth. In this way, the enamel ridges are left sharp. As the anterior teeth are worn away, the whole premaxilla elongates in an occlusal direction to overlap the lower jaw. In aged specimens, the anterior teeth of the maxilla and dentary may be completely worn away and replaced by sharp eburnated edges of bone.

The **eastern bearded lizard** (*Pogona barbata*) (Fig. 6.73) eats mainly insects as a juvenile but, as an adult, is omnivorous and eats leafy greens and vegetables as well. There are three to four teeth on the premaxilla, one of which is caniniform. Each maxilla and dentary bears up to 20 teeth that increase in size posteriorly. This is the only iguanian in which the venom system has been studied. It retains labial glands in both upper and lower jaw, and only 2 of the 13 basal venoms have been identified (Fry et al., 2012).

(A)

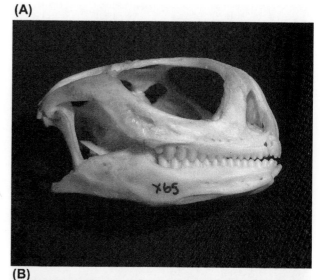

(B)

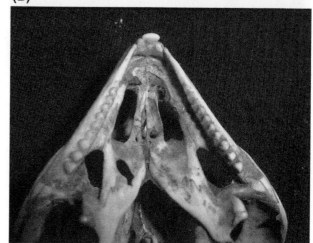

FIGURE 6.72 Dentition of Hardwicke's spiny-tailed lizard (*Uromastyx hardwickii*). (A) Side view. Image width=6 cm. (B) Ventral view of upper jaw, showing the anterior beak. Image width=4.5 cm. *Courtesy MoLSKCL. Catalog no. X65.*

The **Amboina sailfin lizard** (*Hydrosaurus amboinensis*), the largest of the agamid lizards, is herbivorous and has a dentition similar to the eastern bearded lizard, with caniniform teeth anteriorly (Fig. 6.74).

The **peninsular horned tree lizard** (*Acanthosaura armata*) (Fig. 6.75) has many of the dental features typical of agamids, including laterally compressed, acrodont posterior teeth. Unusually, however, it shows an enlarged anterior caniniform tooth in each jaw quadrant, the upper representing the first maxillary tooth.

The **butterfly lizard** (*Leiolepis belliana*) (Fig. 6.76) similarly has a pair of enlarged, caniniform teeth at the front of the jaw, the lower biting in front of the upper and moving in a bony groove between the premaxilla and maxilla.

The dentition of the **eastern garden lizard** (*Calotes versicolor*) is illustrated in Fig. 6.77. There are three teeth

FIGURE 6.73   Dentition of the eastern bearded lizard (*Pogona barbata*). Image width = 7 cm. *Courtesy RCSOMA/ 407.2.*

FIGURE 6.74   Dentition of the Amboina sailfin lizard (*Hydrosaurus amboinensis*). *Courtesy RCSOMA/ 407.1.*

on the premaxilla, of which the middle is the largest. Each maxilla bears up to about 17 mostly tricuspid teeth. The anterior teeth are more heavily worn than the larger posterior teeth. Teeth on each dentary number up to 19; the first 2 teeth are very small, the third tooth is more caniniform and the remainder tricuspid, with wear in the more anterior teeth.

Hatchlings of the **spotted toad-head agama** (*Phrynocephalus melanurus*) have no premaxillary teeth and nine small, acrodont teeth in each jaw quadrant. Two teeth develop later on the premaxilla. Only the premaxillary

FIGURE 6.75   Dentition of peninsular horned tree lizard (*Acanthosaura armata*). Note the enlarged anterior caniniform tooth in each jaw quadrant. Image width = 6.5 cm. *Courtesy Professor S.E. Evans.*

teeth and the first two teeth on each maxilla and dentary are replaced and change from acrodont to pleurodont in the process. With the addition of teeth at the back of the tooth row, the adult dentition consists of four pointed teeth at the front that, in the upper jaw, represent two premaxillary teeth and the first two maxillary teeth, with the latter tooth being the largest. These teeth are followed by eight to nine laterally compressed, broad teeth (and, rarely, a 10th tooth). The age of these agamid lizards can be determined by the presence of resting (hibernation) lines in the bone, and it has been shown that body length does not always correlate with age (Ananieva et al., 2003).

Chamaeleonidae

Chameleons are distinguished by their zygodactylous feet, their prehensile tail, their separately mobile stereoscopic eyes, their long and rapidly protrusible ballistic tongue and their legendary ability to change color. There are more than 180 species in 12 genera (http://reptile-database.org/). They are carnivorous and their diet consists mainly of insects, although the larger species can be cannibalistic. Some small species of chameleons can extend their tongue by more than their body length. Tongue projection is assisted by forward movement of the hyoid apparatus and is very rapid. For example, during a strike at prey 20 cm in front of a **Malagasy giant chameleon** [*Furcifer* (*Chameleo*) *oustaleti*], the tongue reached its full extent in 43 ms and achieved a maximum velocity of 4.9 m/s. Chameleons can adjust the speed of tongue projection: the tongue contacted prey at a distance of 35 cm in only 71 ms and the maximum velocity reached 5.8 m/s (Wainwright et al., 1991). Prey is immobilized by adhesion to the sticky mucus coating the tongue surface, assisted by muscular activity at the tip of the tongue.

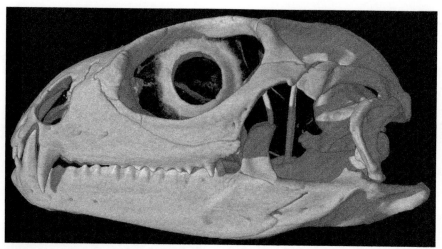

FIGURE 6.76 Dentition of the butterfly lizard (*Leiolepis belliana*): micro-CT image. Image width=3.8 cm. *Courtesy Digimorph.org and Dr. J.A. Maisano.*

FIGURE 6.77 Eastern garden lizard (*Calotes versicolor*). *Courtesy Life S MoLSKCL. Catalog no. X77.*

The teeth of chameleons are acrodont (Fig. 6.78) and are not replaced. There are no teeth on the palate. The premaxilla is reduced in size and may bear up to 2 small teeth, whereas each maxilla and dentary typically bear between 15 and 22 teeth (Edmund, 1969). Tooth size increases from front to back of the tooth row. At hatching there are about 10 small teeth in each maxilla and mandible. Subsequently, larger, triangular, compressed, and tricuspid teeth are added to the back of the jaws. Many of these features can be seen in the dentition of **Owen's chameleon** (*Trioceros oweni*) (Fig. 6.79).

The **veiled chameleon** (*Chamaeleo calyptratus*) has one tooth on the reduced premaxilla. Near hatching, there

(A)

(B)

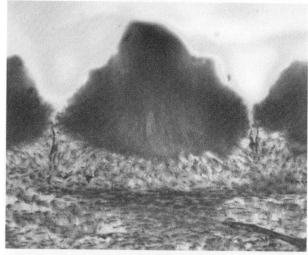

FIGURE 6.78 Ground longitudinal section of the lower jaw of a chameleon (*Chamaeleo* sp.). (A) Low power view showing 16 acrodont teeth. (B) Higher power showing acrodont attachment to the crest of the jaw. © *Museums at the Royal College of Surgeons, Tomes slide collection. Catalog no. 582.*

are 11 mineralized teeth in each upper jaw quadrant and 9 in each lower jaw quadrant. After hatching, larger additional teeth are added to the back of the jaws as room is made available until there are up to 23 maxillary teeth and 21 dentary teeth in each jaw quadrant (Buchtová et al., 2013). The first 9 teeth in the lower jaw and 10 in the upper jaw are small and mineralize during embryonic development. The following five teeth are larger and develop in the juvenile dentition (Fig. 6.80A). The most posterior teeth (nine teeth in the maxilla and seven in the dentary) are the largest and broadest in the dentition and possess additional side cusps (Buchtová et al., 2013). The acrodont teeth fuse to each other at their bases. Near hatching there are 11 in the upper jaw and 9 in the dentary, with a larger egg tooth in the midline flanked on each side by the very small premaxillary tooth (Buchtová et al., 2013). There are no teeth on the palate.

With continuous wear and a lack of replacement, the teeth may be worn away so that in old chameleons the anterior part of the biting surface may be formed in part by the jaw bone (Fig. 6.80B).

Pleurodonta

The 12 families that make up this group include carnivorous, omnivorous, and herbivorous lizards (Cooper and Vitt, 2002).

Teeth can be round in section or flattened, with mesial and distal cutting edges. Cusps are often present on the cutting edges and the number, size, and definition of the cusps vary considerably. The dentitions are generally heterodont: tooth form tends to be simple at the front of the jaw and more complex toward the back. There is a broad correlation between the marginal dentition and the diet, first described by Hotton (1955) and Montanucci (1968) in American pleurodonts. The correlation is outlined below in a simplified form and includes observations on additional species. The classification was based on the morphology of the posterior teeth (degree of lateral tooth flattening and cusp development), the evenness of the profile formed by the tooth tips, and the point in the tooth row that marks the transition between pointed anterior teeth and teeth with more complex crown form.

1. **Herbivores**: The teeth are highly flattened with blade-like mesial and distal cutting edges that bear cusps. The first cusped tooth is located more anteriorly than in the other groups. The posterior teeth of the herbivorous **common** or **green iguana** (*Iguana iguana*) have serrated mesial and distal cutting edges, with numerous small, indistinct cusps (Fig. 6.81). The teeth of the more omnivorous **black iguana**

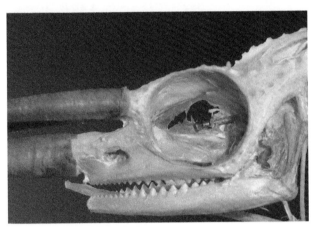

FIGURE 6.79 Dentition of Owen's chameleon (*Trioceros oweni*). Image width=6.2 cm. *Courtesy Horniman Museum & Gardens and Dr. Paolo Viscardi. Catalog no. LDHRN-NH 34.54.*

(A) **(B)**

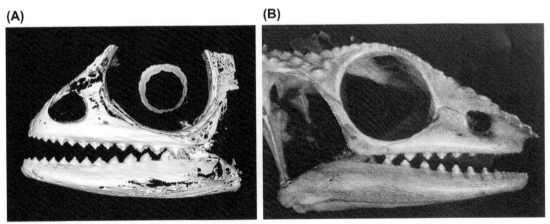

FIGURE 6.80 (A) Dentition of 1.5-year-old veiled chameleon (*Chamaeleo calyptratus*). Micro-CT image. Image width=3.5 cm. *(From Buchtová, M., Zahradníček, O., Balková, S., Tucker, A.S., 2013. Odontogenesis in the veiled chameleon (Chamaeleo calyptratus). Arch. Oral. Biol. 58, 118–133. Courtesy Editors Archives of Oral Biology.)* (B) Dentition of common chameleon (*Chameleo chameleo*), showing wear of teeth in the anterior region of the jaw.

(*Ctenosaura*) are more rounded in section and have less pronounced cutting edges, with three to eight well-defined cusps (Fig. 6.82A and B). Hotton (1955) found teeth similar to those of *Ctenosaura* in the **iguana** (*Dipsosaurus*), the **rock iguanas** (*Cyclura*) (Fig. 6.83), and the **chuckwalla** (*Sauromalus*), and they are also found in other herbivorous members of the Iguanidae, such as the **Galapagos marine iguana** (*Amblyrhynchus*) (Fig. 6.84) and the **Fijian banded iguana** (*Brachylophus*).

2. **Predators on ants and other slow-moving insects**: Teeth are blunt, stout, conical, or peg-shaped and lack

cusps. The crown profile is uneven. Examples include *Phrynosoma* spp. (Phrynosomatidae).

3. **Predators on more active arthropods.** Teeth are sharp and cusp development is variable. Cusped teeth are further back in tooth row than in herbivores.

(A)

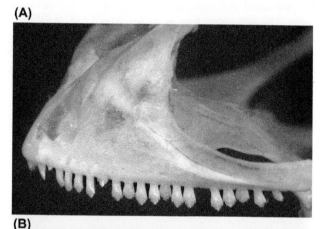

(B)

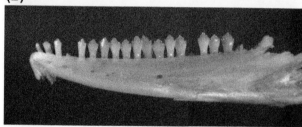

FIGURE 6.82 Dentition of young black iguana (*Ctenosaura similis*), skull length 21.1 mm. (A) Upper jaw. (B) Lower jaw. The anterior teeth are conical and curved, whereas the posterior teeth have three to five cusps on mesial and distal cutting edges. *From Torres-Carvajal, O., 2007. Heterogeneous growth of marginal teeth in the black iguana* Ctenosaura similis *(Squamata: Iguania). J. Herpetol. 41, 528–531. Courtesy Editors Journal of Herpetology.*

(A)

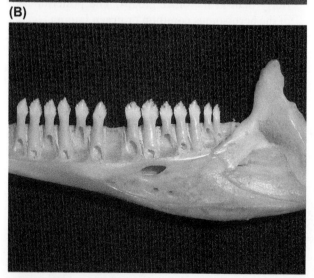

(B)

FIGURE 6.81 (A) Dentition of *Iguana* sp. (probably common iguana, *Iguana iguana*). Image width=4 cm. *(Courtesy Horniman Museum & Gardens and Dr. Paolo Viscardi. Catalog no. LDHRN-NH 2. 1425.)* (B) Lingual aspect of lower jaw of common iguana. In both, note serrated mesial and distal cutting edges on teeth. Note the resorption pits indicating the position of replacement teeth at the bases of the functional teeth. Image width=3 cm. *(Courtesy of Professor S.E. Evans.)*

FIGURE 6.83 Dentition of rhinoceros iguana (*Cyclura cornuta*). Image width=12 cm. *Courtesy © Trustees of the Natural History Museum, London.*

Crown profile is even, with the posterior teeth level with the front teeth. Teeth of species taking arthropods with heavy integuments (eg, beetles, weevils, dragonflies) tend to be slender and somewhat peg shaped with weak-to-moderate cusp development. Examples given by Hotton (1955): **zebra-tailed lizard** (*Callisaurus*) and **tree** or **brush lizards** (*Urosaurus*) (both Phrynosomatidae). Other possible members of this group are the Madagascan **sand iguana** (*Chalarodon*) and **collared iguana** (*Oplurus*) (both Opluridae). In species with a diet including a

high proportion of arthropods with medium-weight integuments (eg, crickets, grasshoppers, flies), teeth tend to be sharp and stoutly conical, with moderately to highly developed cusps. Examples given by Hotton: **collared lizard** (*Crotaphytus*) and **leopard lizard** (*Gambelia*) (both Crotaphytidae), **side-blotched lizard** (*Uta*: Phrynosomatidae). The **basilisk lizard** (*Basiliscus*: Iguanidae) (Montucci, 1968) also seems to be typical of this group.

It is not always possible to assign a particular species to one of these groups with confidence, as Hotton (1955) noted in the case of the **earless lizard** (*Holbrookia*) and two species of the **fence lizard** (*Sceloporus*).

Montanucci (1968) also observed variations in the pterygoid teeth among Pleurodonta with different diets. In the herbivorous iguanids *Ctenosaura* and *Iguana* (see Fig. 6.85), the pterygoid teeth are well developed and seem to play a major role in the gripping function of the dentition. Members of the predominantly insectivorous *Basiliscus* (Corytophanidae) have few pterygoid teeth, probably because the marginal teeth suffice to grasp and crush the small prey.

Iguanidae

All of the iguanids for which quantitative data are available are omnivorous or herbivorous: plant material makes up on average 85% of the diet of the family as a whole (Cooper and Vitt, 2002). Their digestive tract is specialized for digesting plant material, because the microflora of the hindgut is capable of breaking down cellulose.

The teeth tend to be conical at the front of the jaws and are probably used for grasping or tearing. Toward the

FIGURE 6.84 Dentition of the marine iguana (*Amblyrhynchus cristatus*). *Courtesy Dr. J. Seitz.*

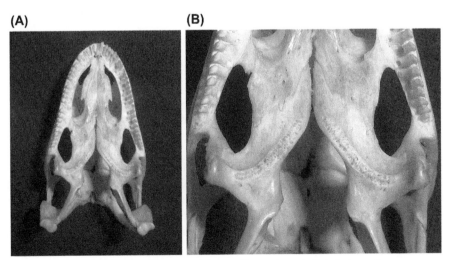

FIGURE 6.85 (A) Ventral view of dentition of the common (green) iguana (*Iguana iguana*). Note that very few spaces are present in the functional dentition. Image width = 6.2 cm. (B) Higher power view of the teeth on the pterygoid. Image width = 3.7 cm. *Courtesy Professor S.E. Evans.*

TABLE 6.7 Tooth Numbers in Various Genera of Iguanian Lizards

Family	Genus	Pterygoid	Premaxilla	Maxilla	Dentary
Iguanidae	*Amblyrhynchus*	0–7	7	17–21	17–24
Iguanidae	*Brachylophus*	1–8	6–7	16–19	18–21
Iguanidae	*Conolophus*	None	7–8	15–20	17–21
Iguanidae	*Ctenosaura*	3–14	7–8	20–26	20–33
Iguanidae	*Cyclura*	9–10	6–10	19–23	22–28
Iguanidae	*Dipsosaurus*	None	7–8	16–19	20–23
Iguanidae	*Iguana*	8–52	6–7	20–26	19–30
Iguanidae	*Sauromalus*	0–7	5–6	16–20	15–25
Opluridae	*Chalarodon*	2–4	6	16–18	19–21
Opluridae	*Oplurus*	4–9	6	15–19	16–23

Apart from the premaxilla the figures shown are from one side of the skull.
From Avery, D.F., Tanner, W.W., 1971. Evolution of the Iguanine Lizards (Sauria, Iguanidae) as Determined by Osteological and Mycological Characters. Brigham Young University Science Bulletin, vol. 12, pp. 1–79. Courtesy Editors Brigham Young University Science Bulletin.

back of the tooth row there is a transition to more specialized teeth that may have a cutting or crushing function. The upper teeth close outside the lowers. The teeth are replaced periodically and there is usually little, if any, evidence of wear or abrasion. There are usually more teeth than among the agamids and tooth numbers vary between genera: some examples are shown in Table 6.7 (Avery and Tanner, 1971).

Small teeth on the posterior margin of the pterygoids are very common in iguanids, but the number and size of the teeth vary considerably. Teeth are less common on the palatines and are variably expressed. Pterygoid teeth are arranged as single, loosely organized rows, although sometimes there are two rows (Mahler and Kearny, 2006). Montanucci (1968) suggested that the double row of pterygoid teeth in the common iguana provides an important function in grasping food while it is being processed by the shearing action of the posterior teeth. Food can also be reduced by being crushed between the tongue and the pterygoid teeth (Reilly et al., 2001).

The **common** or **green iguana** (*I. iguana*) is exclusively herbivorous. The anterior teeth are sharp, pointed, and recurved, whereas the posterior teeth are compressed laterally and the crowns look diamond-shaped when viewed from the side. Their sloping serrated edges are used for cropping leaves and for efficient shredding of leaves, fruits, and flowers (Fig. 6.81B). Tooth replacement is rapid so that, although a tooth row usually shows some gaps at positions where teeth are being replaced, the gaps rarely involve two adjacent tooth positions (Figs. 6.81B and 6.85 and see Chapter 10). Each tooth may be replaced

four or five times per year (Kline and Cullum, 1984). The teeth show few signs of wear, either because of frequent tooth replacement or the lack of tooth-to-tooth contact during feeding or both.

There are usually seven teeth on the premaxilla, including one median tooth. Tooth numbers on the maxillae and dentaries increase with age and size, eventually numbering about 30 on each. The common iguana has a large number of pterygoid teeth (Mahler and Kearny, 2006), usually arranged in two rows (Fig. 6.85). Like the marginal teeth, the number of pterygoid teeth increases with age, eg, from 20 to 67 in individuals between 33 and 93 mm in length (Montanucci, 1968).

The single species of **marine iguana** (*Amblyrhynchus cristatus*) is found only in the Galápagos archipelago. This iguana feeds by diving in the sea for marine algae. Its short snout allows it to make close contact with the marine substrate. Because of the effort required to dive up to a depth of 10 m into the cold sea, marine iguanas only feed once per day. All of the teeth have well-developed lateral cusps (Fig. 6.84). It seems likely that this tricuspid form would increase the grip of the teeth on seaweed during browsing.

The dentition of the **black spinytail iguana** (*Ctenosaura similis*), like that of *Iguana*, consists of pointed, conical, anterior teeth and laterally flattened, multicusped posterior teeth, but the posterior teeth are rounder in section and less flattened than those in *Iguana*. The pterygoids bear a single row of up to 30 large, slightly curved, conical teeth. In the related species, *Ctenosaura* (*Enyaliosaurus*) *clarki*, there are only about 14 much smaller pterygoid teeth.

The number and morphology of the teeth change with age. In black iguanas (Torres-Carvajal, 2007) the premaxillary teeth remain stable at 7, whereas the number on each maxilla increases from 18 to about 27, and on the dentary from 20 to about 36. The pterygoid tooth number also increases with age, from 12 to 30 in individuals between 37 and 95 mm long. In juveniles, there are fewer conical teeth at the front of the dentition than in adults. In addition, the posterior teeth have a greater number of cusps (up to six) (Fig. 6.82A and B) than in adults (usually three, sometimes four) (Fig. 6.86A and B). These changes seem to reflect a change in diet: juvenile black iguana feed primarily on insects, whereas adults eat predominantly soft plant material such as buds, flowers, and fruits, together with insects, eggs and, occasionally, small vertebrates.

Mahler and Kearney (2006) reported that in many specimens of *Ctenosaura* (as well as in *Iguana*, *Basilicus*, *Crotaphytus*, and *Cyclura*), one row of pterygoid teeth is flanked by a second, more medial row, which may be complete.

Dactyloidae and Polychrotidae

The Dactyloidae contains about 400 species, usually assigned to the single genus *Anolis*. This would make *Anolis* lizards the world's most species-rich amniote genus. However, it has been suggested that seven clades could be of generic status. The Polychrotidae (bush anoles) are no longer grouped with *Anolis*. The members of both families are insectivorous; only a small proportion of species consume plant material.

The **green anole** (*Anolis carolinensis*) has a heterodont dentition. There are around 50 teeth on each side of the skull. The front teeth are conical and sharp, whereas those in the back of the jaws have suggestions of multiple cusps (Fig. 6.87).

Allison's (blue-headed) anole (*Anolis allisoni*) clearly displays heterodonty, with the anterior teeth being unicuspid and the posterior teeth being tricuspid (Figs. 6.88 and 6.89).

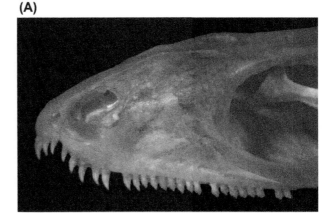

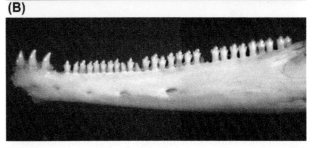

FIGURE 6.86 Dentition of adult black iguana (*Ctenosaura similis*), skull length 72.6 mm. (A) Upper jaw. (B) Lower jaw. Note that the number of conical anterior teeth is much smaller than in the juvenile in Fig. 6.82 and that the posterior teeth have a total of three to five cusps. *From Torres-Carvajal, O., 2007. Heterogeneous growth of marginal teeth in the black iguana Ctenosaura similis (Squamata: Iguania). J. Herpetol. 41, 528–531. Courtesy Editors of Journal of Herpetology.*

FIGURE 6.87 Dentition of green anole (*Anolis carolinensis*): micro-CT image. Image width = 2.2 cm. *Courtesy Digimorph.org and Dr. J.A. Maisano.*

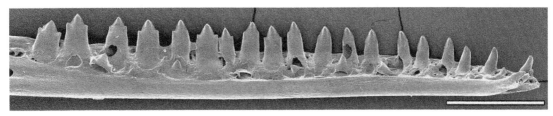

FIGURE 6.88 Scanning electron micrograph of dentition of lower jaw of Allison's anole (*Anolis allisoni*), showing heterodonty between anterior and posterior teeth. Scale = 1 mm. *From Zahradníček, O., Buchtová, M., Dosedelová, H., Tucker, A.S., 2014. The development of complex tooth shape in reptiles. Front. Physiol. 5, 1–7. Courtesy Editors Frontiers of Physiology.*

(A) **(B)**

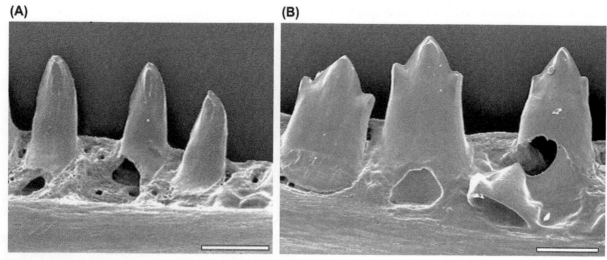

FIGURE 6.89 Higher power scanning electron micrograph of anterior teeth seen in Fig. 6.88. (A) Unicuspid morphology of anterior teeth. (B) Tricuspid morphology of posterior teeth. Scale = 200 μm. *From Zahradníček, O., Buchtová, M., Dosedelová, H., Tucker, A.S., 2014. The development of complex tooth shape in reptiles. Front. Physiol. 5, 1–7. Courtesy Editors Frontiers of Physiology.*

ONLINE RESOURCES

Digimorph. Collection of MicroCT images of reptilian skulls: www.digimorph.org.

Gans Collections and Charitable fund. Provides access to all volumes of Biology of Reptilia (ed. C Gans): http://carl-gans.org/biology-reptilia-online/.

Reptile database. Taxonomic and biogeographich information on over 10,000 reptilian species: www.reptile-database.com.

The Texan Herper. General information on reptiles: www.thetexanherper.com.

University College London Museum. Collection of Amphisbaenia: http://www.ucl.ac.uk/museums-static/obl4he/vertebratediversity/amphisbaenia.html.

REFERENCES

Ananieva, N.B., Orlov, N.L., 2013. Egg teeth of squamate reptiles and their phylogenetic significance. Biol. Bull. 40, 600–605.

Ananieva, N.B., Smirina, E.M., Nikitina, N.G., 2003. Dentition of *Phrynocephalus melanurus*: does tooth number depend on body size and/or age? Russ. J. Herpetol. 10, 1–6.

Avery, D.F., Tanner, W.W., 1971. Evolution of the Iguanine Lizards (Sauria, Iguanidae) as Determined by Osteological and Mycological Characters, vol. 12. Brigham Young University Science Bulletin. pp. 1–79.

Bauer, A.M., Pauwels, O.S.G., Chanome, L., 2002. A new species of cave-dwelling *Cyrtodactylus* (Squamata: Gekkonidae) from Thailand. J. Nat. Hist. Chulalongkorn Univ. 2, 19–29.

Bauer, A.M., Russell, A.P., 1988. Dentitional diversity in *Rhacodactylus* (reptilia: Gekkonidae). Mem. Qld. Mus. 29, 311–321.

Bauer, A.M., Russell, A.P., 1989. A systematic review of the genus *Uroplatus* (Reptilia: Gekkonidae), with comments on its biology. J. Nat. Hist. 23, 169–203.

Bellairs, Ad'A., 1969. The Life of Reptiles. Chapter 4: Feeding and Cranial Mechanics, vol. 1. Weidenfeld and Nicholson, London, pp. 116–183.

Bels, V.L., Chardon, M., Kardong, K.V., 1994. Biomechanics of the hyolingual system in Squamata. In: Bels, V.L., Chardon, M., Vandewalle, P. (Eds.), Biomechanics of Feeding in Vertebrates: Adv. Comp. Env. Physiol., vol. 18. Springer, Berlin, pp. 197–241.

Benton, M.J., 2015. Vertebrate Palaeontology, fourth ed. Wiley-Blackwell, Chichester.

Berkovitz, B.K.B., 2013. Nothing But the Tooth. Elsevier, London.

Bhullar, B.A.S., 2011. The power and utility of morphological characters in systematics: a fully resolved phylogeny of *Xenosaurus* and its fossil relatives (Squamata: Anguimorpha). Bull. Comp. Zool. 160, 65–181.

Bogert, C., Martin del Campo, R., 1956. The gila monster and its allies. Bull. Am. Mus. Nat. Hist. 109, 1–142.

Brizuela, S., Albino, A.M., 2009. The dentition of the neotropical lizard genus *Teius* Merrem 1820 (Squamata Teiidae). Trop. Zool. 22, 183–193.

Buchtová, M., Zahradníček, O., Balková, S., Tucker, A.S., 2013. Odontogenesis in the veiled chameleon (*Chamaeleo calyptratus*). Arch. Oral. Biol. 58, 118–133.

Conrad, J.L., 2004. Skull, mandible, and hyoid of *Shinisaurus crocodilurus* Ahl (Squamata, Anguimorpha). Zool. J. Linn. Soc. 141, 399–434.

Cooper, J.S., 1963. Dental Anatomy of the Genus *Lacerta* (Ph.D. thesis). University of Bristol.

Cooper, J.S., 1966. Tooth replacement in the slow worm (*Anguis fragilis*). J. Zool. Lond. 150, 235–248.

Cooper, J.S., Poole, D.F.G., 1973. The dentition and dental tissues of the agamid lizard, Uromastyx. J. Zool. Lond. 169, 85–100.

Cooper, J.S., Poole, D.F.G., Lawson, R., 1970. The dentition of agamid lizards with special reference to tooth replacement. J. Zool. Lond. 162, 85–98.

Cooper, W.E., Vitt, L.J., 2002. Distribution, extent and evolution of plant consumption by lizards. J. Zool. Lond. 257, 487–517.

Dalrymple, G.H., 1979. On the jaw mechanism of the snail-crushing lizards, *Dracaena* Daudin 1802 (Reptilia, Lacertilia, Teiidae). J. Herpetol. 13, 303–311.

D'Amore, D.C., Blumenschine, R.J., 2009. Komodo Monitor (*Varanus komodoensis*) feeding behavior and dental function reflected through tooth marks on bone surfaces, and the application to ziphodont paleobiology. Paleobiology 35, 525–552.

Delgado, S., Davit-Béal, T., Sire, J.-Y., 2003. The dentition and tooth replacement pattern in *Chalcides* (Squamata; Scincidae). J. Morphol. 256, 146–159.

De Vree, F., Gans, C., 1994. Feeding in tetrapods. In: Bels, V.L., Chardon, M., Vandewalle, P. (Eds.), Biomechanics of Feeding in Vertebrates: Adv. Comp. Env. Physiol., 18. Springer, Berlin, pp. 93–118.

Edmund, A.S., 1969. Dentition. In: Gans, C. (Ed.), Biology of the Reptilia, vol. 1. Academic Press, London, pp. 117–200.

Estes, R., Williams, E.E., 1984. Ontogenetic variation in the molariform teeth of lizards. J. Vert. Palaeontol. 4, 96–107.

Evans, S.E., 2008. The skull of lizards and tuatara. In: Gans, C., Gaunt, A.S., Adler, K. (Eds.), Biology of the Reptilia. Morphology H. The Skull of Lepidosauria, vol. 20. Society for the Study of Amphibians and Reptiles, Ithaca, NY, pp. 1–347.

Fry, B.G., Vidal, N., Norman, J.A., Vonk, F.J., Scheib, H., Ramjan, S.F.R., Kuruppu, S., Fung, K., Hedges, S.B., Richardson, M.K., Hodgson, W.C., Ignjatovic, V., Summerhayes, R., Kochva, E., 2006. Early evolution of the venom system in lizards and snakes. Nat. Lond. 439, 584–588.

Fry, B.G., Wroe, S., Teeuwisse, W., van Osch, M.J., Moreno, K., Ingle, J., McHenry, C., Ferrara, T., Clausen, P., Scheib, H., Winter, K.L., Greisman, L., Roelants, K., van der Weerd, L., Clemente, C.J., Giannakis, E., Hodgson, W.C., Luz, S., Martelli, P., Krishnasamy, K., Kochva, E., Kwok, H.F., Scanlon, D., Karas, J., Citron, D.M., Goldstein, E.J., McNaughtan, J.E., Norman, J.A., 2009. A central role for venom in predation by *Varanus komodoensis* (Komodo Dragon) and the extinct giant *Varanus (Megalania) priscus*. Proc. Natl. Acad. Sci. U.S.A. 106, 8969–8974.

Fry, B.G., Winter, K., Norman, J.A., Roelants, K., Nabuurs, R.J.A., van Osch, M.J.P., Teeuwisse, W.M., van der Weerd, L., McNaughtan, J.E., Kwok, H.F., Scheib, H., Greisman, L., Kochva, E., Miller, L.J., Gao, F., Karas, J., Scanlon, D., Lin, F., Kuruppu, S., Shaw, C., Wong, L., Hodgson, W.C., 2010. Functional and structural diversification of the Anguimorpha lizard venom system. Mol. Cell. Proteomics 9, 2369–2390.

Fry, B.G., Casewell, N.R., Wüster, W., Vidal, N., Young, B., Jackson, T.N.W., 2012. The structural and functional diversification of the Toxicofera reptile venom system. Toxicon 60, 434–448.

Gans, C., 1957. Anguimorph tooth replacement in *Amphisbaena alba* Linnaeus 1758 and *A. fulginosa* Linnaeus 1758. Reptilia: Amphisbaenae. Brevoria 70, 1–12.

Gans, C., 1960. Studies on amphisbaenids (amphisbaenia: reptilia). A taxonomic revision of Trogonophinae. Bull. Am. Mus. Nat. Hist. 119, 133–204.

Gans, C., 1968. Relative success of divergent pathways in amphisbaenian specialization. Am. Nat. 102, 345–362.

Gans, C., 1978. The characteristics and affinities of the Amphisbaenia. Trans. Zool. Soc. Lond. 34, 347–416.

Gorniak, G.C., Rosenberg, H.I., Gans, C., 1982. Mastication in the tuatara, *Sphenodon punctatus* (Reptilia: Rhynchocephalia): structure and activity of the motor system. J. Morphol. 171, 321–353.

Greer, A.R., Chong, J., 2007. Number of maxillary teeth in scincid lizards: lineage characteristics and ecological implications. J. Herpetol. 41, 94–101.

Harrison, H.S., 1901a. Development and succession of the teeth in *Hatteria punctata*. Quart. J. Micr. Sci. 44, 161–213.

Harrison, H.S., 1901b. *Hatteria punctata*, its dentition and its incubation period. Anat. Anz. 22, 145–158.

Herrel, A., Wauters, I., Aerts, P., de Vree, F., 1997. The mechanics of ovophagy in the beaded lizard (*Heloderma horridum*). J. Herpetol. 31, 383–393.

Herrel, A., Schaerlaeken, V., Meyers, J.J., Metzger, K.A., Ross, C.F., 2007. The evolution of cranial design and performance in squamates: consequences of skull-bone reduction on feeding behaviour. Integr. Comp. Biol. 47, 107–117.

Hotton, N., 1955. A survey of adaptive relationships of dentition to diet in the North American Iguanidae. Am. Midl. Nat. 53, 88–144.

Hutchinson, M.N., 1993. Family Scincidae. In: Glasby, C.G., Ross, G.J.B., Beesley, P.L. (Eds.), Fauna of Australia. Amphibia and Reptilia, vol. 2A. AGPS, Canberra, pp. 261–279.

Jones, M.E.H., O'Higgins, P., Fagan, M.J., Evans, S.E., Curtis, N., 2012. Shearing mechanics and the influence of a flexible symphysis during oral food processing in *Sphenodon* (Lepidosauria: Rhynchocephalia). Anat. Rec. 295, 1075–1091.

Kearney, M., 2003. Diet in the amphisbaenian *Bipes biporus*. J. Herpetol. 37, 404–408.

Kearney, M., Rieppel, O., 2006. An investigation into the occurrence of plicidentine in the teeth of squamate reptiles. Copeia 337–350.

Kline, L.W., Cullum, D.R., 1984. A long-term study of the tooth replacement phenomenon in the young green *Iguana iguana*. J. Herpetol. 18, 176–185.

Kluge, A.G., 1962. Comparative osteology of the eublepharid lizard genus *Coleonyx* Gray. J. Morphol. 110, 299–332.

Kluge, A.G., 1976. Phylogenetic relationships in the lizard family Pygopodidae: an evaluation of theory, methods and data. Misc. Publ. Mus. Zool. Univ. Mich. 152, 1–80.

Mahler, I., Kearney, M., 2006. Fieldiana Zool. N.S. 108, 1–61.

Martín, J., Ortega, J., López, P., Pérez-Cembranos, A., Pérez-Mellado, V., 2013. Fossorial life does not constrain diet selection in the amphisbaenian *Trogonophis wiegmanni*. J. Zool. 291, 226–233.

Mateo, J.A., Lopez-Jurado, L.F., 1997. Dental ontogeny in *Lacerta lepida* and its relationship to diet. Copeia 461–463.

McDowell, S.B., Bogert, C.M., 1954. The systematic position of *Lanthanotus* and the affinities of the anguinomorphan lizards. Bull. Am. Mus. Nat. Hist. 105, 1–142.

Mertens, R., 1942. Die Familien der Warane (Varanidae) Part 2. Schadel. Abh. Senckenberg. Naturforsch. Ges. 465, 117–234.

Metzger, K., 2002. Cranial kinesis in Lepidosauria: skulls in motion. In: Aerts, P., D'Août, K., Herrel, A., van Damme, R. (Eds.), Topics in Functional and Ecological Vertebrate Morphology. Shaker Publishing, Aachen, pp. 15–46.

Montanucci, R.R., 1968. Comparative dentition in four iguanid lizards. Herpetologica 24, 305–315.

Moreno, K., Wroe, S., Clausen, P., McHenry, C., D'Amore, D., Rayfield, E.J., Cunningham, E., 2008. Cranial performance in the Komodo dragon (*Varanus komodoensis*) as revealed by high-resolution 3-D finite element analysis. J. Anat. 212, 736–746.

Nikitina, N.G., Ananieva, N.B., 2009. Characteristics of dentition in gekkonid lizards of the genus *Teratoscincus* and other Gekkota (Sauria, Reptilia). Biol. Bull. 36, 237–242.

Patchell, F.C., Shine, R., 1986a. Hinged teeth for hard-bodied prey: a case of convergent evolution between snakes and legless lizards. J. Zool. Lond. 208, 269–275.

Patchell, F.C., Shine, R., 1986b. Feeding mechanisms in pygopodid lizards: how can *Lialis* swallow such large prey? J. Herpetol. 20, 59–64.

Pough, F.H., 1973. Lizard energetics and herbivory. Ecology 54, 837–844.

Pregill, G., 1984. Durophagous feeding adaptations in an amphisbaenid. J. Herpetol. 18, 186–191.

Presch, W., 1974. A survey of the dentition of the macroteiid lizards (Teiidae: Lacertilia). Herpetologica 30, 344–349.

Pyron, R.A., Burbrink, F.T., Wiens, J.J., 2013. A phylogeny and revised classification of Squamata, including 4161 species of lizards and snakes. BMC Evol. Biol. 13, 93.

Reilly, S.M., McBrayer, L.D., White, T.D., 2001. Prey processing in amniotes: biomechanical and behavioural patterns of food reduction. Comp. Biochem. Physiol. Part A 128, 397–415.

Rieppel, O., 1978. Tooth replacement in anguinomorph lizards. Zoomorphologie 91, 77–90.

Rieppel, O., 1979. A functional interpretation of the varanid dentition. Gegenb. Morphol. Jahrb. 125, 797–817.

Rieppel, O., Labhardt, L., 1979. Mandibular mechanics in *Varanus niloticus* (Reptilia: Lacertilia). Herpetologica 35, 158–163.

Reiserer, R.S., Schuett, R.S., Beck, D.D., 2013. Taxonomic reassessment and conservation status of the beaded lizard, *Heloderma horridum* (Squamata: Helodermatidae). Amphib. Reptile Cons. 7, 74–96.

Robinson, P.L., 1976. How *Sphenodon* and *Uromastyx* grow their teeth and use them. In: Bellairs, A.d'A., Cox, B.C. (Eds.), Morphology and Biology of Reptiles. Linnean Society Symposium Series No. 3. Academic Press, London, pp. 43–64.

Savage, J.M., 1963. Studies on the lizard family Xantusiidae IV. Genera. Contrib. Sci. Los Angels Country. Mus. No. 71, 1–38.

Schaerlaeken, V., Holanova, V., Boistel, R., Aerts, P., Velensky, P., Rehak, I., Andrade, D.V., Herrel, A., 2012. Bite to bite: feeding kinetics, bite force, and head shape of a specialised durophagous lizard, *Dracaena guianensis* (Teiidae). J. Exp. Zool. A. Ecol. Genet. Phys. 317, 371–381.

Smith, M.A., Bellairs, A.A., Miles, A.E.W., 1953. Observations on the premaxillary dentition of snakes, with special reference to the egg tooth. J. Linn. Soc. Zool. Lond. 42, 60–68.

Sokol, O.M., 1967. Herbivory in lizards. Evolution 21, 192–194.

Sumida, S.S., Murphy, R.W., 1987. Form and function of the tooth crown structure in gekkonid lizards (Reptilia, Squamata, Gekkonidae). Can. J. Zool. 65, 2886–2892.

Torres-Carvajal, O., 2007. Heterogeneous growth of marginal teeth in the black iguana *Ctenosaura similis* (Squamata: Iguania). J. Herpetol. 41, 528–531.

Townsend, V.R., Akin, J.A., Felgenhauer, B.E., Dauphine, J., Kidder, S.A., 1999. Dentition of the ground skink, *Scincella lateralis* (Sauria, Scincidae). Copeia 783–788.

Vanzolini, P.E., 1951. A systematic arrangement of the family Amphisbaenia (Sauria). Herpetol. J. 7, 113–123.

Wainwright, P.C., Kraklau, D.M., Bennett, A.F., 1991. Kinematics of tongue projection in *Chamaeleo oustaleti*. J. Exp. Biol. 159, 109–133.

Weinstein, S.A., Smith, T.L., Kardong, K.V., 2009. Reptile venom glands. Form, function, and future. In: Mackessy, S.P. (Ed.), Handbook of Venoms and Toxins. CRC Press, Boca Raton, pp. 66–91.

Wellborn, V., 1933. Vergleichende Osteologische Untersuchungen an Geckoniden, Eublephariden und Uroplatiden. Sitzungsber Ges Naturforsch Freunde ZU, Berlin. , pp. 126–199.

Woolrich-Pina, G.A., Smith, G.R., 2012. A new species of *Xenosaurus* from the Sierra Madre Oriental, Mexico. Herpetologica 68, 51–59.

Zahradniček, O., Buchtová, M., Dosedelová, H., Tucker, A.S., 2014. The development of complex tooth shape in reptiles. Front. Physiol. 5, 1–7.

Zangerl, R., 1944. Contributions to the osteology of the skull of the Amphisbaenidae. Am. Midl. Nat. 31, 417–454.

FURTHER READING

Conrad, J.L., Ast, J.A., Montanari, S., Norell, M.A., 2010. A combined evidence phylogenetic analysis of Anguimorpha. Cladistics 26, 1–48.

Gauthier, J.A., Kearney, M., Maisano, J.A., Rieppel, O., Behike, A.D.B., 2012. Assembling the squamate tree of life: perspectives from the phenotype and the fossil record. Bull. Peabody Mus. Nat. Hist. 53, 3–303.

Chapter 7

Reptiles 2: Snakes

INTRODUCTION

There have been two theories about the origin of snakes: that they evolved from aquatic lizards such as mosasaurs; or that they evolved from burrowing lizards. Recent molecular evidence (Vidal and Hedges, 2004) suggests that the latter hypothesis may be correct. Living snakes are contained within the suborder Serpentes of the order Squamata and total over 3500 species. The suborder Serpentes is divided into two infraorders: the smaller infraorder Scolecophidia and the larger infraorder Alethinophidia. In this chapter, the classification into families follows Pyron et al. (2013).

Because of their narrow bodies, the position and size of organs are greatly distorted. Paired organs, such as the kidneys, are situated one in front of the other instead of side by side, while only the right lung is usually present. The requisite flexibility of the body is achieved by multiplying the number of vertebra, which varies from under 100 in vipers to more than 400 in pythons and over 600 in the **arboreal blind snake** (*Typhlops anguisticeps*). The correspondingly large number of ribs (almost 300 ribs in pythons) support and protect the internal organs.

Most snakes are **oviparous**. However, some species are **viviparous** and retain the eggs inside the body until they hatch. Usually, the hatchlings receive no nourishment from the mother (a mode of reproduction sometimes termed **ovoviparity**) but, in a few species, there is formation of a placenta.

Snakes are carnivorous and most catch live prey, although many eat eggs. As they lack limbs, they are mainly ambush predators and can strike quickly. Their dentitions are adapted for grasping and not for cutting, so they swallow their prey whole. The prey is reoriented after capture so that it enters the mouth head first and thus presents the smallest possible cross-section. Nonvenomous snakes swallow their prey alive. Insectivory is not widespread among snakes as it is among lizards and no snakes are herbivorous.

In addition to the usual senses of sight and sound (picked up as vibrations by the lower jaw and transmitted to the inner ear by the columella), snakes have an acute sense of odor detection (chemoreception) and combine this with extensive use of **Jacobson's organ** in the roof of the mouth. The forked tongue is flicked in and out of the mouth frequently and samples particles from the air, which are presented to Jacobson's organ to provide chemosensory information. At the front of the mouth of virtually all snakes there is a gap in the center of the anterior tooth rows which allows the tongue to be protruded frequently without opening the mouth. In addition, some snakes, such as rattlesnakes and pythons, have heat-sensitive receptors of infrared radiation (pit organs) in the head that help them detect prey in their surroundings. The number and position of pit organs differs in different types of snakes (Fig. 7.1). In boas and pythons there are three or more per side on the lips, whereas crotalines (eg, rattlesnakes) have a single pair between the eyes and nostrils.

Snakes have homodont dentitions, although tooth size varies considerably. The number of teeth on any tooth-bearing bone of a snake remains more constant than in lizards throughout postembryonic life and does not increase with age. The fused premaxillae lack teeth except in pythons and a few other primitive snakes. In general, teeth are present on the maxillae, dentaries, palatines, and pterygoids but Scolecophidia and some other snakes lack teeth on the palate (Mahler and Kearney, 2006). Blind worms can have a very restricted number of teeth on only a few bones. The vomers always lack teeth. Whereas, in lizards, teeth on the roof of the mouth tend to be small and suited for gripping, in snakes they are large and morphologically similar to those on the maxilla, effectively providing a double row of teeth on each side of the upper jaw.

Snake teeth are recurved, sharply pointed, and mainly acrodont, although some are pleurodont, as in scolecophidians (Zaher and Rieppel, 1999). At least five genera of colubrid snakes (*Liophidium*, *Scaphiodontophis*, *Sibynophis*, *Lycophidion* and *Mehelya*), plus *Xenopeltis* (Xenopeltidae), have numerous small, hinged teeth (Savitzky, 1981, 1983). This unusual feature is related to the nature of the prey, which consists largely of hard-scaled scincid lizards. The hinged teeth are thought to be less liable to break against the skink's armor than ankylosed teeth. Hinged teeth are also found in lizards with a similar diet (see pages 165–166). Teeth of snakes are continually replaced, from the inner aspect of marginal teeth and the outer aspect of palatine and

The Teeth of Non-Mammalian Vertebrates. http://dx.doi.org/10.1016/B978-0-12-802850-6.00007-2

pterygoid teeth. Typically, the replacement teeth develop in a horizontal position (Figs. 7.19B and 7.43) and, as explained in Chapter 10, teeth are replaced in waves passing from back to front through alternate positions (Figs. 7.16, 7.19B, and 7.43).

As prey is often wide in relation to the gape of the snake, intraoral transport of large prey to the esophagus is associated with a risk of damage to the cranium, and must be accomplished solely by the teeth and jaws, because of the specialization of the tongue as a sensory organ and the absence of limbs. The snake skull displays a number of modifications which overcome these problems. Both temporal arches, including the jugal and squamosal bones, have been lost. The temporal region of the skull is reinforced by downgrowths of the parietal and frontal bones, and the dorsal sutures are fused. The consolidated structure strengthens the cranium and improves protection for the brain. It also eliminates mesokinetic and metakinetic movements, although the nasal-prefrontal joint is mobile in some taxa. While the cranium is consolidated, the elements of both jaws acquire considerable mobility in most species. The quadrate is elongated and can rotate about its articulation with the supratemporal bone (**streptostyly**). The maxillae, palatines, and pterygoids are connected rather loosely with the cranium and articulate with each other by ligaments. In the lower jaw, the prearticular,

articular, and angular bones are fused. The midline symphysis is replaced by a ligamentous connection, so that the two rami can be separated. In addition, each ramus can rotate about its axis and can move with a high degree of independence.

The mobility of the jaws is important both in biting and in transferring the prey to the digestive system. During biting, the pterygoid is drawn forward and upward by its levator and protractor muscles. This motion is transferred to the maxilla by way of the ectopterygoid which is toothless and acts as a connecting link. This sequence of events produces the most dramatic effects in viperine and crotaline snakes, where it acts to erect the venom fangs during a strike at prey. As in these snakes the maxillae are reduced to small bones bearing only the venom fangs, pressure from the ectopterygoids causes them to rotate around their articulation with the prefrontal bones, so that the fangs swing downward and forward into the erect position (Boltt and Ewer, 1964).

The pterygoids are equally important in the transfer of prey to the esophagus. The quadrates and pterygoids form a four-pivot frame which is raised and lowered alternately to grip the prey with the pterygoid and palatine teeth, drag it backward, then release it and renew the grip further forward. The left and right pterygoids act independently and alternately to drag the prey further into the mouth until it enters the esophagus and can be swallowed. This remarkable action was called the **pterygoid walk** by Boltt and Ewer (1964).

The role of the lower jaw is complementary to that of the upper jaw, as its properties allow the prey to pass through the mouth. The size of prey that can be swallowed is limited in the first instance by the distance between the quadrate-articular joints, which can be moved laterally only to a small extent. The ultimate limit on prey size is determined by (a) the length of the mandibular rami and (b) the extent to which the interramus ligaments can be stretched. These two lengths, together with the distance between the jaw joints, define a trapezium-shaped opening through which the prey can pass (Gans, 1961).

Having caught and swallowed its prey, especially if it is large, a snake may need to rest and digest it over a period of days or even weeks.

VENOM AND FANGS

In snakes, the oral secretions contain more potentially toxic substances than in lizards (Fry et al., 2006, 2012), but most snakes are nonvenomous since they lack glands for production and storage of venom and also lack specialized venom fangs. After grasping their prey with their sharp teeth, such snakes either swallow it alive or kill it by constriction. However, about 200 genera of snakes kill prey by injecting venom produced in specialized glands

FIGURE 7.1 Pit organs in snakes: python (top); rattlesnake (bottom). *Arrows* illustrating the position and number of the pit organs are *red*; a *black arrow* points to the nostril. *Courtesy Wikipedia.*

through specialized teeth (**fangs**) on the maxillae. Thereby the prey dies, or is at least immobilized more quickly so that the snake is less vulnerable to injury, and the use of venom means that relatively large prey animals can be captured. Of the four families of venomous snakes, the largest (Colubridae) contains both venomous and nonvenomous genera, while the Viperidae, Atractaspidinae, and Elapidae are exclusively venomous.

Similar to unmodified teeth, fangs are present and functional at birth. In adults they are replaced about every 6–10 weeks. Snakes can be placed in four groups which are named from tooth (fang) morphology (Fig. 7.2), and also from the position and mobility of the fangs, the presence or absence of venom glands, and gland structure and physiology (Young and Kardong, 1996; Deufel and Cundall, 2006; Weinstein et al., 2009).

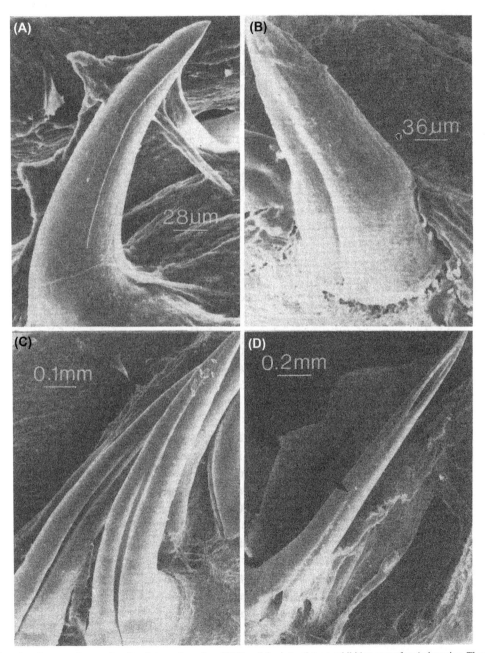

FIGURE 7.2   Scanning electron micrographs of tooth types in snakes. (A) Type I: basic tooth type exhibiting no surface indentation. The raised dental ridge is indicated. Palatine tooth of the viperid, *Azemiops feae*. (B) Type 2: furrowed tooth showing low depression characteristic of this tooth type. Dentary teeth of the colubrid, *Stenorrhina freminvillii*. (C) Type 3: grooved teeth wherein the depression along the tooth is deeply recessed but open its entire length. Posterior maxillary teeth of the colubrid, *Boiga irregularis*. (D) Type 4: hollow tooth produced by surface folds of the tooth that meet and fuse leaving a superficial secondary furrow (*arrow*) to enclose a channel within the tooth with entrance and exit openings at the ends. Fang of the viperid, *A. feae*. *From Young B.A., Kardong K.V., 1996. Dentitional surface features in snakes (Reptilia: Serpentes). Amphibia-Reptilia 17, 261–276. Courtesy Editors Amphibia-Reptilia.*

Aglyphous Snakes

In most of these snakes, there are no specialized fangs and all the teeth are similar, although in some species there is some variation in size and some teeth may be fang-like (Young and Kardong, 1996) (Fig. 7.2A). The teeth are not grooved. About 85% of snakes are aglyphous and are considered nonvenomous. Aglyphous snakes include basal snakes, such as boas, pythons, and also many nonvenomous colubroids. A venom gland is absent in most of these species but in a number of aglyphous species, such **as rat snakes** and **egg-eating snakes** (Colubridae), which evidently are descended from opisthoglyphous ancestors, remnants of atrophied venom glands are present. A few are reported to have venom glands and to be mildly venomous, and may possess ungrooved rear fangs, eg, **garter snakes** (*Thamnophis*: Colubridae) but for envenomation the snake must bite its prey several times.

Opisthoglyphous Snakes

These are the "rear-fanged" snakes. The maxillae are long bones and bear a full row of teeth. The fangs are enlarged, backwardly curved teeth in the middle of the row or at the back (Deufel and Cundall, 2006) (Fig. 7.3). They have a groove on the mesial or lateral tooth surface along which venom flows by capillarity into the prey (Young and Kardong, 1996) (Fig. 7.2B). The anterior teeth and the fangs are differentiated from each other (Jackson, 2003); the fangs have ridges on their palatal and lingual surfaces, whereas the teeth in front have ridges on the mesial and distal surfaces. Moreover, the fangs are often separated from the anterior teeth by a diastema. The venom gland in opisthoglyphs is homologous with those in the tubular-fanged snakes

and, like them, is located behind and below the eye. However, it shows histological and physiological differences and is usually given a separate name: **Duvernoy's gland** (Weinstein et al., 2009). It has no central lumen for storage of venom and is a low-pressure gland lacking a muscular system to expel venom, although adhesion of the gland capsule to the skin and a ligamentous connection to the quadrate may provide some pressurization. The duct of Duvernoy's gland discharges either into a soft-tissue cuff surrounding the fangs or there may be several ducts supplying a number of teeth. Most colubroids and many lamprophiids are opisthoglyphs.

Proteroglyphous Snakes

These are "front-fanged" snakes with fixed fangs and include all members of the Elapidae. Except for an opening near the tip, the venom groove in the fang is enclosed completely by the side walls and is thus converted into a tube, but a "seam" is visible on the surface (Fig. 7.2D). The shortened maxilla bears a small fang at the front and one or more additional, nongrooved small teeth at the back (Fig. 7.4C). Rotation of the maxilla is limited so the fangs are not erected to the same extent as in solenoglyphs. The venom glands have two components which are immediately adjacent to each other: a main gland and an accessory gland (Weinstein et al., 2009). In the main gland, tubular, secretory cisternae empty into a central lumen. The gland is pressurized by a muscle: the superficialis muscle, derived from the adductor superficialis. Venom is expelled through the accessory gland next to the main gland and thence via a secondary duct which leads to the base of the fang. The accessory gland may have the function of conditioning or activating venom as it passes through. Venom finally travels

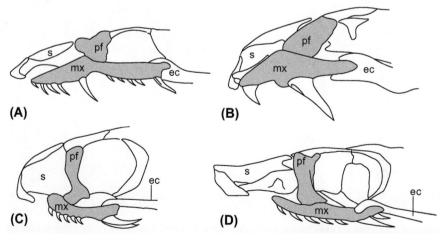

FIGURE 7.3 Variations in maxillary shape and fang position in some opisthoglyphs. (A) *Dryophis*: only the last maxillary tooth is grooved and functions as a fang. (B) *Polemon*: the large rear fang is situated in the middle of the maxilla, opposite to the joint with the prefrontal. (C) *Xenodon*: shortening of the anterior portion of the maxilla. (D) *Heterodon*. Abbreviations: *ec*, Ectopterygoid; *mx*, maxilla; *pf*, prefrontal; *s*, snout. *From Deufel A., Cundall D., 2006. Functional plasticity of the venom delivery system in snakes with a focus on the poststrike prey release behaviour. Zool. Anz. 245, 249–267. Courtesy Editors Zoology.*

down the tube at the center of the fang and is injected into the prey through an opening near the tip. Pressure generated by the superficialis muscle is sufficient to allow some elapids to "spit" venom forward from front-facing openings near the fang tips.

Solenoglyphous Snakes

These are front-fanged snakes with a single pair of large, tubular fangs. The central lumen of the fang is completely enclosed and there is little or no external evidence of a seam (Figs. 7.2D, 7.5 and 7.6A). As in proteroglyphous fangs, the venom duct opens slightly below the tip; this prevents blockage during use, as in a hypodermic needle. The fangs are ankylosed to the maxillae, which are reduced and can be rotated around the prefrontals (see page 202) to erect the fang for use and to fold them into the roof of the mouth when not in use. The venom gland is similar to that in proteroglyphs but the accessory gland lies near the opening of the venom duct into the mouth.

Moreover, the main gland is pressurized by contraction of a different muscle: the compressor glandulae, which is derived from the adductor externus superficialis. This type of fang is found in all viperids (Fig. 7.4A). The mouth opens to almost 180°, so the erected fangs face forward and can strike deeply into the prey. Because of the wide gape, the fangs can be very large; for instance, the fangs of a Gaboon viper may approach 50 mm in length. A similar type of fang is also found in the two genera belonging to the Atractaspidinae (Fig. 7.4B). In *Atractaspis*, the maxilla can rotate around a longitudinal axis, so that the fang can be protruded from the side of the mouth (see page 215). The venom gland of *Atractaspis* is highly elongated and extends backward from the head. It lacks an accessory gland and is compressed by a muscle derived from the adductor externus medialis.

The distribution of teeth according to the type of fang in a wide range of snakes is listed in Table 7.1 (Young and Kardong, 1996).

The development of fangs in front-fanged and rear-fanged snakes have fundamental features in common (Vonk et al., 2008). In basal snakes, there is a single dental lamina running the length of the maxilla, whereas the maxillary dental lamina of rear-fanged snakes has posterior and anterior segments. The anterior segment gives rise only to aglyphous teeth, but the posterior lamina is the origin, not only of the fangs, but also of the primordium of the venom (Duvernoy's) gland (Weinstein et al., 2009). In the front-fanged snakes, the anterior dental lamina is reduced to a

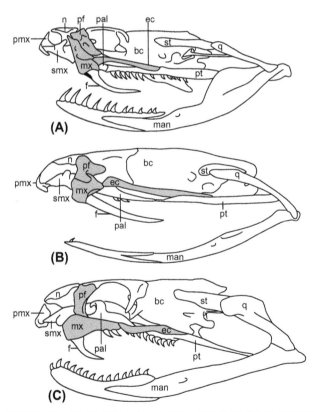

FIGURE 7.4 Skulls of front-fanged snakes. (A) viper; (B) *Atractaspis*; (C) elapid. In both vipers and *Atractaspis*, the maxilla is short and bears only fangs. In elapids there are often additional short, solid teeth on the maxilla posterior to the fangs. Abbreviations: *bc*, Braincase; *ec*, ectopterygoid; *f*, fang; *man*, mandible; *mx*, maxilla; *n*, nasal; *pal*, palatine; *pf*, prefrontal; *pmx*, premaxilla; *pt*, pterygoid; *q*, quadrate; *smx*, septomaxilla; *st*, supratemporal. *From Deufel A., Cundall D., 2006. Functional plasticity of the venom delivery system in snakes with a focus on the poststrike prey release behaviour. Zool. Anz. 245, 249–267. Courtesy Editors Zoology.*

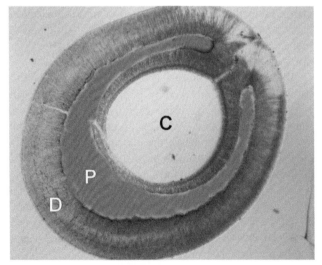

FIGURE 7.5 Cross-section of the venomous fang of a venomous bushmaster snake (*Lachesis muta*) from Central America. The normally central pulp chamber (P) is displaced to one side as a result of the invagination which created the separate central channel (C) that delivers all the venom without any waste into the victim. The rest of the tooth is composed of dentine (D). Image width = 2.1 mm. © *Museums at the Royal College of Surgeons, Tomes slide collection. Catalog no. 606.*

(A)

(B)

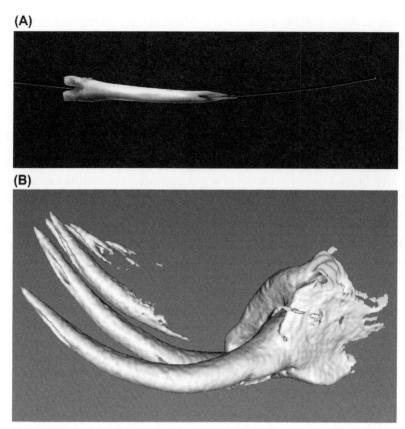

FIGURE 7.6 (A) Fang of a viper, with no external evidence of grooving. A bristle shows that the venom canal extends through the fang. (B) Micro-CT scan of maxilla, showing presence of functional fang and eight successive fangs at varying stages of development. *Courtesy of Professor A. S. Tucker.*

dental ridge and has lost the capacity to form teeth. The posterior lamina gives rise to both fangs and the venom gland, as in rear-fanged snakes. However, during subsequent development, through differential growth of the parts of the maxilla the fangs arrive ultimately at an anterior position (Vonk et al., 2008).

During the development of tubular venom fangs, hard tissue formation is preceded by invagination of the dental epithelium, so that the dental papilla acquires a crescent shape in cross-section (Zahradnicek et al., 2008). In the tooth germs of viper fangs, the extremities of the papilla curve round to the extent that the overlying layers of dental epithelium meet and fuse, forming a tube. The tubular fang is consolidated by deposition of dentine. The epithelium and other tissue remaining in the venom channel is removed by apoptosis. The grooved fangs of opisthoglyphs are formed by the same process, except that the crescent-shaped dental papilla remains open and does not form a tube.

The differences in venom delivery outlined earlier have marked effects on the method of prey capture. In the front-fanged snakes, the possession of high-pressure venom glands, and the tubular fangs, make envenomation very efficient. Thus, the prey is overcome in a short time and this reduces the risk to the snake. Venom delivery

is less efficient in rear-fanged snakes because of the low-pressure venom gland and the position of the fangs, which requires the prey to be maneuvered toward the rear of the mouth and bitten repeatedly to inject a sufficient dose of venom.

Venom has been mainly studied from the point of view of treatment for snake bites in humans, which kill considerable numbers of people (estimated at 20,000–94,000 per year worldwide). For these purposes, venom is broadly categorized into two types: hemotoxic and neurotoxic. Hemotoxic venom affects blood and organs, producing effects such as a drop in blood pressure and inhibition of blood clotting, while neurotoxic venom affects the nervous system and helps to immobilize the prey. Although these toxins are of prime importance in allowing the snake to capture its prey, venom contains a large number of proteins and other bioactive molecules (Fry et al., 2012) which may also contribute to eventual successful prey capture. Venom may also prepare the captured prey for digestion.

A detailed account of tooth numbers in snakes has been reported by Marx and Rabb (1972), who list values for 292 species of the Colubridae, Elapidae, Hydrophiidae, Viperinae, and Crotalinae.

TABLE 7.1 Summary of Tooth Types (0–4, as shown in Fig. 7.2) Within Families of Snakes

Family	N	Dentary	Pterygoid	Palatine	Maxilla (Anterior)	Maxilla (Posterior)
Typhlopidae	1	0	0	0	1	1
Aniliidae	3	1	1	1	1	1
Cylindriophiidae	4	1	1	1	1	1
Uropeltidae	1	1	1	1	1	1
Boidae	54	1	0 (2)	1	1	1
			1 (98)			
Pythonidae	45	1	1	1	1	1
Xenopeltidae	7	1	0 (14)	1	1	1
			1 (86)			
Bolyeriidae	1	1	1	1	1	1
Tropidophiidae	9	1	1	1	1	1
Acrochordidae	7	1	1	1	1	1
Colubridae	716	1 (95)	0 (1)	1 (99)	1	1 (71)
		2 (5)	1 (98)	2 (1)		2 (1)
			2 (1)			3 (28)
Lamprophiidae	45	1 (98)	0 (15)	1 (98)	1 (91)	0 (18)
		2 (2)	1 (85)	2 (2)	4 or 5 (9)	1 (40)
						3 (42)
Elapidae	113	0 (1)	0 (1)	0 (2)	4	0 (15)
		1 (66)	1 (67)	1 (55)		1 (28)
		2 (33)	2 (32)	2 (43)		2 (54)
Viperidae	100	1	1 (98)	1 (98)	4	0
			2 (2)	2 (2)		

N=number of specimens. Prevalence within specimens (%) shown after tooth type in brackets, but when prevalence=100% no percentage is given. Type 5 shown for a few groups is a hollow tooth like Type 4 except no secondary surface groove is present.
Data from Young B.A., Kardong K.V., 1996. Dentitional surface features in snakes (Reptilia: Serpentes). Amphibia-Reptilia 17, 261–276. Prevalences recalculated to harmonize data with classification of Pyron R.A., Burbrink F.T., Wiens J.J., 2013. A phylogeny and revised classification of Squamata, including 4161 species of lizards and snakes. BMC Evol. Biol. 13, 93.

SNAKE DENTITIONS

Scolecophidia

This infraorder comprises three families, containing about 400 species of generally very small and mainly burrowing snakes commonly known as worm, thread, or blind snakes. In conformity with their burrowing habit, they have stout skulls and often greatly reduced eyes, which might only be able to discriminate between light and shade. The mouth has a ventral position. The premaxilla is edentulous and any teeth on the maxilla or dentary are pleurodont. These diminutive snakes prey on small insects, such as ants, termites, and their larva. They show some of the most specialized dentitions among snakes (Kley, 2001, 2006; Rieppel et al., 2009). They also lack teeth on the palatine, vomer, and pterygoid (Mahler and Kearney, 2006).

Anomalepididae

Each maxilla of the **white-nosed blindsnake** (*Liotyphlops albirostris*) carries five sharp, curved teeth and is obliquely oriented, which differs from the typhlopids, where the maxilla and teeth are oriented nearly transversely (Rieppel, 2009). The anterior end of the dentary is expanded transversely and carries two small teeth (Fig. 7.7).

Leptotyphlopidae

The **Texas blind snake** (*Leptotyphlops dulcis*), a wormlike creature up to 20 cm long, has a unique dentition and mode of feeding. The lower jaw is subterminal and extremely short, measuring only 35–40% of the total length of the skull and a single row of four to five teeth is present on each dentary (Figs. 7.8 and 7.9). The cranium and upper jaw are immobile. The maxillae are edentulous and play no part in engulfing prey. Because of the alignment of the elongated quadrate bone, food is transported using a mandibular raking mechanism involving bilateral synchronous

movements of the dentate lower jaw in the anteroposterior axis (Kley, 2006).

Typhlopidae

Movements of the pterygoids in typhlopids serve to protract and retract the toothed maxillae. Prey transport is thus brought about exclusively through rotational movements of the highly mobile maxillae. The interramal connection of the lower jaw in typhlopids is quite rigid.

The skull and dentition of the **lineolate blind snake** (*Typhlops lineolatus*) and the closely related **Jamaican blind**

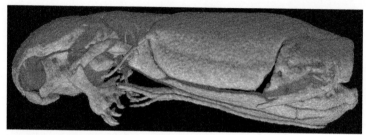

FIGURE 7.7 The dentition of the white-nosed blind snake (*Liotyphlops albirostris*). CT scan. Image width = 4.5 mm. *Courtesy Digimorph.org and Dr. J.A. Maisano.*

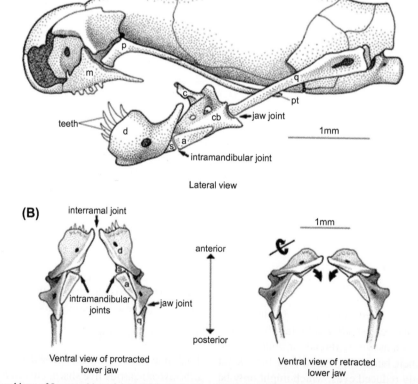

FIGURE 7.8 Skull and and jaws of *Leptotyphlops dulcis*. (A) Lateral view of skull. Teeth are absent from the palatines, pterygoids, and maxillae (which does, however, possess projections of bone). (B) Ventral view of the lower jaw. During protraction (left), the dentary–splenial complex swings forward and rotates about the intramandibular joint, bringing the teeth into a forward-facing position. Abbreviations: *a*, Angular; *c*, coronoid; *cb*, compound bone; *d*, dentary; *m*, maxilla; *p*, palatine; *pt*, pterygoid; *q*, quadrate; *s*, splenial. *From Kley N.J., 2001. Prey transport mechanisms in blindsnakes and the evolution of unilateral feeding system in snakes. Am. Zool. 41, 1321–1337. Courtesy Editors American Zoologist.*

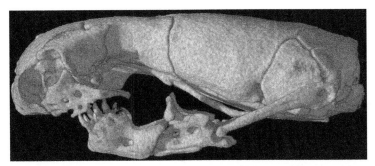

FIGURE 7.9 The dentition of the Texas blind snake (*Leptotyphlops dulcis*). CT image. Image width=5.5 mm. *Courtesy Digimorph.org and Dr. J. A. Maisano.*

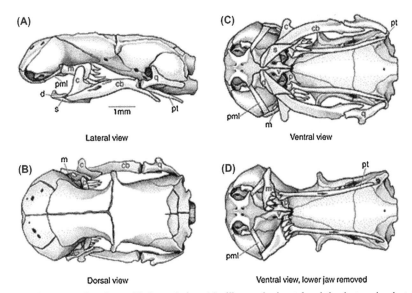

FIGURE 7.10 Skull and jaws of *Typhlops lineolatus*. (A) Lateral view. Maxillae are horizontal and the dentary is edentulous. (B) Dorsal view. The lateral process of the palatine articulates with the dorsal surface of the maxilla. (C) Ventral view. The medial process of the palatine articulates with the posterior edge of the vomer. The premaxilla and the anterior end of the maxilla are connected by thick, obliquely oriented ligaments. (D) Ventral view (lower jaw removed). Asynchronous retraction of the maxillae engaged in raking. Abbreviations: *c*, Coronoid; *cb*, compound bone; *d*, dentary; *m*, maxilla; *p*, palatine; *pml*, premaxillo-maxillary ligament; *pt*, pterygoid; *q*, quadrate; *s*, splenial; *v*, vomer. *From Kley N.J., 2001. Prey transport mechanisms in blindsnakes and the evolution of unilateral feeding system in snakes. Am. Zool. 41, 1321–1337. Courtesy Editors American Zoologist.*

snake (*Typhlops jamaicensis*) are seen in Figs. 7.10 and 7.11. Among the specialized features are the absence of teeth on the lower jaw and the reduced size and rigidity of the lower jaw. Indeed, *Typhlops* is the only genus of snake without teeth on the dentary. Four or five teeth are present on each maxilla. The maxillae are unusual in that they lie horizontally against the roof of the mouth with their transversely oriented tooth rows directed posteriorly. As the pterygoids are long, slender, and widely separated from the quadrate, the upper and lower jaws are functionally decoupled. This enables a maxillary raking mechanism, in which asynchronous ratcheting movements of the highly mobile upper jaws are used to drag prey through the oral cavity (Kley, 2001).

Alethinophidia

The main infraorder of snakes, Alethinophidia, contains 15 families and over 3000 species, of which over 1900 belong to the Colubridae. Pipe snakes, pythons, and boas have a vestigial pelvic girdle that is partially visible as a pair of cloacal spurs.

The vast majority of alethinophidian snakes have, on the palatines and pterygoids, a continuous inner row of relatively long, recurved teeth parallel with the outer maxillary row (Mahler and Kearney, 2006). Reciprocating translational movements of the toothed palatopterygoid arches are primarily responsible for transporting prey through the oral cavity, as explained in the introductory section. In alethinophidian snakes, the distal tips of the dentaries are quite separate from one another and are joined only by connective tissue.

Aniliidae

The **false coral snake** or **pipe snake** (*Anilius scytale*), the only species in this family, has retained the primitive feature of teeth on the premaxilla, in addition to teeth on the main dentigerous bones, including the palatine and pterygoids (Fig. 7.12).

Boidae

This family contains the boa constrictors and anacondas and consists of over 40 species in 8 genera. They are nonvenomous snakes which kill their prey by constriction. Boids lack premaxillary teeth.

The **common boa constrictor** (*Boa constrictor*) is a large snake that can reach lengths of up to 4 m. Its teeth are sharp and recurved (Fig. 7.13). With four rows of teeth in the upper jaw and two in the lower, there are well over 100 teeth in the dentition. The anterior teeth on the maxillae and palatine bones tend to be the largest teeth in the row.

The **green anaconda** (*Eunectes murinus*) is the largest of the boas and usually has 15 teeth on the maxilla (Fig. 7.14). This highly aquatic boa will drag its prey under water to kill it by drowning.

Pythonidae

This family contains 8 genera of constrictor snakes with 26 species. The prey usually consists of birds and mammals, which are captured by a swift strike of the jaws using the long, sharp anterior teeth. Coils of the muscular body then quickly surround the prey and kill it by asphyxiation. Unlike boas, pythons have teeth on the premaxilla.

There are seven species of *Python*. **Pythons** have two teeth on each premaxilla, separated from each other by a small gap. These teeth are shorter and more sharply recurved than the anterior maxillary teeth. There are usually 17 teeth on each maxilla (Figs. 7.15 and 7.16), of which the second or third tooth is the longest. The posterior teeth decrease in size. The recurved teeth have a broad base and end in a sharp point. The tips of the teeth point medially: the tilt increases from the anterior to the posterior teeth. There are six palatine

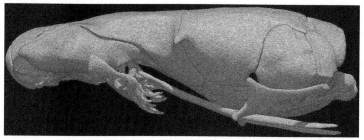

FIGURE 7.11 CT image of the skull of the Jamaican blind snake (*Typhlops jamaicensis*). Image width=9 mm. *Courtesy of Digimorph.org and Dr. J.A. Maisano.*

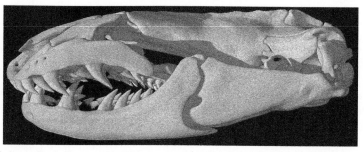

FIGURE 7.12 Dentition of red pipe snake (*Anilius scytale*). CT image. Image width=3 cm. *Courtesy Digimorph.org and Dr. J.A. Maisano.*

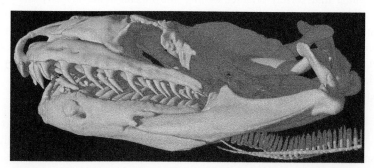

FIGURE 7.13 Dentition of common boa constrictor (*Boa constrictor*). CT image. Image width=11 cm. *Courtesy Digimorph.org and Dr. J.A. Maisano.*

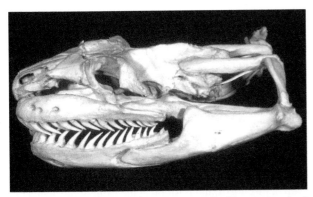

FIGURE 7.14 Dentition of the green anaconda (*Eunectes murinus*). *Courtesy Dr. Mark O'Shea.*

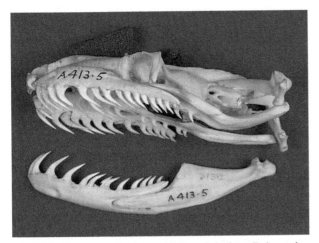

FIGURE 7.15 Dentition of the African rock python, *Python sebae*. Image width=7 cm. *Courtesy RCSOMA/413.5.*

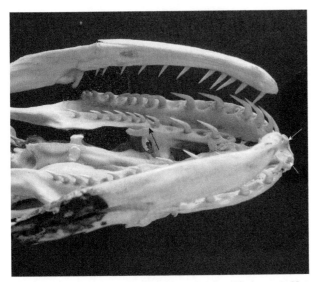

FIGURE 7.16 Ventral view of dentition of python (*Python* sp.). Note teeth on the premaxilla (*white arrows*) and the suture between the palatine and pterygoid bones (*black arrow*). Tooth replacement occurring in waves passing through alternate tooth positions indicated by the position of empty tooth loci. Image width=8 cm. *Courtesy Horniman Museum & Gardens and Dr. Paolo Viscardi. Catalog no. LDHRN_NH 27. 69.*

teeth, of which the first two are largest, and 6–10 small, hook-shaped, pterygoid teeth which project backwards and inwards. There are 18–19 recurved teeth on each dentary. The third or fourth tooth is longest and the size decreases posteriorly (Frazzetta, 1966). The six rows of recurved, pointed teeth prevent the prey from escaping and the prey is engulfed by the ability of the upper jaws of the python to ratchet forwards alternately, "walking over" the prey.

Viperidae

All members of this family are venomous. They are front-fanged, solenoglyphous snakes and have long, recurved, tubular fangs. The fangs are carried on short, highly mobile maxillae, which pivot on the prefrontals and carry no other teeth (Fig. 7.4A). As the mouth opens wide the fangs are erected and point forward (Fig. 7.17). When the mouth is closed, the fangs are folded back along the roof of the mouth and covered by a sheath of soft tissue. Teeth are present on each dentary and pterygoid bone. A small number of teeth are usually also present on the palatines, but are absent in some species (eg, *Ophryacus*) (Mahler and Kearney, 2006). Because the gape is very wide and the fangs are erected by rotation of the maxillae, the fangs of Viperidae are much longer than those of the Elapidae.

The left and right fangs can be rotated together or independently. There are two fang loci on each maxilla that are occupied alternately by a functional tooth. The replacement teeth develop behind the functional teeth and may number up to eight (Fig. 7.6B).

During a strike by a viperid, a large dose of venom is injected into the prey under pressure. Usually viperids strike

FIGURE 7.17 Great Lakes bush viper (*Atheris nitschei*) with mouth open, showing erected fangs surrounded by a protective sheath of soft tissue. *Courtesy Shutterstock.*

and release the prey immediately, then search for it after the venom has had time to kill or immobilize. This method reduces the risk of injury from struggling prey and fewer marginal teeth are required. Arboreal vipers may hold on to their prey after the strike, rather than release it and risk losing it to other predators on the ground. The strike itself can produce considerable damage to the prey which, together with the hemotoxic venom, results in very rapid death (Deufel and Cundall, 2006). The venom also contains digestive juices, which act very quickly. There is evidence that prey is rapidly repositioned, which suggests the existence of (1) fine-scale sensory detection of fang penetration depth; (2) rapid modulation of contraction of antagonistic muscles; and (3) possibly neurological modifications to shorten transmission time between sensory input and motor output (Cundall, 2009).

Although under revision, vipers can be classified under four subfamilies, with 30–40 genera. Over 90% of the more than 220 species belong to either the Viperinae (true or pitless) vipers or the Crotalinae (pit) vipers.

Subfamily Viperinae

The dentition of the **puff adder** (*Bitis arietans*) is illustrated in Figs. 7.18 and 7.19A. The openings of the venom canal

are clearly visible at the base and near the tip of the fang. There are 3 teeth on each palatine, 10 on each pterygoid, and 14 on each dentary. Fig. 7.19B shows the alternating pattern of tooth replacement.

Multiple fangs and pterygoid teeth in the related **rhinoceros horned viper** (*Bitis nasicornis*) are illustrated in Fig. 7.20.

Replacement teeth develop in a horizontal position, presumably to prevent damage to the overlying mucosa when it is stretched during swallowing of the prey, and rotate on erupting into the mouth. Fig. 7.21 shows some replacing teeth in this horizontal position at the front of the dentary of the **rhinoceros horned viper** (*B. nasicornis*).

The **Gaboon viper** (*Bitis gabonica*) has the longest fangs (up to about 5 cm) and the highest venom yield of any venomous snake (Figs. 7.22 and 7.23).

Russell's viper (*Daboia russelii*) is regarded as being responsible for the greatest number of snakebite incidents and deaths, mainly in India. It carries 2–3 palatine teeth, 7–9 pterygoid teeth, and 12–13 dentary teeth (Mao, 1993).

Adders (*Vipera* sp.) (Fig. 7.24) have less prominent fangs than *Bitis* but have more teeth on the dentary and pterygoid bones.

In addition to the fangs on the maxillae, the **asp viper** (*Vipera aspis*) has 3–4 teeth on each palatine, 10–12 on each pterygoid, and 14–15 on each dentary. There is little difference in tooth numbers between four of the

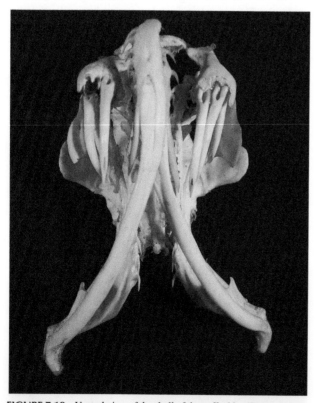

FIGURE 7.18 Ventral view of the skull of the puff adder (*Bitis arietans*). Note the openings of the venom channel at the base and tip of the venom fangs. On the right side of the image two fangs are present while on the left side only one fang (the inner) is present. Image width = 7 cm. *Courtesy RCSOMA/415.12.*

(A)

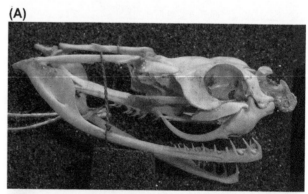

(B)

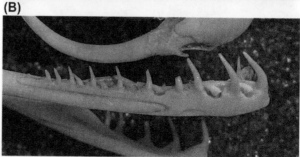

FIGURE 7.19 Puff adder (*Bitis arietans*). (A) Dentition. Image width = 5.2 cm. (B) High power view of lower jaw from Fig. 7.19A, showing unerupted replacement teeth, oriented obliquely backward, in alternate tooth positions. Image width = 2 cm. *Courtesy Horniman Museum and Garden and Dr. P. Viscardi. Catalog no. NH. 27. 64.*

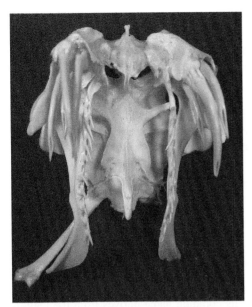

FIGURE 7.20 Ventral view of the upper dentition of the rhinoceros horned viper (*Bitis nasicornis*). Image width=7 cm. *Courtesy RCSOMA/415.14.*

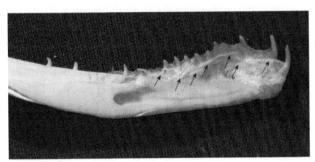

FIGURE 7.21 Lower jaw of the rhinoceros horned viper (*Bitis nasicornis*), showing the horizontal position of replacing teeth (*arrows*). Image width=4 cm. *Courtesy RCSOMA/415.14.*

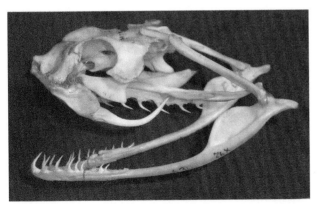

FIGURE 7.22 Dentition of the Gaboon viper (*Bitis gabonica*). Image width=7.5 cm. *Courtesy MoLS KCL. Catalog no. X91.*

FIGURE 7.23 Ventral view of the upper jaw of the Gaboon viper (*Bitis gabonica*). Three tooth positions are present on the palatine bone. Image width=4.2 cm. *Courtesy MoLS KCL. Catalog no. X91.*

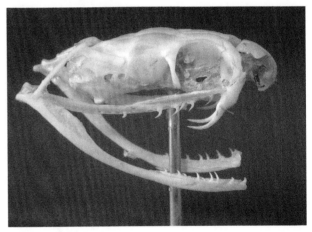

FIGURE 7.24 Dentition of *Vipera* sp. Image width=2.6 cm. *Courtesy Horniman Museum and Gardens and Dr. Paolo Viscardi. Catalog no. LDHRN-NH 88. 37. 213.*

subspecies and there is no evidence of sexual dimorphism (Zuffi, 2014).

Night adders have more teeth than other species of vipers. In addition to the single maxillary fang, **Lichtenstein's night adder** (*Causus lichtensteini*) has 9 palatine teeth, 32 pterygoid teeth, and 29 on each dentary (Marx and Rabb, 1972).

Subfamily Crotalinae

This subfamily contains the rattlesnakes, pit vipers, and lanceheads. There are 18 genera and over 150 species of Crotalinae. The heat-sensing pit organ is located between the eye and nostril on either side of the head and is more sensitive than those in other snakes.

The dentition of a **rattlesnake** (*Crotalus* sp.) is illustrated in Fig. 7.25. Individual counts vary between species but there are usually 3 palatine teeth, up to 10 pterygoid teeth, and 10 dentary teeth. **Barbour's pit viper** *Cerrophidion barbouri* may have up to 12 pterygoid teeth.

Tooth numbers have been reported from three species in the *Cerrophidion godmani* group of pit vipers (Campbell, 1988). Numbers vary from 3 to 5 for palatine teeth, from 8–10 to 13–16 for pterygoid teeth, and from 8–10 to 13–16 for dentary teeth. For the **sharp-nosed pit viper** (*Deinagkistrodon acutus*) there are 4 palatine teeth, 13–16 pterygoid teeth, and 15–17 dentary teeth. The **spotted pit viper** (*Trimeresurus mucrosquamatus*) has 6–7 pterygoid teeth, 11–12 dentary teeth, and lacks palatine teeth. **Stejneger's pit viper** (*Trimeresurus stejnegeri*) has 5 palatine teeth, 13–15 pterygoid teeth, and 13–14 dentary teeth. The tooth count for **Barbour's montane pit viper** (*Mixcoatlus* [*Porthidium*] *barbouri*)

is 3 palatine, 10 dentary, and 10–12 pterygoid (Campbell, 1988). Tooth numbers for some Asian pit vipers are given in Table 7.2.

The generic name of the **water or swamp moccasin** (*Agkistrodon piscivorus*) translates as "hooked-tooth fish eater"; it is the world's only semiaquatic viper. It is usually found in or near water and, although potentially omnivorous, usually feeds on fish and frogs. The related **broad-banded copperhead** (*Agkistodon contortrix*) is illustrated in Fig. 7.26.

Lamprophiidae

This family of snakes includes aglyphous, opisthoglyphous, and solenoglyphous species.

Subfamily Lamprophiinae

The **Cape wolf snake** (*Lycophidion capucinus*) is nonvenomous. Its common name derives from the long, recurved teeth of both upper and lower jaws (Fig. 7.27). The greatly enlarged anterior maxillary teeth are followed by a large diastema, where the maxilla is significantly arched. The diastema is followed by a row of small, closely packed teeth

(A)

(B)

FIGURE 7.25 (A) Micro-CT scan of the dentition of a rattlesnake (*Crotalus* sp.). *Courtesy Dr. R Hardiman*; (B) Dentition of a timber rattlesnake *(Crotalus horridus)*. Image width = 5.6 cm. *Courtesy Horniman Museum and Gardens and Dr. Paolo Viscardi. Catalog no. LDHRN-NH 36.76.*

TABLE 7.2 Tooth Numbers in Nine Species of Asian Pit Vipers

Species	Palatine Teeth	Dentary Teeth	Pterygoid Teeth
Zhaoermia mangshanensis	0	11	8/9
Ovophis monticola	4 (3–4)	17 (17–18)	14 (13–15)
Protobothrops flavoriridis	0	13–14	11
Protobothrops muscrosquamatus	0	9 (8–11)	5 (4–7)
Protobothrops xiangchengensis	0	11 (10–12)	8 (7–9)
Protobothrops jerdonii	0,1,2	11 (10–11)	8 (6–9)
Viridovipera stejnegeri	5 (4–5)	12 (10–14)	12 (9–14)
Viridovipera yunnanensis	5	13 (11–14)	13 (12–14)
Cryptelyttrops albolabris	4 (3–5)	12 (11–13)	10 (9–11)

Figures in parentheses indicate range. Some specimens of *Protobothrops jerdonii* had one or two palatine teeth, while others had 0.
From Guo P., Zhao E.-M., 2006. Comparison of skull morphology in nine Asian pit vipers (Serpentes: Crotalinae). Herpetol. J. 16, 305–313. Courtesy Herpetology Journal.

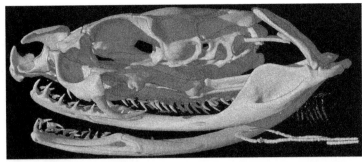

FIGURE 7.26   Dentition of broad-banded copperhead snake (*Agkistrodon contortrix*). Image width=3.2 cm. *Courtesy of Digimorph.org and Dr. J.A. Maisano.*

FIGURE 7.27   Dentition of the Cape wolf snake (*Lycophidion capense*). CT image. Image width=3 cm. *Courtesy Digimorph.org and Dr. J.A. Maisano.*

terminating after a short second diastema in one or two enlarged, but ungrooved, posterior maxillary fangs, with the posterior surfaces modified into blades. The anterior teeth of the dentary are also enlarged, followed by a diastema opposite the arched portion of the maxillary bone, then by a row of smaller teeth. Numerous small teeth are also present on the palatine and pterygoid bones. The Cape wolf snake is durophagous and preys on hard-bodied skinks. The specializations of the dentition can be related to dealing with this type of prey. During prey capture, the enlarged anterior teeth and arched maxilla encircle the prey, which is then killed by constriction. The blade-like posterior fangs slice through the hard scales of the prey (Jackson and Fritts, 2004).

Subfamily Atractaspidinae

In the classification of Pyron et al. (2013), this subfamily consists of two genera: the burrowing asps (*Atractaspis*) and the harlequin snakes (*Homoroselaps*), all of which are venomous. Other genera previously included in the family Atractaspidae are included in other lamprophoriid subfamilies. Hoser (2012) suggested that some species of *Atractaspis* should be transferred to a new genus (*Hoseraspea*) but this is not recognized by Pyron et al. (2013).

Burrowing asps (*Atractaspis*) have enormous venom glands extending back from the head and possess very long, solenoglyphous fangs on a shortened maxilla which bears no other teeth (Fig. 7.4B). In this feature, they superficially resemble vipers. However, *Atractaspis* has but few other teeth: a maximum of four on the palatine and dentary and none on the pterygoid (Marx and Rabb, 1972). Although

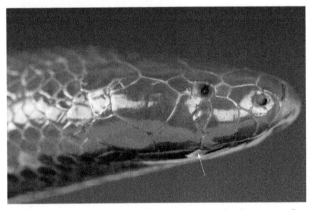

FIGURE 7.28   Burrowing asp (*Atractaspis* sp.), showing venom fang (*arrow*) protruding from side of mouth. *Courtesy Matthijs Kuijpers.*

the fangs of vipers and *Atractaspis* share some similarities, there are major differences that relate to the burrowing habit of *Atractaspis* and to methods of prey capture (Deufel and Cundall, 2003). *Atractaspis* does not launch strikes from a distance like vipers. Instead, it envenomates prey by stabbing downwards and backward with a single, long, caudally directed fang that is extruded from one side of the closed mouth while the snake crawls alongside its prey (Fig. 7.28). This is made possible by the ability of the maxilla to rotate around a longitudinal axis. A slit along the mouth-line is opened and provides enough space for the fang to protrude from the mouth (Shine et al., 2006). Maxillary rotation is limited by two ligaments that span the maxillary-prefrontal joint caudally. The loss of pterygoid teeth and maxillary movement present in most other snakes means that

Atractaspis cannot swallow prey by the usual "pterygoid walk." Instead *Atractaspis* transports prey through the oral cavity using movement cycles in which mandibular adduction, anterior trunk compression and ventral flexion of the head alternate with mandibular abduction and extension of head and anterior trunk over the prey.

Elapidae

This family contains over 350 species of venomous snakes, including cobras, coral snakes, and sea snakes, distributed in about 60 genera. Elapids range in length from less than 20 cm (**crowned snakes**, *Drysdalia*) to 6 m (**king cobra**, *Ophiophagus hannah*). The majority are ground dwellers, but there are some tree dwellers and aquatic species. Some species have the capacity to "spit" venom (Wuster and Thorpe, 1992). Detailed information on tooth numbers in cobras and related elapids have been reported by Bogert (1943) and Marx and Rabb (1972).

The elapids are the proteroglyphous or "front-grooved" snakes with fixed fangs (Fig. 7.4C). The shortened maxilla bears a small but enlarged fang at the front and a variable number of small, solid teeth behind. The fangs of elapids are shorter than those of viperids, so do not deliver venom quite so efficiently. Thus, elapids do not use the strike-and-release method of killing prey. Instead they hold on to their victims for a short period after striking, to ensure enough venom enters the prey (Deufel and Cundall, 2006).

The Elapidae usually have two fangs in each maxilla, one anchored and functional, the other incompletely anchored and nonfunctional (Fig. 7.29). Both fangs are rarely anchored equally well and functional. Thus, functional fangs alternately occupy the inner and outer sockets of each maxilla. The fangs are angled backward and slot into grooves in the floor of the mouth.

Behind each fang there are at least two or three replacement fangs in progressive stages of growth. As the growing fang nears completion, the functional fang in the maxillary socket is shed and the base of the developing fang moves into the vacated socket. The replacement fang rotates from its initial horizontal position until it reaches the functional position, with the main axis of the fang at an angle of 30–40 degrees to the horizontal axis of the maxilla, and finally becomes fully ankylosed. The fangs may be symmetrical, with the functional fangs on both maxillae occupying the inner or outer socket, or asymmetrical, with the functional fang occupying the outer socket on one maxilla and the inner socket on the other. Snakes at hatching may have symmetrical fangs, but they may become asymmetrical during life.

The smaller, solid, recurved teeth behind the fangs on the maxilla, as well as similar teeth on the other dentigerous bones, may show weak grooving, the significance of which is unknown (Vaeth et al., 1985; Young and Kardong, 1996). *Hemachatus*, *Aspidelaps*, and *Walterinnesia* lack

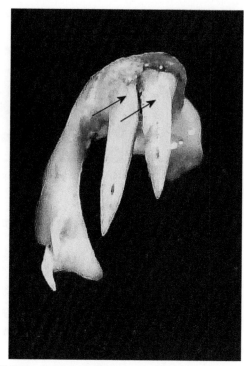

FIGURE 7.29 Isolated maxilla of the black-necked spitting cobra (*Naja nigricollis*) showing two erupted fangs with forward-facing apertures. Note also the presence of a groove running along the anterior surface of the solenoglyphous fangs (*arrows*). Image width = 8 mm. *Courtesy. RCSOMA/414.21.*

solid maxillary teeth. *Naja* species have none, one or two such teeth, *Pseudohaje* may possess from two to four, while *Boulengerina* and *Hamadryas* both normally possess three. *Elapsoidea* normally has three or rarely four, while in other elapid genera (eg, *Apistocalmus* and *Pseudechis*), there may be up to seven or more teeth. Among the Elapidae, the number of teeth on the dentary ranges from four in Hediger's coral snake (*Parapistollcalamus hedigeri*) to 35 in Muller's crowned snake (*Aspidomorphus muelleri*) (Marx and Rabb, 1972).

The **red-bellied black snake** (*Pseudechis porphyriacus*) grows up to about 2 m in length and its prey is mainly frogs, although it will also eat other snakes. Behind the fangs, the maxilla also carries six to seven ungrooved teeth (Fig. 7.30). There are about 12 palatine and 27 pterygoid teeth forming an inner row on each side of the upper jaw, while each dentary has about 19 teeth.

Reaching lengths of up to 2 m, the venom of the **Eastern brown snake** (*Pseudonaja* [*Demansia*] *textilis*) is one of the deadliest found in snakes (Fig. 7.31). In the upper jaw, in addition to the fangs, each maxilla carries eight ungrooved teeth, with up to eight on each palatine and 20 on each pterygoid. The dentary has up to 25 teeth.

The **Cape cobra** (*Naja nivea*) is a generalist feeder and will eat rodents, lizards, birds, and carrion as well as other snakes. There are two grooved fangs on the shortened

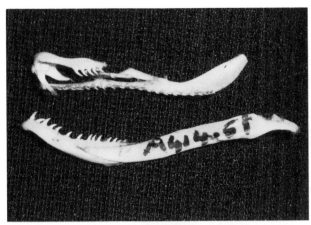

FIGURE 7.30 Dentition of the red-bellied black snake (*Pseudechis porphyriacus*). Five solid maxillary teeth separated by a diastema from the fang, and the suture separating the palatine bone from the pterygoid bone (*arrow*). Image width=3.3 cm. *Courtesy RCSOMA/414.51.*

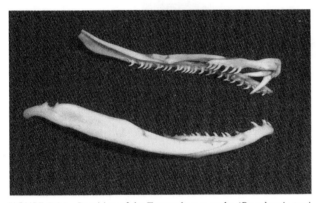

FIGURE 7.31 Dentition of the Eastern brown snake (*Pseudonaja textilis*). Image width=4.2 cm. *Courtesy RCSOMA/414.7.*

FIGURE 7.32 Dentition of the Cape cobra (*Naja nivea*). Note the replacement fangs. Image width=3.2 cm. *Courtesy RCSOMA/414.3.*

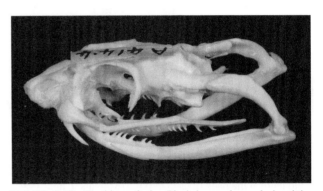

FIGURE 7.33 Dentition of the Ringhals or ring-necked spitting cobra (*Hemachatus haemachatus*). Image width=4 cm. *Courtesy RCSOMA/414.4.*

maxilla as well as usually two small, solid teeth behind, separated from the fang by a diastema (Fig. 7.32). There are 6–8 teeth on each palatine, 14–20 on each pterygoid, and 13–15 on each dentary. The Chinese cobra (*Naja altra*) has one solid maxillary tooth, 7–8 palatine teeth, 16–17 pterygoid teeth, and 14 dentary teeth (Mao, 1993).

The **ringhals** or the **ring-necked spitting cobra** (*Hemachatus haemachatus*) is unique among African cobras in being ovoviviparous. In addition to sharing the hooded threat posture with other cobras, this cobra is also one of the few species of cobra that can defend itself by "spitting" venom at an enemy from its fangs (Fig. 7.33). The orifice of the venom channel is on the anterior surface of the fang and shows varying degrees of reduction in discharge orifice size so that the venom may be forced out at high pressure. Only the fangs are present on the maxillary bone, while there are 5–6 teeth on each palatine, 15–18 on each pterygoid, and 12–16 on each dentary. In some species of spitting cobra, the venom channel has spiral grooves that help to direct the jet of venom. Some rear their heads well off the ground to

spit (Wuster and Thorpe, 1992). In the **black-necked spitting cobra** (*Naja nigricollis*), the range may be up to 3 m.

The **Eastern coral snake** (*Micrurus fulvius*) (Fig. 7.34) feeds on a wide range of prey, including insects, fish, frogs, lizards, and other snakes. Its venom is injected using the pair of short, fixed, hollow fangs at the front of the mouth. To achieve significant transfer of venom, the snake must bite and hold onto its victim for a brief period.

The color pattern of coral snakes is particularly beautiful, consisting of a series of rings, with wide red and black rings separated by narrow yellow rings. A number of similarly colored species of nonvenomous snakes, such as the colubrid scarlet snake (*Cemophora coccinea*) mimic that of the Eastern coral snake, giving rise to the folk rhyme: "Red touches black, friend of Jack, red touches yellow, kill a fellow."

The **sea snakes** are venomous elapid snakes that live in a marine or lake habitat for most of their lives. They are distinguished from land snakes by their laterally compressed, fin-like tail, but they continue to breathe air. They have limited ability to move on land. Some species have among the most potent venoms of all snakes. Some have gentle dispositions and bite only when provoked, but others are much more aggressive. There are just over 60 species of sea snake distributed in 17 genera. In addition to the short, paired fangs at

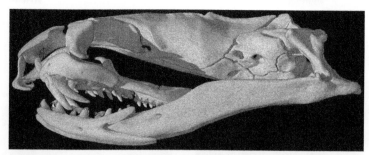

FIGURE 7.34 Dentition of the Eastern coral snake (*Micrurus fulvius*). CT image. Image width=2.7cm. *Courtesy Digimorph.org and Dr. J.A. Maisano.*

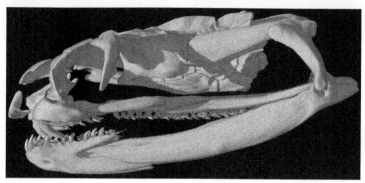

FIGURE 7.35 Dentition of the yellow-lipped sea krait (*Laticauda colubrina*), a species of sea snake. CT image. Image width=3.7cm. *Courtesy Digimorph.org and Dr. J.A. Maisano.*

the front of the maxillae there is usually a variable number of solid teeth at the back of each maxilla (Mao and Chen, 1980).

The **yellow-lipped sea krait** (*Laticauda colubrina*) possesses two solid maxillary teeth behind the paired fangs. In the roof of the mouth there are five to six teeth on each palatine, followed by a longer row of 14–17 teeth on each pterygoid, and each dentary bears 11–12 teeth (Fig. 7.35). The related **blue-lipped sea krait** (*Laticauda laticauda*) has a similar dentition, while the **black-banded sea krait** (*Laticauda semifasciculata*) has only one solid maxillary tooth behind the fang. The dentition of **Ijima's sea snake** (*Emydocephalus ijimae*), however, differs significantly in that the dentary and palatine are toothless, the maxilla has no solid teeth behind the small fangs, while the pterygoid has 18–19 teeth. The **slender-necked sea snake** (*Hydrophys melanocephalus*) has a greater number of teeth: 7 solid maxillary teeth, 7 palatine teeth, 14–15 pterygoid teeth and 16 dentary teeth (Mao and Chen, 1980). The **many-banded krait** (*Bungarus multicinctus*) has 3–4 solid maxillary teeth, 12–13 palatine teeth, 11–13 pterygoid teeth, and 16–18 dentary teeth (Mao, 1993).

Colubridae

With over 300 genera, the Colubridae is the largest snake family and includes about two-thirds of all living snake species. Most species are aglyphous and nonvenomous, while others are opisthoglyphous and can be venomous to varying extents. In some species, eg, rat snakes, the venom system

appears to have regressed (Fry et al., 2008). Examples of variations in the maxillary dentition of this group of snakes are seen in Fig. 7.3. Among the Colubridae, the number of teeth on the dentary ranges from just one or two in the common egg-eating snake (*Dasypeltis scabra*) to over 40 in the Western forest file snake (*Mehelya poenisis*) (Marx and Rabb, 1972).

Venom delivery is less efficient in rear-fanged snakes, and the prey must be moved toward the rear of the mouth to bring the fangs into function. Thus, the size of prey is limited and the anterior marginal teeth on the maxillae are required to catch and hold on to the prey (Fig. 7.3). Some rear-fanged snakes, such as the **boomslang** (*Dispholidus typus*), are more dangerous than others, because the fangs are further forward in the mouth and Duvernoy's gland has some compressive musculature.

Subfamily Natricinae

The **grass (water) snake** (*Natrix natrix*) often lives close to water and feeds almost exclusively on amphibians. The diet of grass snakes varies according to the season. In the spring they take fish when they are spawning and easier to catch. During the summer, they favor newts for the same reason. From July onwards, they move on to the land and catch mostly frogs and toads. They eat mice and voles but only rarely. The recurved small teeth of the grass snake form two rows in the upper jaw and one row on the dentary (Fig. 7.36).

(A)

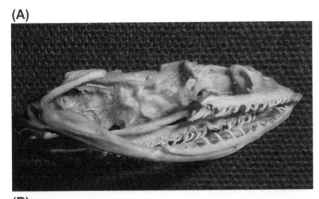

(B)

FIGURE 7.36 Dentition of the grass (water) snake (*Natrix natrix*). (A) Lateral view. Image width=3 cm *(Courtesy Grant Museum, University College London. Catalog no. LDUCZ.X254.).* (B) Ventral view showing teeth in the roof of the mouth. Image width=2.6 cm. *(Courtesy Horniman Museum & Gardens and Dr. P. Viscardi. Catalog no. LDHRN-NH Z.873.)*

FIGURE 7.37 Dentition of buff-striped keelback (*Amphiesma stolata*). Image width=2.5 cm. *Courtesy of Digimorph.org and Dr. J.A. Maisano.*

Keelback snakes show the typical enlargement of the posterior maxillary teeth and the presence of numerous teeth on all tooth-bearing bones as represented by the **buff-striped keelback** (*Amphiesma stolata*) (Fig. 7.37).

In the Japanese **tiger keelback snake** or **yamakagashi** (*Rhabdophis tigrinus*), the two rearmost maxillary teeth or fangs are separated by a prominent diastema from the teeth anterior to them. They possess a sharp posterior cutting edge and, though grooveless, are prominently recurved and about 2.25 times longer than the teeth immediately preceding them. The yamakagashi is unique in that, although its teeth are ungrooved, it is venomous and

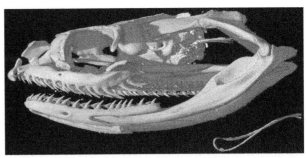

FIGURE 7.38 Dentition of the checkered garter snake (*Thamnophis marcianus*). Notice the enlarged teeth at the back of the maxilla. Image width=3.7 cm. *Courtesy of Digimorph.org and Dr. J.A. Maisano.*

stores steroidal toxins known as bufadienolides in specialized glands on its neck. However, it does not produce these toxins itself but sequesters them from venomous toads it consumes as prey. Furthermore, females containing high levels of bufadienolides can provision their offspring with toxins (Hutchinson et al., 2007).

Many of the garter snakes (*Thamnophis*) are aquatic. Although thought to be nonvenomous, they are now known to produce a mild, neurotoxic venom. The **checkered garter snake** (*Thamnophis marcianus*) has typically enlarged teeth at the back of the maxilla (Fig. 7.38) and subsists on earthworms, frogs, toads, and fish.

Some garter snakes (*Thamnophis*) are **malacophagous**, ie, they feed exclusively on gastropods (slugs and snails). Malophagous snakes have various techniques to first extricate the snail from the shell. Laporta-Ferreira and Salomao (2004) associated this diet with the presence of longer teeth on the anterior dentary and posterior maxilla. To test this hypothesis, the morphology of the teeth of four species of *Thamnophis* with different diets were compared (Britt et al., 2009):

The **Northwester garter snake** (*Thamnophis ordinoides*), a specialist on banana slugs.

The **Sierra garter snake** (*Thamnophis couchii*), a piscivorous aquatic species.

Two subspecies of the **terrestrial garter snake** (*Thamnophis elegans*), a prey generalist: an inland population (*T. elegans elegans*) which eats mainly fish and anurans, and a coastal population (*T. elegans terrestris*) which eats slugs.

Three basic tooth types were observed in all populations, namely straight, curved, and recurved. Teeth on the dentary were straight or curved. In the maxilla, teeth were curved at the anterior end, recurved or curved in the middle and straight or curved at the posterior end. In all four populations, the two most posterior maxillary teeth were the longest in the mouth and had posterior ridges that gave the teeth a blade-like appearance. These are the first teeth to contact prey during intraoral transport. Furthermore, there was no indication that the teeth of the slug eaters (*T. ordinoides* and *T. elegans terrestris*) were longer or more slender than those of the nonslug-eating populations, although they possessed a more pronounced ridge on the trailing edge of the posterior maxillary tooth. However, the piscivorous *T. couchii* had a higher density of teeth per jawbone and more curved teeth than the other populations.

Subfamily Dipsadinae

The **ringneck snake** (*Diadophis punctatus*) is a mildly venomous snake whose diet consists of worms, small amphibians, lizards, and smaller snakes. They are opisthoglyphous and the enlarged, nongrooved teeth at the back of the maxillary tooth row (Fig. 7.39) deliver venom. A combination of constriction and envenomation is used to kill the prey. The venom from Duvernoy's gland drains out of an opening at the rear of the last maxillary tooth. The prey is maneuvered backward to ensure that this tooth punctures the skin and allows venom transfer.

A distinguishing feature of the **Eastern hognose snake** (*Heterodon platirhinos*) is the upturned snout that is used for digging up its prey, particularly toads, in sandy soils (Fig. 7.40). This is a rear-fanged snake, but is considered only mildly venomous. Its generic name refers to the large size difference along the tooth row: the two ungrooved posterior teeth are two to three times larger than the maxillary teeth in front, from which they are separated by a diastema.

Subfamily Colubrinae

The **Western ground snake** (*Sonora semiannulata*) is an opisthoglyphous or rear-fanged snake. Its diet is chiefly invertebrates. The front, recurved teeth on the maxilla,

about 10 in number, are small, whereas the posterior two or three are enlarged, face backward and have shallow grooves on their outer sides (Fig. 7.41). As its envenomation system is inefficient, it retains numerous other teeth in prominent rows on the pterygoid, palatine, and dentary bones to deal with struggling prey.

The **lyre snake** (*Trimorphodon biscutatus*) (Fig. 7.42) is a venomous rear-fanged snake. Its name refers to the three distinct kinds of recurved teeth in the maxilla. At the front are three to four larger teeth followed by a similar number of smaller teeth. The back teeth in the maxilla are large grooved fangs. Fig. 7.43 illustrates the alternating pattern of tooth replacement.

FIGURE 7.41 Dentition of the Western ground snake (*Sonora semiannulata*). CT image. Image width=1.2 cm. *Courtesy Digimorph.org and Dr. J.A. Maisano.*

FIGURE 7.39 Dentition of the ringneck snake (*Diadophis punctatus*). CT image. Image width=1.4 cm. *Courtesy Digimorph.org and Dr. J.A. Maisano.*

FIGURE 7.42 Dentition of the lyre snake (*Trimorphodon biscutatus*). CT image. Image width=3.7 cm. *Courtesy Digimorph.org and Dr. J.A. Maisano.*

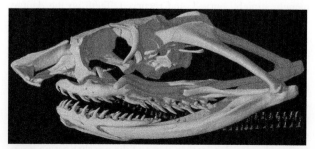

FIGURE 7.40 The Eastern hognose snake (*Heterodon platirhinos*). CT image. Image width=9 mm. *Courtesy Digimorph.org and Dr. J.A. Maisano.*

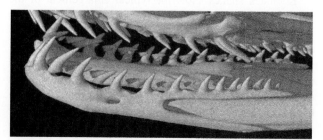

FIGURE 7.43 Higher power view of Fig. 7.42 showing alternation of replacing teeth in dentary of the lyre snake (*Trimorphodon biscutatus*). CT image. *Courtesy Digimorph.org and Dr. J.A. Maisano.*

Dasypeltis (11 species) and *Oligodon* (75 species) are egg-eating snakes. The **common egg-eating snake** (*Dasypeltis scabra*) (Fig. 7.44) has no teeth on the maxillae or dentaries, but retains some small teeth on the palatine bone. It has extremely flexible jaws and neck for eating eggs much larger than its head. Once the egg has been swallowed, strong muscular contractions squeeze it against the underside of the neck vertebra where a series of 25–35 downwardly projecting bony spines crack the shell. The soft contents of the egg are absorbed and the crushed shell is eventually regurgitated.

The diet of *Oligodon* consists of reptile eggs rather than bird eggs. Unlike rigid, brittle-shelled avian eggs that are crushed, the larger, leathery-shelled eggs of most oviparous reptiles require a different approach for access. *Oligodon* uses repeated slashes with blade-like teeth to enlarge a slit in the egg through which the snake can push its head and swallow the yolk and embryo within. Jaw movements require extreme displacement of the maxillary bone. In the **Taiwan snake** (*Oligodon formosanus*), teeth are present on the premaxilla, palatine, pterygoid, and dentary, none of which are grooved. The teeth are small, except for those on the posterior end of the maxilla, where one or two of the teeth are longer, laterally compressed and with the posterior margin forming a sharp cutting edge (Coleman et al., 1993; Fig. 7.45).

The **rat snake** (*Ptyas* [*Zaocys*] *dhumnades*) has 25–26 maxillary teeth, 18–20 palatine teeth, 24–28 pterygoid teeth and 25–27 dentary teeth (Mao, 1993).

One genus of colubrine snakes, exceptionally, contains many species that can be classed as insectivorous. **Dwarf snakes** (*Eirenis*) rarely exceed a maximum length of 60 cm. Their diet mainly consists of spiders, scorpions, centipedes, beetles, other insects, and worms. The dentitions of 18 species derived from 331 specimens have been described in detail by Mahlow et al. (2013) (Table 7.3). Tooth numbers are high in this genus. Tooth size in the maxilla generally increases posteriorly while in the dentary it decreases posteriorly. Differences in tooth numbers can be seen as well as subtle differences in tooth size and spacing. Three examples are illustrated in Figs. 7.46–7.48.

Subfamily Homalopsinae

This group contains a number of piscivorous snakes, such as the **masked water snake** (*Homalopsis buccata*). It characteristically possesses numerous, long, sharp, highly curved, or recurved teeth. The **keel-bodied water snake** (*Bitia hydroides*) unusually has greatly enlarged palatine teeth (Savitzky, 1983). The **rice paddy snake** (*Enhydris plumbea*) has about 15 maxillary teeth, 9–10 palatine teeth, 24 pterygoid teeth, and 25–27 dentary teeth (Mao, 1993).

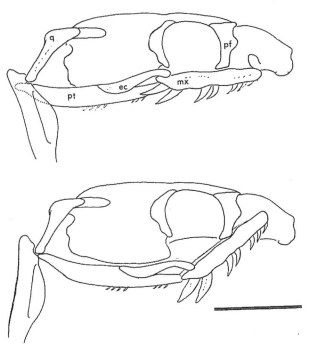

FIGURE 7.45 Diagram of skull of egg-eating Taiwan snake (*Oligodon formosanus*), showing maxillae at rest (upper diagram) and movement of maxillae (lower diagram), which brings the enlarged posterior maxillary teeth in a ventral and anterolateral arc into contact with the surface of the egg. *ec*, ectopterygoid; *mx*, maxilla; *pf*, prefrontal; *pt*, pterygoid; *q*, quadrate. Bar=5 mm. *From Coleman K., Rothfuss L.A., Ota H., Kardong K.V., 1993. Kinematics of egg-eating by specialised Taiwan snake* Oligodon formosanus *(Colubridae). J. Herpetol. 27, 320–327. Courtesy Editors of Herpetology,*

FIGURE 7.44 Common egg-eating snake (*Dasypeltis scabra*) after recently swallowing an egg. *Courtesy Wikipedia.*

TABLE 7.3 Tooth Numbers in Various Species of the Genus *Eirenis,* Derived From a Total of 331 Specimens

Species	Maxilla	Palatine	Pterygoid	Dentary
Eirensis aurolineatus	16–21	9–12	16–19	18–22
Eirensis modestus	14–20	8–13	14–21	14–21
Eirensis decemlineatus	19–26	10–14	18–24	19–24
Eirensis africanus	16–18	9–12	15–19	18–22
Eirensis barani	16–20	8–12	14–20	17–21
Eirensis collaris	17–20	8–11	17–21	16–20
Eirensis coronella	12–13	7–9	13–15	12–15
Eirensis coronelloides	11–14	7–9	12–17	10–15
Eirensis eiselti	15–18	8–10	17–21	17–20
Eirensis hakkariensis	16–20	9–12	18–23	17–21
Eirensis levantinus	16–22	9–13	15–22	14–22
Eirensis lineomaculatus	9–14	6–10	11–17	11–17
Eirensis medus	14–16	8–11	17–20	17–20
Eirensis punctatolineatus	17–21	10–12	16–22	17–22
Eirensis rechingeri	17	11	25	18–19
Eirensis rothii	14–15	9–11	13–18	13–17
Eirensis thospitis	16–17	9–11	18–19	18–20
Eirensis persicus-complex[a]	13–17	8–11	12–17	12–17

[a]*Including* Eirensis nigrofasciatus, Eirensis persicus, Eirensis occidentalis, *and* Eirensis walteri.
From Mahlow K., Tillack F., Schmidtler J.S., Muller J., 2013. An anotated checklist, description and key to the dwarf snakes of the genus *Eirenis* JAN, 1863 (Reptilia: Squamata: Colubridae), with special emphasis on the dentition. Vert. Zool. 63, 41–85; K Mahlow (personal communication).

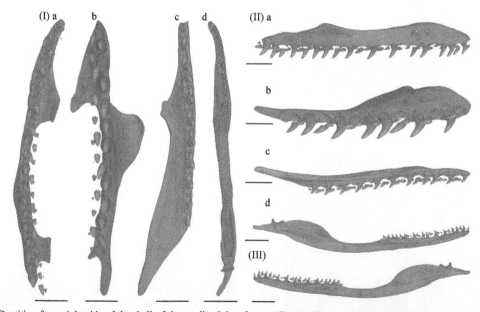

FIGURE 7.46   Dentition from right side of the skull of the ten-lined dwarf racer (*Eirenis* [*Eoseirenis*] *decemlineatus*). (I) ventral views of (a) maxilla, bar=1 mm; (b) palatine, bar=0.65 mm; (c) pterygoid, bar=1 mm; (d) dorsal view of lower jaw, bar=2 mm. (II) outer lateral view of (a) maxilla, bar=0.9 mm; (b) palatine, bar=0.65 mm; (c) pterygoid, bar=1 mm; (d) lower jaw, bar=2 mm. (III) inner lateral view of lower jaw, bar=2 mm. (*Courtesy Museum für Naturkunde, Berlin. Catalog no. ZMB 26677a.). From Mahlow K., Tillack F., Schmidtler J.S., Muller J., 2013. An anotated checklist, description and key to the dwarf snakes of the genus* Eirenis *JAN, 1863 (Reptilia: Squamata: Colubridae), with special emphasis on the dentition. Vert. Zool. 63, 41–85. Courtesy Editors Vertebrate Zoology.*

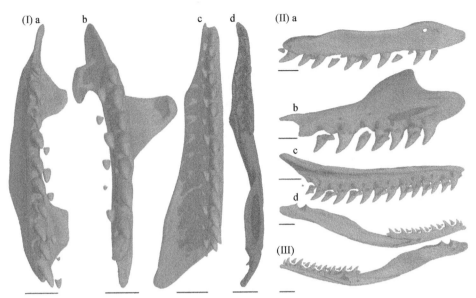

FIGURE 7.47 Dentition from right side of the skull of the dwarf racer (*Eirenis* [*Pediophis*] *lineomaculatus*). (I) ventral views of (a) maxilla, bar=0.55 mm; (b) palatine, bar=0.4 mm; (c) pterygoid, bar=0.65 mm; (d) dorsal view of lower jaw, bar=1 mm. (II) outer lateral view of (a) maxilla, bar=0.45 mm; (b) palatine, bar=0.3 mm; (c) pterygoid, bar=0.65 mm; (d) lower jaw, bar=1 mm. (III) inner lateral view of lower jaw, bar=1 mm. (*Courtesy Museum für Naturkunde, Berlin. Catalog no. ZMB 19912.*). *From Mahlow K., Tillack F., Schmidtler J.S., Muller J., 2013. An anotated checklist, description and key to the dwarf snakes of the genus* Eirenis *JAN, 1863 (Reptilia: Squamata: Colubridae), with special emphasis on the dentition. Vert. Zool. 63, 41–85. Courtesy Editors Vertebrate Zoology.*

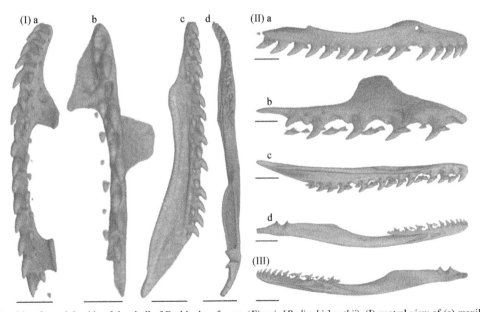

FIGURE 7.48 Dentition from right side of the skull of Roth's dwarf racer (*Eirenis* [*Pediophis*] *rothii*). (I) ventral view of (a) maxilla, bar=0.45 mm; (b) palatine, bar=0.4 mm; (c) pterygoid, bar=0.65 mm; (d) dorsal view of lower jaw, bar=0.9 mm. (II) outer lateral view of (a) maxilla, bar=0.4 mm; (b) palatine, bar=0.3 mm; (c) pterygoid, bar=0.55 mm; (d) lower jaw, bar=0.9 mm. (III) inner lateral view of lower jaw, bar=0.9 mm. (*Courtesy Museum für Naturkunde, Berlin. Catalog no. ZMB 77666.*). *From Mahlow K., Tillack F., Schmidtler J.S., Muller J., 2013. An anotated checklist, description and key to the dwarf snakes of the genus* Eirenis *JAN, 1863 (Reptilia: Squamata: Colubridae), with special emphasis on the dentition. Vert. Zool. 63, 41–85. Courtesy Editors Vertebrate Zoology.*

REFERENCES

Bogert, C.M., 1943. Dentitional phenomena in cobras and other elapids with notes on adaptive modifications of fangs. Bull. Am. Mus. Nat. Hist. 81, 285–360.

Boltt, R.E., Ewer, R.F., 1964. The functional anatomy of the head of the puff adder, *Bitis arietans* (Merr.). J. Morphol. 114, 83–106.

Britt, E.J., Clark, A.J., Bennett, A.F., 2009. Dental morphologies in garter-snakes (*Thamnophis*) and their connection to dietary preferences. J. Herpetol. 43, 252–259.

Campbell, J.A., 1988. The distribution, variation, natural history, and relationships of *Porthidium barbouri* (Viperidae). Acta Zool. Mex. 26, 1–32.

Coleman, K., Rothfuss, L.A., Ota, H., Kardong, K.V., 1993. Kinematics of egg-eating by specialised Taiwan snake *Oligodon formosanus* (Colubridae). J. Herpetol. 27, 320–327.

Cundall, D., 2009. Viper fangs: functional limitations of extreme teeth. Physiol. Biochem. Zool. 82, 63–79.

Deufel, A., Cundall, D., 2003. Feeding in *Atractaspis* (Serpentes: Atractaspididae). Zoology 106, 43–61.

Deufel, A., Cundall, D., 2006. Functional plasticity of the venom delivery system in snakes with a focus on the poststrike prey release behaviour. Zool. Anz. 245, 249–267.

Frazzetta, T.H., 1966. Studies on the morphology and function of the skull in the Boidae (Serpentes). J. Morphol. 118, 217–296.

Fry, B.G., Vidal, N., Norman, J.A., Vonk, F.J., Scheib, H., Ramjan, S.F.R., Kuruppu, S., Fung, K., Hedges, S.B., Richardson, M.K., Hodgson, W.C., Ignjatovic, V., Summerhayes, R., Kochva, E., 2006. Early evolution of the venom system in lizards and snakes. Nature 439, 584–588.

Fry, B.G., Scheib, H., van der Weerd, L., Young, B., McNaughtan, J., Ramjan, S.F.R., Vidal, N., Poelmann, R.E., Norman, J.A., 2008. Evolution of an arsenal. Structural and functional diversification of the venom system in the advanced snakes (Caenophidia). Mol. Cell Proteomics 7, 215–246.

Fry, B.G., Casewell, N.R., Wüster, W., Vidal, N., Young, B., Jackson, T.N.W., 2012. The structural and functional diversification of the Toxicofera reptile venom system. Toxicon 60, 434–448.

Gans, C., 1961. The feeding mechanism of snakes and its possible evolution. Am. Zool. 1, 217–227.

Guo, P., Zhao, E.-M., 2006. Comparison of skull morphology in nine Asian pit vipers (Serpentes: Crotalinae). Herpetol. J. 16, 305–313.

Hoser, R.T., 2012. A reassesment of the burrowing asps, *Atractaspis* Smith 1849, with the erection of a new genus and two tribes (Serpentes: Atractaspidae). Austral J. Herpetol. 11, 56–58.

Hutchinson, D.A., Mori, A., Savitzky, A.H., Burghardt, G.M., Wu, X., Meinwald, J., Schroeder, F.C., 2007. Dietary sequestration of defensive steroids in nuchal glands of the Asian snake *Rhabdophis tigrinus*. Proc. Nat. Acad. Sci. 104, 2265–2270.

Jackson, K., Fritts, T.H., 2004. Dentitional specialisations for durophagy in the Common Wolf snake, *Lycodon aulicus Capucinus*. Amphibia-Reptilia 25, 247–254.

Jackson, K., 2003. The evolution of venom-delivery systems in snakes. Zool. J. Linn. Soc. 137, 337–354.

Kley, N.J., 2001. Prey transport mechanisms in blindsnakes and the evolution of unilateral feeding system in snakes. Am. Zool. 41, 1321–1337.

Kley, N.J., 2006. Morphology of the lower jaw and suspensorium in the Texas blindsnake, *Leptotyphlops dulcis* (Scolecophidia: Leptotyphlopidae). J. Morphol. 267, 494–515.

Laporta-Ferreira, I.L., Salomao, M.G., 2004. Reptilian predators of terrestrial gastropods. In: Barker, G.M. (Ed.), Natural Enemies of Terrestrial Gastropods. CAB International, Wallingford, Oxfordshire, pp. 427–481.

Mahler, I., Kearney, M., 2006. The palatal dentition in squamate reptiles. Fieldiana Zool. NS 108, 1–61.

Mahlow, K., Tillack, F., Schmidtler, J.S., Muller, J., 2013. An anotated checklist, description and key to the dwarf snakes of the genus *Eirenis* JAN, 1863 (Reptilia: Squamata: Colubridae), with special emphasis on the dentition. Vert. Zool. 63, 41–85.

Mao, S.-H., Chen, B.-Y., 1980. Sea Snakes of Taiwan. Publication No 4. National Science Council, Taipei.

Mao, S.-H., 1993. Common Terrestrial Venomous Snakes of Taiwan. Special Publication No. 5. National Museum of Natural History, Taipei.

Marx, H., Rabb, G.B., 1972. Phyletic analysis of fifty characters of advanced snakes. Fieldiana Zool. 63, 1–321.

Pyron, R.A., Burbrink, F.T., Wiens, J.J., 2013. A phylogeny and revised classification of Squamata, including 4161 species of lizards and snakes. BMC Evol. Biol. 13, 93.

Rieppel, O., Kley, N.J., Maisano, J.A., 2009. Morphology of the skull of the white-nosed blindsnake *Liotyphlops albirostris* (Scolecophidia: Anomalepididae). J. Morphol. 270, 536–557.

Savitzky, A.H., 1981. Hinged teeth in snakes: an adaptation for swallowing hard-bodied prey. Science 212, 346–349.

Savitzky, A.H., 1983. Coadapted character complexes among snakes: fossoriality, piscivory and durophagy. Am. Zool. 23, 397–409.

Shine, R., Branch, W.R., Harlow, P.S., Webb, J.K., Shine, T., 2006. Biology of burrowing asps (Atractaspididae) from southern Africa. Copeia 103–115.

Vaeth, R.H., Rosssman, D.A., Shoop, W., 1985. Observations of the tooth surface morphology in snakes. J. Herpetol. 19, 20–26.

Vidal, N., Hedges, S.B., 2004. Molecular evidence for a terrestrial origin of snakes. Proc. R. Soc. Lond. B 271 (Suppl.), S226–S229.

Vonk, F.J., Admiraal, J.F., Jackson, K., Reshef, R., de Bakker, M.A.G., Vanderschoot, K., van den Berge, I., van Atten, M., Burgerhout, E., Beck, A., Mirtschin, P.J., Kochva, E., Witte, F., Fry, B.G., Woods, A.E., Richardson, M.K., 2008. Evolutionary origin and development of snake fangs. Nature 454, 630–633.

Weinstein, S.A., Smith, T.L., Kardong, K.V., 2009. Reptile venom glands. In: Mackessy, S.P. (Ed.), Handbook of Venoms and Toxins of Reptiles. CRC Press, Boca Raton, pp. 65–91.

Wuster, W., Thorpe, R.S., 1992. Dentitional phenomena in cobras revisited: spitting and fang structure in the Asiatic species of *Naja* (Serpentes: Elapidae). Herpetologica 48, 424–434.

Young, B.A., Kardong, K.V., 1996. Dentitional surface features in snakes (Reptilia: Serpentes). Amphibia-Reptilia 17, 261–276.

Zaher, H., Rieppel, O., November 1999. Tooth implantation and replacement in squamates, with special reference to mososaur lizards and snakes. Am. Mus. 3271, 1–19.

Zahradnicek, O., Horacek, I., Tucker, A.S., 2008. Viperous fangs: development and evolution of the venom canal. Mech. Dev. 125, 786–796.

Zuffi, M.A.L., 2014. Teeth number variations and cranial morphology within *Viper aspis* group. Basic Appl. Herpetol. 8, 87–97.

FURTHER READING

Bellairs, Ad'A., 1969. The Life of Reptiles, vol. 1. Weidenfeld and Nicholson, London (Chapter 4): Feeding and cranial mechanics, pp. 116–183. (Chapter 5): The venom apparatus and venom, pp. 184–216.

Edmund, A.G., 1969. Dentition. In: Gans, C. (Ed.), Biology of the Reptilia, vol. 1. Academic Press, London, pp. 117–200.

Gower, D., Garrett, K., Stafford, P., 2012. Snakes. Natural History Museum, London.

Chapter 8

Reptiles 3: Crocodylia

INTRODUCTION

The order Crocodylia comprises three families: Alligatoridae, Crocodylidae, and Gavialidae. They are mainly large, powerful, semiaquatic reptiles with short webbed feet, powerful tails, and tough, armored scales covering the dorsal surface. They range in length from 1 to 1.5 m (dwarf caiman) to 7 m (Australian saltwater crocodile). Males are larger than females. The eggs are protected by the mother and the young hatch using a caruncle at the tip of the upper jaw. Sex determination is temperature related. At higher temperatures during egg development, more males develop, and at lower temperatures, more females.

Crocodylians are exclusively carnivorous and their diet consists of fish, birds, other reptiles, such as snakes and lizards, and mammals. The prey of forms such as the Australian freshwater crocodile (*Crocodylus johnsoni*) and the Indian gharial (*Gavialis gangeticus*), which have extremely slender snouts with needle-like teeth, comprises mainly smaller prey, such as fish and crustaceans.

FEEDING

Crocodylians are ambush predators and lie in wait for terrestrial prey by floating mostly submerged; this strategy is facilitated by the placement, high up on the head, of the nose, eyes, and ears. Unlike other living reptiles, the Crocodylia have a bony secondary palate, formed by lateral processes of the premaxillae, maxillae, palatines, and pterygoids, which extends far back in the mouth. The respiratory and feeding pathways are therefore separated. The mouth can be isolated from the respiratory tract by a palatobuccal valve at the back of the oral cavity, and ingress of water into the respiratory tract can be prevented by closure of the external nares. These adaptations enable crocodylians to kill prey by holding them under water until they drown.

The skull is flattened and solidly built, with no movement between the bones, unlike the kinetic skulls of lizards and snakes. The snout of crocodylians is robust. Alligators have rounded (U-shaped) snouts; crocodiles have less rounded, tapering (V-shaped) snouts; and gharials have elongated jaws. Crocodylians generate the highest bite forces and tooth pressure known for any living animal species (Erickson et al., 2012). The bite force in the posterior region of the jaws ranges from 900 to 8983 N (*Paleosuchus palpebrosus* and *Crocodylus porosus*, respectively). The latter value is greater than the highest recorded value for a mammalian carnivore: 4500 N in the spotted hyena (*Crocuta crocuta*) (Binder and Valkenburgh, 2000). When bite force and tooth pressure are scaled isometrically with body mass, there is surprisingly no correlation with the shape of the crocodylian snout.

DENTITION

The crocodylian dentition, reviewed by Edmund (1962, 1969) and Grigg and Gans (1993), consists of single marginal rows of teeth in both jaws, with no palatal teeth. The teeth are large, conical, and pointed, but among alligators and crocodiles they vary in size and shape along the length of the jaw. The upper and lower jaws have undulating profiles and the positions of the largest teeth coincide with the crests of the jaw profile. Gharials have teeth of uniform size with a straight jaw line. Alligators have a larger upper jaw which completely overlaps the lower jaw, hiding the lower teeth in small depressions or sockets. In alligators the largest (fourth) tooth on the dentary fits into a pit in the upper jaw. The teeth of crocodiles and gharials interdigitate and the jaws are of equal size. As the large fourth mandibular tooth of crocodiles rests in an indentation in the upper jaw in the region of the premaxilla–maxilla junction, the teeth are prominent when the mouth is closed.

All living crocodilians have similar numbers of teeth: 5 premaxillary and 13–16 maxillary teeth. Most crocodiles have only 15 mandibular teeth, while alligators have 17–22. This gives total numbers of teeth of 60–72 in crocodiles and 72–82 in alligators. Due to its long slender snout, the gharial has the largest number of teeth: five premaxillary, 23–24 maxillary, and 25–26 mandibulars (making 106–110 in total). Tooth numbers are stable throughout life with only one or two additional teeth added to the back of the jaws of a young crocodilian through the rest of its life, although individual teeth will increase significantly in size with subsequent tooth replacement.

The Teeth of Non-Mammalian Vertebrates. http://dx.doi.org/10.1016/B978-0-12-802850-6.00008-4

The anterior (caniniform) teeth in crocodylians are longer, more slender, and generally rounder in cross-section than the posterior teeth. Besides being utilized for seizing prey, the caniniform teeth are also used in fighting, defense, aggression, and display. The posterior (molariform) teeth are blunt-tipped, shorter, and used for crushing and grasping prey before swallowing.

The teeth of the Crocodylia are unable to slice or masticate food but are very effective in grasping and killing prey. While crocodylians simply swallow small prey whole, they may firmly grasp large mammalian prey and rapidly spin round and rip off large portions. Large prey may also be trapped under water and left until soft enough to dismember.

In the crocodylians, the premaxillae are paired, and in the following sections tooth numbers refer to just one side.

As in most non-mammalian vertebrates the teeth are continuously replaced and, in the case of the Nile crocodile, there may be in the order of 50 or more replacements during its life (Poole, 1961). Tooth replacement is considered further in Chapter 10.

TOOTH ATTACHMENT

Crocodylians are the only non-mammalian vertebrates in which the teeth are attached to the jaw by **gomphosis**, as in mammals. The teeth have roots which are considerably longer than the crown. The roots are covered with a bonelike tissue (**cementum**) and are embedded in sockets within **alveolar bone** on the crest of the jaw bones. The tooth is retained within the socket by the **periodontal ligament**, which is a densely fibrous, vascular, and innervated connective tissue occupying the narrow space (about $200 \mu m$ in mammals) between the tooth and the socket wall. The collagen fibers are embedded in the cementum at one end and the bone of the socket wall at the other. This mode of attachment may be better adapted than ankylosis to the great mechanical demands on the dentition.

In crocodylians, replacement teeth develop within the same socket as their predecessors and in due course erupt into this same socket (Fig. 8.1; see also Fig. 8.11; **thecodont gomphosis**). Among mammals, in contrast, replacement permanent teeth come to occupy their own individual sockets and the periodontal ligament and socket of the deciduous predecessor disappear during growth and remodeling.

The available studies, conducted on the caiman (*Caiman sclerops* and *Caiman crocodilus*) and the American alligator (*Alligator mississippiensis*) (Kvam, 1960; Soule, 1967; Miller, 1968; Berkovitz and Sloan, 1979; Tadokoro et al., 1998; McIntosh et al., 2002) show that the tooth attachment structures in crocodylians and mammals share many similarities. Therefore, we present a general description of crocodylian tooth attachment and note important differences from mammals (for an account of the mammalian ligament, see Berkovitz et al., 2009).

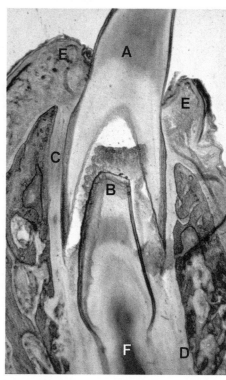

FIGURE 8.1 Caiman (*Caiman sclerops*). Transverse ground section of jaw embedded in methacrylate to retain soft tissues. Functional tooth (A) with successional tooth (B) forming basally within the same socket. (C) Periodontal ligament of functional tooth; (D) periodontal ligament of successional tooth; (E), gingiva; (F), pulp cavity of successional tooth. Note the apparent continuity of the periodontal ligament of functional and successional teeth. Image width = 6 mm. *From Berkovitz, B.K.B., Sloan, P., 1979. Attachment tissues of the teeth in Caiman sclerops (Crocodilia). J. Zool. Lond. 187, 179–194. Courtesy Editors of the Journal of Zoology.*

In young crocodylians, the posterior teeth are not initially separated from each other and lie in a common trough of bone until the interdental bony septa are formed (Miller, 1968). The alveolar bone consists of bundle bone, which is simpler than the complex Haversian bone of mammals. The portions of the periodontal ligament fibers embedded in the bone (**Sharpey's fibers**) do not appear to be as deep as in mammals (Berkovitz and Sloan, 1979). As in mammals, there is a layer of acellular cementum on the root surface and a layer of cellular cementum on top of that in the basal two-thirds of the root. Thick nodules of cellular cementum also occur at the alveolar crest (Kvam, 1960; Berkovitz and Sloan, 1979), whereas in mammals cellular cementum is absent from this region.

As in mammals, the collagen fibers form dense, interlacing bundles radiating from the roots and connecting the cementum to the alveolar bone. In longitudinal sections, the fibers connecting to the bone form dentoalveolar, horizontal, and oblique groups (Fig. 8.2). Toward the root apex, circular fibers have been described in the speckled caiman (*C. crocodilius*) (Tadokoro et al., 1998). In transverse sections, the fibers show a radial orientation (Fig. 8.3).

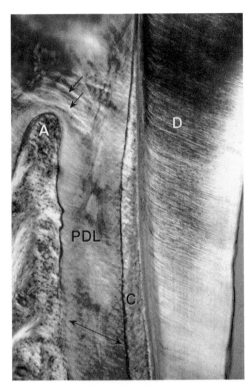

FIGURE 8.2 Caiman (*Caiman sclerops*). Longitudinal ground section of methacrylate-embedded tooth, examined between crossed polars and showing mainly obliquely orientated periodontal ligament fibers (*double-headed arrow*), plus some fibers crossing the alveolar crest (*small arrows*). *A*, alveolar bone; *PDL*, periodontal ligament; *C*, cementum; *D*, dentine. Image width = 1.5 mm.

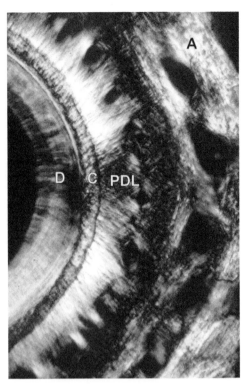

FIGURE 8.3 Caiman (*Caiman sclerops*). Transverse ground section of methacrylate-embedded tooth, examined between crossed polars and showing the radial orientation of the periodontal fibers. *A*, alveolar bone; *PDL*, periodontal ligament; *C*, cementum; *D*, dentine. Image width = 1.5 mm.

Transseptal collagen fibers run in mesiodistal sections from the cementum of one tooth over the alveolar crest to the cementum of the neighboring tooth (Kvam, 1960), or in buccolingual sections into the gingiva (Fig. 8.1). Most of the collagen fibers of the ligament have an average diameter of 55 nm, similar to those of mammals (45 nm), but some giant fibers have diameters up to 250 nm (Fig. 8.4; Berkovitz and Sloan, 1979). Oxytalan fibers, passing from the cementum up the periodontal ligament without attaching to bone, are less numerous than in mammals (Soule, 1967). Numerous small foci of mineralization have been observed in the young caiman ligament (Soule, 1967); these do not occur in mammals.

The rate of collagen turnover in the mammalian periodontal ligament is among the highest in the body, although the reasons for this are poorly understood. This may also be true of the caiman periodontal ligament, even though it is less cellular and more fibrous than its mammalian counterpart. The fibroblasts possess all the organelles associated with protein synthesis and secretion and a high level of collagen secretion is suggested by the presence of intracellular procollagen vesicles.

The central part of the ligament in the caiman contains a rich supply of nerves and blood vessels. Both free and encapsulated nerve endings are present (Tadokoro et al., 1998). In the proximal half of the caiman ligament and around the apex there occur a number of circular bodies, up to 30 μm in diameter, which contain nuclei and consist of concentric, horseshoe-shaped membrane profiles (Fig. 8.5; Berkovitz and Sloan, 1979; Tadokoro et al., 1998). They are devoid of cytoplasmic organelles, the main constituents being microfilaments and the occasional vesicle. Fig. 8.5 shows what appears to be an unmyelinated nerve fiber at the center of one of these bodies. Berkovitz and Sloan (1979) suggested that these structures are specialized pressure receptors, possibly resembling Herbst corpuscles, although they show some cytological differences from those bodies. Mechanoreceptors in the periodontal ligament may be utilized in reflex jaw activity, perhaps conferring the delicate control necessary for large, lumbering crocodylians to carry their young safely in their mouths. No similar bodies have been seen in the mammalian periodontal ligament.

CROCODYLIAN DENTITIONS

Alligatoridae

This family comprises the alligators and caimans. There are four genera containing eight species In the upper jaw, the fourth premaxillary tooth, and the ninth and eleventh

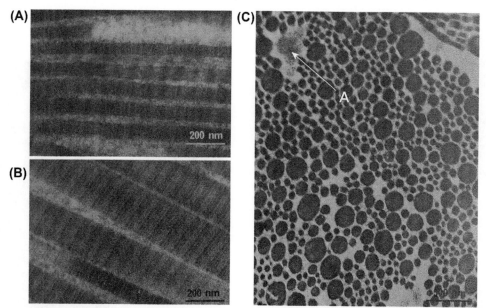

FIGURE 8.4 Caiman (*Caiman sclerops*). Transmission electron micrograph of periodontal ligament. (A) Longitudinal view of 55 nm fibers; (B) Longitudinal view of giant 250 nm fibers; (C) Transverse section of fibers showing both 55 and 250 nm fibers, together with an oxytalan fiber (*arrow A*). *From Berkovitz, B.K.B., Sloan, P., 1979. Attachment tissues of the teeth in Caiman sclerops (Crocodilia). J. Zool. Lond. 187, 179–194. Courtesy Editors of the Journal of Zoology.*

maxillary teeth are the largest. In the lower jaw, the first, fourth, eleventh, and twelfth teeth are the largest (Chiasson, 1962; Grigg and Gans, 1993).

There are two species of alligators: the American alligator and the Chinese alligator.

The **American alligator** (*A. mississippiensis*) can reach a length of about 5 m. Hatchlings feed mainly on invertebrates, such as insects, snails, and worms. As they grow, they take larger prey, including fish, amphibians, snakes, birds, and larger mammals such as wild pigs. The large, conical teeth are pointed anteriorly and are slightly blunter posteriorly. There are 5 teeth on each premaxilla, 13–15 on each maxilla, and 19–20 on each dentary (Chiasson, 1962; Westergaard and Fergusson, 1987, 1990; Fig. 8.6). In the upper jaw the third, ninth, and the last six teeth are enlarged. In the lower jaw, the largest teeth are the first and fourth, and the teeth around positions 13 and14.

The **Chinese alligator** (*Alligator sinensis*) (Fig. 8.7) is smaller than the American alligator and more heavily armored. It has a similar dentition. Snails may form a large part of its diet and its teeth may appear blunter.

Caimans inhabit Central and South America. There are three genera, comprising six species, of which the **black caiman** (*Melanosuchus niger*) is the largest. Indeed, at 5 m it is similar to the American alligator. Compared with other caiman species, it has distinctly larger eyes and a relatively narrow snout. Its dentition is similar to that of alligators, with 5 teeth on each maxilla, 13–14 on each maxilla, and 18–19 on each dentary, giving a total tooth count of 72–76 (Fig. 8.8).

The **spectacled caiman** (*C. crocodilus*) is a small- to mid-sized crocodylian: males grow up to a length of 2 m,

while females are smaller. It derives its name from the bridge of bone underneath its eyelids that makes it look as if it is wearing spectacles (Fig. 8.9). Some details of its dentition, which is similar to that of the black caiman, have been provided by Miller and Radnor (1970) and Berkovitz and Sloan (1979). The crowns are conical in shape and waisted at the junction with the root (Fig. 8.10). The roots are flared and thin walled. Varying degrees of resorption of the roots are seen depending on the size of the replacing tooth (Fig. 8.10). Radiological examination of the jaws reveals the presence of the underlying replacing teeth (Fig. 8.11).

Ősi and Barrett (2011) noted extensive wear on the posterior teeth of two specimens of *Caiman latirostris* and suggested that this was due to a large proportion of molluscs, crustaceans, and turtles in the diet. They also suggested that the shape of these teeth was rounded rather than pointed. However, as no definitely unworn teeth were present, this suggestion must await confirmation.

There are two species of dwarf caiman. **Cuvier's dwarf caiman** (*Paleosuchus palpebrosus*) is the smallest extant crocodylian with males reaching a length of 1.5 m. It has heavily ossified armor on both dorsal and ventral surfaces. Its head is unusual in having a dome-shaped skull and a short, smooth, concave snout with an upturned tip. Juvenile dwarf caimans feed mainly on invertebrates, small fish, and frogs, while adults eat similar but larger prey as well as molluscs. The upper jaw extends markedly further forward than the lower jaw. Unlike other members of the Alligatoridae, the premaxilla contains only four teeth. There are 14 upper teeth and 21–22 lower teeth, giving a total of 78–82 teeth (Fig. 8.12).

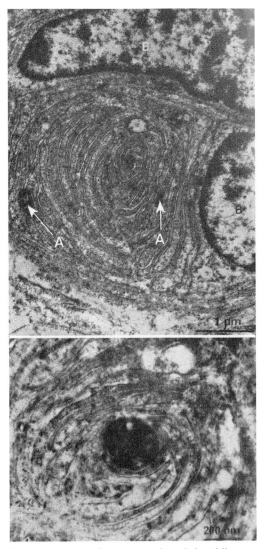

FIGURE 8.7 Chinese alligator (*Alligator sinensis*). Note that the fourth lower caniniform tooth on the dentary is hidden from view. *Courtesy Wikipedia.*

FIGURE 8.8 Dentition of black caiman (*Melanosuchus niger*). © *Sergy Skieznev/Dreamstime.*

FIGURE 8.5 Putative mechanoreceptors in periodontal ligament of the caiman (*Caiman sclerops*). Top: nerve ending showing a number of concentric membranes. *A*, attachment plaque; *B*, nucleus. Bottom: central part of a nerve ending showing horseshoe-shaped cell membranes surrounding a possible nerve fiber. *From Berkovitz, B.K.B., Sloan, P., 1979. Attachment tissues of the teeth in Caiman sclerops (Crocodilia). J. Zool. Lond. 187, 179–194. Courtesy Editors of the Journal of Zoology.*

FIGURE 8.9 Dentition of spectacled caiman (*Caiman crocodilus*). © *Mikelane45/Dreamstime.*

Crocodylidae

This family contains three genera with 14 species. In the upper jaw, the fifth premaxillary tooth, the ninth, and eleventh maxillary teeth are usually the largest. In the lower jaw, the first, fourth, eleventh, and twelfth teeth are the largest.

FIGURE 8.6 Dentition of the American alligator (*Alligator mississippiensis*). *Courtesy Wikipedia.*

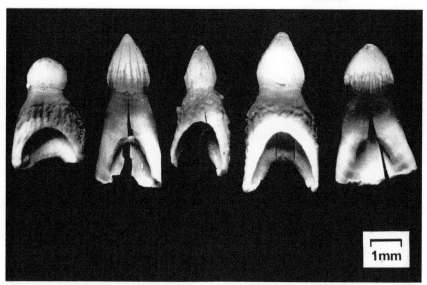

FIGURE 8.10 Individual teeth of spectacled caiman (*Caiman crocodilus*), showing resorption of root. *From Berkovitz, B.K.B., Sloan, P., 1979. Attachment tissues of the teeth in Caiman sclerops (Crocodilia). J. Zool. Lond. 187, 179–194. Courtesy Editors of the Journal of Zoology, London.*

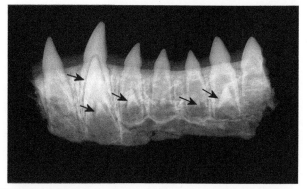

FIGURE 8.11 Radiograph of teeth of spectacled caiman (*Caiman crocodilus*), showing replacing teeth (*arrows*). Note that the second tooth in from the left border, which is about to be shed, shows two replacing teeth.

FIGURE 8.12 Dentition of dwarf caiman (*Paleosuchus palpebrosus*). © *Mgkuijpers/Dreamstime.*

FIGURE 8.13 Dentition of American crocodile (*Crocodylus acutus*) with mouth open. © *Steve Byland/Dreamstime.*

The **American crocodile** (*Crocodylus acutus*) is one of the largest crocodiles, growing to a length of 5 m. It has 30–40 teeth in the upper jaw and 28–32 teeth in the lower (Fig. 8.13). The teeth are conical, sharply pointed, but vary in size along the length of the jaws. The third and ninth teeth in the upper jaw and the first and fourth in the lower jaw are particularly enlarged. As mentioned previously, when the jaws are closed the lower teeth are clearly visible (Fig. 8.14). Hunting for small mammals, turtles, and birds, the American crocodile can be found in both fresh water and marine environments.

The **Nile crocodile** (*Crocodylus niloticus*) is a large, carnivorous, ambush predator growing up to a length of over 5 m. It is a generalist feeder and takes fish, reptiles, birds, and mammals. Its dentition has been described by a number of authors (Poole, 1961; Kieser et al., 1993; Putterill and Soley, 2003; Peterka et al., 2010; Fruchard, 2013). It has a typical crocodylian dentition, with 5 teeth on each premaxilla, 13–14 on each maxilla, and 14–15 on each dentary, giving a total of 64–68 teeth (Figs. 8.15 and 8.16). From a study of tooth development, and discontinuities of the dental lamina and tooth spacing, Kieser et al., (1993) interpreted the dentition of the Nile

FIGURE 8.14 Dentition of American crocodile (*Crocodylus acutus*) with mouth closed. *Courtesy Wikipedia.*

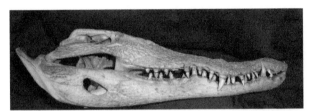

FIGURE 8.15 Dentition of Nile crocodile (*Crocodylus niloticus*). Image width = 38 cm. *Courtesy RCSOMA/403.2.*

FIGURE 8.16 Dentition of Nile crocodile (*Crocodylus niloticus*). *Courtesy Wikipedia.*

crocodile as being heterodont rather than homodont. They regarded the first five upper and first three lower teeth as equivalent to incisors, the next five upper and lower teeth as caniniform, and the last six upper and lower seven teeth as molariform. Detailed observations of feeding are required to establish whether these groupings have any functional significance.

The **Australian saltwater crocodile** (*C. porosus*) is the largest and most aggressive of all the crocodiles and the one most responsible for attacking and killing humans. Although most crocodylians are social animals, even sharing food, saltwater crocodiles are more territorial and less tolerant of their own kind. Although usually no more than 5-m long, lengths of up to 6 m have been recorded. As the name implies, this saltwater crocodile inhabits marine and brackish environments. Its teeth are large, with the fourth tooth from the front on the dentary being up to 9-cm long (Fig. 8.17A and B).

The **Australian freshwater crocodile** (*C. johnsoni*) is smaller than the more dangerous saltwater crocodile. Its snout is slenderer and its teeth are smaller. Its diet is mainly fish and its teeth are more sharply pointed than in other crocodiles (Fig. 8.18). It has the usual complement of 68–72 teeth (ie, 5 premaxillary, 14–16 maxillary, and 15 dentary).

The **Siamese crocodile** (*Crocodylus siamensis*) is a freshwater crocodile with males reaching up to 4 m in length. Unusually for a crocodile, it generally has only four teeth on each premaxilla. It bears the usual number of teeth on each maxilla (13–14) and dentary (15). Its teeth appear to be more pointed, perhaps reflecting its diet of fish (Figs. 8.19).

The **New Guinea crocodile** (*Crcodylus novaeguineae*) has slightly narrowed jaws. It has 66–68 thecodont teeth, with five on the premaxilla, 13–14 on the maxilla, and 15 on the dentary (Fig. 8.20).

The **dwarf crocodile** (*Osteolaemus tetraspis*) exists as two subspecies. These small crocodiles are nocturnal feeders. Like *C. siamensis*, it has only four teeth on each premaxilla. There are 12–13 teeth on each maxilla and 14–15 on each dentary (Fig. 8.21).

(A)   **(B)**

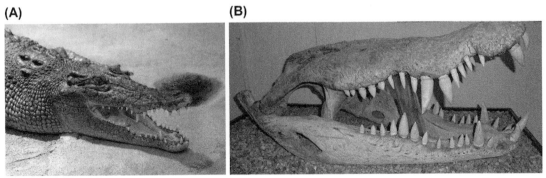

FIGURE 8.17 Saltwater crocodile (*Crocodylus porosus*). (A) Dentition. *(Courtesy Wikipedia.)*; (B) Model of skull. *(Courtesy The Hunterian, University of Glasgow. Catalog no. GLAHM:159542.)*

FIGURE 8.18   Dentition of Australian freshwater crocodile (*Crocodylus johnstonia*). *Courtesy Wikipedia.*

(A)   **(B)**

FIGURE 8.19   Dentition of Siamese crocodile (*Crocodylus siamensis*). *(A) Courtesy Wikipedia; (B) © Panuruangjan/Dreamstime.*

FIGURE 8.20   Dentition of New Guinea crocodile (*Crocodylus novaeguineae*). *Courtesy Wikipedia.*

FIGURE 8.21   Dentition of dwarf crocodile (*Osteolaemus tetraspis*). Image width = 16 cm. *Courtesy RCSOMA/403.32.*

Ghavialidae

The **gharial** (*G. gangeticus*) is the only member of this crocodile family and is found in the Indian Subcontinent. Mature males may be up to 4.5-m long. The gharial is a critically endangered species.

The gharial is distinguished from all other crocodiles by the shape of its skull and by its dentition, both of which are highly specialized for the capture of fish. The snout is elongated and narrow, although it becomes proportionally shorter and thicker with age. The teeth are more numerous than any other crocodylian species: there are 5 teeth on each premaxilla, 23–24 on each maxilla, and 25–26 on each dentary, giving a total of about 110

FIGURE 8.22 Dentition of male gharial (*Gavialis gangeticus*). Note the bulbous nasal protuberance. *Courtesy Misad/Dreamstime.*

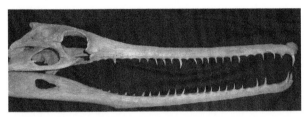

FIGURE 8.23 Dentition of gharial (*Gavialis gangeticus*). Image width=45 cm. *Courtesy Prof. P T Sharpe.*

FIGURE 8.24 Dentition of male gharial (*Gavialis gangeticus*). *Courtesy Wikipedia.*

(Figs. 8.22 and 8.23). Upper and lower teeth both project outwards and the two sets interdigitate. The teeth all have a similar sharp and conical shape, the largest teeth being at the front, with the first three teeth on the dentary fitting into notches in the upper jaw.

Gharials are the only living crocodylian with visible sexual dimorphism; its name derives from the fact that males develop a hollow bulbous nasal protuberance at sexual maturity that resembles the shape of an earthenware pot known locally as a "ghara" (Fig. 8.24). The function of the

nasal boss is unknown, but it is possibly used as a visual sex indicator, as a sound resonator, or for bubbling or other associated sexual behaviors.

ONLINE RESOURCES

Crocodilian species list (written by Adam Britton, hosted by the Crocodile Specialist Group): http://crocodilian.com/cnhc/csl.html.

REFERENCES

Berkovitz, B.K.B., Holland, G.R., Moxham, B.J., 2009. Oral Anatomy, Histology and Embryology, fourth ed. Elsevier, London, pp. 179–204.

Berkovitz, B.K.B., Sloan, P., 1979. Attachment tissues of the teeth in *Caiman sclerops* (Crocodilia). J. Zool. Lond. 187, 179–194.

Binder, W.J., van Valkenburgh, B., 2000. Development of bite strength and feeding behavior in juvenile spotted hyenas (*Crocuta crocuta*). J. Zool. Lond. 252, 273–283.

Chiasson, R.B., 1962. Laboratory Anatomy of the Alligator. WM.C. Brown Co, Dubuque, Iowa.

Edmund, A.G., 1962. Sequence and rate of tooth replacement in the crocodile. R Ontario Mus. Life Sci. Contr. 56, 1–42.

Edmund, A.S., 1969. Dentition. In: Gans, C. (Ed.), Biology of the Reptilia, vol. 1. Academic Press, London, pp. 117–200.

Erickson, G.M., Gignac, P.M., Steppan, S.J., Lappin, A.K., Vliet, K.A., Brueggen, J.D., Inouye, B.D., Kledzik, D., Webb, G.J.W., 2012. Insights into the ecology and evolutionary success of crocodilians revealed through bite-force and tooth-pressure experimentation. PLoS One 7, e31781. http://dx.doi.org/10.1371/journal.pone.0031781.

Fruchard, C., July 2013. The Nile crocodile, a new model for investigating heterodonty and dental continuous renewal in vertebrates. Biosci. Master Rev. 1–10.

Grigg, G., Gans, C., 1993. Morphology and physiology of the Crocodylia. In: Glasby, C.J., Ross, G.J.B., Beesley, P.L. (Eds.), Fauna of Australia. Amphibia and Reptilia, vol. 2A. Australian Government Publishing Service, pp. 326–343.

Kieser, J.A., Klapsidis, C., Law, L., Marion, M., 1993. Heterodonty and patterns of tooth replacement in *Crocodilus niloticus*. J. Morphol. 218, 195–201.

Kvam, T., 1960. The teeth of *Alligator mississippiensis* Daud: VI. Periodontium. Acta Odont. Scand. 18, 67–82.

McIntosh, J.E., Anderton, X., Flores de Jacoby, L., Carlson, D.S., Shuler, C.F., Diekwisch, T.G.H., 2002. Caiman periodontium as an intermediate between basal vertebrate ankylosis-type attachment and mammalian "true" periodontium. Microsc. Res. Tech. 59, 449–459.

Miller, W.A., 1968. Periodontal attachment apparatus in the young *Caiman sclerops*. Arch. Oral Biol. 13, 735–743.

Miller, W.A., Radnor, C.J.P., 1970. Tooth replacement patterns in young *Caiman sclerops*. J. Morphol. 130, 501–510.

Ősi, A., Barrett, P.M., 2011. Dental wear and oral food processing in *Caiman latirostris*: analogue for fossil crocodylians with crushing teeth. N. Jb. Geol. Paläont. Abh. 261, 201–207.

Peterka, M., Sire, J.-V., Hovorakova, M., Prochazka, J., Fougeirol, L., Peterkova, R., Viriot, L., 2010. Prenatal development of *Crocodylus niloticus niloticus* Laurenti, 1768. J. Exp. Zool. B Mol. Dev. Evol. 314B, 353–368.

Poole, D.F.G., 1961. Notes on tooth replacement in the Nile crocodile *Crocodilus niloticus*. Proc. Zool. Soc. Lond. 136, 131–140.

Putterill, J.F., Soley, J.T., 2003. General morphology of the oral cavity of the Nile crocodile *Crocodylus niloticus* (Laurenti, 1768). 1. Palate and gingivae. Onderstepoort J. Vet. Res. 70, 281–297.

Soule, J.D., 1967. Oxytalan fibres in the periodontal ligament of the caiman and alligator (Crocodilia, Reptilia). J. Morphol. 122, 169–174.

Tadokora, O., Mishima, H., Maeda, T., Kozawa, Y., 1998. Innervation of the periodontal ligament in the alligatorid Caiman crocodilius. Eur. J. Oral. Sci. 106 (Suppl. 1), 519–523.

Westergaard, B., Ferguson, M.W.J., 1987. Development of the dentition in *Alligator mississippimsis*. Later development in the lower jaws of hatchlings and young juveniles. J. Zool. Lond. 212, 191–222.

Westergaard, B., Ferguson, M.W.J., 1990. Development of the dentition in *Alligator mississippiensis*: upper jaw dental and craniofacial development in embryos. hatchlings, and young juveniles, with a comparison to lower jaw development. Am. J. Anat. 187, 393–421.

Chapter 9

Tooth Formation

INTRODUCTION

The hard dental tissues which make up a tooth are each formed by a two-part process, in which an organic matrix is first secreted by the formative cells and is then hardened by deposition of mineral (calcium phosphate). This process occurs between juxtaposed layers of epithelium and mesenchyme which together form a **tooth germ**. Like many organs, such as hair, feathers, or salivary glands, teeth develop through a series of reciprocal epithelial–mesenchymal interactions. Among non-mammalian vertebrates, the epithelium in the tooth germ of oral teeth is usually ectodermal in origin, but among amphibians the posterior teeth may be derived wholly or partly from foregut endoderm, as may the pharyngeal tooth germs of bony fish (Soukup et al., 2008). It has been suggested that in mice posterior teeth are also formed from the endoderm (Ohazama et al., 2010) but this is controversial, and evidence suggests that mouse molars are derived from oral ectoderm (Rothová et al., 2012a). The mesenchymal component consists of **neural crest** cells, which migrate from the region of the presumptive hindbrain and midbrain to the jaw regions during embryogenesis. Neural crest cells give rise to many structures besides teeth in the head region (Miletich and Sharpe, 2004). The tissue derived from neural crest cells is frequently distinguished as **ectomesenchyme**. The complex sequence of exchange of signaling molecules between the epithelium and the ectomesenchyme will be summarized later in this chapter.

TOOTH DEVELOPMENT

Tooth Germ Initiation

The site of initial tooth germ formation varies between the classes of vertebrates. In elasmobranchs, an ingrowth from the oral epithelium results in formation of a sheet-like **dental lamina** parallel with the surface of the jaw cartilage. The lamina persists throughout life and both the initial generation of tooth germs and all successor generations are initiated at the inner margin (Fig. 9.1). The developing teeth move forward through the dental lamina until they erupt into function. A dental lamina is also formed in amphibians (Davit-Béal et al., 2007) and reptiles (Whitlock and Richman, 2013). In these groups, only the first-generation tooth germs are formed at the margin of the primary dental lamina (Fig. 9.2),

while later generations bud from a successional lamina—an outgrowth from the outer dental epithelium of the predecessor tooth (Fig. 9.3). Typically, the dental lamina in these groups persists throughout life and generally retains a connection with the surface epithelium. The lamina forms on the lingual aspect of the marginal tooth rows in amphibians and reptiles. The lamina for a single row of pterygoid teeth on the palate is labially placed. Where two rows of palatal teeth are present the lamina is labially placed on the outer row and lingually on the medial row (Mahler and Kearney, 2006). The dental and successional laminae consist of two epithelial layers with some interstitial cells. Successional teeth develop from the aspect of the successional dental lamina which faces the predecessor tooth. Among reptiles, a successional dental lamina is observed even in the chameleon, which is monophyodont, but this ceases to proliferate and undergoes apoptosis (see Fig. 10.26; Buchtová et al., 2013).

Mammals also possess a dental lamina (Fig. 9.4; Berkovitz et al., 2009). The deciduous tooth germs form in association with the first-formed, anterior portion of the dental lamina. These are succeeded by permanent teeth which develop from a successional lamina, as in reptiles. The dental lamina continues to grow backward after the deciduous teeth have started to form and permanent molars, which have no predecessors, develop from this posterior portion of the dental lamina. The second permanent molar develops as a downgrowth of dental lamina from the first permanent molar, while the third permanent molar develops similarly from the dental lamina from the second permanent molar. Thus, culture of an early, isolated first (permanent) mouse molar tooth germ can give rise to all three molars (Lumsden, 1988). Lineage tracing has established that stem cells from the first molar and overlying epithelium give rise to the epithelial cell lineages of the second and third molars (Juuri et al., 2013; Gaete et al., 2015).

Bony fish differ from other vertebrates in that the teeth do not develop from a single dental lamina. The first generation of tooth germs bud directly from the oral epithelium (Fig. 9.5), but the origin of later generations of teeth varies. First, the locus of tooth initiation can be either superficial to the dentigerous bone (extraosseous) or within a trough or crypt inside the bone (intraosseous; Trapani, 2001). Second, the relationship of the tooth germ to precursor teeth varies. **Extraosseous tooth germs** are formed as an invagination into the epithelium: often, perhaps usually, in the outer layer of

The Teeth of Non-Mammalian Vertebrates. http://dx.doi.org/10.1016/B978-0-12-802850-6.00009-6

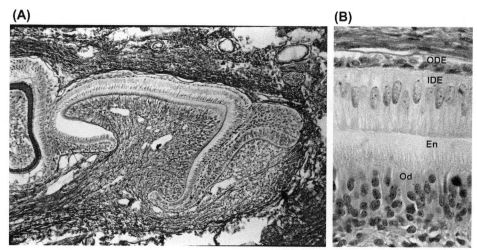

FIGURE 9.1 Dental lamina in an elasmobranch, the thornback ray (*Raja clavata*). Three tooth germs are visible. In the youngest (right), the dental epithelium has formed a cup-shaped rudiment enclosing a mass of mesenchyme. In the next oldest germ, the enameloid layer is completely formed, while in the oldest dentine formation is advanced. Gomori reticulin method. Field width = 1.03 mm. Inset: detail of tooth germ with complete enameloid matrix, showing the outer dental epithelium (ODE), consisting of a layer of small cuboidal cells immediately adjacent to the columnar inner dental epithelial cells (IDE), with no intervening layer. *En*, developing enameloid layer; *Od*, odontoblast layer. Field width = 14 μm. Hematoxylin and eosin.

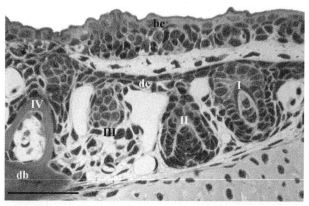

FIGURE 9.2 Transverse section of mandible of an amphibian, the Iberian ribbed newt (*Pleurodeles waltl*), showing dental lamina linking four adjacent tooth positions, of which tooth family I is closest to the symphysis and IV (functional tooth) is furthest from the symphysis. *be*, buccal epithelium; *db*, dentary bone; *dl*, dental lamina. Marker bar = 100 μm. *From Davit-Béal, A.F., Sire J.-Y., 2006. Morphological variations in a tooth family through ontogeny in* Pleurodeles waltl *(Lissamphibia, Caudata). J. Morphol. 267, 1048–1065. Courtesy Editors of the Journal of Morphology.*

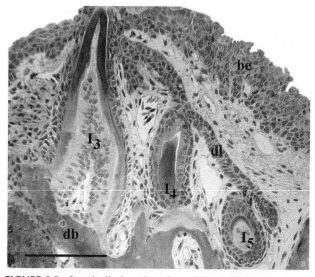

FIGURE 9.3 Longitudinal section of mandible of Iberian ribbed newt (*Pleurodeles waltl*), showing a single tooth family consisting of a functional tooth (I_4) and two successional teeth (I_5 and I_6), linked by a common successional lamina. *be*, buccal epithelium; *db*, dentary bone; *dl*, dental lamina. Marker bar = 100 μm. *From Davit-Béal, A.F., Sire J.-Y., 2006. Morphological variations in a tooth family through ontogeny in* Pleurodeles waltl *(Lissamphibia, Caudata). J. Morphol. 267, 1048–1065. Courtesy Editors of the Journal of Morphology.*

the epithelial sheath covering the shaft of the predecessor tooth (Fig. 9.6). Sometimes, the forming successional tooth remains embedded within the epithelium of the predecessor tooth until an advanced stage of development (Fig. 9.7). Alternatively, the dental epithelium of the developing tooth may become separated from that of the predecessor, while retaining a connection with the oral epithelium (Fig. 9.8). Finally, successional teeth can also develop from placodes formed within the oral epithelium, without formation of a dental lamina, and with no connection with predecessor teeth (Figs. 9.9 and 9.10).

Intraosseous tooth germs are, by definition, initiated at a distance from the oral epithelium. In some species,

intraosseous tooth germs are initiated at the tip of an epithelial strand, equivalent to a successional lamina, which buds from the epithelial sheath of a functional tooth (Fig. 9.11; Vandervennet and Huysseune, 2005; Fraser et al., 2013). As the replacement tooth erupts, the epithelial strand may break down (Fraser et al., 2013). In piranhas, the first generation of tooth germs presumably develops in this fashion. However, later generations of successional tooth germs originate from the outer dental epithelium of their predecessors as the

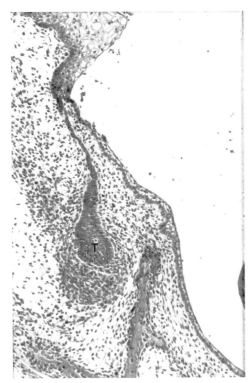

FIGURE 9.4 Dental lamina in a mammal, the ferret (*Mustela putorius*), with a tooth germ (T) at the bud stage. Masson's trichrome. Image width=4.5 cm. *From Berkovitz B.K.B., Holland G.R., Moxham B.M., 2009. Oral Anatomy, Histology and Embryology, fourth ed. Mosby Elsevier, London.*

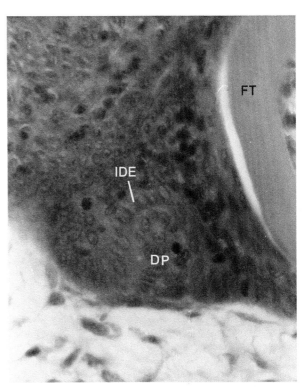

FIGURE 9.6 Early pharyngeal tooth germ in a bony fish, the European eel (*Anguilla anguilla*). The tooth germ consists of an invagination into the epithelium adjacent to a functional tooth (FT). The invaginated epithelium constitutes the inner dental epithelium (IDE), which surrounds a condensed mass of mesenchymal cells forming the dental papilla (DP). Hematoxylin and eosin. Field width=82 μm.

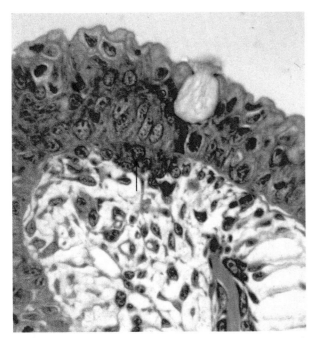

FIGURE 9.5 Precursor of tooth germ (placode) in a bony fish, the rainbow trout (*Oncorhynchus mykiss*). Note the condensation of mesenchyme beneath the epithelial placode (*arrow*). Masson's trichrome. Magnification ×1000. *From Berkovitz B.K.B., 1978. Tooth ontogeny in the upper jaw and tongue of the rainbow trout (Salmo gairdneri). J. Biol. Buccale 6, 205–215. Courtesy Editors of the Journale de Biologie Buccale.*

latter start to erupt (Berkovitz and Shellis, 1978). As the predecessor tooth erupts, it remains connected with its successor tooth germ by a lengthening epithelial strand (Fig. 9.12). Replacement teeth are thus initiated autonomously in this species and the epithelial strand is not exactly equivalent to a successional lamina, as it does not initiate tooth germs.

Crown Formation

In elasmobranchs, the earliest stage of tooth development is an invagination at the inner margin of the dental lamina (Fig. 9.1). In other vertebrates which form a dental lamina, tooth germs are first visible as swellings (**tooth buds**) at specific sites on the inner margin of the lamina. Ectomesenchymal cells proliferate and form a condensed mass around the tooth bud. The epithelial bud next develops into a hollow structure (the **cap** stage) by proliferation and ingrowth of epithelial cells around the central part of the mesenchymal aggregate (Fig. 9.13). The epithelial component is referred to as the **dental epithelium**. In relation to mammals, the dental epithelium is called the enamel epithelium and the epithelial cap is called the enamel organ. This terminology would also be valid for reptiles and adult amphibians, but for other vertebrates, which possess enameloid rather than enamel, it is inappropriate. In this book, we will refer to "dental epithelium" in all vertebrates.

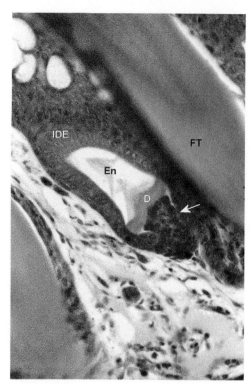

FIGURE 9.7 European eel (*Anguilla anguilla*). Pharyngeal tooth germ at stage of enameloid mineralization still embedded in epithelial sheath of functional tooth (FT). *IDE*, inner dental epithelium; *En*, remains of enameloid matrix; *D*, dentine. *Arrow* indicates boundary between dental epithelium and papilla, marking future position of tooth shaft. Hematoxylin and eosin. Field width = 123 μm.

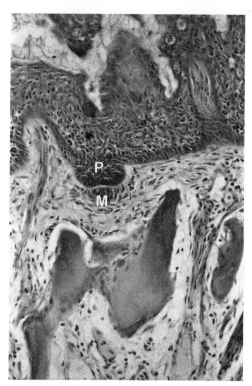

FIGURE 9.9 Sea bass (*Dicentrarchus labrax*). Initial pharyngeal tooth germ, consisting of a condensation of mesenchyme (M) around a placode (P) formed from the basal layer of the oral epithelium. Masson trichrome. Field width = 203 μm.

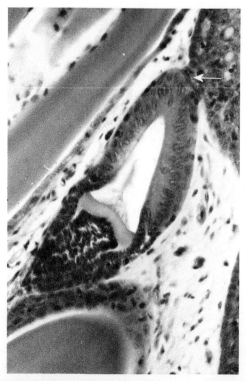

FIGURE 9.8 Black bream (*Spondyliosoma cantharus*). Pharyngeal tooth germ at same stage of development as in Fig. 9.7, but detached from epithelial sheath of predecessor tooth while retaining connection with oral epithelium (*arrow*). Masson trichrome. Field width = 123 μm.

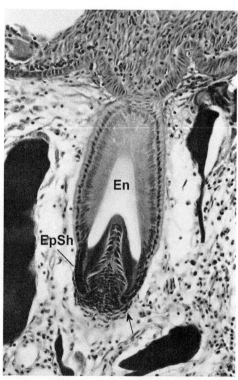

FIGURE 9.10 Sea bass (*Dicentrarchus labrax*). Advanced pharyngeal tooth germ. Enameloid cap is fully mineralized and is represented by an enameloid space (En) left after demineralization during histological processing. The tooth shaft is beginning to form on the inner surface of a two-layered epithelial sheath (EpSh). Note condensation of mesenchymal cells around basal extremity of shaft rudiment (*arrow*). Masson's trichrome. Field width = 810 μm.

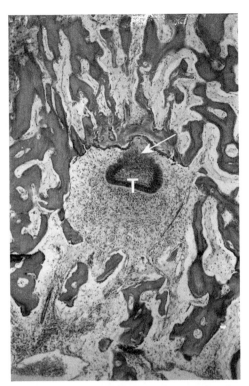

FIGURE 9.11 Ballan wrasse (*Labrus bergylta*): pharyngeal tooth. Very early intraosseous tooth germ (T) at the cap stage. The epithelial strand connecting this tooth germ with the oral epithelium is outside the plane of section, but its origin at the top of the epithelial cap is visible (*arrow*). Hematoxylin and eosin. Field width=810 μm.

The mesenchymal component comprises two regions: the **dental papilla** and the **dental follicle**. The dental papilla is the future source of the dental pulp and of the odontoblasts, which form dentine. In mammals the follicle is responsible for formation of the tooth attachment tissues (cementum, periodontal ligament, and alveolar bone; Diekwisch, 2001; Diep et al., 2009), and the same is probably true in other vertebrates. The dental papilla has been localized (Rothová et al., 2012b) to a relatively small volume of condensed mesenchyme immediately adjacent to the inner dental epithelium and enclosed by the epithelial cap. The majority of the dental mesenchyme belongs to the follicle. In mammals, after the cap stage, the follicle forms a prominent layer of mesenchymal cells embedded in a loose connective tissue, which encloses the tooth germ. A distinct layer of this kind is also present in association with reptilian tooth germs (Delgado et al., 2005; Buchtová et al., 2013), but in most amphibians and bony fish, is either absent or is confined to the basal region of the tooth germ (Figs. 9.10 and 9.12; Davit-Béal et al., 2006, 2007). In the light of the revised concept of the dental follicle due to Rothová et al. (2012b), the absence of a histologically obvious follicle is probably related to the fact that formation of the tooth attachment in amphibians and fishes takes place at the base of the tooth germ, not at the sides. A well-developed follicle is, however, present around developing thecodont actinopterygian teeth (Fig. 9.20).

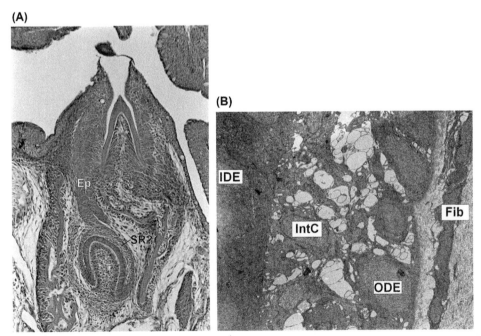

FIGURE 9.12 (A) Two developing teeth in piranha (*Serrasalmus rhombeus*). The shaft is forming in the tooth germ nearer the oral cavity, while the deeper tooth germ is in the earliest stages of enameloid formation. The two are connected by an epithelial strand (Ep), which passes through the gubernaculum, an opening in the bone. In the younger tooth germ there is a layer (SR) between the inner and outer dental epithelia (IDE, ODE), which is reminiscent of a stellate reticulum (cf. Figs. 9.15 and 9.16). Hematoxylin and eosin. Magnification ×155; (B) European eel (*Anguilla anguilla*): tooth germ during late enameloid matrix formation (TEM). IDE and ODE separated by thin intermediate layer of interconnected stellate cells (IntC). *Fib*, fibroblast. Field width=25.3 μm.

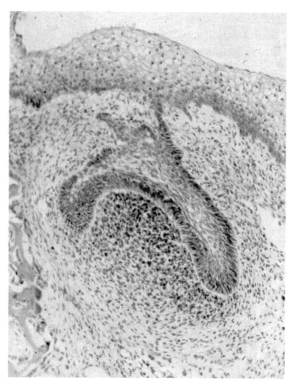

FIGURE 9.13   Cap stage of tooth development in ferret (*Mustela putorius*). Growth of dental epithelium has resulted in enlargement and formation of a cap shape enclosing the condensed mesenchyme of the dental papilla. Masson's trichrome. Magnification ×150.

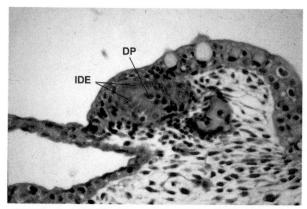

FIGURE 9.14   Tooth germ of rainbow trout (*Oncorhynchus mykiss*) prior to formation of enameloid matrix. The boundary between the inner dental epithelium (IDE) and the dental papilla (DP), marked by a green-stained basement membrane, has assumed the final form of the mature tooth tip. Masson's trichrome. Magnification ×750. *From Berkovitz B.K.B., 1978. Tooth ontogeny in the upper jaw and tongue of the rainbow trout (*Salmo gairdneri*). J. Biol. Buccale 6, 205–215.*

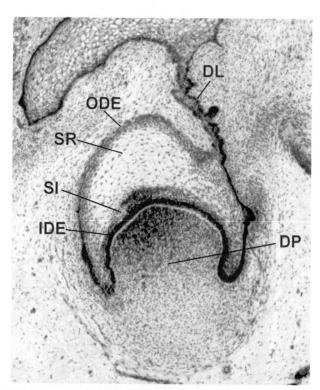

FIGURE 9.15   Bell stage of tooth development in ferret (*Mustela putorius*). The large volume between the inner (IDE) and outer dental epithelia (ODE) is occupied by the stellate reticulum (SR) and stratum intermedium (SI). The tooth germ is still connected with the oral epithelium by the dental lamina (DL). *DP*, dental papilla. Hematoxylin and eosin. Magnification ×100.

From the cap stage onwards, mesodermal cells from outside the tooth germ migrate into the ectomesenchyme of the dental papilla and form the blood vessels which supply the tooth with nutrients and function in gas exchange (Rothová et al., 2011). Blood vessels are not a universal feature of developing teeth in non-mammalian vertebrates, or may appear only at an advanced stage of development. This seems to be the case with small tooth germs, probably because the diffusion distances are short enough for exchange of gases, nutrients, and metabolic waste products to be maintained without vascularization. Innervation, commencing during the bud stage, is co-ordinated by interactions between the tooth germ and nerves, mediated by TGFβ, WNT and FGF (Luukko and Kettunen, 2014).

As the tooth germ expands, the epithelial–mesenchymal interface develops a specific form through a process of localized cell proliferation and folding within the dental epithelium. The shape of this interface is the primary determinant of the external crown morphology. Where the teeth will be covered with enamel, the epithelial–mesenchymal interface defines the future enamel–dentine junction, and the external shape of the crown is defined by the distribution of enamel upon this junction. In teeth that will be covered with enameloid, the shape of the interface is more or less the same as that of the final tooth crown (Fig. 9.14), because enameloid matrix is deposited mainly on the mesenchymal surface of the interface. However, in bony fish, the inner dental epithelium may deposit some matrix on the outer

aspect of the interface, thereby modifying the external tooth shape (Chapter 12).

In mammals, prior to the establishment of crown form, the inner and outer dental epithelia become separated by additional epithelial components (Figs. 9.15 and 9.16). The **stratum intermedium** forms a layer covering the inner

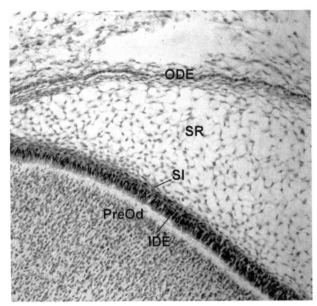

FIGURE 9.16 Part of a bell stage tooth germ in ferret (*Mustela putorius*), to show the organization of the epithelial component. From the outside, the outer dental epithelium (ODE), stellate reticulum (SR) characterized by widely spaced, stellate cells, the stratum intermedium (SI), and the inner dental epithelium (IDE). The outer layer of cells of the dental papilla have differentiated into preodontoblasts (PreOd). Hematoxylin and eosin. Magnification ×200.

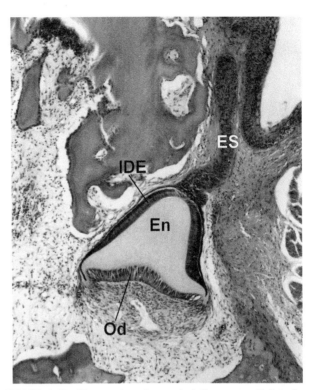

FIGURE 9.17 Pharyngeal tooth germ in ballan wrasse (*Labrus bergylta*), in which the enameloid matrix (En) is fully formed but has not started to mineralize. The tooth germ remains connected to the oral epithelium by an epithelial strand (ES) running toward the top of the field. Both the odontoblasts (Od) and inner dental epithelial (IDE) cells are active during enameloid formation in bony fishes and both appear columnar. Hematoxylin and eosin. Field width = 322 μm.

dental epithelium and is characteristically rich in the enzyme alkaline phosphatase. According to Harada et al. (2006), the stratum intermedium originates from the inner dental epithelium before the cells in that layer differentiate into ameloblasts. The stratum intermedium in turn is separated from the outer dental epithelium by the **stellate reticulum**: a thick zone of widely spaced stellate cells, embedded in a glycosaminoglycan-rich ground substance and connected with each other via cell processes (Fig. 9.16). Because of the presence of the stellate reticulum, the tooth germ at this stage of development has a rounded shape, so is referred to as the **bell stage** (Fig. 9.15). Tooth germs of reptiles appear to develop a stellate reticulum, but the presence of a stratum intermedium is uncertain (Delgado et al., 2005; Handrigan and Richman, 2010a; Zahradnicek et al., 2012; Buchtová et al., 2013). In bony fish, a stellate reticulum is often absent (eg, Figs. 9.7–9.11) but evidence of cells possibly equivalent to this layer has been seen in some fish (Fig. 9.12) (Ishiyama et al., 1999; Vandervennet et al., 2005; Fraser et al., 2013) and this layer might even harbor stem cells (Fraser et al., 2013).

The epithelial cap expands by cell proliferation at the margins, specifically within the structure known as the **cervical loop**, where the dental epithelium is folded and the inner and outer dental epithelia enclose a region of stellate reticulum. The cervical loop has been studied most intensively in rodent incisors, which grow continuously, but the loop probably has the same organization and role in teeth of limited growth. The loop forms a niche for a population of epithelial

stem cells, located at the boundary of the stellate reticulum and basal epithelium (Harada et al., 1999). The stem cells generate both inner and outer dental epithelium, stellate reticulum, and stratum intermedium (Juuri et al., 2012).

Following the morphogenetic changes during the bell stage, the inner dental epithelium and the adjacent outer layer of cells of the dental papilla differentiate and acquire the ability to synthesize and secrete the organic matrices of the hard tissues. In all vertebrates, the outer layer of cells of the dental papilla differentiates into a layer of **odontoblasts**, which produce dentine. The differentiated inner dental epithelial cells of mammals, reptiles, and adult amphibians are termed **ameloblasts** and are responsible for production of enamel. The role of the corresponding inner dental epithelial cells varies among fish and larval urodeles, so the term ameloblast is inappropriate in this context.

Hard tissues are laid down at the interface between the inner dental epithelium and the dental papilla. In fish and larval amphibians, the first tissue matrix to be laid down is that of enameloid, which is fully formed before deposition of dentine matrix begins to be deposited on its inner aspect (Fig. 9.17). In adult amphibians, and in reptiles and mammals, a thin layer of dentine matrix is secreted first in the region of the cusp tips or incisal edges and enamel matrix formation begins

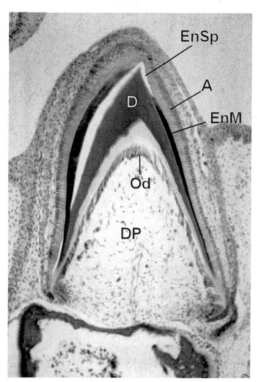

FIGURE 9.18 Late bell stage tooth germ in green lizard (*Lacerta viridis*). *EnM*, enamel matrix. At the tip of the tooth germ, the enamel is fully mineralized, so the organic matrix is reduced to a low level and in this demineralized section this part of the enamel is represented by a space (EnSp), *A*, Ameloblasts layer (inner dental epithelium); *D*, dentine; *Od*, odontoblast layer; *DP*, dental pulp. Hematoxylin and eosin. Image width = 140 μm.

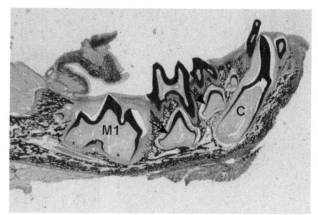

FIGURE 9.19 Longitudinal section of the demineralized lower jaw of a kit of the ferret (*Mustela putorius*). In the tooth germs of premolars enamel formation is incomplete: while the enamel overlying the cusps is fully mineralized and represented by a space, partly mineralized, yellowish enamel matrix (*arrows*) is still present in the lateral regions. The enamel at the coronal surfaces of the canine (C) and first molar (M1) is completely mineralized and represented by an enamel space. All the permanent teeth will need to erupt vertically to reach their functional positions. Basally, the rudiments of the roots are forming within epithelial root sheaths (one root on the canine; two roots, marked by asterisks, on the molar). Van Gieson stain. Magnification ×8. *From Berkovitz B.K.B., Moxham B.J., 1990. The development of the periodontal ligament with special reference to collagen fibre ontogeny. J. Biol. Buccale 18, 227–236. Courtesy Editors of the Journale de Biologie Buccale.*

almost immediately afterward (Fig. 9.18). As enameloid or enamel mineralizes, the matrix is lost progressively, so that mature enamel contains only a small residue of organic matrix and is represented as a space in demineralized histological sections (Figs. 9.7, 9.10 and 9.19). As the tooth germ enlarges, formation of dentine and enamel is initiated at the margins while the first-formed enamel increases in thickness. The final morphology of the crown is established by the pattern of secretion of coronal hard tissue (Fig. 9.19), as detailed further in Chapters 11 and 12.

Tooth Eruption

After the formation of coronal hard tissues is well advanced, the more basal portion of the tooth begins to form. In elasmobranchs, the attachment tissues begin to form directly but in most vertebrates, there is some elongation of the tooth as the shaft or root forms.

During this phase of tooth formation, the tooth moves from its site of formation to its functional position (**tooth eruption**) and becomes attached to the jaw skeleton. In elasmobranchs this is accomplished by forward movement of the whole array of developing and mature teeth, which brings new completed teeth into the functional position. In other vertebrates the process of tooth eruption involves vertical movement. In addition,

teeth of bony fish and reptiles may develop in a recumbent position and have to rotate as they come into function. Among amphibians, most reptiles and bony fishes, in which the teeth develop extraosseously, growth in length of the tooth shaft seems to be an important component of eruption, as it brings the tooth tip toward the epithelium. However, once the base of the tooth becomes fixed to the jaw bone, movement ceases and the final phase of eruption may be brought about by remodeling of the epithelium so that the tooth tip is exposed (Huysseune et al., 1998; van der Heyden et al., 2000; Vandervennet and Huysseune, 2005). In mammals, crocodylians, and many bony fish, the teeth form intraosseously and have to move a considerable vertical distance during eruption (Figs. 9.19 and 9.20) (Berkovitz and Moxham, 1990). The mechanism by which the teeth reach their functional position in this situation is not fully known, but experimental evidence supports the hypothesis that in mammals the eruptive force is generated within the periodontal ligament and its precursor, the dental follicle (Berkovitz and Moxham, 1995; Cho and Garant, 2000).

Tooth Attachment

Whereas development of the distal ("crown") portion of a tooth is the product of the inner dental epithelium and the dental papilla, the tooth attachment tissues originate from the dental papilla, the dental follicle, and, possibly, mesenchymal tissue from outside the tooth germ.

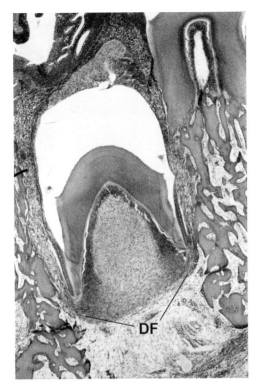

FIGURE 9.20 Ballan wrasse. Tooth erupting from the intrabony crypt where it formed (cf. Figs. 9.11 and 9.17) toward the oral cavity (top), after removal of overlying bone. *DF*, dental follicle. Ehrlich's hematoxylin and eosin. Field width=1.6 mm.

The first rudiment of the basal plate of **elasmobranchs** is a meshwork of collagen fibers which extends out of the dental lamina from the base of the tooth "crown" (Fig. 9.21A). This rudiment is formed between odontoblasts (continuous with those with the dental papilla) and mesenchymal cells, the origin of which is not known. During further development, the tissues of the basal plate, consisting of an inner layer of dentine and an outer layer of mineralized, atubular mineralized tissue, are formed respectively by odontoblasts and external mesenchymal cells (Fig. 9.21B and C; Shellis, 1982).

Among other vertebrates the dental epithelium extends beyond the crown as an **epithelial sheath**, within which is formed the shaft of the tooth or, in crocodylian teeth, the root (Fig. 9.22). The epithelial sheath presumably regulates morphogenesis of the shaft or root. It also has the important function of controlling the deposition of the tooth attachment tissues. When the sheath remains intact, the attachment tissues are confined to the tooth base but when the sheath retracts or breaks down, attachment tissues can be formed on the lateral surfaces of the tooth sheath (Howes, 1978; Shellis, 1982; Cho and Garant, 2000; Diekwitsch, 2001).

Except for crocodylians, the teeth of reptiles, amphibians, and actinopterygians are attached by **ankylosis** of the tooth to the bone, whether the tooth is pedicellate or not. Ankylosis begins as the base of the elongating tooth germ

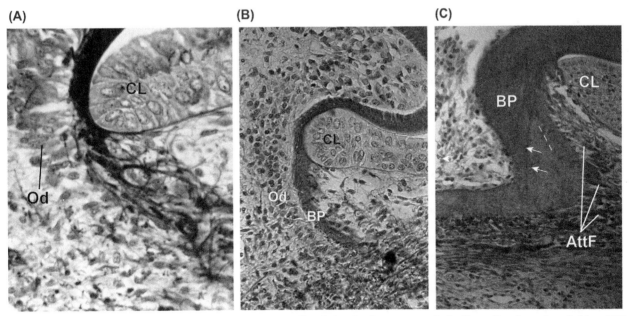

FIGURE 9.21 Thornback ray (*Raja clavata*). (A) Earliest rudiment of basal plate forming as a network of collagen fibers beyond cervical loop (CL). Odontoblast (Od), continuous with layer inside dental epithelium, differentiating on inner aspect of fibrous rudiment. Other mesenchymal cells on outer aspect of fiber network and some enclosed within it. Azan. Image width=290 μm. (B) More advanced rudiment of basal plate (BP) than in (A). Fine-textured collagenous tissue laid down on fibrous rudiment. Well-defined layer of odontoblasts on inner aspect of plate tissue. Other mesenchymal cells on outer aspect. Celestin blue-hemalum-van Gieson. Image width= μm. (C) Almost mature BP, consisting of inner layer of dentine, with tubules (*arrows*) and an atubular outer layer. *Dashed line* marks approximate boundary between the two layers. Attachment fibers (AttF) connect BP to connective tissue sheet running around and beneath the teeth. Celestin blue-hemalum-van Gieson. Image width=180 μm.

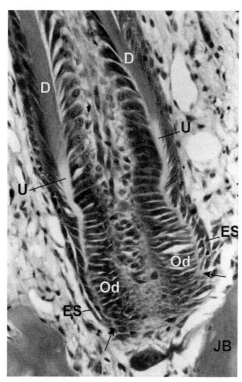

FIGURE 9.22 European eel (*Anguilla anguilla*). Late stage of shaft formation, showing shaft tissues forming between epithelial sheath (ES) and odontoblast layer (Od). In the upper part of the figure, the distal shaft tissue consists of mineralized dentine (D), which is stained. In the lower part, the shaft tissues (U) are not mineralized (unstained) and comprise the basal dentine and the rudiments of the dividing zone and the pedicel. *JB*, jaw bone. Celestin blue-hemalum-eosin. Image width = 130 μm.

approaches the jaw bone. Outgrowths of dentine from the base of the tooth and bone from the adjacent bone meet and form a composite tissue in which bone and dentine are in intimate contact (Fig. 9.23A; Shellis, 1982; David-Béal et al., 2007; Zahradniček et al., 2012). Sometimes, the term cementum is applied to all or part of the mineralized tissue connecting tooth and bone in ankylosis (David-Béal et al., 2007; Zahradniček et al., 2012). However, as cementum is a bone-like tissue distinguishable histologically only when it forms a layer on roots of thecodont teeth, there seems to be no justification for this. In amphibians and most reptiles, pleurodont attachment is consolidated by deposition of bone between the labial surface of the tooth and the lingual surface of the jaw bone, without involvement of odontoblasts (Fig. 9.23B). Bone formation on the lateral tooth surfaces also occurs in thecodont teleost teeth (Figs. 4.13 and 4.14). In both pleurodont and thecodont ankylosis, bone formation at the lateral tooth surfaces is correlated with loss of the epithelial sheath from this region (Figs. 4.13, 4.14, and 9.23B).

In pedicellate teeth, the dividing zone and **pedicel** start to form after the rudiment of the tooth shaft has grown to its full length. Both tissues are laid down between the epithelial sheath and odontoblasts in amphibian teeth

(Fig. 9.23A; Parsons and Williams, 1962; Davit-Béal et al., 2007), although the sheath retracts somewhat in late stages of pedicel formation and osteoblasts gain access to the outer surface (Davit-Béal et al., 2007). The involvement of the epithelial sheath in formation of the attachment tissues in pedicellate teeth of teleosts is controversial. Some workers (Huysseune et al., 1998; Vandervennet and Huysseune, 2005) considered that the sheath never extends beyond the dividing zone, and considered that the pedicel is an outgrowth of the bone. However, it has also been observed (Shellis, 1982) that the epithelial sheath covers the earliest rudiment of the pedicel and dividing zone (a stage apparently not observed by Huysseune et al.), but then retracts or is broken down and in later stages reaches only the dividing zone, as seen in Figs. 9.22 and 9.24A–B. This sequence suggests that the epithelial sheath is involved in morphogenesis of the pedicel, but not in its formation. In both teleosts and amphibians, odontoblasts differentiate in relation to both the dividing zone and pedicel and actively contribute to both tissues (Parsons and Williams, 1962; Shellis, 1982 and papers cited therein; Huysseune et al., 1998; Davit-Béal et al., 2007). As the epithelial sheath covers the pedicel until a late stage, it is likely that the amphibian pedicel is produced almost entirely by odontoblasts (Parsons and Williams, 1962; Davit-Béal et al., 2007). Whatever the role of the epithelial sheath proves to be, it is clear that the teleost pedicel is a joint product of the internal odontoblasts and the osteoblasts which cover the outer surface and that the latter make a greater contribution to the structure than in amphibians (Shellis, 1982; Huysseune et al., 1998).

The **gomphosis** of mammals develops from the dental follicle, which gives rise to the cementum, periodontal ligament and alveolar bone (Cho and Garant, 2000; Diep et al., 2009). Since the follicle is derived from the same neural crest cell population as the dental papilla, the tooth and its attachment system clearly form a developmental unit (Diep et al., 2009). Alveolar bone begins to form early, at the bell stage of tooth development, and is later continuously remodeled in response to tooth growth, eruption, and jaw growth (Cho and Garant, 2000; Lungová et al., 2011). During root development, the epithelial sheath (here known as **Hertwig's epithelial root sheath**) begins to break down and this allows access of mesenchymal follicle cells, which differentiate into cementoblasts (Cho and Garant, 2000; Diekwisch, 2001). The fragments of Hertwig's sheath remain as epithelial rests within the periodontal ligament, but do not differentiate into cementoblasts, as was once suggested (Diekwisch, 2001; Yamamoto et al., 2015). Development of the crocodylian gomphosis has not been studied in the same detail as its mammalian counterpart, but it is known that breakdown of Hertwig's sheath is crucial for initiation of cementogenesis (McIntosh et al., 2001).

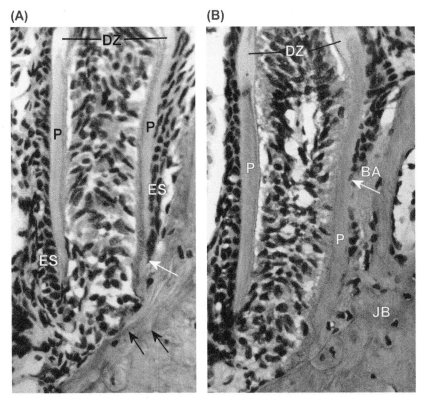

FIGURE 9.23 Great crested newt (*Triturus cristatus*). (A) Earliest stage of pleurodont ankylosis of pedicel (P) to jaw bone (right). Epithelial sheath covers the whole pedicel except for the terminal basal portion where its lower limit is marked by a *white arrow*. Basal to this, ankylosis is formed by outgrowths from pedicel and bone. *Black arrows* indicate reversal lines in jaw bone left by previous teeth. *DZ*, dividing zone; (B) Late stage of pleurodont ankylosis, with bone of attachment (BA) being deposited between jaw bone (JB) and pedicel (P). *White arrow* marks lower limit of epithelial sheath (ES); *DZ*, dividing zone. (A, B) Harris's hematoxylin, image width = 270 μm.

CONTROL OF TOOTH DEVELOPMENT

The complex processes of histogenesis and morphogenesis that occur during tooth ontogeny require a continuous and reciprocal exchange of information between the epithelial and ectomesenchymal components of the tooth germ. It has been estimated that more than 300 genes are involved with tooth development. As detailed treatment of this topic is beyond the scope of this book, we provide here only a brief summary. Most of the research in this field has been carried out on the mouse, which has a very simple dentition, consisting of one continuously growing incisor and three molars in each quadrant. Apart from rudiments of deciduous teeth, which are resorbed, there is only one generation of teeth. Therefore, we first summarize odontogenesis in the mouse, and then review the more scanty evidence on tooth formation in non-mammalian vertebrates. These accounts are followed by a discussion of the phenomenon of tooth loss during evolution, which may shed indirect light on tooth formation. Aspects of tooth development are also considered in Chapters 10–13.

Tooth Formation in Mice

Communication between the epithelium and ectomesenchyme, which ensures correct morphogenesis of the tooth crown and coordinated differentiation of the secretory cells, is mediated by many paracrine cell signaling agents. The principal signaling molecules belong to the bone morphogenetic protein (BMP), fibroblast growth factor (FGF), sonic hedgehog (SHH), WNT, and tumor necrosis factor (TNF) families of growth factors. Signaling agents are active throughout tooth development and are conserved in both mammalian and non-mammalian teeth. They control communication between epithelium and mesenchyme and also between cells within these layers by regulating transcription factors, such as PITX2, MSX1, MSX2, and PAX9. The changes in gene expression at each stage influence the response of the target cells to subsequent signaling events, so that a large variety of responses can be elicited by a limited number of signaling molecules, and their activities interact in a variety of ways.

A simplified illustration of some of the growth factors and transcription factors is provided in Fig. 9.25. Tabulated data on the genes and gene products involved in tooth formation are given at http://bite-it.helsinki.fi/, and concise descriptions of their possible roles are given by Jernvall and Thesleff (2000), Thesleff (2003), Zhang et al. (2005), and Catón and Tucker (2009).

The fate of the neural crest cells is not determined on arrival at the presumptive oral cavity. During the early

(A)   **(B)**

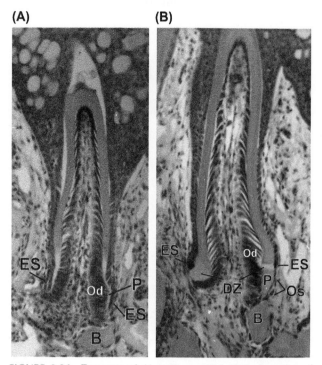

FIGURE 9.24 European eel (*Anguilla anguilla*). Early formation of attachment (slightly later than in Fig. 9.22). Rudiment of pedicel (P) recognizable between epithelial sheath (ES) and odontoblasts (Od). Pedicel close to bone (B); B. Later stage of attachment process. Pedicel (P) well formed at right and contacting bone. It is lined on the inside by odontoblasts (Od) and on the outside by osteoblasts (Os). Epithelial sheath now extends only to the level of the dividing zone (DZ). This is an incomplete pedicel (cf. Fig. 4.10B); at the left, the fibers of the DZ will connect directly to the supporting bone (B). A, B: Celestin blue-hemalum-eosin, image width = 200 μm.

stages of odontogenesis (in the mouse, up to embryonic day E11.5), the potential to initiate tooth germs, and to determine their locations, is located in the oral epithelium and its signals include SHH, BMP4, FGF8, and PITX2. These initiate the expression of transcription factors, such as MSX1 and PAX9. The earliest markers of odontogenesis are the expression, well in advance of histological differentiation, of *Pitx2* in the epithelium and of *Pax9* in the mesenchyme (Zhang et al., 2005), followed by expression of other factors, such as Shh.

As with a broad range of other tissues, *Shh* is a major determinant of (or at least marks) the sites at which teeth develop. *Shh* gene expression is restricted to the dental epithelium at sites of tooth development and also appears to be involved in epithelial–mesenchymal interactions. Recent research has shown a role for SHH in shaping the developing placode, while FGF signaling has a role in stratification of the epithelium (Li et al., 2016). Later on, SHH may induce cell proliferation in the enamel organ and dental papilla, and may regulate cell survival within the stellate reticulum. Restriction of expression of *Shh* to tooth-forming regions is important and this seems to be accomplished by interactions with WNT. These interactions thus determine the boundary between presumptive dental and oral epithelium. When WNT signaling is inhibited, tooth morphogenesis is arrested at an early stage. Conversely, stimulation of WNT signaling in the oral epithelium, by stabilizing β-catenin, results in the production of extra teeth. In the molar region of mice, these supernumerary teeth are in the form of unicuspid cones (Järvinen et al., 2006).

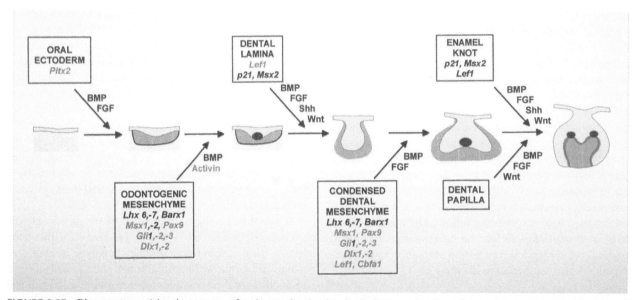

FIGURE 9.25 Diagram summarizing the sequence of exchange of molecular signals, between dental ectoderm and mesenchyme, involved in tooth formation, from the tooth primordium to the advanced tooth germ. *Courtesy Prof. I. Thesleff.*

The diastema region in the mouse, which lacks teeth, has been used as an experimental model for manipulation of genes to produce supernumerary teeth (Wang et al., 2009). Causative factors for toothlessness in the diastema region are the repression of SHH, FGF, and WNT signaling pathways. However, supernumerary teeth can be produced in the mouse diastema region by genetic manipulation to alter the expression of such pathways, which have both positive and negative feedback loops (Porntaveetus et al., 2012; Nakamura and Fukumoto, 2013; Lan et al., 2014). As WNT signaling is so important in regulating tooth development, Porntaveetus et al. (2012) studied WNT signaling in *Lrp4* mutant mice that develop supernumerary teeth in the diastema region. In the normal diastema, *Wnt5a* expression was present in the mesenchyme, whereas *Wnt4* and *Wnt10b* were present in the epithelium. Both tissues expressed *Wnt6* and *Wnt11*. The WNT coreceptor *Lrp6* was weakly expressed in the diastema overlapping with weak *Lrp4* expression, a coreceptor that inhibits WNT signaling. Secreted WNT inhibitors *Dkk1*, *Dkk2*, and *Dkk3* were also expressed in the diastema. In the *Lrp4* mutant, which develop supernumerary teeth, there was upregulation of WNT signaling and *Lrp6* expression. The authors concluded that WNT signaling is usually attenuated in the diastema by these secreted and membrane-bound WNT inhibitors.

In response to the initial epithelial signals, the epithelium first thickens locally and then invaginates to form the early tooth bud, while mesenchyme proliferates and condenses around the bud. After the early bud stage (in the mouse, from E11.5 onwards), the expression domains of the ectomesenchyme are fixed and the control of odontogenesis is transferred from the epithelium to the mesenchyme. While odontogenic potential is lost by the epithelium, epithelial signaling remains essential for maintenance of mesenchymal competence.

A very important development during the bud stage is the induction by the mesenchyme of the **primary enamel knot**: a localized, nonproliferating cluster of cells at the center of the inner dental epithelium, which is central to establishing the form of the tooth crown (Fig. 9.26; Thesleff et al., 2001). The primary enamel knot is an important signaling center expressing 10 or more growth factors, including members of the BMP, FGF, SHH, and WNT families (Fig. 9.27). The primary enamel knot is not a permanent feature of the tooth germ (Berkovitz, 1967): it is removed by apoptosis at the end of the cap stage. In monocuspid teeth (incisors, canines), the primary knot is the only one to form and it marks the site of the tooth tip. However, in premolars and molars, additional **secondary knots** are formed, possibly with contributions from the primary knot, at the sites of the future cusps and in a sequence that matches the relative importance of the cusps in the mature tooth. Although the primary and secondary knots consist of nondividing cells, it is thought that they secrete mitogenic factors that diffuse outwards to stimulate proliferation and folding

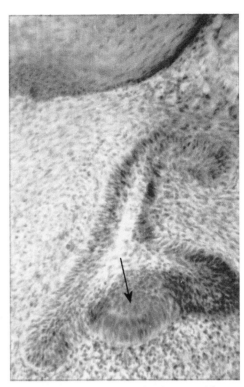

FIGURE 9.26 Cap-stage tooth germ, showing enamel knot (*arrow*) in the common brushtail possum (*Trichosurus vulpecula*). Hematoxylin and eosin. Magnification ×180. *From Berkovitz B.K.B., 1967. An account of the enamel cord in* Setonix brachyurus (Marsupialia) *and of the presence of an enamel knot in* Trichosurus vulpecula. *Arch. Oral Biol. 12, 49–59. Courtesy Editors of the Archives of Oral Biology.*

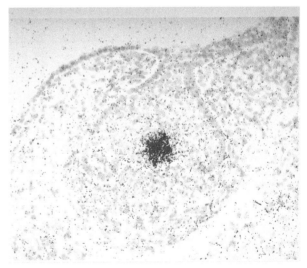

FIGURE 9.27 The primary enamel knot in the mouse molar characterized by intense expression and localization of *Fgf4* mRNA (red stain) in a cap stage (E14) by in situ hybridization. To aid interpretation of gene expression using ^{35}S-labeled in situ hybridization, both dark and bright field images were taken. Silver grains in the dark field image were selected, colored red, and then superimposed on to the bright field image. Magnification ×120. *Courtesy Dr. T. Aberg.*

of the adjacent internal enamel epithelial and mesenchymal cells. This results in expansion of the epithelial cap and also in folding of the inner dental epithelium; downgrowth of the expanding dental epithelium on the flanks of the enamel knots result in the formation of cusps during the bell stage. Many mutations affecting signaling factors associated with enamel knots result in altered tooth morphology (Jernvall and Thesleff, 2012). As many genes participate in the signaling activities of the enamel knots, the system is capable of very fine control of crown morphology.

Enamel knots have been identified in many mammals other than mice (Jernvall and Thesleff, 2012). The occurrence of knots in an opossum (*Monodelphis domestica*) indicates that marsupials share this signaling mechanism with placental mammals.

Regulation of proliferation of stem cells in the cervical loop, and the subsequent differentiation of ameloblasts is central to growth and morphogenesis of the tooth. It is clear from studies on the continuously growing mouse incisor (Wang et al., 2007; Hu et al., 2014) that this system is influenced by a large number of signaling molecules, including FGF, BMP, Notch, Activin, and hedgehog (HH), under the control of a network of interacting genes. This network is illustrated in Fig. 9.28.

Tooth Formation in Non-Mammalian Vertebrates

Research on fish and reptiles has shown that many features of the mechanisms controlling tooth formation in

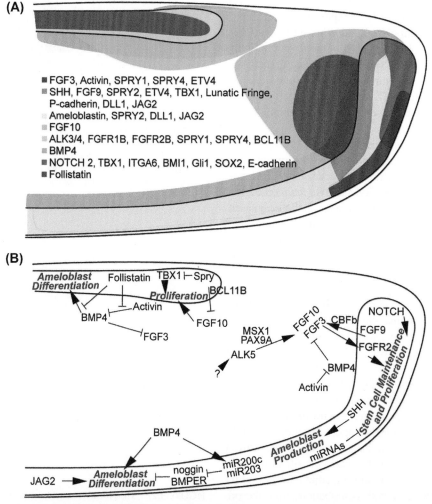

FIGURE 9.28 Diagrams showing the gene networks which control cell proliferation and differentiation at the proximal end of the continuously growing rodent incisor. In both diagrams, the upper portion shows the cervical loop (CL) region of the cementum-covered lingual aspect of the incisor and the lower portion shows the CL region of the enamel-covered labial aspect. (a) Fields of expression of signaling molecules and transcription factors that regulate the formation and maintenance of the CLs; (b) The gene network regulating stem cell maintenance, proliferation, and differentiation, and thereby maintenance of the homeostasis of the stem cell niche. Promotional activities of signaling molecules and transcription factors are indicated by → and inhibitory activities by ⊣. *From Hu J.K.-H., Vagan Mushegyan V., Ophir D. Klein O.D., 2014. On the cutting edge of organ renewal: identification, regulation and evolution of incisor stem cells. Genesis 52, 79–92 Diagram by courtesy Dr. J.K.-H. Hu. By permission from the Editors Genesis.*

mammals are shared with other vertebrates, and it is generally accepted that a core of regulatory systems is conserved among all vertebrates. The patterns of expression of *shh*, *pitx2*, *wnt*, *fgf*, and *bmp* in fish and reptiles are on the whole similar to those in the mouse (Jackman et al., 2004, 2010; Fraser et al., 2004, 2006, 2013; Handrigan and Richman, 2010a; Richman and Handrigan, 2011; Rasch et al., 2016). Handrigan and Richman (2010a) obtained evidence that the first null-generation of rudimentary teeth in the **bearded dragon** (*Pogonia itticeps*) showed impaired responsiveness to SHH signaling.

There are, however, some notable differences in tooth ontogeny between non-mammalian vertebrates and mammals. For instance, *bmp4* is not expressed in the odontogenic epithelium during tooth initiation in cichlids and rainbow trout (Fraser et al., 2004, 2006). There are differences in expression of *dlx* genes between **zebrafish** (*Danio rerio*) and the mouse (Borday-Birraux et al., 2006). Fraser et al. (2004) found that *pitx2* is expressed strongly during initiation of oral teeth, but not of pharyngeal teeth in the rainbow trout. In an *edar* mutant of the **medaka** (*Oryzias latipes*), tooth number is reduced considerably but tooth morphology is not affected, suggesting that *edar* is involved with regulating tooth number (Atukorala et al., 2010). In the mouse and in humans, it is associated with regulating molar crown morphology and tooth number. The role of this gene in regulation of tooth number is therefore conserved. This study provided a further example of a difference between the oral and pharyngeal dentitions, in that the reduction of tooth number in the mutant strain was much greater in the pharyngeal dentition (74% vs. 44%).

In the **veiled chameleon**, *Chamaeleo calypratus*, a cluster of epithelial cells is distinguishable from early bud to early bell stage (Buchtová et al., 2013). As the chameleon has multicuspid teeth, Buchtová et al. raised the possibility that this cluster may be homologous to the enamel knots in mammalian teeth. Structures resembling enamel knots have not been observed in most non-mammalian vertebrates (Jernvall and Thesleff, 2012), and it has not been established that the structure in the veiled chameleon has a signaling function (Buchtová et al., 2013). However, the inner dental epithelium expresses several of the growth factors and transcription factors observed in the enamel knot, but in a less localized pattern (Handrigan and Richman, 2010a,b; Richman and Handrigan, 2011; Jernvall and Thesleff, 2012; Fraser et al., 2013). Furthermore, in reptiles there is no functional requirement for apoptosis in the inner dental epithelium and these two factors may explain why there is no localized swelling of cells in reptiles akin to a mammalian enamel knot.

Two studies have explored the role of growth factors in control of tooth crown shape in bony fish. Fraser et al. (2013) found that inhibition of Fgf and Bmp resulted in an increase in the number of cusps in cichlid teeth. Jackman et al. (2013) confirmed that Bmp inhibition converted unicuspid teeth into bicuspid teeth in the zebrafish and the Mexican tetra. However, similar results were obtained with overexpression, rather than inhibition, of Fgf. Besides the effects on tooth form, Fraser et al. (2013) observed that tooth replacement was prevented when the pathways of either Bmp, Notch, or Wnt/β-catenin were interrupted, while Jackman et al. (2013) observed the presence of supernumerary teeth in the zebrafish when *fgf* was overexpressed or Bmp was inhibited. Their results allowed Jackman et al. (2013) to conclude that, despite their apparent complexity, the evolutionary origin of multicuspid teeth in teleosts is positively constrained, likely requiring only slight modifications of a preexisting mechanism for patterning the number and spacing of individual teeth.

Tooth development has been studied in reptiles to determine whether enamel crests and multiple cusps result from simple variations of enamel thickness or involve folding at the enamel–dentine junction. Enamel crests were seen to be due to asymmetrical deposition of enamel. The same was true of the cusps on the bicuspid teeth of the **Madagascan gecko** (*Paroedura picta*; Fig. 9.29). However, in the tricuspid teeth of anoles and chameleons, there was folding of the inner dental epithelium, with a separate mineralization center for each cusp (Fig. 9.30; Zahradniček et al., 2014).

MISSING TEETH

A large number of vertebrates have lost teeth during their evolution. Davit-Béal et al. (2009) estimated that about a third of living tetrapod species are edentulous: 350 amphibians (toads), 257 reptiles (turtles), 10,000 birds, and 27 mammals. In addition, there is partial edentulism in many living vertebrates, following tooth loss from some elements of the jaw system. For instance, among the amphibians, all frogs except one (**Guenther's marsupial tree-frog**, *Gastrotheca guentheri*: see Fig. 5.69) lack mandibular teeth while, among the bony fish, cypriniforms have lost teeth from the oral cavity while retaining pharyngeal teeth. These cases raise several important questions. How are teeth lost without compromising feeding efficiency? What are the underlying molecular controls? To what extent is tooth loss reversible? With regard to the first question, Davit-Béal et al. (2009) found that tooth loss among tetrapods was possible when there was a pre-adapted secondary feeding mechanism that could become progressively dominant in feeding and could eventually supplant the teeth. For instance, during the evolution of birds and turtles, the role of feeding was progressively taken over by a keratinized beak and a muscular gizzard as teeth were gradually lost. Experimental embryological studies have shed some light on the molecular changes

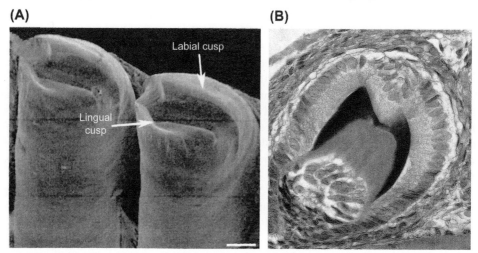

FIGURE 9.29 Madagascan gecko (*Paroedura picta*). (A) Bicuspid tooth of adult. SEM. Scale bar = 50 μm. (B) Developing tooth, showing the dome of dentine (DE) capped by enamel (E) which is deposited asymmetrically to form two cusps. Hematoxylin and eosin/Alcian blue. Scale bar = 10 μm. *From Zahradniček O., Horacek I., Tucker A.S., 2012. Tooth development in a model reptile: functional and null generation teeth in the gecko* Paroedura picta. *J. Anat. 221, 195–208. Courtesy Editors Journal of Anatomy.*

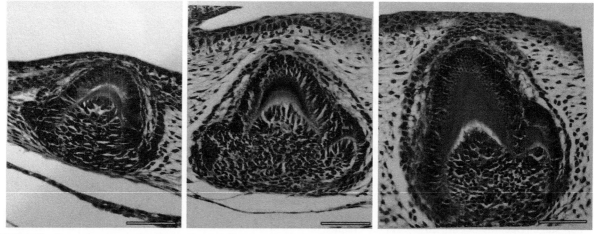

FIGURE 9.30 Successive stages of tooth formation in Allison's anole (*Anole allisoni*). In contrast to the gecko tooth germ in Fig. 9.29, there is folding of the inner dental epithelium, with a separate mineralization center for each cusp. Hematoxylin and eosin. Scale bar = 50 μm. *From Zahradniček O., Buchtova M., Dosedelova H., Tucker A.S., 2014.The development of complex tooth shape in reptiles. Front. Physiol. 5, 2–7. Courtesy Editors of the Journal of Anatomy.*

underlying tooth loss and on the question of whether loss is reversible. These studies have mostly addressed tooth loss in birds and have been performed on the domestic chicken (*Gallus domesticus*).

Hen's Teeth

Teeth are absent from all species of modern birds, which use a hard beak instead of teeth to collect food. The toothless condition has been established for 80 million years or more. However, the ancestors of birds, which first appeared about 150 million years ago, did have teeth (Davit-Béal et al., 2009). The absence of teeth in modern birds has given rise to the phrase "as rare as hen's teeth."

Research on the developmental reason for the complete disappearance of teeth would clearly have important implications for understanding the mechanism of major evolutionary changes.

During the development of the chick, transient thickenings resembling the placodes that precede the dental lamina in the mouse form in the oral epithelium but do not develop further. Several studies show that much of the odontogenic signaling system is present in the chick. It is not functional but can be reactivated under the right conditions to form primordia of hen's teeth.

Cross-species recombination experiments (Wang et al., 1998) suggested that chick oral mesenchyme retains the ability to participate in tooth germ formation when exposed to the right stimulus. Thus, combination of mouse odontogenic epithelium with chick oral mesenchyme (but not the reverse combination) resulted in formation of tooth buds and mesenchymal expression of Msx-1 and Bmp-4. After culture in vivo for six days, the tooth buds progressed to the point of secreting enamel and dentine matrix.

There is evidence that a lack of *BMP4* expression accounts for the inability of chick oral epithelium to initiate tooth formation. This factor is not expressed in the lateral regions of the developing chick mouth where the tooth rudiment-like epithelial thickenings are located, even though several other markers for odontogenic tissue are expressed (Chen et al., 2000). In support of this concept, expression of *MSX1*, *MSX2*, and *BMP4* is induced by culture of chick mesenchyme in the presence of exogenous BMP4 (Wang et al., 1998; Chen et al., 2000). A further defect in the signaling mechanism could be a failure for transfer of *BMP4* expression from the epithelium to the mesenchyme. This is normally mediated by mesenchymal *MSX1*, which is not expressed by chick oral mesenchyme (Chen et al., 2000). As *MSX1* can be bypassed by exogenous BMP4, Chen et al. (2000) tested the effect of culturing chick mandibles in which mesenchymal *BMP4* overexpression was induced by viral infection. This resulted in epithelial invagination, which supports the hypothesis that the epithelium has a latent odontogenic potential. Further support came from experiments in which chick oral epithelium was recombined with mesenchyme from the dorsal skin, which actively expresses *BMP4*. This combination resulted in formation of primordia resembling cap-stage tooth germs (Chen et al., 2000).

A third study in which tooth germs were produced experimentally in the chick was carried out by Mitsiadis et al. (2003, 2006). Replacement of endogenous neural crest of the chick by neural crest transplanted from the mouse resulted in formation of tooth germs which developed to the stage of secreting predentine-like extracellular matrix, which was said to mineralize (Mitsiadis et al., 2003).

Together, these experiments suggest that defects in the odontogenic signaling system in the chick do not appear to be located exclusively in the epithelium or the mesenchyme.

The previous three examples of hen's teeth were produced by experimental manipulation. In a fourth example, teeth were formed naturally, but in a chick mutant. This condition also sheds light on why the teeth appear and helps us to interpret the preceding experiments. The mutation in question is talpid[2], in which disruptive changes in some important signaling molecules (especially SHH) cause many abnormalities in the limbs and head, and the chick dies before hatching. Although this mutation was discovered 50 years ago, it is only recently that a number of small, unmineralized conical tooth germs were

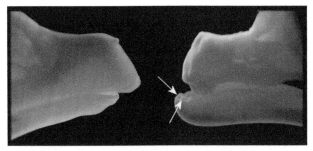

FIGURE 9.31 Rostra of normal and talpid[2] mutant chicks, after removal of rhamphotheca. Normal chick embryo on left. Talpid[2] mutant on right, showing small teeth projecting from the beak region (*arrow*). Note the abnormal growth with foreshortening of the beak in the mutant. *From Harris M.P., Hasso S.M., Ferguson M.W.J., Fallon J.F., 2006. The development of archosaurian first-generation teeth in a chicken mutant. Curr. Biol. 16, 371–377. Courtesy Dr. M.P. Harris.*

discovered in the developing beak (Fig. 9.31). These tooth germs resemble the first-generation teeth of reptiles in that they do not progress to later stages of development (Harris et al., 2006).

The formation of tooth germs in this chick mutant was explained by Harris et al. (2006) as follows. In the normal chick, the epithelium at the junction between the oral and aboral regions acts as a signaling center but the immediately underlying mesenchyme is not competent to respond. However, mesenchyme located more medially expresses signaling molecules and has the potential to produce tooth germs, but cannot interact with the epithelium because the two are spatially separated. Hence, normal chicks do not develop teeth. In the mutant chick, on the other hand, the junction between the oral and aboral epithelia is displaced medially, so the signaling center is juxtaposed with competent oral mesenchyme and this allows the interactions necessary for tooth formation.

This interpretation of the data from the talpid[2] mutant suggests that a major phenotypic change—loss of teeth—does not require a genetic or developmental change of similar complexity. It could simply have arisen as the result of a minor change in the relative positions of normally interacting epithelial and mesenchymal cell layers. It should be noted that Louchart and Viriot (2011) have proposed an alternative interpretation of the findings of Harris et al. (2006). They suggested that it is possible for teeth to form in the talpid[2] mutation because keratinization of the oral surface of the ramphotheca (the horny sheath of a bird's beak) occurs later than in the wild type. They proposed that, during the evolution of the beak, the genes responsible for odontogenesis in the toothed ancestors of the birds were diverted into keratinization, which shares many of the same signaling pathways with tooth formation.

Tooth rudiments induced in the chick without the involvement of mouse tissue (Chen et al., 2000; Harris et al., 2006) did not progress beyond formation of a small amount of dentine matrix, which was probably entirely composed of collagen (Davit-Béal et al., 2009). As a result of changes to the genome during the long history of the birds, the genes necessary to produce dental hard tissue matrices are no longer functional. Two enamel genes have disappeared and, while the genes for amelogenin and dentine sialophosphoprotein persist, they have long been invalidated, so it will be impossible for teeth to reevolve in birds (Sire et al., 2008; Kawasaki and Amemiya, 2014). The tooth rudiments in the cross-species recombination experiments of Mitsiadis et al., (2003) developed to the point where predentine-like extracellular matrix was laid down, because the mouse tissue possessed the requisite genes, including dentine sialoprotein.

Oral Tooth Loss in Bony Fish

Among bony fish there are several groups that lack oral teeth, but they are not completely edentulous as they retain the pharyngeal dentition. The major group with this type of dentition is the Cypriniformes, and two studies have addressed the question of how the oral teeth have been lost. The results indicate that the causes of tooth loss differ between these fishes and birds.

Stock et al. (2006) compared the expression of odontogenesis-related growth factors and transcription factors in the **zebrafish** (*D. rerio*: Cypriniformes), which lacks oral teeth; the related **Mexican tetra** (*Astyanax mexicanus*: Characiformes); and the unrelated **medaka** (*O. latipes*) and a catfish (*Synodontis multipunctatus*), all three of which have oral teeth. Their results suggested that a possible factor in tooth loss from zebrafish is a lack of mesenchymal Fgf signaling to induce epithelial dlx2a and dlx2b.

From a similar cross-species comparison, Aigler et al. (2014) identified lack of *eda* expression as a cause of tooth loss. An *eda* mutant of the zebra fish did not form pharyngeal teeth. In contrast, overexpression of *eda* in a transgenic zebrafish was associated with supernumerary and bicuspid pharyngeal teeth, such as occurs after activation of Fgf or inhibition of Bmp signaling (Jackman et al., 2013). Strikingly, teeth were formed on the normally edentulous upper pharyngeal bones, but oral teeth were not induced. The atavistic upper pharyngeal teeth are maladapted, as they are attached to bony plates that do not replace the keratinized chewing pad, so the upper teeth do not occlude properly with the lower dentition. Aigler et al. (2014) showed that the downstream *eda* signal transduction pathway was intact in the oral cavity, indicating that loss of oral teeth must involve multiple mutations. The results of this study show that regain of oral teeth (which could be advantageous for some cypriniforms, such as fish-eaters) and restoration of an ancestral pharyngeal dentition with both upper and lower teeth would both require complex genetic changes to achieve a functional, adaptive result.

REFERENCES

Aigler, S.R., Jandzi, D., Hattac, K., Uesugid, K., Stock, D.W., 2014. Selection and constraint underlie irreversibility of tooth loss in cypriniform fishes. Proc. Natl. Acad. Sci. 111, 7707–7712.

Atukorula, A.D.S., Inohaya, K., Baba, O., Tabata, M.J., Ratnayake, R.A.R.K., Abduweli, D., Kasugei, S., Mitani, H., Takano, Y., 2010/2011. Scale and tooth phenotypes in medaka with a mutated ectodysplasin-A receptor: implications for the evolutionary origin of oral and pharyngeal teeth. Arch. Histol. Cytol. 73, 139–148.

Berkovitz, B.K.B., Moxham, B.J., 1990. The development of the periodontal ligament with special reference to collagen fibre ontogeny. J. Biol. Buccale 18, 227–236.

Berkovitz, B.K.B., Moxham, B.J., 1995. The periodontal ligament and physiological tooth movements. In: Berkovitz, B.K.B., Moxham, B.J., Newman, H.N. (Eds.), The Periodontal Ligament in Health and Disease, second ed. Mosby-Wolfe, London, pp. 183–214.

Berkovitz, B.K.B., Shellis, R.P., 1978. A longitudinal study of tooth succession in piranhas (Pisces: Characidae), with an analysis of the tooth replacement cycle. J. Zool. 184, 545–561.

Berkovitz, B.K.B., Holland, G.R., Moxham, B.M., 2009. Oral Anatomy, Histology and Embryology, fourth ed. Mosby Elsevier, London.

Berkovitz, B.K.B., 1967. An account of the enamel cord in *Setonix brachyurus* (Marsupialia) and of the presence of an enamel knot in *Trichosurus vulpecula*. Arch. Oral Biol. 12, 49–59.

Berkovitz, B.K.B., 1978. Tooth ontogeny in the upper jaw and tongue of the rainbow trout (*Salmo gairdneri*). J. Biol. Buccale 6, 205–215.

Borday-Birraux, V., van der Heyden, C., Debiais-Thibaud, M., Verreijdt, L., Stock, D.W., Huysseune, A., Sire, J.Y., 2006. Expression of Dlx genes during the development of the zebrafish pharyngeal dentition: evolutionary implications. Evol. Dev. 8, 130–141.

Buchtová, M., Zahradníček, O., Balková, S., Tucker, A.S., 2013. Odontogenesis in the veiled chameleon (*Chamaeleo calyptratus*). Arch. Oral Biol. 58, 118–133.

Catón, J., Tucker, A.S., 2009. Current knowledge of tooth development: patterning and mineralization of the murine dentition. J. Anat. 214, 502–515.

Chen, Y., Zhang, Y., Jiang, T.-X., Barlow, A.J., St Amand, T.R., Hu, Y., Heaney, S., Francis-West, P., Chuong, C.-M., Maas, R., 2000. Conservation of early odontogenic signalling pathways in Aves. Proc. Natl. Acad. Sci. 97, 10044–100049.

Cho, M.I., Garant, P.R., 2000. Development and general structure of the periodontium. Periodontology 2000 24, 9–27.

Davit-Béal, T., Chisaka, H., Delgado, S., Sire, J.-Y., 2007. Amphibian teeth: current knowledge, unanswered questions, and some directions for future research. Biol. Rev. 82, 49–81.

Davit-Béal, T., Tucker, A.S., Sire, J.-Y., 2009. Loss of teeth and enamel in tetrapods: fossil record, genetic data and morphological adaptations. J. Anat. 214, 477–501.

Davit-Béal, A.F., Sire, J.-Y., 2006. Morphological variations in a tooth family through ontogeny in *Pleurodeles waltl* (Lissamphibia, Caudata). J. Morphol. 267, 1048–1065.

Delgado, S., Davit-Béal, T., Allizard, F., Sire, J.-Y., 2005. Tooth development in a scincid lizard, *Chalcides viridanus* (Squamata), with particular attention to enamel formation. Cell Tissue Res. 319, 71–89.

Diekwisch, T.G.H., 2001. Developmental biology of cementum. Int. J. Dev. Biol. 45, 695–706.

Diep, L., Matalova, E., Mitsiadis, T.A., Tucker, A.S., 2009. Contribution of the tooth bud mesenchyme to alveolar bone. J. Exp. Zool. Mol. Dev. Evol. 312B, 510–517.

Fraser, G.J., Graham, A., Smith, M.M., 2004. Conserved deployment of genes during odontogenesis across osteichthyans. Proc. R. Soc. Lond Biol. Sci. 271, 2311–2317.

Fraser, G.J., Graham, A., Smith, M.M., 2006. Developmental and evolutionary origins of the vertebrate dentition: molecular controls for spatio-temporal organisation of tooth sites in osteichthyans. J. Exp. Zool. Mol. Dev. Evol. 306B, 183–203.

Fraser, G.J., Bloomquist, R.F., Streelman, J.T., 2013. Common developmental pathways link tooth shape to regeneration. Dev. Biol. 377, 399–414.

Gaete, M., Fons, J.M., Mădălina Popa, E., Chatzeli, L., Tucker, A.S., 2015. Epithelial topography for repetitive tooth formation. Biol. Open 4, 1625–1634.

Handrigan, G.R., Richman, J.M., 2010a. Autocrine and paracrine Shh signaling are necessary for tooth morphogenesis, but not tooth replacement in snakes and lizards (Squamata). Dev. Biol. 337, 171–186.

Handrigan, G.R., Richman, J.M., 2010b. A network of Wnt, hedgehog and BMP signaling pathways regulates tooth replacement in snakes. Dev. Biol. 348, 130–141.

Harada, H., Kettunen, P., Jung, H.-S., Mustonen, T., Wang, Y.A., Thesleff, I., 1999. Localization of putative stem cells in dental epithelium and their association with Notch and FGF signaling. J. Cell Biol. 147, 105–120.

Harada, H., Ichimori, Y., Yokohama-Tamaki, T., Ohshima, H., Kawano, S., Katsube, K., Wakisaka, S., 2006. Stratum intermedium lineage diverges from ameloblast lineage via Notch signaling. Biochem. Biophys. Res. Comm. 340, 611–616.

Harris, M.P., Hasso, S.M., Ferguson, M.W.J., Fallon, J.F., 2006. The development of archosaurian first-generation teeth in a chicken mutant. Curr. Biol. 16, 371–377.

Howes, R.I., 1978. Root formation in ectopically transplanted teeth of the frog, *Rana pipiens* II. Comparative aspects of the root tissues. Acta Anat. 100, 461–470.

Hu, J.K.-H., Vagan Mushegyan, V., Ophir, D., Klein, O.D., 2014. On the cutting edge of organ renewal: identification, regulation and evolution of incisor stem cells. Genesis 52, 79–92.

Huysseune, A., van der Heyden, C., Sire, J.Y., 1998. Early development of the zebrafish (*Danio rerio*) pharyngeal dentition (Teleostei, Cyprinidae). Anat. Embryol. 198, 289–305.

Ishiyama, M., Inage, T., Shimokawa, H., 1999. An immunocytochemical study of amelogenin proteins in the developing tooth enamel of the gar-pike, *Lepisosteus oculatus* (Holostei, Actinopterygii). Arch. Histol. Cytol. 62, 191–197.

Jackman, W.R., Draper, B.W., Stock, D.W., 2004. Fgf signaling is required for zebrafish tooth development. Dev. Biol. 274, 139–157.

Jackman, W.R., Yoo, J.J., Stock, D.W., 2010. Hedgehog signaling is required at multiple stages of zebrafish tooth development. BMC Dev. Biol. 10, 119–132.

Jackman, W.R., Davies, S.H., Lyons, D.B., Stauder, C.K., Denton-Schneider, B.R., Jowdry, A., Aigler, S.R., Vogel, S.A., Stock, D.W., 2013. Manipulation of Fgf and Bmp signaling in teleost fishes suggests potential pathways for the evolutionary origin of multicuspid teeth. Evol. Dev. 15, 107–118.

Järvinen, E., Salazar-Ciudad, I., Birchmeier, W., Taketo, M.M., Jernvall, J., Thesleff, I., 2006. Continuous tooth generation in mouse is induced by activated epithelial Wnt/β-catenin signalling. Proc. Natl. Acad. Sci. 103, 18627–18632.

Jernvall, J., Thesleff, I., 2000. Reiterative signaling and patterning during mammalian tooth morphogenesis. Mech. Dev. 92, 19–29.

Jernvall, J., Thesleff, I., 2012. Tooth shape formation and tooth renewal: evolving with the same signals. Development 139, 3487–3497.

Juuri, E., Saito, K., Ahtiainen, L., Seidel, K., Tummers, M., Hochedlinger, K., Klein, O.D., Thesleff, I., Michon, F., 2012. Sox2+ stem cells contribute to all epithelial lineages of the tooth via Sfrp5+ progenitors. Dev. Cell 23, 317–328.

Juuri, E., Jussila, M., Seidel, K., Holmes, S., Wu, P., Richman, J., Heikinheimo, K., Chuong, C.-M., Arnold, K., Hochedlinger, K., Klein, O., Michon, F., Thesleff, I., 2013. Sox2 marks epithelial competence to generate teeth in mammals and reptiles. Development 140, 1424–1432.

Kawasaki, K., Amemiya, C.T., 2014. SCPP genes in the coelacanth: tissue mineralization genes shared by sarcopterygians. J. Exp. Zool. Mol. Dev. Evol. 322B, 390–402.

Lan, Y., Jia, S., Jiang, R., 2014. Molecular patterning of the mammalian dentition. Semin. Cell Dev. Biol. 25–26, 61–70.

Li, J., Chatzeli, L., Panousopoulou, E., Tucker, A.S., Green, J.B.A., 2016. Epithelial stratification and placode invagination are separable functions in early morphogenesis of the molar tooth. Dev. Adv. Online Artic. http://dev.biologists.org/lookup/doi/10.1242/dev.130187.

Louchart, A., Viriot, L., 2011. From snout to beak: the loss of teeth in birds. Trends Ecol. Evol. 26, 663–673.

Lumsden, A.G.S., 1988. Spatial organization of the epithelium and the role of neural crest cells in the initiation of the mammalian tooth germ. Development 103 (Suppl.), 155–169.

Lungová, V., Radlanski, R.J., Tucker, A.S., Renz, H., Míšek, I., Matalová, E., 2011. Tooth-bone morphogenesis during postnatal stages of mouse first molar development. J. Anat. 218, 699–716.

Luukko, K., Kettunen, P., 2014. Coordination of tooth morphogenesis and neuronal development through tissue interactions: lessons from mouse models. Exp. Cell Res. 325, 72–77.

Mahler, D.L., Kearney, M., 2006. The palatal dentition in squamate reptiles: morphology, development, attachment, and replacement. Fieldiana Zool. 1–61 New Series No.108.

McIntosh, J.E., Anderton, X., Flores-de-Jacoby, F., Carlson, D.S., Shuler, C.F., Diekwisch, T.G.H., 2001. Caiman periodontium as an intermediate between basal vertebrate ankylosis-type attachment and mammalian "true" periodontium. Microsc. Res. Tech. 59, 449–459.

Miletich, I., Sharpe, P.T., 2004. Neural crest contribution to mammalian tooth formation. Birth Def. Res. C 72, 200–212.

Mitsiadis, T.A., Chéraud, Y., Sharpe, P.T., Fontaine-Pérus, J., 2003. Development of chick embryos after mouse neural crest transplantations. Proc. Natl. Acad. Sci. U.S.A. 100, 6541–6545.

Mitsiadis, T.A., Caton, J., Cobourne, M., 2006. Waking up the sleeping beauty: recovery of the ancestral bird odontogenic program. J. Exp. Zool. Mol. Dev. Evol. 306B, 227–233.

Nakamura, T., Fukumoto, S., 2013. Genetics of supernumerary tooth formation. J. Oral Biosci. 55, 180–183.

Ohazama, A., Haworth, K.E., Ota, M.S., Khonsari, R.H., Sharpe, P.T., 2010. Ectoderm, endoderm, and the evolution of heterodont dentitions. Genesis 48, 382–389.

Parsons, T.S., Williams, E.E., 1962. The teeth of amphibia and their relation to amphibian phylogeny. J. Morphol. 110, 375–389.

Porntaveetus, T., Ohazama, A., Choi, H.Y., Herz, J., Sharpe, P.T., 2012. Wnt signalling in the murine diastema. Eur. J. Orthod. 34, 518–524.

Rasch, L.J., Martin, K.J., Cooper, R.L., Metscher, B., Underwood, C., Fraser, G.J., 2016. An ancient dental gene set governs development and continuous regeneration of teeth in sharks. Dev. Biol. http://dx.doi.org/10.1016/j.ydbio.2016.01.038.

Richman, J.M., Handrigan, G.R., 2011. Reptilian tooth development. Genesis 49, 247–260.

Rothová, M., Feng, J., Sharpe, P.T., Peterková, R., Tucker, A.S., 2011. Contribution of mesoderm to the developing dental papilla. Int. J. Dev. Biol. 55, 59–64.

Rothová, M., Peterková, R., Tucker, A.S., 2012a. Fate map of the dental mesenchyme: dynamic development of the dental papilla and follicle. Dev. Biol. 366, 244–254.

Rothová, M., Thompson, H., Lickert, H., Tucker, A.S., 2012b. Lineage tracing of the endoderm during oral development. Dev. Dyn. 241, 1183–1191.

Shellis, R.P., 1982. Comparative anatomy of tooth attachment. In: Berkovitz, B.K.B., Moxham, B.J., Newman, H.N. (Eds.), The Periodontal Ligament in Health and Disease. Pergamon, Oxford, pp. 3–24.

Sire, J.-Y., Delgado, S.C., Girondot, M., 2008. Hen's teeth with enamel cap: from dream to impossibility. BMC Evol. Biol. 8, 246–257.

Soukup, V., Eppeerlein, H.-H., Horacek, I., Cerny, R., 2008. Dual epithelial origin of vertebrate oral teeth. Nature 445, 795–798.

Stock, D.W., Jackman, W.R., Trapani, J., 2006. Developmental genetic mechanisms of evolutionary tooth loss in cypriniform fishes. Development 133, 3127–3137.

Thesleff, I., Keranen, S., Jernvall, J., 2001. Enamel knot as signalling centers linking tooth morphogenesis and odontoblast differentiation. Adv. Dent. Res. 15, 14–18.

Thesleff, I., 2003. Epithelial-mesenchymal signalling regulating tooth morphogenesis. J. Cell Sci. 116, 1647–1648.

Trapani, J., 2001. Position of developing replacement teeth in teleosts. Copeia 35–51.

Van der heyden, C., Huysseune, A., Sire, J.Y., 2000. Development and fine structure of pharyngeal replacement teeth in juvenile zebrafish (*Danio rerio*) (Teleostei, Cyprinidae). Cell Tiss. Res. 302, 205–219.

Vandervennet, E., Huysseune, A., 2005. Histological description of tooth formation in adult *Eretmodus* cf. *Cyanostictus* (Teleostei, Cichlidae). Arch. Oral Biol. 50, 635–643.

Wang, Y.-H., Upholt, W.B., Sharpe, P.T., Kollar, E.J., Mina, M., 1998. Odontogenic epithelium induces similar molecular responses in chick and mouse mandibular mesenchyme. Dev. Dyn. 213, 386–397.

Wang, X.-P., Suomalainen, M., Felszeghy, S., Zelarayan, L.C., Alonso, M.T., Plikus, M.V., Maas, R.L., Chuong, C.-M., Schimmang, T., Thesleff, I., 2007. An integrated gene regulatory network controls stem cell proliferation in teeth. PLoS Biol. 5, e159.

Wang, X.-P., O'Connell, D.J., Lund, J.J., Saadi, I., Kuraguchi, M., Turbe-Doan, A., Cavallesco, R., Kim, H., Park, P.J., Harada, H., Kucherlapati, R., Maas, R.L., 2009. Apc inhibition of Wnt signaling regulates supernumerary tooth formation during embryogenesis and throughout adulthood. Development 136, 1939–1949.

Whitlock, J.A., Richman, J.M., 2013. Biology of tooth replacement in amniotes. Int. J. Oral Sci. 5, 66–70.

Yamamoto, T., Yamada, T., Yamamoto, T., Hasegawa, T., Hongo, H., Oda, K., Amizuka, N., 2015. Hertwig's epithelial root sheath fate during initial cellular cementogenesis in rat molars. Acta Histochem. Cytochem. 48, 95–101.

Zahradniček, O., Horacek, I., Tucker, A.S., 2012. Tooth development in a model reptile: functional and null generation teeth in the gecko *Paroedura picta*. J. Anat. 221, 195–208.

Zahradniček, O., Buchtova, M., Dosedelova, H., Tucker, A.S., 2014. The development of complex tooth shape in reptiles. Front. Physiol. 5, 2–7.

Zhang, Y.D., Chen, Z., Song, Y.Q., Liu, C., Chen, Y.P., 2005. Making a tooth: growth factors, transcription factors, and stem cells. Cell Res. 15, 301–316.

Chapter 10

Tooth Replacement and Ontogeny of the Dentition

INTRODUCTION

With few exceptions, such as acrodont lizards, non-mammalian vertebrates replace their teeth throughout life (polyphodonty). In contrast, most toothed mammals have only a single tooth replacement (diphyodonty) or, as in many rodents and cetaceans, have no tooth replacement and retain a single set throughout life (monophyodonty). The limited number of tooth generations in mammals is correlated with the evolution of precise occlusion, through the acquisition of tribosphenic molars and other adaptations, such as socketted teeth and a jaw joint with greater freedom of movement. The functioning of this type of dentition is not compatible with periodic replacement of partly worn teeth by unworn teeth. In addition to the continual succession of teeth in non-mammalian vertebrates, we also consider how the number, position, and form of teeth are established, a process in which tooth replacement plays an integral part.

FUNCTIONS OF POLYPHYODONTY

Continual tooth replacement has several interlinked functions. Formation of a tooth has a metabolic cost and it must be assumed that, in any given species, natural selection will ensure that the rate and pattern of replacement are such as to balance this cost against the functional requirements of the dentition. The functions of tooth replacement can be described under several headings: growth; adaptation to changes of diet during the life cycle; and compensation for wear.

Growth

As part of the general growth process, the jaws and the dentition must enlarge to keep pace with increasing nutritional requirements. This is achieved by two processes. First, the number of teeth may increase by addition of further teeth as the jaws grow. Second, the tooth at each locus in the dentition may become progressively larger by incremental increases in size of successive replacement teeth. Usually, growth of the dentition involves both processes (eg, *Pomatomus*; Bemis et al., 2005), although some species such as piranhas have a fixed number of teeth and growth is achieved entirely by tooth replacement (Fig. 10.8; Berkovitz and Shellis, 1978).

Age-Related Changes of Tooth Form

Juvenile forms often have a different diet than adults. Probably the most common reason is that the size and strength of the dentition of young animals are insufficient to deal with foodstuffs consumed by adults. There is also a selective advantage if individuals of different ages are not competing for the same food resources. Age-related changes in diet often involve changes in tooth form as well as size and these modifications are brought about by tooth replacement, over one or more tooth generations. Sire et al. (2002) found that among non-mammalian vertebrates in which the embryonic phase of the life cycle is short (actinopterygians, dipnoans, urodeles), the first-generation functional teeth are very small and avascular, with atubular dentine. Such "Type 1" teeth are easily formed, are appropriate for larval feeding and are replaced by the adult dentition through tooth replacement. Among elasmobranchs, squamates, and crocodylians, which have a prolonged embryonic phase, Type 1 teeth are suppressed during development of the dentition, or appear only as nonfunctional rudiments, and the first generation of functional teeth are more substantial than Type 1 and resemble miniature versions of the adult teeth.

Tooth form usually becomes more complex during development (eg, unicuspid to tricuspid, as in some cichlid fish; Streelman et al., 2003). However, tooth form can also become simpler by replacement (as in the Mexican burrowing caecilian, *Dermophis mexicanus*), or may change more than once, as in the tigerfish (*Hydrocynus*) (Gagiano et al., 1996).

Compensation for Wear

All teeth wear during use. Even though the teeth of non-mammalian vertebrates do not wear against their counterparts in the opposite jaw, as do mammalian teeth, they wear by interaction with the food. The rate of wear varies greatly and can be considerable, as in the bristle-like teeth of rock gobies, which are used to scrape food from rock surfaces. Even relatively light wear will blunt pointed tooth tips or the sharp edges of cutting teeth and hence increase the force required to penetrate prey. The necessity to compensate for wear will increase the rate of tooth replacement beyond that required by growth alone. There are too few direct

The Teeth of Non-Mammalian Vertebrates. http://dx.doi.org/10.1016/B978-0-12-802850-6.00010-2

measurements of replacement rate to quantify the effect directly, but the number of successional teeth associated with each functional tooth can be used as an indicator of the rate. Only a single successor is generally present in bony fish and crocodylians, through two or three in most reptiles, about eight in sharks and fangs of some snakes, 15 in rays, and 45 in some gobies. The high numbers in the latter two groups are correlated with crushing (rays) and scraping (gobies). The rate of replacement, being a genetic adaptation, is not affected by whether the animal feeds or not.

ONTOGENY OF THE DENTITION

The dentition is established initially by the eruption of the first teeth in a specific pattern and then continued by formation of additional teeth, sometimes behind the existing teeth and sometimes between them. The dentition continues to develop by replacement of the first teeth.

The pattern and sequence of tooth replacement are important aspects of the organization of a polyphyodont dentition. When teeth are continually replaced, there are usually temporary gaps within the functional tooth row, where replacing teeth have not yet erupted into function. Clearly, if many adjacent teeth were lost at the same time and eruption were relatively slow, the function of the dentition would be impaired but, as will be seen, this rarely occurs because of the pattern of tooth replacement. A common pattern, seen in many vertebrates, is that replacement tends to involve alternate teeth rather than adjacent teeth, which ensures that more than 50% of teeth are present at any one time.

In this work, we describe the patterns of tooth replacement before the sequence of initiation. This is because tooth replacement was the first of these aspects of the dentition to be studied extensively and the formation of the first dentition was studied later to test theories of tooth

replacement. We then summarize recent research on the systems of molecular signaling which control and integrate the formation of the dentition.

TOOTH REPLACEMENT
CHONDRICHTHYES

In sharks and rays (Chapters 2 and 3) the dentition comprises a number of tooth families, each forming a file of one or more functional teeth plus several replacement teeth in the process of development. Chondrichthyan teeth become attached to a common sheet of connective tissue (Chapters 2 and 3, and 13) as the base of the tooth forms. Although the details of the mechanism of tooth replacement are poorly understood, the interconnected families of teeth are carried en masse over and around the jaws before being shed at the front (Poole and Shellis, 1976). Teeth being shed are not resorbed as they would be in other vertebrates, but by breakdown of the attaching collagen fibers.

Within tooth rows of sharks, adjacent teeth may overlap with each other to a variable extent, as shown in Fig. 10.1 (Strasburg, 1963; Moss, 1967). The pattern of imbrication influences the sequence of tooth shedding, since the outward movement of a particular tooth can be physically blocked by adjacent, overlapping neighbors. Consequently, teeth may be replaced either in groups of individual teeth or even as whole rows (Strasburg, 1963). In the simplest arrangement (Fig. 10.1A), where there is no overlap (independent dentition), any tooth can be replaced independently. This type of dentition is seen typically in great white sharks (*Charcharodon*) and thresher sharks (*Alopias*), while bramble sharks (*Echinorhinus*) and six-gill sharks (*Hexanchus*) have only a few teeth that are slightly overlapped.

There are two potential patterns of tooth overlap: both ends of teeth in alternate positions (alternate overlap; Fig. 10.1B)

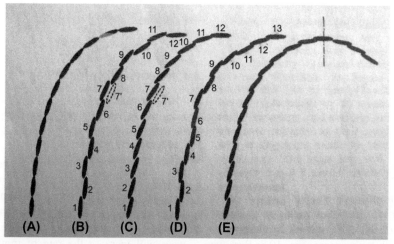

FIGURE 10.1   Tooth arrangements in sharks. (A) No overlap (independent dentition); (B) alternate overlap; (C) imbricate overlap; (D) mixed alternate and imbricate overlap; (E) modified imbricate overlap. The dashed outlines in (B) and (C) at tooth position 7 represent a replacement tooth. *From Strasburg, D.W., 1963. The diet and dentition of* brasiliensis, *with remarks on tooth replacement in other sharks. Copeia 33–40. Courtesy Editors of Copeia.*

or the same edge for the whole tooth row (imbricate overlap; Fig. 10.1C). However, whole dentitions in which there is either one or the other pattern are never encountered. Instead, overlap of both types occurs in patches of the same dentition (mixed dentition; Fig. 10.1D). The latter arrangement is seen in the dentitions of requiem sharks (*Carcharhinus*) and hammerhead sharks (*Sphryna*). Here, the independent loss of many teeth may be prevented. However, with tooth replacement, the nature of the overlap may change. Fig. 10.1E shows a modified imbricate overlap where the two halves of the jaw have imbricated teeth but in opposite directions. This type of dentition is seen in cookie-cutter sharks (*Isistius*), where the whole tooth row is replaced simultaneously, although it would seem necessary for the mesial tooth to fall out first. A number of the shed teeth end up in the stomach of the shark (Strasburg, 1963).

Assuming that semierect teeth in a shark are about to become functional, an analysis of the dentitions of numerous sharks leads to the conclusion that, with exceptions, such as kitefin and cookie-cutter sharks (see Fig. 2.40), in general only a few teeth are replaced at any one time along the jaws (in the order of 15–30%). There is also evidence that the rates of tooth replacement may differ between upper and lower jaws (Springer, 1960; Strasburg, 1963).

The teeth of rays are much more closely packed than those of sharks, within both rows and files, and form a pavement with teeth packed in an alternating pattern. The forward movement of the dental pavement, therefore, results in simultaneous loss of all members of a tooth row.

Rates of Replacement

The large numbers of successional teeth indicate that tooth replacement in sharks is rapid. Direct measurements have shown that tooth rows are replaced every 10–14 days in young lemon sharks (*Negaprion brevirostris*; Moss, 1967; Boyne, 1970; Reif et al., 1978) and immature dogfish (*Mustelus canis*; Ifft and Zinn, 1948), and 28 days in the nurse shark (*Ginglymostoma cirratum*; Reif et al., 1978). When young lemon sharks were deprived of food for a period of about 3 weeks, the overall rate of replacement was not affected but the time taken to shed teeth was slightly increased (Moss, 1967).

Luer et al. (1990) investigated replacement times in three juvenile nurse sharks (*G. cirratum*) over 3 years. They found that tooth replacement (every 9–21 days) was more rapid during the summer, when water temperature was 27–29°C, and slowest (every 51–70 days) during winter, when the water temperature was 19–22°C. Somewhat surprisingly, the rate of tooth replacement for this shark did not slow down as the animals aged. This latter observation conflicts with that of Wass (1973), who observed that the time taken to replace a tooth row in young sandbar sharks (*Carcharhinus milberti*) was 18 days, whereas for older animals it was 38 days.

By relating the width of teeth in successive generations to total body length, as in Fig. 10.8, Strasburg (1963) calculated that in the cookie-cutter shark (*Isistius brasiliensis*) there were 15 replacements during growth from 140 to 501 mm body length. From the fact that the width of each successive tooth in young lemon sharks (*N. brevirostris*) increased by 0.14 mm, Moss (1967) calculated that roughly 125 replacements occurred during growth to a maximum size of 250 cm. Similar calculations for young dogfish (*M. canis*), suggested that six tooth row replacements occurred during an increase of 10 cm in body length Moss (1972). Reif (1976) estimated, conservatively, that the Pacific bull-head shark (*Heterodontus*) produces between 2000 and 4000 teeth in the course of 10 years.

OSTEICHTHYES

Development of Replacement Teeth

The teeth of bony fish are attached individually to the jaws. With very few exceptions (eg, piranhas), the teeth are spaced apart and do not overlap, so each tooth can be replaced independently. Usually, a successional tooth develops by the side of, or beneath, its predecessor and replacement involves resorption and loss of the predecessor and

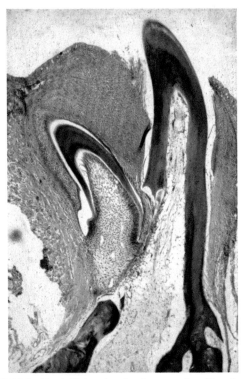

FIGURE 10.2 Decalcified section showing resorption of functional tooth by lingually positioned, developing successional tooth in the rainbow trout (*Oncorhynchus mykiss*). Masson trichrome. ×55. *From Berkovitz, B.K.B., 1977b. The order of tooth development and eruption in the rainbow trout (*Salmo gairdneri*). J. Exp. Zool. 201, 221–226. Courtesy Editors of the Journal of Experimental Zoology.*

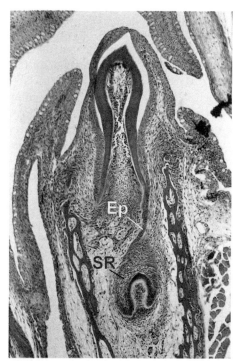

FIGURE 10.3 Two developing teeth in piranha (*Serrasalmus rhombeus*). The shaft is forming in the tooth germ nearer the oral cavity, while the deeper tooth germ is in the earliest stages of enameloid formation. The two are connected by an epithelial strand (Ep), which passes through an opening in the bone. In the younger tooth germ there is a layer (SR) between the inner and outer dental epithelia which is reminiscent of a stellate reticulum. Hematoxylin and eosin. ×135. *From Berkovitz, B.K.B., 1975. Observations on tooth replacement in piranhas (Characidae). Arch. Oral Biol. 20, 53–56. Courtesy Editors of the Archives of Oral Biology.*

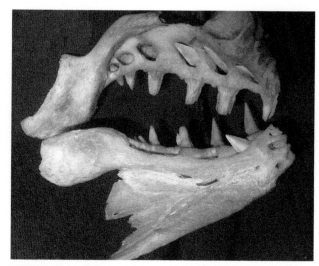

FIGURE 10.4 Upper and lower jaws of tigerfish (*Hydrocynus* sp.). Lingual view, showing replacement teeth developing intra-osseously in crypts beneath the functional teeth. The replacement teeth are all in the same stage of development. They also develop in a recumbent position, pointing backwards: this is most obvious in the mandible. Actual image width = 12 cm. *Courtesy MoLS KCL. Catalogue no. V171.*

its attachment tissues (Fig. 10.2). The successional tooth then erupts into position in the tooth row and is cemented to the jaw bone by development of its attachment tissues. Replacement teeth may develop superficially in the soft tissue outside the bone, to which they will subsequently attach (extraosseous development), as in trout (Fig. 10.2), or within a crypt or groove within the bone (intraosseous development; Chapter 9). Teeth which have developed intraosseously must erupt a considerable distance into function, as in piranhas (Fig. 10.3).

Trapani (2001) investigated the location of replacement teeth in nearly 130 teleost species. Extraosseous development is plesiomorphic for bony fish and is the rule among the basal groups. Among the Elopiformes, intraosseous development was observed only in *Albula*. However, the oral teeth of the Characiformes show intraosseous development, but pharyngeal teeth develop extraosseously. Among Percomorphaceae intraosseous development is predominant (Trapani, 2001; Bemis et al., 2005; Hilton and Bemis, 2005). Both types of development can coexist, eg, among Cyprinodontids (killifish, guppies) oral teeth often develop extraosseously while pharyngeal teeth develop intraosseously (Trapani, 2001),

or both types can be found in the oral dentition (Hilton and Bemis, 2005).

In species with large teeth, eg, the tigerfish (*Hydrocynus*; Characidae), the replacement teeth may develop in a recumbent position and become erect during eruption (Fig. 10.4).

Patterns of Replacement

Tooth replacement studies in bony fish have tended to be limited to species where the teeth are arranged in discrete single rows. We begin with a description of longitudinal studies of tooth replacement in a protacanthopterygian fish, the rainbow trout (*Oncorhynchus mykiss*) by Berkovitz and Moore (1974, 1975). The pattern of replacement illustrates several important features shared with other dentitions, not only in bony fish but also in amphibians and reptiles.

In addition to single rows of teeth on the marginal jaw bones, the rainbow trout also possesses teeth on the tongue (Fig. 4.46). The presence or absence of teeth at each locus was recorded from impressions in dental wax taken from anesthetized fish twice weekly for up to 9 months. Charts were constructed to show the times of eruption and loss for each locus and hence to visualize the pattern of tooth replacement. Fig. 10.5 shows such a chart covering 7 months for 20 tooth loci in the left dentary and five lingual teeth on each side of the tongue (Berkovitz and Moore, 1974). Gaps between the solid vertical lines indicate when a tooth was absent. At any one time approximately 70% of teeth were present in the jaw and in all fish in the sample this proportion varied between 57% and 77%. Virtually no instances occurred over the experimental period when two adjacent teeth were absent simultaneously. This indicates that the trout dentition is always maintained in a highly functional state (DeMar, 1973; Osborn, 1975).

The generation time of a tooth (the time between its first appearance in the mouth and its replacement by a successor)

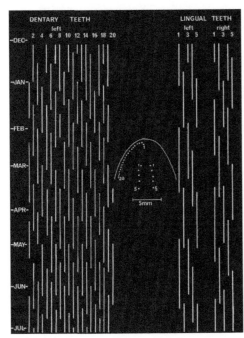

FIGURE 10.5   Tooth chart for the lower jaw (dentary teeth) and tongue (lingual teeth) of a rainbow trout (*Oncorhynchus mykiss*). *From Berkovitz, B.K.B., Moore, M.H., 1974. A longitudinal study of replacement patterns of teeth on the lower jaw and tongue in the rainbow trout* Salmo gairdneri. *Arch. Oral Biol. 19, 1111–1119. Courtesy Editors of the Archives of Oral Biology.*

was on average 8 weeks for small trout (12–15-cm long) and 12–14 weeks for larger trout (20–23-cm long). The ratio of functional time (period when a tooth was present in the mouth) to nonfunctional time (when a tooth was absent) varied between individual fish from 1.3 to 3.3.

On the chart in Fig. 10.5, it is possible to draw lines linking the eruption times of teeth in adjacent or alternate tooth positions, or indeed any other series of regularly spaced points (eg, every third tooth; Fig. 10.6). However, it is convenient to focus on the set of lines through alternate tooth positions, which illustrate the functional state of the dentition as a whole at a given moment. The slope of the lines reflects the fact that, along each line, teeth toward the back of the jaw erupt earlier than those at the front. In other words, the lines can be thought of as representing waves of replacement passing from the back to the front of the jaw. In the dentaries of all the trout examined by Berkovitz and Moore (1974, 1975), the replacement waves passing through alternate tooth positions passed steeply from back to front. A similar pattern was seen in the premaxillary/maxillary tooth row, although the replacement waves were flatter (Fig. 10.7) (Berkovitz and Moore, 1975).

Huysseune and Witten (2006) found that in the related Atlantic salmon (*Salmo salar*), teeth in every third position are at the same stage of development in most specimens but a few showed irregularities.

Patterns similar to those seen in these salmoniforms occur frequently among amphibians and reptiles, but are by no means universal among bony fish, among which teeth can be replaced in a great variety of ways. The descriptions

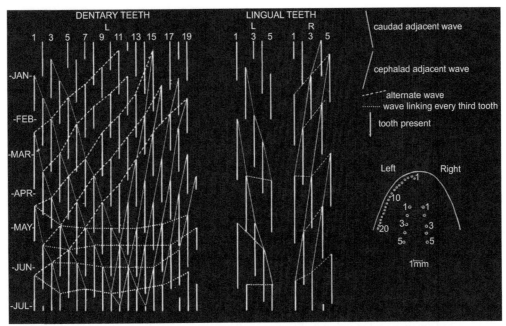

FIGURE 10.6   Same tooth chart as in Fig. 10.7 but various possible waves of tooth replacement indicated by linking adjacent teeth, alternate teeth, and every third tooth. *From Berkovitz, B.K.B., Moore, M.H., 1974. A longitudinal study of replacement patterns of teeth on the lower jaw and tongue in the rainbow trout* Salmo gairdneri. *Arch. Oral Biol. 19, 1111–1119. Courtesy Editors of the Archives of Oral Biology.*

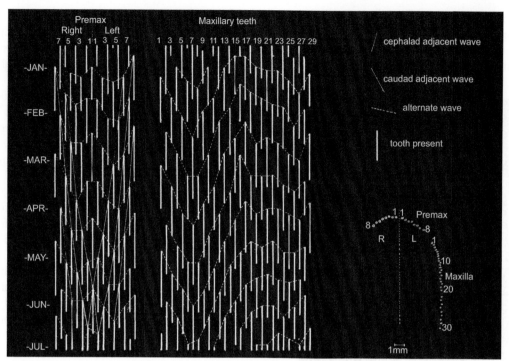

FIGURE 10.7 Tooth chart for the premaxillary and maxillary teeth on one side of the upper jaw of a rainbow trout (*Oncorhynchus mykiss*). *From Berkovitz, B.K.B., Moore, M.H., 1975. Tooth replacement in the upper jaw of the rainbow trout* (Salmo gairdneri). *J. Exp. Zool. 193, 221–234. Courtesy Editors of the Journal of Experimental Zoology.*

in the remainder of this chapter are arranged according to the phylogeny of Betancur-R et al. (2014).

Holostei

In skeletal material of the bowfin (*Amia calva*), Miller and Radnor (1973) found that the teeth on different bones were often distinctly different in tooth morphology and replacement pattern, and bilateral symmetry was rare. Although tooth replacement sometimes occurred in waves affecting alternate teeth, random replacement was common and became more so with age. In several bones, particularly the maxillae, series with every third or fourth tooth at a similar stage of development were observed.

Characiformes

The remarkably diverse dentitions of these fish show some unusual features associated with tooth replacement. Intraosseous tooth development is the rule and in many species all the teeth in a quadrant develop and are replaced in synchrony (Roberts, 1967; Berkovitz and Shellis, 1978; Trapani et al., 2005).

Adult tigerfish (*Hydrocynus*; Alestidae) have a few large, conical teeth. The successional teeth lie at an angle to the functional teeth and are enclosed in individual crypts (Fig. 10.4). Gagiano et al. (1996) observed, in captive tigerfish, that all the teeth were replaced simultaneously over

about 5 days: the rapidity explained why replacement had not been observed previously.

The large multicuspid rostral teeth of the Mexican tetra (*Astyanax mexicanus*; Characidae) develop intraosseously. They form and erupt in synchrony. The small caudal teeth on the dentary are, however, replaced at random (Trapani et al., 2005). During the initial 169 days of their study, Trapani et al. (2005) observed that four generations of teeth developed and they calculated that over a 6-year lifespan there would be about 40 replacements of the teeth.

Most of the teeth in the unique slicing dentition of piranhas (*Serrasalmus*; *Pygocentrus*; Characidae) are interlocked, so they cannot be shed individually (Fig. 4.34; Shellis and Berkovitz, 1976). Examinations of skeletal material indicated that teeth within each quadrant must be shed simultaneously, that upper and lower quadrants of the same side are synchronized, and that right and left halves of the dentition are out of phase (Roberts, 1967; Berkovitz, 1975). A 13-month longitudinal study (Berkovitz and Shellis, 1978) confirmed that the teeth within each jaw quadrant were shed and replaced simultaneously, as also were the smaller ectopterygoid teeth. However, the pattern of tooth replacement proved to be more variable than was suggested by skeletal material. All four quadrants may be in phase (so that for a few days no teeth at all were present in the mouth); one side of the jaws may be out of synchrony with the other side; diagonally opposing quadrants (eg, upper left and lower right) may be in synchrony; or all four quadrants may be out of phase with each other. The teeth remained functional for about 100 days and the time between exfoliation of one tooth row and its replacement by the succeeding row was very rapid (taking about 5 days), so teeth were absent from each quadrant for only about 5% of the cycle: this explained why tooth replacement had not been observed in skeletal material, as for the tigerfish (see page 260). The period when each quadrant was nonfunctional was also short enough not to risk starvation, even though only two out of the eight piranhas studied by Berkovitz and Shellis (1978) would retain one functional side of the dentition at all times. However, 5 days seems an adequate time for the bone of attachment to be renewed after tooth loss.

To investigate the stability of the patterns of tooth replacement, two piranhas used in the initial study of Berkovitz and Shellis (1978) were kept for 3 more years and tooth replacement patterns noted for a further 10-month period (Berkovitz, 1980). Although the teeth in each quadrant continued to be replaced simultaneously, the timing of replacement had changed so that the patterns seen 3 years earlier were no longer recognizable. In addition, the functional life of teeth in the older specimens increased to about 125 days compared with about 100 days three years earlier.

A Strasburg plot of tooth width against jaw length from cross-sectional data (Fig. 10.8) indicated that teeth would have been replaced about 27 times in the largest piranha studied (jaw length 45 mm). In the largest specimen, no replacement teeth were present in one jaw and the functional teeth showed unusually heavy wear. This suggests that in very old specimens replacement teeth may eventually fail to develop.

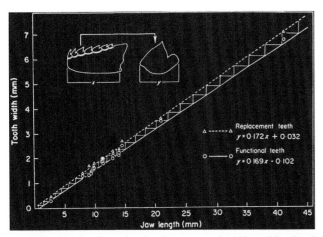

FIGURE 10.8 Strasburg plot of replacement tooth widths and functional tooth width against jaw lengths for piranhas (*Serrasalmus*). The length of the mandible, and the widths of the third lower functional tooth and of its developing replacement were measured on radiographs as indicated. *From Berkovitz, B.K.B., Shellis, R.P., 1978. A longitudinal study of tooth succession in piranhas (Pisces: Characidae), with an analysis of the replacement cycle. J. Zool. Lond. 184, 545–561. Courtesy Editors of the Journal of Zoology, London.*

Although tooth replacement within quadrants appears simultaneous to periodic visual examination, there is actually a small anterior–posterior gradient of both initiation and eruption: anterior teeth start to form first and probably erupt first (Berkovitz, 1975; Berkovitz and Shellis, 1978). This seems to be essential for the overlapping arrangement of teeth to be established (Fig. 10.9). The replacement teeth in each quadrant slightly overlap and develop with a backward tilt. As they move upwards and erupt, the overlying bone is resorbed and the functioning teeth shed. Starting with the front tooth, each new tooth erupts and rotates forward, so that the anterior scoop-shaped depression docks with the small rear cusp of the tooth immediately in front. Finally, new bone is formed below to which the fresh set of teeth becomes attached. A new set of replacing teeth then develops and the cycle is repeated.

Paracanthomorphacea

The dentition of the cod (*Gadus morhua*) is unlike that of trout and piranha, in that the oral teeth form several rows and are of different sizes (see Fig 4.58). Replacement teeth continually arise from the epithelium. Holmbakken and Fosse (1973) could find no evidence for a pattern of alternate tooth replacement. The number of nonfunctional tooth germs exceeds the number of functional teeth in a ratio of 2.3:1. As a general rule, the number of functional teeth increases with the total length of the fish, although the number varies between fish of the same length. Thus, in the lower jaw a fish about 30-cm long had about 20 teeth in function, whereas one 100-cm long had over 50 functional teeth.

Percomorphaceae

The gobiid fish *Sicyopterus japonicus* and *Sicydium plumieri* (Mochizuki and Fukui, 1983; Kakizawa et al., 1986; Mochizuki et al., 1991) scrape algae from rocks in fast-flowing streams and the resulting rapid tooth wear is correlated with rapid turnover of teeth. Beneath each functional tooth in the upper jaw there may be a graduated series of up to 45 successional teeth, arranged in a complex spiral. Replacement teeth form rows with an alternating zigzag arrangement except in the row immediately behind the functional row where there is no alternation. Teeth are thus replaced more or less simultaneously (Mochizuki and Fukui, 1983). Tooth replacement is very frequent (about every 9 days). Members of each tooth family are arranged in the form of a compressed circle, with each replacing tooth being at a progressively younger stage of development than the one above it. In Figs. 10.10 and 10.11 a tooth family of the rock-climbing monk goby (*S. japonicus*) with 26 sets of replacing teeth is illustrated. The youngest tooth is at the base of the first arm of the "n" and the oldest and sole functioning tooth lies at the base of the second arm of the

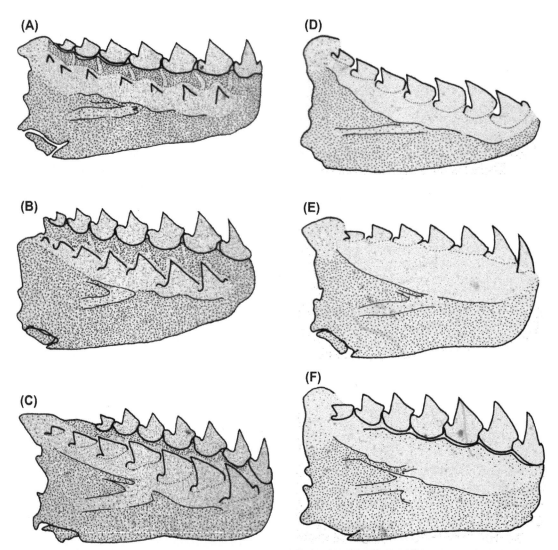

FIGURE 10.9 Tracings from radiographs of six different lower jaws of piranhas representing a complete cycle of tooth replacement and illustrating the development, overlapping, eruption, and rotation of the teeth necessary to establish the interlocked dentition. (A) New set of teeth in position and attached to newly formed attachment bone. Replacement teeth within an intraosseous crypt are at an early stage of development with just the tips of the teeth mineralized. (B) Replacement teeth more advanced, with main blade and tip of subsidiary cusp mineralized. Functional and replacement teeth separated by a reasonable thickness of bone. (C) Replacement teeth now tilted backward so that their open bases face forward. The sixth and seventh teeth lie behind the last functional tooth. The replacement teeth are erupting, as shown by thinning of the bone separating them from the functioning teeth. (D) Functional teeth have been shed and the attachment bone completely resorbed. Replacement teeth have erupted further, as shown by the increased height above the base of the crypt. Replacement teeth still tilted backward. (E) The replacement teeth erupt forward diagonally and rotate into the upright position, thereby ensuring that they interlock correctly. (F) New functional teeth become attached to new bone forming immediately beneath them. The new set of replacing teeth are unmineralized, so are invisible in radiographs. *From Berkovitz, B.K.B., Shellis, R.P., 1978. A longitudinal study of tooth succession in piranhas (Pisces: Characidae), with an analysis of the replacement cycle. J. Zool. Lond. 184, 545–561. Courtesy Editors of the Journal of Zoology, London.*

"n." Whenever this functional tooth is shed, all the replacing sets move up one position along the compressed circle and a new tooth is budded off in the deepest position. The effect resembles an escalator. From Figs. 10.10 and 10.11, it can be seen that the tooth base develops comparatively late (from about tooth generation 18–26). This base quickly fuses to the jaw only as the tooth becomes functional. It appears that the functional teeth are not shed but resorbed and subsequently degraded (Fig. 10.12), perhaps allowing minerals to be reutilized.

The monk goby shares several characteristics with other unrelated percomorph fish which also scrape algae from rocks. An example is the opaleye (*Girella nigricans*; Eupercaria; Kyphosidae), in which numerous replacement teeth are associated with each functional tooth on the premaxilla and dentary. Tooth replacement is rapid: the youngest teeth require only 22–32 days to be replaced and shed. There is a high degree of convergence in the dentitions of related species with a similar feeding habit, such as the bumphead damselfish (*Microspathodon bairdi*; Ovalentaria; Pomacentridae)

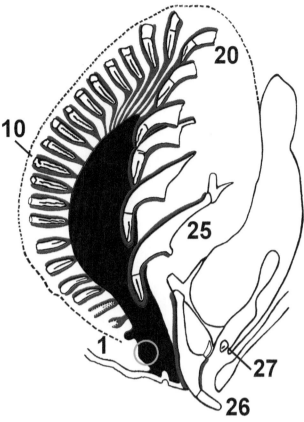

FIGURE 10.10 Section of the upper jaw of a monk goby (*Sicyopterus japonicus*), showing the generations of replacing teeth beneath each functional tooth. *From Moriyama, K., Watanabe, S., Lida, M., Fukui, S., Sahara, N., 2009. Morphological characteristics of upper jaw dentition in a gobiid fish (Sicyopterus japonicus): a micro-computed tomography study. J. Oral Biosci. 51, 81–90. Courtesy Editors of the Journal of Oral Biosciences.*

FIGURE 10.11 Line drawing of Fig. 10.10. In this 3-cm specimen of a monk goby, a total of 26 generations of teeth are visible from the youngest (number 1) to the oldest (number 26), which has erupted into the mouth and become attached to the jaw. Each developing tooth generation is outlined in red. The green open circle represents the site at which all new teeth form from the blue colored zone. As each new tooth starts to grow, it moves upwards. The 10th and 20th generation of teeth are labeled, while the 25th generation tooth is almost complete and lies immediately beneath the functional tooth. As soon as the functional tooth is lost, it will be replaced by tooth number 25. All the other teeth will move up one position and a new tooth will be budded off beneath tooth 1, which will then become tooth number 2. The remnants of a previously erupted tooth labeled number 27 can be seen enclosed by the lining of the mouth. An older, 10 cm specimen had 45 generations of teeth. *From Moriyama, K., Watanabe, S., Lida, M., Fukui, S., Sahara, N., 2009. Morphological characteristics of upper jaw dentition in a gobiid fish (Sicyopterus japonicus): a micro-computed tomography study. J. Oral Biosci. 51, 81–90. Courtesy Editors of Cell and Tissue Research.*

and the large-banded blenny (*Ophioblennius steindachneri*; Ovalentaria; Blenniidae; Norris and Prescott, 1959).

Morgan and King (1983) identified waves of tooth replacement passing through alternate tooth positions in skeletal material of king mackerel (*Scomberomorus cavalla*; Pelagiaria; Scombridae). However, there is little symmetry and waves are irregular, in that some waves pass from back to front, others from front to back while others are relatively flat. Over 40% of tooth loci are occupied by developing replacement teeth in both jaws. Tooth shedding is unusual in that resorption occurs, not only in the basal region associated with the replacing tooth, but also at a second site at the mesial and distal edges just above the jaw line, producing an arrowhead shape. Morgan and King suggested that this may reduce the amount of tooth substance which must be resorbed from the base prior to shedding. In another pelagiarian, the bluefish (*Pomatomus saltatrix*), flat replacement waves pass regularly from back to front and there is an almost perfectly alternating replacement pattern (Bemis et al., 2005), so that functional teeth are widely spaced, as in *Hydrocynus* among teleosts (Fig. 4.30B) and in snakes (Chapter 7). In anguimorph lizards (Figs. 6.54–6.69), similar spacing is achieved by the varanid mode of replacement. The spacing of the teeth must increase the depth of the bite.

Synchronous tooth development, as seen among characiforms, also appears to be synchronized in the unrelated **plaice** (*Pleuronectes*; Carangaria; Pleuronectidae), and replacement appears to be seasonal as well. Kerr (1960) stated that teeth are shed during the winter and that the new replacing teeth move up in unison to occupy their loci by spring.

Tooth replacement patterns among cichlids (Ovalentaria, Cichlidae) seems, from the very limited data available, to be variable, being highly irregular in *Hemichromis bimaculatus*, but regular in eretmodine species (Huysseune

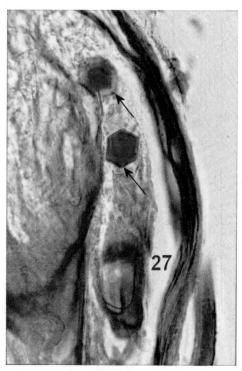

FIGURE 10.12   High-power view of inset in the region near the base of the functional tooth labeled number 27 in Fig. 10.11. It shows evidence of fragments of older functional teeth (*arrows*) that have been engulfed and are being digested, perhaps allowing important constituents such as calcium to be reutilized. *Courtesy of Dr. N. Sahara.*

and Witten, 2006). The rate of tooth replacement in cichlids is dependent on age and species, with adults replacing their teeth every 30–100 days (Tuisku and Hildebrand, 1994; Streelman et al., 2003).

Wakita et al. (1977) studied tooth replacement in the premaxillae and dentaries of the sawtail surgeonfish (*Prionurus microlepidotus*; Eupercaria; Acanthuridae). From histological material, they could observe no symmetry either between right and left sides or between upper and lower jaw quadrants. The authors concluded that tooth replacement proceeded independently in each quadrant. Unusually, the functional tooth at each locus is accompanied by two developing replacing teeth at different stages of development, each connected to the surface by a dental lamina. Whereas it would be expected that the two replacing teeth represented members of the same series, this is not the case. One replacing tooth was located lingually and the other labially. At each tooth locus, replacement alternated so that a functional tooth that originally developed lingually was replaced by a tooth which had developed labially and vice versa. The predecessor teeth being replaced are therefore resorbed alternately on the lingual and labial aspects. Thus, the randomness of replacement within each quadrant contrasts with the regular alternation of replacement at each locus. This suggested to the authors that some control was effected through inhibition of a

developing tooth germ by its neighbor. Wakita et al. (1977) also observed that when a functional tooth was assessed to have been lost earlier than expected (eg, through trauma), tooth replacement at the affected locus proceeded faster than at the two adjacent loci.

Lobotes and *Datnioides* (Eupercaria; Lobotiformes) have two rows of oral teeth: a medial field of small, extraosseous teeth and a lateral row of larger, intraosseous teeth. Unusually, while the medial teeth are replaced individually, those in the lateral row seem to be replaced in groups (Hilton and Bemis, 2005). In the tripletail (*Lobotes surinamensis*), the 50 or so teeth forming the lateral row appear to be arranged in groups of up to 5 teeth. Within each group, there is an anterior–posterior gradient of development, the most anterior tooth being the oldest. Beneath the functional teeth, the replacement teeth are arranged in corresponding groups, in each of which there is the same anterior–posterior gradient in development as in the functional teeth. The implication of these findings is that tooth replacement occurs in sequence from front to back within each group of teeth.

This account has concerned tooth replacement in the oral dentition. A few studies have focused on the pharyngeal dentition, particularly in cypriniforms, which lack oral teeth. Evans and Deubler (1955) and Nakajima (1979, 1984, 1987) showed that in cyprinid and cobitid fish, tooth replacement occurs in waves passing through alternate tooth positions. The teeth are replaced in an alternating manner in the young zebrafish (*Danio rerio*; Cyprinidae) after functioning for 8–10 days (Van der Heyden and Huysseune, 2000). A replacing tooth develops by the side of its functional predecessor and when the base of the latter has been resorbed, the replacing tooth erupts and becomes functional with minimum delay. Sometimes two functional teeth are present at the same time in one tooth locus.

In the medaka (*Oryzias latipes*; Beloniformes; Adrianichthyidae), the pharyngeal teeth are organized in families of up to five generations of teeth, with two functional and three developing successional teeth (Fig. 10.13). There is no obvious dental lamina. Longitudinal studies using calcein to label mineralizing tissues revealed that the life cycle is approximately 4 weeks: 3 weeks of growth and development and only 1 week in function, followed by rapid shedding (Abduweli et al., 2014).

AMPHIBIANS

Caecilians

In this group (see pages 121–123), the morphology of teeth in fetuses may differ significantly from those of the adult (Parker and Dunn, 1964; Wake, 1976). In many viviparous species, the fetuses use their teeth to obtain nutrition by scraping off the oviduct epithelium. In some oviparous

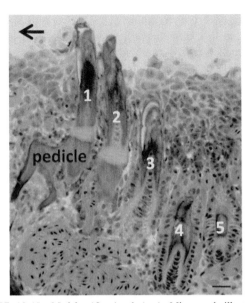

FIGURE 10.13 Medaka (*Oryzias latipes*). Micrograph illustrating a pharyngeal tooth family of five generations, of which the first two generations are functional. Scale bar = 10 μm. *From Abduweli, D., Baba, O., Tabata, M.J., Higuchi, K., Mitani, H., Takano, Y., 2014. Tooth replacement and putative odontogenic stem cell niches in pharyngeal dentition of medaka* (Oryzias latipes). *Microscopy 63, 141–153. Courtesy Editors of Microscopy.*

Urodeles

Lawson et al. (1971) observed waves of replacement passing through alternate tooth positions in skeletal material of the red-backed salamander (*Plethodon cinereus*), although again there was variation in the pattern. Some specimens showed almost perfect alternation, some had waves passing from back to front, while in others the waves passed from front to back.

There are seasonal variations in tooth replacement patterns in adult mudpuppies (*Necturus maculosus*; Miller and Rowe, 1973). Teeth are not replaced during hibernation or for much of the spring and summer (or at least it is very slow at these latter times of the year). This was demonstrated by experimental studies in which animals ceased to replace their teeth when maintained in the dark and at low temperatures to simulate hibernation conditions. Tooth replacement is most active in late summer and autumn. Replacement patterns are very irregular and classical waves passing through alternate tooth positions are rarely seen. Indeed, in a number of animals, replacement waves affect every third tooth.

If a segment of the mandible and its associated teeth are excised in adult urodeles, regeneration will occur. Furthermore. this process can occur more than once and is more rapid during re-regeneration (Graver, 1978). New teeth are added in an alternate sequence at the anterior end of the tooth row in the regenerate and re-regenerate jaw. Regenerated teeth are of normal adult size and can be replaced without the presence of surrounding bone, which grows in later and is often incomplete (Howes, 1978). If a digit is also transplanted to the regeneration site in the maxilla of frogs, ankylosis takes place earlier but does not affect the rate of regeneration of teeth, which still become ankylosed to a separately induced bony base rather than to the transplanted bone of the digit (Howes and Eakers, 1984).

Anurans

In the African clawed frog (*Xenopus laevis*) tooth replacement during the period leading up to metamorphosis occurs in flat waves passing through alternate tooth positions and was calculated to take place about every 16 days (Shaw, 1979). The odd-numbered series is about 8 days out of phase with the even-numbered series. In a longitudinal study of adult *Xenopus*, Shaw (1985) reported a complete development cycle of between 60 and 70 days, of which the tooth is in a functional position for about 25 days. There is a gap of 10–15 days between replacements and the remaining 25–30 days is occupied by the preeruptive, developmental stages.

In the Sumaco horned treefrog (*Hemiphractus proboscideus*), functional teeth in general are present in alternate loci, while between them there is either no functional tooth or a tooth undergoing resorption with its successional tooth

species the hatchlings use their teeth to peel off and digest the mother's nutritious skin. The teeth of fetal viviparous caecilians differ markedly from those of the adults, which are often bicuspid and arranged in single rows, whereas in the fetal dentition, the crowns are separated from the pedicel by a hinge joint, which allows some slight movement of the teeth. Fetal teeth are often polystichous, usually forming 3–4 rows, although there may be 12 rows in *Chthonerpeton petersi*. This results from the retention of replacement series on the dentigerous elements. Tooth rows are lost in subsequent development, so that in adults only a single row is left (Parker and Dunn, 1964; Wake, 1976).

While the adult dentition in the Kenyan caecilian (*Boulengerula taitanus*) has pointed teeth with either one or two cusps (Chapter 5, Fig. 5.18A and B), the dermatophagous young have very different teeth (Fig. 5.18C–F). The vomeropalatine teeth and the anteriormost three to four teeth of the premaxilla and dentary teeth are monocuspid. The remaining teeth are multicuspid and combine a pronounced blade-like labial cusp with a lingual cusp that has two or three subsidiary cusps (Fig. 5.18D), which may be short and blunt (Fig. 5.18E) or may have more elongated, pointed processes resembling grappling hooks (Fig. 5.18F) (Kupfer et al., 2006). Wake (1976, 1980) showed that, in viviparous caecilians, some of the first-formed teeth were rudimentary and underwent resorption without becoming functional. Histological examination revealed evidence of alternation in the pattern of tooth replacement.

beginning to erupt (Shaw, 1989). Gillette (1955) reported similar results from skeletal material of the Northern leopard frog (*Rana pipiens*), observing that teeth in alternate tooth positions were at similar stages of development. Furthermore, functional teeth generally alternated with empty loci where teeth were in the process of replacement. Thus, the replacement waves are flat. Gillette deduced that total life cycle time for teeth was about 90 days.

Lawson (1966) found a similar alternating pattern of tooth replacement in the common frog (*Rana temporaria*), but the proportion of tooth positions lacking a functional tooth was lower (26%) than in true alternation (50%). Lawson explained this partial obscuring of a true alternating pattern as being a consequence of the rapidity with which the final stages of resorption and loss of the functional teeth occur. This condition could result if only specimens with the majority of teeth in the terminal stages in the tooth life cycle are examined.

Goin and Hester (1961) also observed alternating waves of replacement in skulls of the green tree frog (*Hyla cinerea*), but these were more irregular than those seen by Gillette (1955). Some replacement waves sloped backward, others forward, and others remained flat. Groups of functional teeth in even-numbered tooth loci alternated with groups in the odd-numbered positions, the number of groups varying between 4 and 13. Similar variation was seen in three other types of frog.

REPTILES

Mode of Tooth Replacement

Three modes of tooth replacement patterns can be recognized in reptiles with regard to the position of the replacement tooth in respect to the tooth being replaced (McDowell and Bogert, 1954; Edmund, 1960; Rieppel, 1978), namely the varanid, iguanid, and intermediate modes.

In the **varanid** mode, seen in the majority of anguimorph lizards, the replacement tooth develops interdentally distal to the tooth that it replaces. No resorption pits occur in relation to the replacing tooth, and the functional tooth is lost as a result of independent resorption at the site of the bone of attachment. This mode of replacement maintains a row of widely spaced teeth (Figs. 6.54–6.69).

In the **iguanid** mode, present in nonanguimorph lizards, the replacement tooth lies directly lingual to its predecessor and is associated with the appearance of a resorption pit on the functional tooth that enlarges and is occupied by the growing replacing tooth. The ankylosis surface on the lingual aspect of the jaw bone is the last part to be resorbed, so a functional tooth can remain in function to a very late stage of resorption (eg, Fig. 6.81B: tooth locus 5).

In the **intermediate** mode, found in anguinid lizards, the replacement teeth arise distolingually, rather than directly lingually, but move into resorption pits toward the

final stages of the resorption cycle. Delgado et al. (2003) found that in *Chalcides* the iguanid mode found in adults was preceded by the intermediate mode of replacement in young and juvenile specimens.

In almost every modern reptile, each functional tooth is accompanied by at least one or two developing replacement teeth, and in lizards and snakes three or even four developing replacements are not uncommon. In the case of poison fangs, up to eight teeth are present (see Fig 7.6B; Zahradnicek et al., 2008).

Patterns of Tooth Replacement

Comprehensive accounts of the patterns of tooth replacement in reptiles as determined by analysis of skeletal material were presented by Edmund (1960, 1962, 1969). From a survey of a large cross-section of different species (including fossils), he demonstrated that the vast majority showed replacement waves passing through alternate tooth positions in a back-to-front direction (although there is little bilateral synchrony). Examples of this pattern can be seen in Chapter 7 in the dentitions of pythons (Fig. 7.16), viperids (Fig. 7.19B), and colubrids (Fig. 7.43). In elapid snakes, replacement waves pass from front to back, while in the crotaline viperids, there is a tendency for simple alternation. In a large series of crocodile skulls, Edmund (1962) established that, although waves of replacement pass through alternate tooth positions in adult animals from back to front, in the young the direction is reversed. His explanation for this change was that teeth at the back of the jaws remained in function for longer periods compared with anterior teeth. He noted that, with age, the replacement waves became more irregular.

Miller and Radnor (1970) found that 15 skulls of young spectacled caiman (*Caiman sclerops*) all showed evidence of replacement waves affecting alternate tooth positions, but there was considerable variability. Bilateral symmetry was never encountered in the dentary, but was present in three pairs of maxillae. Apparent crossing over of replacement waves was observed, which the authors interpreted in terms of precocious or retarded development of one or two individual teeth.

In the common agama (*Agama agama*), only the nine anterior, pleurodont teeth (five upper and four lower) are replaced, while the remaining posterior, acrodont teeth are not, although they increase in number by addition to the back of the tooth row. In skeletal material of this lizard, Cooper et al. (1970) found that the anterior teeth were replaced in a successive sequence from front to back. Replacement was limited to three (possibly four) generations of teeth. The replacement teeth were very much larger than their predecessors and there was little symmetry of replacement.

Tooth replacement patterns have been studied in the Gran Canaria skink (*Chalcides sexlineatus*) and the West

Canary skink (*Chalcides viridanus*; Delgado et al., 2003). The majority of specimens were of known age (0.5 months to 6 years old) and the stages of development were determined from radiographs. No differences between the two species were noticed in patterns of replacement. For older specimens, replacement waves were seen to pass from back to front of the jaws, affecting alternate tooth positions as in most other reptiles. However, in young and juvenile specimens this pattern was more difficult to establish, perhaps because of the higher rate of tooth replacement. There was little symmetry between either right and left or upper and lower dentitions.

Longitudinal Studies of Replacement

Edmund (1960) studied tooth replacement using radiography in two young specimens of *Alligator mississippiensis* at monthly intervals for up to 3 years and showed that replacement waves passed through alternate tooth positions, in a back-to-front direction. The waves were irregular and became more irregular with age. The average functional life of an anterior tooth was about 9 months, while for a posterior tooth it was about 16 months. The average total time from initial development to exfoliation for a tooth in a young alligator was estimated to be about 2 years.

Erickson (1996) observed much shorter tooth replacement times of 45–150 days (mean 120 days) in two specimens of American alligator (*Alligator mississipiensis*), initially 0.85 m long. Kaye [quoted by Erickson (1996)] observed replacement rates of 110–130 days in the spectacled caiman (*Caiman sclerops*). Using incremental lines, Erickson (1996) estimated replacement times of 111 days for *C. sclerops* and 83–122 days for *A. mississipiensis*.

Tooth replacement in the lizard genus *Lacerta* was investigated by Cooper (1963), utilizing a wax impression technique. He found that the functional life of teeth in the ocellated lizard (*Lacerta lepida*) increased with age from 2 to 3 weeks at hatching to about 38 weeks in the mature animal. As this animal can live for up to 20 years, Cooper calculated (conservatively) that there may be as many as 80 replacements during life. He reported the existence of the classic pattern of waves of replacement passing through alternate tooth positions and that bilateral symmetry was a conspicuous feature of replacement patterns (Fig. 10.14). However, portions of waves showing a flat gradient or even local reversal of wave motion were frequently evident anteriorly. Such biological variation needs to be taken into account when undertaking mathematical modeling of replacement patterns (DeMar, 1972, 1973; Osborn, 1972, 1974).

The form of the replacement wave in *Lacerta* varies between species. In the more primitive species *L. lepida* the acceleration of waves anteriorly is not so pronounced as in the more advanced species *Lacerta muralis*. The difference

in length of functional life between anterior and posterior teeth became more marked in the advanced species (Cooper, 1963).

Cooper (1966) also studied tooth replacement in the slow-worm (*Anguis fragilis*), again using wax impression, combined with direct visual observation through a microscope at weekly intervals. Three adult females were studied following hibernation from May to December. Replacement was close to a pattern of almost perfect alternate replacement throughout the dentition (Fig. 10.15). Thus, about half the teeth in a quadrant were generally absent in either the odd- or even-numbered tooth loci (although the wave sometimes sloped slightly from back to front, other times sloping from front to back). Replacement patterns were bilaterally symmetrical, and there was partial symmetry between patterns in the upper and lower jaws. Such symmetry and synchrony may be important in maintaining a functional dentition. It seems most likely that replacement ceases during hibernation in the slow-worm. The functional life of all teeth appeared similar, irrespective of their size and position in the jaws, and there was little change related to age.

Tooth replacement has been studied in young green iguanas (*Iguana iguana*) over periods of up to two-and-a-half years using weekly wax impressions (Kline and Cullum, 1984, 1985). Waves of tooth replacement were seen to

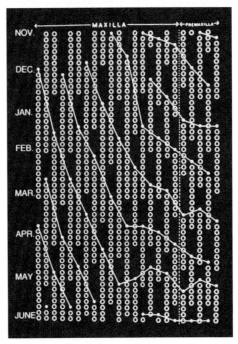

FIGURE 10.14 Tooth replacement graph from a longitudinal study of living ocellated lizard (*Lacerta lepida*), showing alternate waves of tooth replacement passing from back to front. *Round open circles* indicate functioning teeth in the maxilla and premaxilla. The numbers indicate tooth loci from front to back of jaw. The steepness of the replacement waves is due to the longer functioning life cycle of the posterior teeth. *From Cooper, J.S., 1963. Dental Anatomy of the Genus* Lacerta. *(Ph.D. thesis), University of Bristol. Courtesy University of Bristol.*

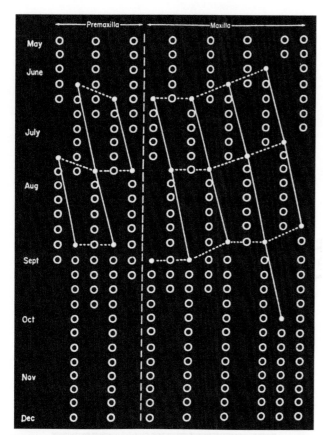

FIGURE 10.15   Chart of tooth replacement in a female three- to four-year-old slow-worm (*Anguis fragilis*). Each horizontal row is a record showing the circumstances at every tooth position, ie, functional tooth present (*open circles*), tooth newly erupted (*black dots*), or vacant space. The records were taken at weekly intervals and are arranged consecutively from the top to the bottom of the page. The suture between maxilla and premaxilla is indicated by a broken vertical line. Zahnreihen are marked by continuous lines and the replacement sequence by *dotted line*. *From Cooper, J.S., 1966. Tooth replacement in the slow-worm* (Anguis fragilis). *J. Zool. Lond. 150, 235–248. Courtesy Editors of the Journal of Zoology, London.*

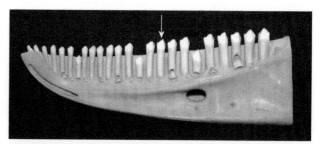

FIGURE 10.16   Model of the lower jaw of a green lizard (*Lacerta viridis*). Tooth locus 16 from front is *arrowed. Courtesy Dr. J. S. Cooper.*

The theory is conveniently described on the basis of a model of the lower jaw in the lizard genus *Lacerta* (Figs. 10.16–10.18; Cooper, 1963). At first sight the arrangement of developing teeth appears random (Fig. 10.16). However, if alternate teeth are compared, a recognizable pattern emerges. Thus, starting for instance at tooth position 16 (*arrow*), the tooth has recently erupted and, as yet, no successor is evident. Progressing anteriorly through tooth positions 14, 12, 10, 8, and 6, the replacing teeth are at younger stages of development. Thus, these teeth will erupt successively in an alternating sequence from the back to the front of the jaw. A similar pattern exists for the teeth in the odd-numbered tooth positions. Fig. 10.17 shows the same model as in Fig. 10.16, but lines showing the waves of tooth replacement passing through alternate tooth positions have been superimposed. According to Edmund's theory, this pattern is set up by stimuli passing from front to back along the jaw, initiating tooth formation at successive tooth loci. The teeth initiated during the progress of a stimulus form a set (each group connected by an oblique solid line in Fig. 10.18), known as a **Zahnreihe** (German: tooth row or series; pl. Zahnreihen).

In Edmund's theory, the interval between successive stimuli, measured in terms of the number of tooth positions apart, can determine the number of teeth in each Zahnreihe and the direction of the subsequent replacement wave. Even though teeth are initiated in a front to back sequence, if the Zahnreihen spacing is greater than two (and less than three) then, after a few generations, replacement waves passing through alternate tooth positions and running from back to front will be established (solid black teeth in Fig. 10.18, passing sequentially through tooth loci 14, 12, 10, 8, 6, and 4). This pattern is by far the most common pattern in reptiles. This is also seen in the dentition of some fish, eg, the rainbow trout (Berkovitz and Moore, 1974). With a Zahnreihe spacing of 2.0, the developing successors will all be at about the same stage of development and almost perfect alternation of waves of replacement would be seen, as in the slow-worm (*A. fragilis*) (Cooper, 1966). With a Zahnreihe spacing of less than two (and greater than one), alternate waves of replacement would pass from front to back of the jaws.

pass through alternate tooth positions from back to front. Unlike Edmund (1962) in alligators and Cooper (1963) in lizards, Kline and Cullum (1984) did not find any differences in the life span of anterior compared with posterior teeth. During the course of the study, mean tooth width in the dentary increased from 0.85 to 1.31 mm, and in the maxilla from 0.88 to 1.54 mm.

The Zahnreihen Theory

In a series of papers, Edmund (1960, 1962, 1969) extended earlier work by Woerdeman (1919, 1921) to formulate an explanation as to how the alternating patterns of tooth replacement in reptiles are established. This became known as the **Zahnreihen** theory. He postulated that, throughout life, tooth-generating impulses pass along the jaws from front to back at regular intervals, stimulating the development of teeth at each successive tooth locus.

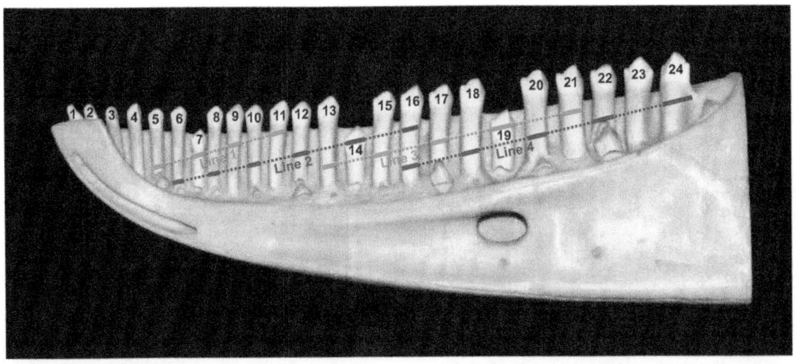

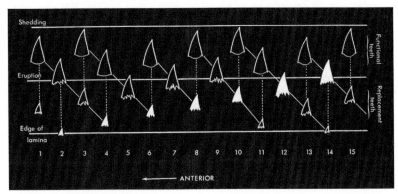

FIGURE 10.17 Same model as in Fig. 10.18, but waves of tooth replacement passing through alternative tooth positions are indicated by *oblique lines*. See text for explanation. *Courtesy Dr. J. S. Cooper.*

FIGURE 10.18 Schematic diagram of a reptilian jaw showing oblique rows of teeth on the dental lamina. The first tooth of the dentition is formed at position 1 and has moved a little way up the dental lamina by the time the second tooth is initiated at position 2. These two teeth both continue to move up the lamina and the third forms at position 3, and so on. In this way, an oblique row of teeth at successive stages of development is formed on the dental lamina, and further similar rows (Zahnreihen), marked by continuous oblique lines, follow one behind the other until the whole dentition is built up. Eventually, a replacement 'wave', proceeding forwards through alternate tooth positions, develops (solid white teeth). *From Cooper, J.S., 1965. Tooth replacement in amphibians and reptiles. Br. J. Herpetol. 6, 214–217. Courtesy Editors of the British Journal of Herpetology.*

Edmund (1960) assumed that, during embryogenesis, the first tooth would be formed at the most anterior position and later teeth in an alternating pattern. Information on the order of tooth development was not available to Edmund but is fundamental to testing the veracity of the Zahnreihen theory. However, subsequent studies in reptiles (described in the next section) have shown that Edmund's theory fails to pass this test (Osborn, 1970). There are examples of dentitions in which teeth are initiated sequentially from front to back of the jaws (eg, elasmobranchs, piranhas) but in most non-mammalian vertebrates this is not the case. Moreover, there is no evidence from molecular biological studies for "waves" stimulating tooth initiation.

ONTOGENY OF THE DENTITION

REPTILES

Studies on the sequence of tooth initiation in lizards (Osborn, 1971; Westergaard, 1988), alligators (Westergaard and Ferguson, 1986, 1987, 1990), slow-worms (Westergaard, 1986, 1988), and the tuatara (Westergaard, 1986) show that

teeth never develop in a continuous sequence from front to back.

A frequent feature of development of the reptilian dentition is the initial formation of numerous rudimentary teeth, some only containing dentine, which are resorbed or which function for only a very short time. These null-generation teeth arise directly from the oral epithelium, before the appearance of a dental lamina (Zahradníček et al., 2012; Fig. 10.19). The first functional teeth develop from a dental lamina and later generations of teeth develop from successional laminae: ingrowths of the dental lamina which usually maintain contact with the surface epithelium.

Osborn (1971) found that the first tooth to be initiated in the lower jaw of the common or viviparous lizard (*Zootoca* [*Lacerta*] *vivipara*) is situated toward the back, at tooth locus 11. Osborn termed this tooth the **dental determinant**. Further teeth were budded off in succession behind this tooth as space became available. In front, however, interstitial growth between teeth allowed for the later development of an intervening row. Thus, an initial row of teeth formed in the odd-numbered positions (ie, 9, 7, 5, and 3), was closely followed by those in the even-numbered positions 10, 8, 6,

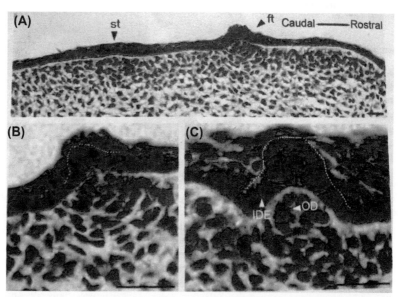

FIGURE 10.19 Development of the null-generation teeth from the oral epithelium in the Madagascan ground gecko (*Paroedura picta*). (A) The first tooth (ft) and second tooth (st) start to form at the oral surface, and the underlying mesenchyme starts to condense at 13 days postovulation; (B) High-power view of the first tooth. The cells of the dental papilla are organized into layers adjacent to the oral epithelium; (C) During later development these teeth shift deeper into mesenchyme and no longer project into the oral cavity. The dental papilla cells adjacent to the inner dental epithelium (IDE) mature into odontoblasts. *IDE*, inner dental epithelium (outlined by *dotted line*); *od*, odontoblast. Scale bar = 10 μm. *From Zahradníček, O., Horacek, I., Tucker, A.S., 2012. Tooth development in a model reptile: functional and null generation teeth in the gecko* Paroedura picta. *J. Anat. 221, 195–208. Courtesy Editors of the Journal of Anatomy.*

4, and 2. A number of these first-generation teeth were rudimentary, developed directly from the surface epithelium and disappeared without becoming functional.

Because his study was based on somewhat incomplete material, Osborn (1993) later acknowledged that a minor change in the initial sequence of development was necessary following constructive comments from Westergaard and Ferguson (1987).

Osborn proposed that the teeth were evenly spaced from each other because a regular zone of inhibition was set up around each developing tooth germ, which temporarily prevented initiation of adjacent teeth. The sequential (as opposed to alternate) initiation of tooth buds behind the dental determinant reflected "tip" growth (rather than interstitial growth). Whereas Osborn considered zones of inhibition to be regular, Weishampel (1991) and Lubkin (1997) produced mathematical models with more variable zones of inhibition. The concept of zones of inhibition during the establishment of the dentition has been utilized in some models (Huysseune and Witten, 2006; Fraser et al., 2008).

In very detailed studies of reptile tooth development in the lower jaw, Westergaard and Ferguson (1986, 1987, 1990) demonstrated that in the American alligator (*A. mississippiensis*), although there are later areas where there is clear evidence of alternation, the overall sequence does not show perfect regularity, especially where the first-formed, rudimentary teeth (which number approximately 19) are concerned. Thus, the initial three tooth germs in the lower

jaw arise in tooth loci 3, 6, and 12, respectively. Westergaard (1980, 1983, 1986) proposed a tooth position theory of tooth development, taking into account the important regional differences in interstitial growth, together with the concepts of inhibition and fields of form determination. Up to five generations of teeth may develop prior to hatching. Although there is no very close correlation, the initiation pattern in the jaws is an approximate indicator of the eruption pattern, which showed waves passing through more or less alternate tooth positions. The first four waves are almost horizontal, the following ones more typically sloping from back to front (Fig. 10.20; Westergaard and Ferguson, 1987, 1990). Kieser et al. (1993) concluded that Zahnreihen with a periodicity of one are easily identifiable in the dentition of the Nile crocodile (*Crocodylus niloticus*).

In contrast to the description of Z. *vivipara* by Osborn (1971), irregular initiation of rudimentary teeth and an anteriorly placed dental determinant have also been observed in the sand lizard (*Lacerta agilis*; Fig. 10.21B), the slow-worm (*A. fragilis*; Fig. 10.21C), and *A. mississipiensis* (Fig. 10.21D). There was evidence of more regular alternation, but from front to back, in the tuatara (*Sphenodon punctatus*; Fig. 10.21A; Westergaard, 1986, 1988). These studies emphasized that interstitial growth was variable and may create space for one or more intervening tooth positions to develop, and it was only with the subsequent formation of the dental lamina that replacement teeth developed at the same positions as their

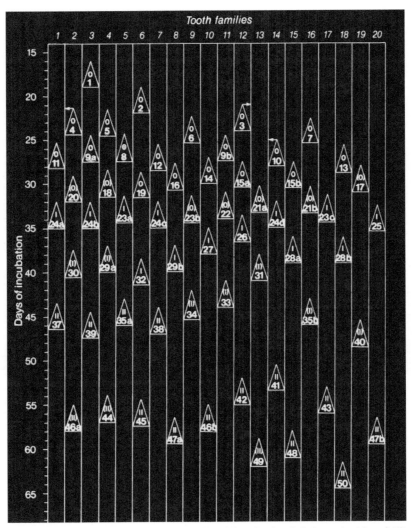

FIGURE 10.20 Chart of the pattern of tooth initiation at functional tooth positions for the whole lower dentition of *Alligator mississippiensis* up to hatching. *Tips of triangles* mark initiation times. Small numbers represent the initiation sequence. 0, resorptive teeth; I, first set of functional teeth; II, second set of functional teeth. *Brackets* indicate variability, as some teeth are either resorptive or transiently functioning teeth. *Arrows* signify that teeth are placed in the direction of the arrow compared to the next tooth generation, tooth families thus being rather arbitrarily defined. *From Westergaard, B., Ferguson, M.W.J., 1987. Development of the dentition in Alligator mississippiensis. later development in the lower jaws of hatchlings and young juveniles. J. Zool. Lond. 212, 191–222. Courtesy Editors Journal Zoology.*

predecessors. Mathematical modeling of the inhibition theory, using data from developmental studies, led Lubkin (1997) to conclude that, in the alligator, the posterior portion of the lower jaw grows much faster than the anterior portion, whereas in the tuatara growth is faster in the anterior portion of the jaw.

Sneyd et al. (1993) proposed a mathematical model for the initiation and spatial positioning in relation to the dental determinant in the alligator. Their model makes use of biological data derived from previous developmental studies (Westergaard and Ferguson, 1986, 1987). The model also incorporates mechanochemical and reaction-diffusion mechanisms and demonstrates that simple, intuitive approaches to pattern formation can be misleading in the study of pattern formation in more complex situations.

As adults, leopard geckos (*Eublepharis macularius*) have bicuspid teeth, but as juveniles they have monocuspid teeth. Handrigan and Richman (2011) found that the change from unicuspid to bicuspid teeth occurred during the third generation of teeth and that it was not the result of folding of the internal enamel epithelium as in mammals, but by a process of differential secretion of enamel.

Formation of a null generation of nonfunctional teeth precedes development of the lower dentition in the Madagascan ground gecko, (*Paroedura picta*; Zahradníček et al., 2012). Fifteen of these teeth develop, the first few directly from the oral epithelium before a dental lamina has formed (Fig. 10.19). Some of the functional teeth form at the positions of the null-generation teeth but the remaining functional tooth positions are established by development of

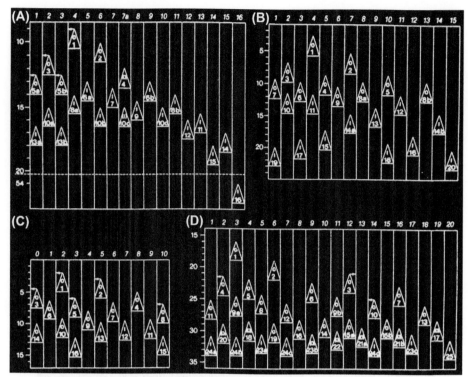

FIGURE 10.21 Diagrams of embryonic tooth initiation (tips of triangles) establishing tooth positions present in the lower jaw at hatching, indicated by *italic numbers* (horizontal axis). *Arrows* signify that teeth are placed in the direction of the *arrow* compared with the next tooth generation, making the tooth family rather arbitrary. Small numbers in triangles represent tooth initiation sequence. Upper numbers: 0 = resorbing teeth; I = first set of functional teeth. *Brackets* indicate variability, as some teeth are either resorbing, transitory/normally functioning, or not formed. (A) Tuatara (*Sphenodon punctatus*). Vertical axis: rough estimate of weeks after egg laying. Position 7a is not present at hatching if tooth 4 is resorbing and tooth 10c is not formed; (B) Sand lizard (*Lacerta agilis*). Vertical axis: days after egg laying; (C) Slow-worm (*Anguis fragilis*). Vertical axis: days after the latest stage without visual odontogenic tissue mobilization. Position 0 is not present at hatching; (D) Alligator (*Alligator mississipiensis*). Vertical axis: days after egg laying. Total tooth initiation pattern up to hatching in Westergaard and Ferguson (1987). *From Westergaard, B., 1988. Early dentition development in the lower jaws of* Anguis fragilis *and* Lacerta agilis. *Soc. Fauna Flora Fenn. Mem. 64, 148–151. Courtesy Editors of Societas pro Fauna et Flora Fannica Memoranda.*

teeth from the dental lamina without any null-generation predecessors (Fig. 10.22). Fig. 10.23 shows how the 31 tooth positions are established by 40 days postovulation. Some of the null-generation teeth are shed as the successional teeth develop and erupt, while others appear to be engulfed by the developing jaw bone. The null-generation teeth have a simple conical shape, while the functional teeth are typically bicuspid.

In the slow-worm an initial series of nonfunctional rudimentary teeth (which undergo resorption) develop in a seemingly complex sequence which does not show alternation. However, the first functional teeth develop rapidly in the odd-numbered positions in an alternating sequence from back to front (Fig. 10.21C; Westergaard, 1986, 1988), which conforms with the alternating pattern of tooth replacement in the adult (Cooper, 1966).

In the monophyodont veiled chameleon (*Chamaeleo calyptratus*) (Buchtová et al., 2013), no rudimentary teeth precede the appearance of the functional dentition. The first tooth to develop is the egg tooth in the middle of the premaxilla. All the teeth arise from a dental lamina and in the maxilla and dentary the teeth arise in an alternating

sequence. Initially the teeth are well spaced, allowing subsequent teeth to develop interstitially. The sequence of ontogeny is similar in the maxilla and dentary. In the maxilla, the first tooth appears at position 9, followed by positions 7, 5, 10, 8, 6, 4, 3, 1, 11, and 2. In the dentary the first tooth to appear is at position 7, followed by positions 5, 3, 8, 6, 4, 2, 1, and 9. A successional lamina develops but is soon resorbed (Fig. 10.24).

In three nonvenomous **snakes**, three generations of teeth were present at hatching, but Buchtová et al. (2008) observed no evidence of the nonfunctional teeth seen in some lizards.

AMPHIBIANS

Aspects of the development of the dentition have already been described for the smooth newt (*Triturus vulgaris meridionalis*) and the axolotl (*Ambystoma mexicanum*; see Chapter 5).

During the development of the dentition of caecilians (Parker and Dunn, 1964; Wake 1976, 1977, 1980), there appear in the fetus rows of specialized, often spoon-shaped

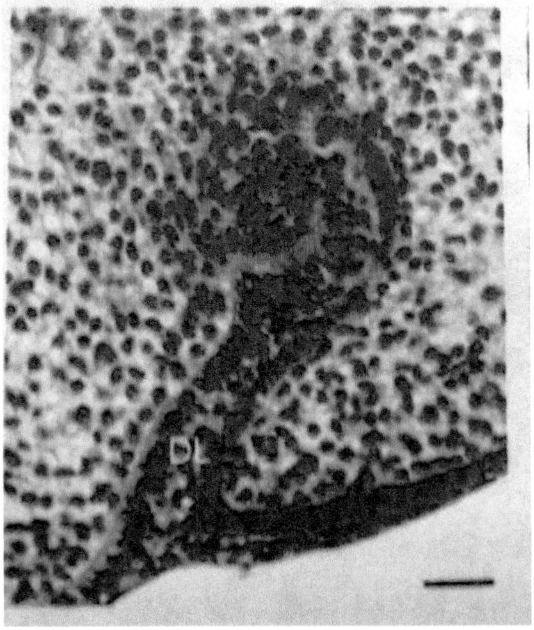

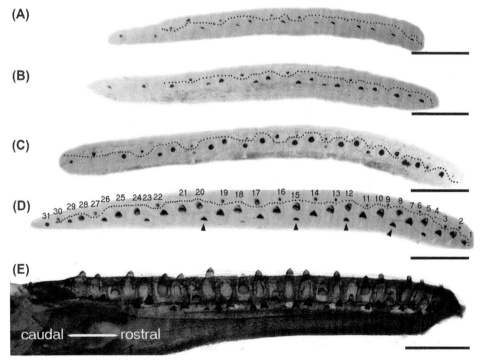

FIGURE 10.22 Developing tooth at the cap stage arising directly from the dental lamina (DL) in the Madagascan ground gecko (*Paroedura picta*). Stain Alcian blue, hematoxylin, and Sirrus red. Scale bar = 100 μm. *From Zahradniček, O., Horacek, I., Tucker, A.S., 2012. Tooth development in a model reptile: functional and null generation teeth in the gecko* Paroedura picta. *J. Anat. 221, 195–208. Courtesy Editors of the Journal of Anatomy.*

teeth used to obtain nutrients from the oviduct wall. In the Mexican burrowing caecilian (*D. mexicanus*), the number of rows on the dentary may increase to eight during fetal development. The development of these functional teeth is preceded by development of a series of tiny, prefunctional teeth with a thin enameloid layer extended into three to seven apical spikes. The crowns are weakly mineralized and the teeth are not ankylosed (Wake, 1980). As the fetus becomes a free-living juvenile, the dentition changes to the adult form and monocuspid teeth appear within the rows and replace the fetal teeth. The rows are reduced to the single rows of the adult.

During the ontogeny of the Asian salamander (*Ranodon sibiricus*), it takes over nine tooth replacements to establish the adult dentition (Vassilieva and Smirnov, 2001).

The *first series* of teeth appear on the palatal bones and coronoids and a little later on the dentaries and premaxillae. The teeth calcify from a single center and become ankylosed to the bone. The teeth are monocuspid, sharply pointed, nonpedicellate and form single rows (Fig. 10.25A).

The teeth of the *second series* are also monocuspid and nonpedicellate. On the premaxillae and dentaries, the teeth of the first and second series alternate, to produce a single row of teeth. On the vomers, palatines, and coronoid, the

FIGURE 10.23 Dental laminae of mandibles at different stages of development in the Madagascan ground gecko (*Paroedura picta*), demonstrating the developing dental pattern. Anterior to the right. Mineralized tooth germs were visualized by alizarin red and viewed from the lingual side, oral side uppermost. (A–D). Dissected dental laminae at (A) 30 days postovulation (dpo), (B) 32 dpo, (C) 35 dpo, and (D) 40 dpo. The dotted line separates the more superficially developing single generation of null-generation tooth germs (above the line) from the deeper functional tooth germs (below the line). *Green arrows* mark functional teeth that will replace the null-generation teeth. *Black arrows* in (D) indicate functional teeth that will replace the first set of functional teeth. The numbering show the tooth positions at hatching. The *black numbers* indicate tooth positions where tooth families are established by functional teeth, and the *blue numbers* indicate tooth positions established by null-generation teeth. (E) Ankylosed teeth in the jaw before hatching at 60 dpo. Scale bar = 500 μm. *From Zahradniček, O., Horacek, I., Tucker, A.S., 2012. Tooth development in a model reptile: functional and null generation teeth in the gecko* Paroedura picta. *J. Anat. 221, 195–208. Courtesy Editors of the Journal of Anatomy.*

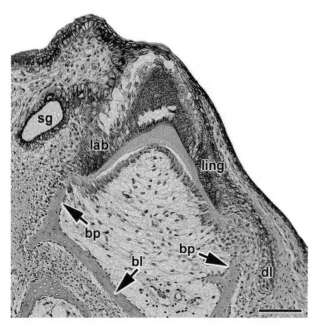

FIGURE 10.24 Tooth in monophyodont dentition of veiled chameleon, showing evidence of a lingual downgrowth that subsequently disappears by apoptosis without developing a successional tooth. *bl*, bone lamella; *bp*, bone pedicles; *dl*, dental lamina; *lab*, labial side of cervical loop; *ling*, lingual side of cervical loop; *sg*, salivary (dental) gland, Scale bar = 0.1 mm. *From Buchtová, M., Zahradniček, O., Balková, S., Tucker, A.S., 2013. Odontogenesis in the veiled chameleon* (Chameleo calypratus)*. Arch. Oral Biol. 58, 118–133. Courtesy Editors of the Archives of Oral Biology.*

second series of teeth lie medial to the first, so these regions bear two rows of teeth.

The *third series* of teeth replace the first series on the premaxillae and dentaries. Although the teeth are still monocuspid, two centers of calcification appear, with the first evidence of a hypomineralized zone, so in this series the teeth are subpedicellate. The tooth row is therefore mixed as it consists of nonpedicellate teeth (second series) alternating with subpedicellate teeth (third series). On the vomers, palatines, and coronoids, extensive resorption occurs until only a single row of nonpedicellate, monocuspid teeth is present (Fig. 10.25B).

The *fourth series* of teeth resemble those of the third series and replace the second series on the premaxillae and dentaries, which now bear rows consisting only of subpedicellate teeth.

The *fifth series* of teeth, which replace the third series on the premaxillae and dentaries, are slightly larger with a complete zone of division and are still monocuspid and subpedicellate. On the vomers, palatines, and coronoid bones, the fifth series of subpedicellate teeth alternate with non-pedicellate teeth of the fourth series.

The *sixth series* of teeth replace the fourth series in the premaxilla and dentary at the start of metamorphosis. They are still monocuspid but have a more defined pedicel at their base (Fig. 10.25B). The palatines and coronoids continue

to resorb while the vomer remodels and contains a row of monocuspid teeth, each with a zone of division.

The *seventh series* of teeth are still monocuspid and replace the fifth series at the loci between the teeth of the sixth series. Similar monocuspid, subpedicellate teeth start to appear for the first time on the maxilla. The vomer dentition is still in the form of a single row of monocuspid teeth with a pronounced hypomineralized zone.

The *eighth series* of teeth on the premaxilla and dentary are bicuspid and pedicellate and are the first of the adult series. The two cusps are the same height and the teeth alternate with monocuspid teeth of the seventh series. The maxillae now bear more teeth than the premaxillae. The vomerine teeth remain monocuspid.

The *ninth series* of teeth on the premaxillae, dentaries (replacing the seventh series) and maxillae are bicuspid but the cusps are of unequal size, the lingual cusp being larger (Fig. 10.25C). Following further remodeling of the vomer, the teeth are aligned transversely and are still monocuspid. With further replacement the full adult dentition of exclusively bicuspid and pedicellate teeth is established (Fig. 10.25D).

Davit-Béal et al. (2006) studied the sequence of tooth initiation and replacement in the Iberian ribbed newt (*Pleuronectes waltl*). Precision was increased by focusing attention on the most anterior tooth on the dentary, the tooth closest to the symphysis. Between its first appearance in the embryo, through its larval and metamorphic phases until sexual maturity at 2 years, six generations of teeth are formed. The total cycle time for each tooth from initiation to replacement increases from 55 days for the first generation to 14 months for the sixth generation (Fig. 10.26). The first-generation tooth develops from the deep surface of the overlying oral epithelium in the absence of a dental lamina (Fig. 10.27). Subsequent tooth generations develop and are connected by a dental lamina arising from the surface epithelium and three or four generations may be present at a time (Fig. 10.28). A dividing zone only becomes clear in the fourth generation, while the originally monocuspid, nonpedicellate tooth changes to a bicuspid form at a similar stage (Fig. 10.29). One unusual finding of this study was that the second-generation tooth does not replace its predecessor but erupts and functions by its side. Although the first-generation tooth is subsequently resorbed, the precise mechanism of this process is unclear.

In the African clawed frog (*Xenopus laevis*), about 20 teeth are present in a single row on the premaxilla and maxilla up to the time of metamorphosis, compared with about 60 teeth in old adults (Shaw, 1979 and personal communication). The teeth develop rapidly in an alternating sequence, commencing at the back of the jaw and passing forward. In the vast majority of cases (55 out of 60), the even-numbered tooth positions develop and erupt first, commencing at position 20. The first set of teeth take 26 days to develop, erupt, and

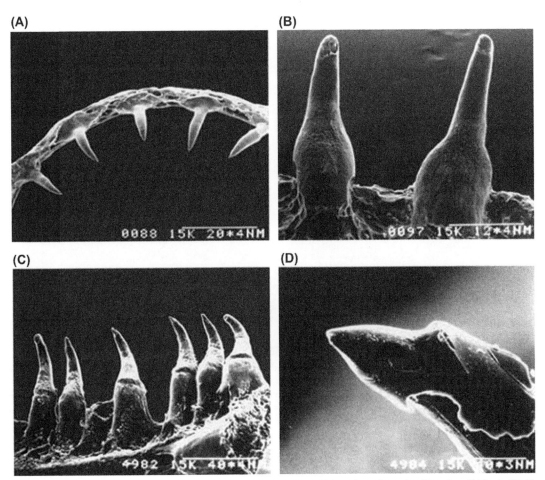

FIGURE 10.25 Gross tooth morphology of the dentary teeth during development of the Asian salamander (*Ranodon sibiricus*). (A) Nonpedicellate monocuspids; (B) subpedicellate monocuspids; (C) pedicellate bicuspids; (D) Enamel cap of the pedicellate bicuspid tooth with large lingual and small labial cusps. *From Vassilieva, A.B., Smirnov, S.V., 2001. Development and morphology of the dentition in the Asian salamander,* Ranodon sibiricus *(Urodela: Hynobiidae). Russ. J. Herpetol. 8, 105–116. Courtesy Editors of the Russian Journal of Herpetology.*

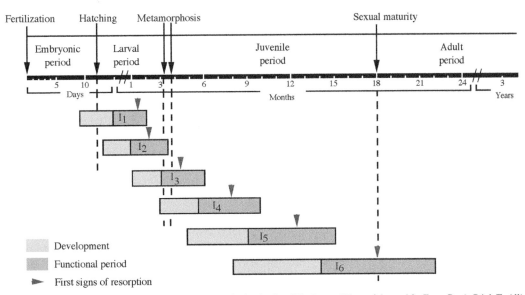

FIGURE 10.26 Table showing stages in life cycle of the first tooth in the Iberian ribbed newt (*Pleurodeles waltl*). *From Davit-Béal, T., Allizard, F., Sire, J.-Y., 2006. Morphological variations in a tooth family through ontogeny in* Pleuronectes waltl *(Lissamphibia, Caudata). J. Morphol. 257, 1048–1065. Courtesy Editors of the Journal of Morphology.*

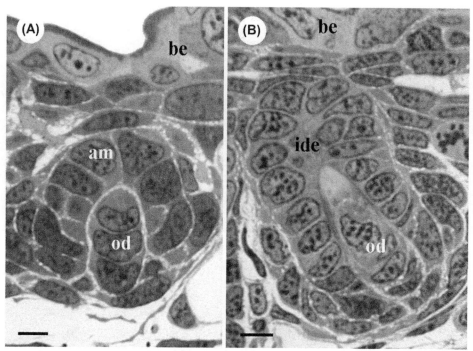

FIGURE 10.27   Micrograph of first-generation tooth in the Iberian ribbed newt (*Pleurodeles waltl*), developing directly from the buccal epithelium (be). In (A) the inner dental epithelium (IDE) is differentiating into ameloblasts (am), while the mesenchyme of the dental papilla is differentiating into odontoblasts (od). At a later stage in (B) the first-formed predentine has been deposited by the odontoblasts (od), while the inner dental epithelial cells (ide) are differentiating into ameloblasts. *be*, buccal epithelium. Scale bar = 10 μm. *From Davit-Béal, T., Allizard, F., Sire, J.-Y., 2006. Morphological variations in a tooth family through ontogeny in* Pleuronectes waltl *(Lissamphibia, Caudata). J. Morphol. 257, 1048–1065. Courtesy Editors of the Journal of Morphology.*

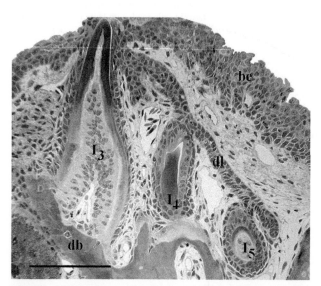

FIGURE 10.28   Micrograph showing three generations of teeth (I₃, I₄, I₅) in the Iberian ribbed newt (*Pleurodeles waltl*), developing directly attached to the dental lamina. *be*, buccal epithelium; *db*, dentary bone. Scale bar = 100 μm. *From Davit-Béal, T., Allizard, F., Sire, J.-Y., 2006. Morphological variations in a tooth family through ontogeny in* Pleuronectes waltl *(Lissamphibia, Caudata). J. Morphol. 257, 1048–1065. Courtesy Editors of the Journal of Morphology.*

become ankylosed, and are functional for 7 days. The main phase of resorption is very rapid, taking only about one day.

BONY FISH

Sequence of Tooth Initiation

Berkovitz (1977a,b) studied tooth development in the **rainbow trout** to determine its relationship with the subsequent development of the alternating pattern of tooth replacement. A true dental lamina is absent, so the teeth arise directly from the oral epithelium, and there are no null-generation teeth like those of reptiles. The first 10 teeth on the dentary develop in an alternate sequence. The teeth in odd-numbered tooth positions develop first and are rapidly followed by the teeth in the even-numbered positions. The teeth beyond tooth position 10 were assumed by Berkovitz (1977b) to appear in succession. The precise sequence of development was 3, 5, 1, 7, 4, 9, 2, 6, 10, 8, 11, and 12. For the upper marginal tooth row (Berkovitz, 1978), the first maxillary tooth appears in position 1 and there is an almost perfect sequence of tooth development passing back through the odd-numbered tooth positions, closely followed by tooth formation at the even-numbered

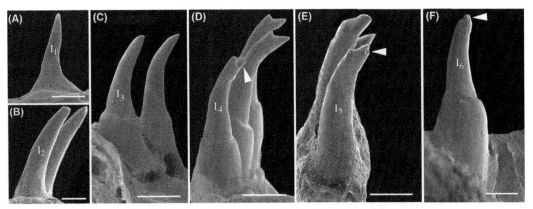

FIGURE 10.29 Scanning electron micrographs showing different morphology of the first tooth closest to the symphysis on the left dentary in the Iberian newt (*Pleuronectes waltl*) at varying stages following tooth succession. (A) Larva, Stage 41 (25 days postfertilization [dpf]); (B) Larva, Stage 48 (50 dpf); (C) Larva, Stage 55a (90 dpf); (D) Juvenile, 6-month old; (E) Juvenile, 12 months old; (F) Adult. The first three generations of teeth (I1 to I3) have only one cusp, while the fourth onwards have two. Scale bars in (A) 20 μm; (B) 50 μm; (C–F) 100 μm. *From Davit-Béal, T., Allizard, F., Sire, J.-Y., 2006. Morphological variations in a tooth family through ontogeny in* Pleuronectes waltl *(Lissamphibia, Caudata). J. Morphol. 257, 1048–1065. Courtesy Editors of the Journal of Morphology.*

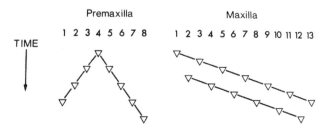

FIGURE 10.30 Diagram showing the order of tooth development in the upper jaw of the rainbow trout (*Oncorhynchus mykiss*). *From Berkovitz, B.K.B., 1978. Tooth ontogeny in the upper jaw and tongue of the rainbow trout (*Salmo gairdneri*). J. Biol. Buccale 6, 205–215. Courtesy Editors of the Journale Biologie Buccale.*

positions (Fig. 10.30). The pattern is closely similar to that in the dentary (although the first lower tooth forms at tooth locus 3) and the initial eruption and replacement pattern follow the sequence of development. However, for the eight teeth in the premaxilla, the sequence of development is entirely different. The first tooth forms at position 4 and the remaining teeth develop in sequence, both in front and behind (Fig. 10.30). As the maxillary and premaxillary teeth both show similar alternating waves of tooth replacement (see Fig. 10.7), it is clear that such replacement waves can be derived from very different sequences of tooth initiation. Whether the difference in initiation sequence between maxillary and premaxillary teeth reflects a different pattern of growth in the underlying bones awaits clarification.

Huysseune and Witten (2006) found that the sequence of tooth initiation in the Atlantic salmon was similar to that seen in the rainbow trout. They also confirmed that the posterior teeth are added sequentially. The sequence is: 5, 3, 1, 7, 4, 6, 2, 8, 9, 10, 11, 12, and 13.

Van der Heyden and Huysseune (2000) have described in detail the development of the pharyngeal dentition of the **zebrafish** (*D. rerio*; Cyprinidae). Of the three tooth rows in the adult, teeth first appear in the ventral row. The order of development for the five teeth in this row alternates. The first tooth to appear is in position 4, followed by position 3 and 5, then position 2, and last of all position 1. Tooth replacement is very rapid, such that the tooth in position 4 is replaced before the tooth in position 2 has developed. The four teeth in the mediodorsal row of the zebrafish start to appear as the tooth in position 1 of the ventral row appears (Van der Heyden and Huysseune, 2000; Huysseune et al., 2005). The first medio-dorsal tooth forms at position 3, followed by positions 4 and 2. The tooth in position 1 is the last tooth to appear in this row, which is at about the same time as that in tooth position 1 of the dorsal row, followed by the second tooth in the dorsal row.

The sequence of tooth initiation in the jewel cichlid (*H. bimaculatus*) resembles that in the rainbow trout and salmon in that, after formation of the first few teeth, additional teeth develop in alternate positions (Huysseune and Witten, 2006). However, the first teeth to form occupy even-numbered rather than odd-numbered positions. The sequence is: 4, 2, 6, 5, 7, 3, 1, 8, 9, and 10.

Huysseune and Witten (2006) traced the establishment of the tooth replacement pattern in three species. In the zebrafish, an alternating sequence appears quickly in the ventral row, with replacement at positions 2 and 4, followed by 1, 3, and 5, symbolized as (2–4)–(1–3–5). However, over time this changes to (5–2)–3–(1–4), so the alternate sequence is lost. In the Atlantic salmon, the initial sequence of tooth initiation leads directly in most cases to a replacement pattern with every third tooth at the same stage of development. In these species, it is important to note that teeth which are replaced at the same time do not all belong to the same generation. In the jewel cichlid the replacement pattern is alternate at first, although somewhat irregular, and becomes increasingly irregular.

Loss of Teeth During Growth

In a number of bony fish, the juvenile forms possess teeth but these are not replaced and the adults are edentulous. Examples include the Curimatidae (Characiformes) (Vari, 1989), armored catfish, eg, *Corydoras* spp. and *Hoplosternum littorale* (Huysseune and Sire, 1997), and the swordfish (*Xiphias*; see Chapter 4). Juvenile stages of the Mekong giant catfish (*Pangasianodon gigas*; Siluriformes) possess a marginal row of simple, conical teeth on the upper and lower jaws, on the palate, and in the pharynx. There is only a single tooth replacement. The maximum number of functional teeth is 25 in a 9.5-cm long fish. By the time the fish has reached a length of 17 cm, only four functional teeth remain and no teeth are present at a length of 24 cm (Kakizawa and Meenakarn, 2003).

In Chilodontidae and Prochilodontidae (Characiformes), which are mud feeders like the curimatids, the first generation of teeth are attached to the jaw bones. These are replaced many times but these numerous later generations remain embedded within the soft tissues of the lips, without being attached to bone (Castro and Vari, 2004).

Changes in Tooth Form During Growth

There are many examples among bony fishes of a correlation between tooth replacement and a change in tooth morphology with age. This phenomenon has been reported most frequently among characiforms.

In the Mexican tetra (*A. mexicanus*; Characidae), the adult dentition comprises two rows of multicuspid teeth on the premaxilla (four to five anteriorly and five to seven posteriorly), one or two multicuspid teeth on the maxilla and on the dentary a single row of four anterior (rostral) teeth, and three to five posterior (caudal) teeth, which are variable in size and may have one or several cusps (Atukorala et al., 2013; Fig. 10.31B). Its development has been studied by Trapani et al. (2005). Maxillary teeth appear much later than those of the premaxilla and dentary. The sequence of tooth initiation is summarized in Fig. 10.32. It is irregular and differs between premaxilla (6, 3, 8, 1, 5, 10, 7, 2, 4, and 9) and dentary (3, 5, 1, 6, 4, 8, 10, 2, 7, and 9). The first two generations of teeth develop extraosseously and are monocuspid (Fig. 10.31A). These teeth are replaced in an irregular sequence which does not correspond to the order of initiation. From the second replacement cycle onwards, replacement teeth develop intraosseously and are multicuspid. Initially successional teeth are bi- or tricuspid but, with further replacements, teeth can have up to six cusps. The multicuspid teeth arise from single enamel organs and not by fusion of discrete enamel organs as had previously been

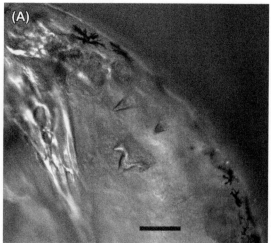

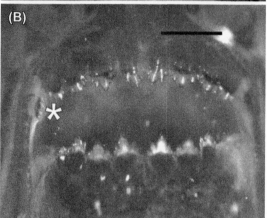

FIGURE 10.31 Transitional and adult dentitions in Mexican tetra (*Astyanax mexicanus*). (A) Live fish 65 days postfertilization (dpf), 16.9 mm total length (TL), showing premaxilla with the first multicuspid teeth developing prior to replacing conical predecessors. Scale bar = 100 μm; (B) cleared-and-stained adult multicusped teeth of fish 163 dpf, 41.8 mm TL. *Asterisk* indicates maxillary tooth. Scale bar = 1 mm. *From Trapani, J., Yamamoto, Y., Stock, D.W., 2005. Ontogenetic transition from unicuspid to multicuspid oral dentition in a teleost fish:* Astyanax mexicanus, *the Mexican tetra (Ostariophysi: Characidae). Zool. J. Linn. Soc. 145, 523–538. Courtesy Editors of the Zoological Journal of the Linnean Society.*

suggested. From the third replacement cycle onwards, the premaxilla has two rows of teeth, and the premaxillary teeth and the rostral dentary teeth are replaced simultaneously, although jaw quadrants are not synchronized. The caudal teeth on the dentary, however, continue to develop extraosseously and are replaced in an irregular pattern. A striking feature in the Mexican tetra is the very large increment in size between replacement teeth and their predecessors. The first generation of multicuspid teeth can be up to five times larger than their monocuspid precursors, and there can be a two- to two-and-a-half-fold increase at other replacements.

In contrast to the Mexican tetra, the earliest teeth in tigerfish (*Hydrocynus*) are conical but are replaced by tricuspid teeth when the fish is about 14 mm long. Young specimens

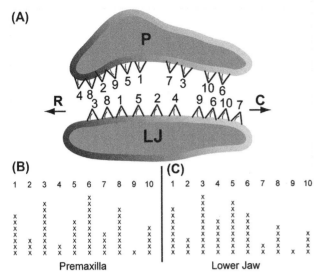

FIGURE 10.32 Order of appearance of the first-generation dentition on one-half (right) of the jaw in *Astyanax mexicanus*. (A) Schematic of the premaxillary and lower jaw. Order is indicated by numbers; (B) graph of tooth appearance patterns on the premaxilla. Numbers represent tooth positions and time runs vertically down; (C) graph of tooth appearance patterns on the lower jaw. *C*, caudal; *LJ*, lower jaw; *P*, premaxilla; *R*, rostral. *From Trapani, J., Yamamoto, Y., Stock, D.W., 2005. Ontogenetic transition from unicuspid to multicuspid oral dentition in a teleost fish:* Astyanax mexicanus, *the Mexican tetra (Ostariophysi: Characidae). Zool. J. Linn. Soc. 145, 523–538. Courtesy Editors of the Zoological Journal Linnean Society.*

have a functional dentition with a seemingly random mixture of tricuspid, bicuspid, and unicuspid teeth (Roberts, 1967; Brewster, 1986). Ultimately all teeth are replaced by a small number of large, conical teeth in the adult (Brewster, 1986; Gagiano et al., 1996).

Roberts (1967) described changes in tooth morphology accompanying tooth replacement in other characid fish. In the gold-banded pencilfish (*Nannostomus harrisoni*), young of 9 mm length have simple conical teeth, while older specimens longer than 30 mm have the characteristic five- or six-cusped teeth. Fishes of intermediate size show a mixture of tooth shapes. In adult sharktail distichodus (*Distichodus fasciolatus*) the adult typically has a double row of bicuspid teeth in both jaws, while the inner row of teeth in the lower jaw of young specimens consists of conical teeth.

Pufferfishes (Tetraodontidae) possess a unique oral dentition of only four stacks of elongated teeth (laminae; Britski et al., 1985), one stack in each quadrant, that are cemented together to form a beak-like structure (Chapter 4). A study of the Mekong puffer (*Monotrete suvattii*) (Fraser et al., 2012) showed that a first generation of tooth germs similar to those seen in other bony fish is formed. In the lower jaw this eventually consists of 14–18 small conical teeth, which form in the sequence 2, 3, 1, 4, 5, 6, 7, 8, 9, 10... The first generation of teeth in the upper jaw is limited to four to

six teeth. The pulps of these first-generation teeth are filled with dentine and fused together at their bases. The laminar teeth that form the typical parrotfish beak only appear as the successors to the rows of first-generation teeth. Fraser et al. suggested that replacement occurs only at the four parasymphysial tooth positions. These teeth enlarge, and successional laminae appear on their lingual side from which develops an elongated laminar tooth of dentine surmounted by enameloid. The successional laminae express the conserved key regulators of tooth initiation, namely sonic hedgehog (SHH), Pitx-2, and Bmp-4. Continued successional lamina downgrowth results in a stacked series of tooth laminae typical of these pufferfishes. In other words, the successor of the small first-generation tooth at the front of each jaw quadrant expands posteriorly to form a single elongated laminar tooth extending the length of the jaw. However, further research on this interesting dentition would be valuable. For instance, whereas Fraser et al. assumed the existence of a successional lamina, Pflugfelder (1930) observed that the tooth germ for each successive element buds off from the oral epithelium basal to the beak.

The reader is referred to Chapter 4 for a description of developmental plasticity in the dentition of a cichlid species.

CHONDRICHTHYES

Tooth development in sharks and rays proceeds with rows of teeth developing usually in alternating positions, although in some sharks (eg, frilled shark) the rows are aligned. There is considerable variation in the speed at which teeth are added along the row. Teeth initially develop in the symphysial region.

Batoidea

Tooth development patterns have been investigated and reviewed by Underwood et al. (2015) and are illustrated diagrammatically in Fig. 10.33. In *Myliobatis*, tooth development is identical in upper and lower jaws. Initially there are two rudimentary parasymphysial teeth in the first row. The second tooth row comprises three teeth, all of which are larger than those in the first row, with the addition of a symphysial tooth. The third row lacks a symphysial tooth and comprises four tooth files. No further proximal tooth files are added, with successional tooth rows gradually enlarging. The dentition, therefore, comprises two alternate series of teeth: one of three teeth, including an enlarged symphysial tooth, and the other of four teeth. The heterodonty in the dentition does not appear until the third generation, when the teeth are still widely spaced.

In the specialized *Aetobatus*, the first generation of teeth comprise two parasymphysial teeth. The second and subsequent generations consist only of a single symphysial tooth,

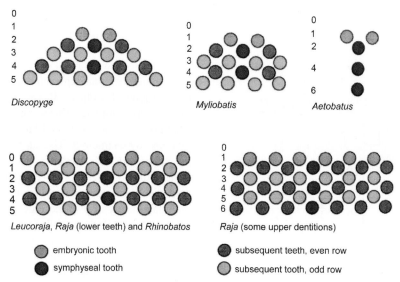

FIGURE 10.33 Diagram showing patterns of development in Batoidea. See text for explanation. *From Underwood, C.J., Johanson, Z., Welten, M., Metscher, B., Rasch, L.J., Fraser, G.J., Smith, M.M., 2015. Development and evolution of dentition pattern and tooth order in the skates and rays (Batoidea; Chondrichthyes). PLoS One 10, e0122553. doi:10.1371. Courtesy Editors of the Public Library of Science (PLoS).*

which subsequently enlarges to form the single file of the adult.

The first teeth to appear in embryos of the torpediniform *Discopyge tschudii*, as in *Myliobatis,* are two parasymphysial teeth. The second row comprises three teeth: a symphysial tooth and two teeth either side of this. Subsequent tooth rows alternate in a similar pattern to the first two, with rows bearing a symphysial tooth alternating with rows lacking this tooth but having a pair of parasymphysial tooth. In the third row, an additional tooth is added on proximally to the end of the row. Each successive tooth row has an additional tooth added to the back end until a final number is reached.

In contrast to *Myliobatis* and *Discopyge*, the first row of teeth in *Rhinobatos* are numerous, extend along the whole length of the jaws, and contain a symphysial tooth. The second row, alternating with the first, lacks a symphysial tooth. As the full adult number of teeth appears to be present at this initial stage, alternation occurs in successive replacements without further addition to the tooth row. Some variation to this pattern was observed in some specimens of *Raja clavata*, where the first row of developing teeth in one jaw had a symphysial tooth while the first row in the opposing jaw lacked a symphysial tooth (Underwood et al., 2015).

Selachii

The general process of tooth replacement in sharks is similar to that in rays, with symphysial teeth appearing very early in development and teeth developing in alternate rows (Reif, 1976, 1982, 1984; Underwood et al., 2015). Initially, there is a row or two of rudimentary teeth and, depending on

the complexity of crown morphology, it may take a number of generations before the adult form appears. This may also apply to the symphysial teeth. Whereas rays have a bilaterally symmetrical dentition, the slope of cusps on one side of shark dentitions is the mirror image of that on the opposite side (Fig. 2.46). In sharks such as the frilled shark (*Chlamydoselachus*), where the files of teeth are spaced apart, the developing and functional teeth in adjacent files are aligned within rows (see Fig. 2.32).

The first generation of teeth in the rostrum of saw-sharks comprises 23 pairs but these are recumbent and become erect and fixed at their bases only after birth. With growth, a second set of smaller teeth (about one-quarter the size of the first set) appears between the members of the initial set. With further growth, a third set appears that is about one-half to two-thirds the size of the initial set of teeth. With eruption of the teeth, a sequence of different-sized teeth appears along the rostrum. Initially this is fairly regular (long-intermediate-short-intermediate-long) but, with loss of teeth and replacement, a less regular sequence of tooth size results (Slaughter and Springer, 1968; Welten et al., 2015).

MOLECULAR CONTROL OF ONTOGENY OF THE DENTITION AND TOOTH REPLACEMENT

The problems of how the number, arrangement, and variations in form of teeth in a dentition are determined have attracted research attention for many decades. Older theories, such as the field model of Butler (1956), have not found supporting evidence. As shown earlier, the predictions made by the Zahnreihen theory of Edmund are not confirmed by observation. The modern molecular biological approach to the problem, although in its early stages, is proving much more fruitful.

The chief experimental model for studying how teeth develop has been the monophyodont rodent dentition, consisting of continuously growing incisors and molars of limited growth, as outlined in Chapter 9. This model offers the hope of understanding pathologies, such as missing teeth (hypodontia), supernumerary teeth (hyperdontia), abnormally large teeth (macrodontia), abnormally small teeth (microdontia), deficiencies of mineralization (amelogenesis imperfecta, dentinogenesis imperfecta), and noneruption (cleidocranial dysostosis; Berkovitz et al., 2009). However, it has limitations for the study of wider aspects of the dentition, such as the integration of morphogenesis and tooth succession. Consequently, studies of patterning and growth of the dentition have extended to a greater number and variety of vertebrates.

Patterning of the Dentition in Mammals

In mammals, tooth number and position seem to be controlled by interaction between the growth factors SHH, EDA, FGF, WNT, and BMP and their respective antagonists (Cobourne and Sharpe, 2010). In the diastema region of the mouse dentition only rudiments of teeth, in which SHH is expressed transiently at low levels, normally develop and are resorbed. However, if SHH is expressed experimentally at higher levels, these rudiments can complete development and to form supernumerary teeth. The activity of SHH in the diastema region seems to be controlled by the epithelium (Cobourne et al., 2004). WNT is important to initiation of tooth development. If the expression of *Wnt* is inhibited, tooth morphogenesis is arrested at an early stage. Conversely, upregulation of *Wnt*, either by depleting endogenous inhibitors such as APC (Wang et al., 2009) or LRP4 (Porntaveetus et al., 2012), or by forced activation of WNT/β-catenin signaling (Järvinen et al., 2006), results in formation of supernumerary teeth.

The distribution of teeth with different morphologies (incisor, molar, etc.) is thought to be determined by regionally differentiated expression of homeobox genes in the mesenchyme, such that different tooth types develop in association with particular combinations of homeobox genes (the **odontogenic homeobox code**; Sharpe, 2001). Non-Hox homeobox genes code for DNA-binding proteins that regulate gene transcription and thus control the expression of other genes necessary for the development of a particular structure (in this case a tooth). Several homeobox genes, including *Msx1*, *Msx2*, *Dlx1*, *Dlx2*, *Barx1*, and *Alx3*, are expressed by tooth germ ectomesenchyme in response to epithelial signals during early odontogenesis. According to the odontogenic

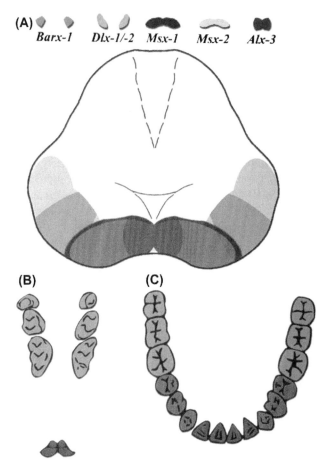

FIGURE 10.34 Diagrams to illustrate the mesenchymal odontogenic homeobox gene code. (A) Lower jaw of mouse embryo. Note the overlap (orange) between the domains of *Msx1* (red) and *Msx2* (yellow). (B) Model representing the mouse dentition. Note that molars develop from cells expressing *Barx1* and incisors from cells expressing *Msx1*, *Msx2*, and *Alx3*. (C) Model representing the human dentition, predicting that incisors develop from cells expressing *Msx1*, *Msx2*, and *Alx3*, canines and premolars from cells expressing *Msx1* and *Msx2*, and molars from cells expressing *Barx1*. *From McCullum, M., Sharpe, P.T., 2001. Evolution and development of teeth. J. Anat. Lond. 199, 153–159. Courtesy Editors of the Journal of Anatomy.*

homeobox code hypothesis, tooth form is dictated by the combination of homeobox genes expressed by the ectomesenchyme of the tooth germ. In the mouse, incisors form in the anterior region of the jaws, where *Msx1,2* (but not *Barx1*) is expressed, while molars form in the posterior region, where *Barx1*, *Dlx1,2*, and *Pitx1* (but not *Msx1,2*) are expressed (Fig. 10.34). The regional differences in the homeobox profile appear to be established by the epithelium by the thickening/placode stage, through the distribution of epithelial growth factors. Thus, *Bmp4*, which upregulates *Msx1* and *Msx2* but inhibits *Barx1*, is expressed anteriorly, while *Fgf8*, which stimulates expression of *Barx1* and *Dlx2*, is expressed in the future molar region.

The homeobox code seems to differ between the mandible and the maxilla. The expression of the different forms of the *Dlx* gene varies between different regions of the jaws (Zhao et al., 2000). When *Dlx1* and *Dlx2* gene expression is disrupted, mandibular molars still develop but maxillary molars do not (Sharpe, 2001).

The odontogenic homeobox code hypothesis is supported by various experiments which explore the effect on the shape of a developing tooth of alterations in activities of the relevant genes (Catón and Tucker, 2009). For instance, knockout of *Dlx1,2* results in loss of maxillary molars without effect on incisor formation. In early incisor tooth germs cultured in the presence of noggin protein (a BMP antagonist), so that epithelial BMP cannot switch on *Msx1* in the ectomesenchyme, not only is *Msx1* downregulated in the mesenchyme but Barx-1 (normally only present in the molar region) is expressed. When these tooth germs are transplanted to renal capsules and allowed to continue development, multicuspid molariform teeth, rather than incisiform teeth, can be produced (Tucker et al., 1998).

Although the homeobox code has been used to explain regional differentiation of heterodont dentitions, a similar code exists in birds, which have not possessed teeth for 100 million years. It appears that the role of the code in patterning the dentition is an extension of its original function, which was the patterning of the bones of the jaws and middle ear (Catón and Tucker, 2009).

Patterning of the Dentition in Non-Mammalian Vertebrates

Fraser et al. (2008), working with cichlid fish, proposed that teeth form within an odontogenic band: a field rendered competent for tooth initiation by expression of shh and pitx. Spacing of teeth is achieved by interactions between shh, eda, and wnt7b, which creates a zone of inhibition around each tooth locus. According to this model, multiple rows of teeth are formed lingual to the first row(s) by a "copy-and-paste" mechanism.

A number of genetic markers seen in the developing first generation of teeth are also present during tooth replacement. For example, in the rainbow trout, in which successional teeth originate extraosseously from the outer layer of the dental epithelial sheath of its predecessor, *pitx2* is expressed in the basal oral epithelium during initiation of the first generation of teeth and also in the outer dental epithelium of this tooth at the site of formation of the successional tooth. *Bmp4* is coexpressed in the adjacent mesemchyme, indicating the importance of these two molecules in tooth initiation and replacement (Fraser et al., 2006). An important finding is the absence of *shh* from the outer dental epithelium of the replacement teeth.

Most studies of the molecular control of tooth replacement have been carried out in amphibians or reptiles, in which replacement teeth develop from a persistent dental lamina or a successional lamina, or in bony fish with intraosseous replacement, where the teeth develop at the tip of an epithelial strand that appears to be equivalent to a successional lamina (see Chapter 9). It is postulated that in each group a population of stem cells in the oral epithelium adjacent to the functional tooth supplies cells for formation of tooth-forming laminae (Tucker and Fraser, 2014). A stem-cell cluster has been identified in a corresponding position in elasmobranchs (Tucker and Fraser, 2014). The variation in the pattern of tooth initiation observed among bony fish (Chapter 9) may require some modification of this general hypothesis.

Replacement teeth develop from the surface of a dental or successional lamina facing the predecessor tooth. As the lamina is usually lingual to the tooth row, the labial surface is usually where teeth form. The opposite (lingual) surface shows a greater rate of cell proliferation. It is, therefore, not surprising that the two epithelial surfaces also show different patterns of gene expression. The early signaling agents seen in mammals are also present in non-mammalian vertebrates. *shh* and its receptor *ptc1* are preferentially expressed on the lingual side of the dental lamina, while *edar* and various *wnt* pathway genes are preferentially expressed on the labial side. There are asymmetries in ectomesenchymal gene expression. Among squamates, *bmp2* is more highly expressed on the lingual side under the oral ectoderm whereas *bmp4* is more evenly expressed (Richman and Handrigan, 2011; Whitlock and Richman, 2013).

In the developing teeth of nonvenomous snakes, expression of *shh* is at first confined to the odontogenic band and defines the position of the future dental lamina (Buchtová et al., 2008). *Shh* is expressed in the inner enamel epithelium and stellate reticulum but not by the external dental epithelium or its derivative, the successional lamina. In squamates, the successional lamina lacks Pct-1, the receptor of SHH, confirming that pathways other than SHH may regulate tooth replacement (Handrigan and Richman, 2010a,b).

Continuous tooth replacement clearly requires a permanent niche of pluripotent stem cells in the dental lamina and/or the associated dental mesenchyme. Slow-cycling stem cells were found in the leopard gecko (*E. macularius*) on the lingual surface of the successional lamina. These stem cells were WNT responsive and expressed several stem-cell markers (eg, *lgr5*, *dkk3*, and *igfbp5*; Handrigan et al., 2010).

Studies of the distribution of Sox-2, a transcription factor essential for self-renewal and a marker for stem cells, have shown that stem cells are present in the dental lamina and successional lamina in snakes, alligators, geckos and iguanas (Juuri et al., 2013). Abduweli et al. (2014) identified stem cells in the pharyngeal epithelium adjacent to the posterior end of each family of pharyngeal teeth in the medaka (*O. latipes*; Fig. 10.35).

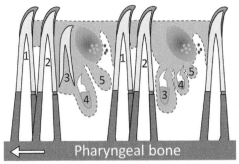

FIGURE 10.35 Diagram illustrating tooth families and putative odontogenic stem-cell niches in the pharyngeal dentition of the medaka (*Oryzias latipes*). The teeth are organized in well-orientated tooth families, each having up to five generations of functional teeth (1 and 2) and successional tooth germs (3–5) at different developmental stages and/or tooth-replacing cycles. Maintenance of slow-cycling dental epithelial and Sox-2-positive cells (*light green dots*) at the posterior end of each tooth family (*dark zone*), and possible interactions between these and adjacent mesoderm-containing putative mesodermal stem cells (*red dots*) may contribute to continuous tooth replacement. The *arrow* indicates the rostral direction. *From Abduweli, D., Baba, O., Tabata, M.J., Higuchi, K., Mitani, H., Takano, Y., 2014. Tooth replacement and putative odontogenic stem cell niches in pharyngeal dentition of medaka* (Oryzias latipes). *Microscopy 63, 141–153. Courtesy Editors of Microscopy.*

There is evidence that the WNT signaling pathway is important in regulating tooth replacement. The very tip of the successional dental lamina in reptiles shows high cell proliferation and also increased WNT and BMP signaling (Richman and Handrigan, 2011). In the ball python (*Python regius*), these WNT-active cells at the extending tip of the dental lamina could represent the immediate descendants of the putative stem cells housed in the lingual face of the dental lamina (Handrigan and Richman, 2010a,b).

During normal development, WNT/β-catenin signaling is concentrated at the lef1-positive successional lamina, with sox2-positive stem cells located slightly further back in the lamina. Evidence for a possible role of active WNT/β-catenin signaling in elongation of the successional lamina and orderly production of the next tooth generation by an increase in proliferation rate was obtained in the **corn snake** (*Pantherophis guttatus*) (Gaete and Tucker, 2013). An increase in the level of signaling in culture resulted in increased cell proliferation and disordered initiation of supernumerary teeth in ectopic positions.

In alligators, putative stem cells become distributed in the distal end of the dental lamina where replacement teeth are initiated (Wu et al., 2013). During development of the teeth, shh is expressed in the surface and lingual oral epithelium, and later in the inner dental epithelium. The distribution of β-catenin (a multifunctional molecule involved in cell–cell adhesion, signaling, and the maintenance of stem cells) is present localized to the nucleus in the distal end of the dental lamina immediately prior to tooth initiation. Increasing WNT activity in explants of tooth-containing segments of the alligator jaw increases cell proliferation

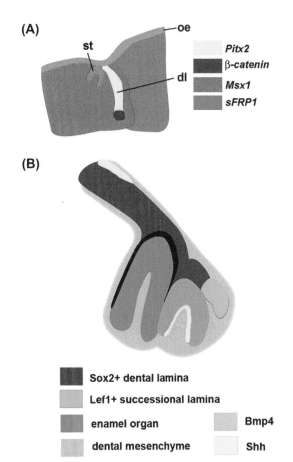

FIGURE 10.36 (A) The distribution of various signaling agents associated with tooth replacement in the alligator (*From Wu, P., Wu, X., Jiang, T.X., Elsey, R.M., Temple, B.L., Divers, S.J., Glenn, T.C., Yuan, K., Chen, M.H., Widelitz, R.B., Chuong, C.M., 2013. Specialized stem cell niche enables repetitive renewal of alligator teeth. Proc. Natl. Acad. Sci. U.S.A. 110, E2009–E2018, Courtesy Editors of the Proceedings of the National Academy of Sciences, USA.*). (B) Diagram showing the distribution of various additional signaling agents for tooth replacement in reptiles (*Courtesy Prof. A. S. Tucker.*).

dramatically. As soluble, frizzled-related protein1 (sFRP1), a WNT pathway antagonist, is present in the preinitiation stage dental lamina, but absent following the appearance of a successional tooth, dental lamina initiation may require reduced *wnt* expression. Deliberate extraction of a functional tooth speeds up development of its successor, with an associated increase in WNT signaling, while inhibition of WNT signaling blocks the growth of replacement teeth by inducing apoptosis in cells of the dental papilla (Wu et al., 2013). As canonical (β-catenin-dependent) WNT signaling pathway is found in the mesenchyme on the buccal side of the dental lamina, these two gene networks may help determine the position of the functional tooth. Fig. 10.36A illustrates the distribution of various signaling agents associated with tooth replacement in the alligator (Wu et al., 2013), while Fig. 10.36B illustrates the distribution of other important signaling agents for tooth replacement known in other reptiles (Tucker, personnel communication).

Although these studies indicate that canonical (β-catenin-dependent) WNT signaling pathway is very likely to be important in regulating tooth replacement, it should be noted that in one study this pathway was reduced in zebrafish without affecting the eruption pattern of the pharyngeal teeth, leading the authors to conclude that WNT control may be quite complex and with possible species differences in its control (Huysseune et al., 2014).

An intriguing sidelight on the control of tooth replacement came from a study on a bony fish. Tuisku and Hildebrand (1994) studied the effect of unilateral section of part of the mandibular nerve in the spotted tilapia (*Tilapia mariae*). The normal interval between tooth shedding and replacement was rapid (5–7 days), while the functional life of a tooth was about 100 days, regardless of size. Following unilateral sectioning of the inferior alveolar nerve, functional teeth were not replaced after 100 days but remained in place. Occasional teeth lost on the denervated side were not replaced and 300 days after denervation, the tooth count was 25% lower on the denervated side. When teeth were extracted on both control and denervated sides, those on the control side continued to be replaced but not those on the denervated side. These results imply that nerves may have

a role in maintaining tooth initiation and replacement. In agreement with this the parasympathetic nervous system has been shown to be a source of mesenchymal stem cells in mouse incisors (Kaukua et al., 2014). Nerves may therefore play an important role in both supporting stem cells and providing a source of stem cells for the tooth.

Morphogenesis and replacement of teeth have several signaling pathways in common (Jernvall and Thesleff, 2012). Fraser et al. (2013) highlighted the central role in these processes of five signaling pathways: BMP, FGF, Hedgehog, Notch, and WNT/β-catenin. These pathways, expressed in intricate temporal and spatial networks, determine the size and spacing of teeth, control one-for-one replacement of teeth, and form new teeth repetitively throughout the life of the animal. However, there is at present no information on how the timing of tooth formation and replacement are controlled. This aspect of the problem is clearly crucial to the adaptation of the dentition to function.

Polyphyodonty Versus Diphyodonty

It is of interest to compare polyphyodont dentitions with the diphyodont or monophyodont dentitions of mammals. Fig.

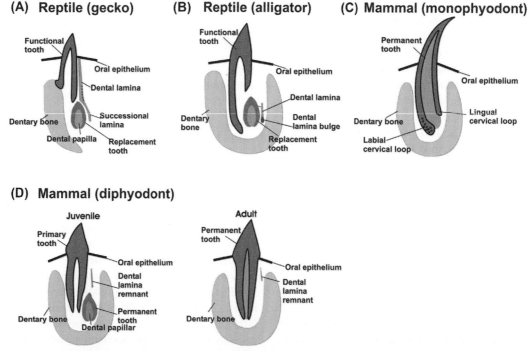

FIGURE 10.37 Comparison of stem-cell niches in adult gecko, alligator, and mammalian teeth. (A) Gecko: dental lamina is linked to the oral epithelium. LRCs (red dots) are diffusely distributed in the lingual portion of the dental lamina but excluded from their distal end (successional lamina). (B) Adult alligator: dental lamina is not linked to the oral epithelium. LRCs are located in the distal tip of the dental lamina at the preinitiation stage. (C) Adult mice lack a dental lamina for tooth renewal. In incisors, stem cells for continual growth are located in the cervical loop. (D) Diphyodont mammals (including humans) have only two sets of teeth. After formation of the crown of a permanent tooth, the dental lamina degenerates. *From Wu, P., Wu, X., Jiang, T.X., Elsey, R.M., Temple, B.L., Divers, S.J., Glenn, T.C., Yuan, K., Chen, M.H., Widelitz, R.B., Chuong, C.M., 2013. Specialized stem cell niche enables repetitive renewal of alligator teeth. Proc. Natl. Acad. Sci. U.S.A. 110, E2009–E2018. Courtesy Editors of the Proceedings of the National Academy of Sciences, USA.*

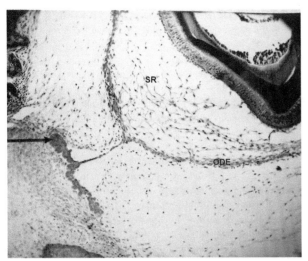

FIGURE 10.38 Micrograph of demineralized section showing a downgrowth (*arrow*) lingual to the upper permanent second premolar of the marsupial quokka (*Setonix brachyurus*). Image width = 1.1 mm. *ODE*, outer dental epithelium; *SR*, stellate reticulum. *From Berkovitz, B.K.B., 1966. The homology of the premolar teeth in* Setonix brachyurus *(Macropodidae: Marsupialia). Arch. Oral Biol. 11, 1371–1384. Courtesy Editors of the Archives of Oral Biology.*

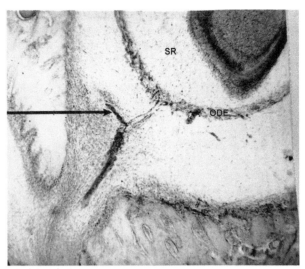

FIGURE 10.39 Micrograph of demineralized section showing a downgrowth (*arrow*) lingual to the developing upper third premolar of a cat. Image width = 2.0 mm. *ODE*, outer dental epithelium; *SR*, stellate reticulum. *From Berkovitz, B.K.B., 1966. The homology of the premolar teeth in* Setonix brachyurus *(Macropodidae: Marsupialia). Arch. Oral Biol. 11, 1371–1384. Courtesy Editors of the Archives of Oral Biology.*

10.37 compares the feature of stem-cell niches in the two types of dentition (Wu et al., 2013). In mammals, the dental lamina connecting teeth to the oral epithelium and connecting the teeth to each other is not persistent but breaks down, leaving the teeth isolated in the mesenchyme. Three hypotheses to explain why no more than one replacing generation of teeth develops can be formulated:

1. A successional lamina does not appear from any of the permanent teeth.
2. A successional lamina appears by the lingual side of the permanent tooth but degenerates because of apoptosis and/or lack of stem cells or signaling molecules necessary to generate a second series of teeth.
3. A successional dental lamina appears by the lingual side of the permanent tooth with the necessary attributes to potentially form a second-generation tooth but there is a lack of responsive ectomesemchyme to form a second series of teeth.

The available evidence indicates that the first hypothesis is incorrect, and that one of the other two hypotheses is correct. At least some of the permanent teeth of mammals, both in marsupials and placentals, possess histological evidence of what appears to be the beginnings of a putative dental lamina (Figs. 10.38 and 10.39; Berkovitz, 1966; Dosedělová et al., 2015). However, this lingual downgrowth remains vestigial and never gives rise to a third generation of teeth. A transient, rudimentary lamina has also been observed on the lingual aspect of the molars of the monophyodont mouse (Dosedělová et al., 2015).

REFERENCES

Abduweli, D., Baba, O., Tabata, M.J., Higuchi, K., Mitani, H., Takano, Y., 2014. Tooth replacement and putative odontogenic stem cell niches in pharyngeal dentition of medaka (*Oryzias latipes*). Microscopy 63, 141–153.

Atukorala, A.D.S., Hammer, C., Dufton, M., Franz-Odendaal, T.A., 2013. Adaptive evolution of the lower jaw dentition in Mexican tetra (*Astyanax mexicanus*). EvoDevo 4, 28.

Bemis, W.E., Giuliano, A., McGuire, B., 2005. Structure, attachment, replacement and growth of teeth in bluefish, *Pomatomus saltatrix* (Linnaeus, 1766), a teleost with deeply socketed teeth. Zoology 108, 317–327.

Berkovitz, B.K.B., Moore, M.H., 1974. A longitudinal study of replacement patterns of teeth on the lower jaw and tongue in the rainbow trout *Salmo gairdneri*. Arch. Oral Biol. 19, 1111–1119.

Berkovitz, B.K.B., Moore, M.H., 1975. Tooth replacement in the upper jaw of the rainbow trout (*Salmo gairdneri*). J. Exp. Zool. 193, 221–234.

Berkovitz, B.K.B., Shellis, R.P., 1978. A longitudinal study of tooth succession in piranhas (Pisces: Characidae), with an analysis of the replacement cycle. J. Zool. Lond. 184, 545–561.

Berkovitz, B.K., Holland, G.R., Moxham, B.J., 2009. Oral Anatomy, Histology and Embryology, fourth ed. Mosby, London.

Berkovitz, B.K.B., 1966. The homology of the premolar teeth in *Setonix brachyurus* (Macropodidae: Marsupialia). Arch. Oral Biol. 11, 1371–1384.

Berkovitz, B.K.B., 1975. Observations on tooth replacement in piranhas (Characidae). Arch. Oral Biol. 20, 53–56.

Berkovitz, B.K.B., 1977a. Chronology of tooth development in the rainbow trout (*Salmo gairdneri*). J. Exp. Zool. 200, 65–70.

Berkovitz, B.K.B., 1977b. The order of tooth development and eruption in the rainbow trout (*Salmo gairdneri*). J. Exp. Zool. 201, 221–226.

Berkovitz, B.K.B., 1978. Tooth ontogeny in the upper jaw and tongue of the rainbow trout (*Salmo gairdneri*). J. Biol. Buccale 6, 205–215.

Berkovitz, B.K.B., 1980. The effects of age on tooth replacement patterns in piranhas (Pisces: Characidae). Arch. Oral Biol. 25, 833–835.

Betancur-R, R., Wiley, E., Bailly, N., Miya, M., Lecointre, G., Ortí, G., 2014. Phylogenetic Classification of Bony Fishes – Version 3. https://sites.google.com/site/guilleorti/home/classification.

Boyne, P.J., 1970. Study of the chronologic development and eruption of teeth in elasmobranchs. J. Dent. Res. 49, 556–560.

Brewster, B., 1986. A review of the genus *Hydrocynus* Cuvier (Teleostei: Chariformes). Bull. Br. Mus. Nat. Hist. (Zool) 50, 163–206.

Britski, H.A., Andreucci, R.D., Menezes, N.A., Carneiro, J.J., 1985. Coalescence of teeth in fishes. Rev. Bras. Zool. 2, 459–482.

Buchtová, M., Handrigan, G.R., Tucker, A.S., Lozanoff, S., Town, L., Fu, K., Diewert, V.M., Wicking, C., Richman, J.M., 2008. Initiation and patterning of the snake dentition are dependent on sonic hedgehog signalling. Dev. Biol. 319, 132–145.

Buchtová, M., Zahradniček, O., Balková, S., Tucker, A.S., 2013. Odontogenesis in the veiled chameleon (*Chameleo calypratus*). Arch. Oral Biol. 58, 118–133.

Butler, P.M., 1956. The ontogeny of molar pattern. Biol. Rev. 31, 30–70.

Castro, R.M.C., Vari, R.P., 2004. Detritivores of the South American fish family Prochilodontidae (Teleostei: Ostariophysi: Characiformes): a phylogenetic and revisionary study. Smithson. Contr. Zool. 622, 1–188.

Catón, J., Tucker, A.S., 2009. Current knowledge of tooth development: patterning and mineralization of the murine dentition. J. Anat. 214, 502–515.

Cobourne, M.T., Sharpe, P.T., 2010. Making up the numbers: the molecular control of mammalian dental formula. Semin. Cell Dev. Biol. 21, 314–324.

Cobourne, M.T., Miletich, I., Sharpe, P.T., 2004. Restriction of sonic hedgehog signalling during early tooth development. Development 131, 2875–2885.

Cooper, J.S., Poole, D.F.G., Lawson, R.A., 1970. The dentition of agamid lizards with special reference to tooth replacement. J. Zool. Lond. 162, 85–98.

Cooper, J.S., 1963. Dental Anatomy of the Genus *Lacerta* (Ph.D. thesis). University of Bristol.

Cooper, J.S., 1966. Tooth replacement in the slow-worm (*Anguis fragilis*). J. Zool. Lond. 150, 235–248.

Davit-Béal, T., Allizard, F., Sire, J.-Y., 2006. Morphological variations in a tooth family through ontogeny in *Pleuronectes waltl* (Lissamphibia, Caudata). J. Morphol. 257, 1048–1065.

Delgado, S., Davit-Béal, T., Sire, J.-V., 2003. Dentition and tooth replacement pattern in *Chalcides* (Squamata; Scincidae). J. Morphol. 256, 146–159.

DeMar, R., 1972. Evolutionary implications of Zahnreihen. Evolution 26, 435–450.

DeMar, R., 1973. The functional implications of the geometrical organizations of dentitions. J. Paleontol. 47, 452–461.

Dosedělová, H., Dumková, J., Lesot, H., Glocová, K., Kunová, M., Tucker, A.S., Veselá, I., Krejčí, P., Tichý, F., Hampl, A., Buchtová, M., 2015. Fate of the molar dental lamina in the monophyodont mouse. PLoS One 10 (5), e0127543. http://dx.doi.org/10.1371/journal.pone.0127543.

Edmund, A.G., 1960. Tooth replacement phenomena in the lower vertebrates. Contr. R. Ont. Mus. Life Sci. 52, 1–190.

Edmund, A.G., 1962. Sequence and rate of tooth replacement in the crocodile. Contr. R. Ont. Mus. Life Sci. 56, 1–42.

Edmund, A.G., 1969. Dentition. In: Gans, C., Bellairs Ad', A., Parsons, T.S. (Eds.), Biology of the Reptiles. Academic Press, New York, pp. 117–200.

Erickson, G.M., 1996. Daily apposition of dentine in juvenile *Alligator* and assessment of tooth replacement rates using incremental line counts. J. Morphol. 228, 189–194.

Evans, H.E., Deubler, E.E., 1955. Pharyngeal tooth replacement in *Semotilus atromaculatus* and *Clinostomus elongates*, two species of cyprinid fishes. Copeia 31–41.

Fraser, G.J., Berkovitz, B.K.B., Graham, A., Smith, M.M., 2006. Gene deployment for tooth replacement in the rainbow trout (*Oncorhynchus mykiss*): a developmental model for evolution of the osteichthyan dentition. Evol. Dev. 8, 446–457.

Fraser, G.J., Bloomquist, R.F., Streelman, J.T., 2008. A periodic pattern generator for dental diversity. BMC Biol. 6, 32.

Fraser, G.J., Britz, R., Hall, A., Johanson, Z., Smith, M.M., 2012. Replacing the first-generation dentition in pufferfish with a unique beak. Proc. Natl. Acad. Sci. U.S.A. 109, 8179–8184.

Fraser, G.J., Bloomquist, R.F., Streelman, J.T., 2013. Common pathways link tooth shape to regeneration. Dev. Biol. 377, 399–414.

Gaete, M., Tucker, A.S., 2013. Organised emergence of multiple generations of teeth in snakes is dysregulated by activation of Wnt/beta catenin signalling. PLoS One 8 (9), e74484. http://dx.doi.org/10.1371/journal.pone.0074484.

Gagiano, C.I., Steyn, G.J., Du Preez, H.H., 1996. Tooth replacement of tigerfish *Hydrocynus vittatus* from the Kruger National Park. Koedoe 39, 117–122.

Gillette, R., 1955. The dynamics of continuous succession of teeth in the frog (*Rana pipiens*). Am. J. Anat. 96, 1–36.

Goin, C.J., Hester, M., 1961. Studies on the development, succession and replacement in the teeth of the frog *Hyla cinerea*. J. Morphol. 109, 279–287.

Graver, H.T., 1978. Re-regeneration of lower jaws and the dental lamina in adult urodeles. J. Morphol. 157, 269–280.

Handrigan, G.R., Richman, J.M., 2010a. A network of Wnt, hedgehog and BMP signaling pathways regulates tooth replacement in snakes. Dev. Biol. 348, 130–141.

Handrigan, G.R., Richman, J.M., 2010b. Autocrine and paracrine Shh signalling are necessary for tooth morphogenesis, but not tooth replacement in snakes and lizards (Squamata). Dev. Biol. 337, 171–186.

Handrigan, G.R., Richman, J.M., 2011. Unicuspid and bicuspid tooth crown formation. J. Exp. Zool. Mol. Dev. Evol. 316, 598–608.

Handrigan, G.R., Leung, K.J., Richman, J.M., 2010. Identification of putative dental epithelial stem cells in a lizard with life-long tooth replacement. Development 137, 3545–3549.

Hilton, E.J., Bemis, W.E., 2005. Grouped tooth replacement in the oral jaws of the tripletail, *Lobotes surinamensis* (Perciformes: Lobotidae), with a discussion of its proposed relationship to Datnioides. Copeia 665–672.

Holmbakken, N., Fosse, G., 1973. Tooth replacement in *Gadus callarias*. Z. Anat. Entwicklungsges 143, 65–79.

Howes, R.I., Eakers, E.C., 1984. Augmentation of tooth and jaw regeneration in the frog with a digital transplant. J. Dent. Res. 63, 670–674.

Howes, R.I., 1978. Regeneration of ankylosed teeth in the adult frog premaxilla. Acta Anat. 101, 179–186.

Huysseune, A., Sire, J.-Y., 1997. Structure and development of teeth in three armoured catfish, *Corydoras aeneus, C. arcuatus* and *Hoplosternum littorale* (Siluriformes, Callichthyidae). Acta Zool. 78, 69–84.

Huysseune, A., Delgado, S., Witten, P.E., 2005. How to replace a tooth: fish(ing) for answers. Oral Biosci. Med. 213, 75–81.

Huysseune, A., Witten, P.E., 2006. Developmental mechanisms underlying tooth patterning in continuously replacing osteichthyan dentitions. J. Exp. Zool. Mol. Dev. Evol. 306B, 204–215.

Huysseune, A., Soenens, M., Elderweirdt, F., 2014. Wnt signaling during tooth replacement in zebrafish (*Danio rerio*): pitfalls and perspectives. Front. Physiol. 5, 386. http://dx.doi.org/10.3389/fphys.2014.00386.

Ifft, J.D., Zinn, D.J., 1948. Tooth succession in the smooth dogfish *Mustelus canis*. Biol. Bull. 95, 100–106.

Järvinen, E., Salazar-Ciudad, I., Birchmeier, W., Taketo, M.M., Jernvall, J., Thesleff, I., 2006. Continuous tooth generation in mouse is induced by activated epithelial Wnt/Beta-catenin signaling. Proc. Natl. Acad. Sci. U.S.A. 103, 18627–18632.

Jernvall, J., Thesleff, I., 2012. Tooth shape formation and tooth renewal: evolving with the same signals. Development 139, 3487–3497.

Juuri, E., Jussila, M., Seidel, K., Holmes, S., Wu, P., Richman, J., Heikinheimo, K., Chuong, C.M., Arnold, K., Hochedlinger, K., Klein, O., Michon, F., Thesleff, I., 2013. Sox2 marks epithelial competence to generate teeth in mammals and reptiles. Development 140, 1424–1432.

Kakizawa, Y., Meenakarn, W., 2003. Histogenesis and disappearance of the teeth of the Mekong giant catfish, *Pangasianodon gigas* (Teleostei). Oral Sci. 45, 213–221.

Kakizawa, Y., Kajiyama, N., Nagai, K., Kado, K., Fujita, M., Kashiwaka, Y., Imai, C., Hirama, A., Yorioka, M., 1986. The histological structure of the upper and lower teeth in the gobiid fish *Sicyopterus japonicus*. J. Nihon Univ. Sch. Dent. 28, 175–187.

Kaukua, N., Shahidi, M.K., Konstantinidou, C., Dyachuk, V., Kaucka, M., Furlan, A., An, Z., Wang, L., Hultman, I., Ahrlund-Richter, L., Blom, H., Brismar, H., Lopes, N.A., Pachnis, V., Suter, U., Clevers, H., Thesleff, I., Sharpe, P.T., Ernfors, P., Fried, K., Adameyko, I., 2014. Glial origin of mesenchymal stem cells in a tooth model system. Nature (London) 513, 551–554.

Kerr, T., 1960. Development and structure of some actinopterygian and urodele teeth. Proc. Zool. Soc. Lond. 133, 401–422.

Kieser, J.A., Klapsidis, C., Law, L., Marion, M., 1993. Heterodonty and patterns of tooth replacement in *Crocodylus niloticus*. J. Morphol. 218, 198–201.

Kline, L.W., Cullum, D.R., 1984. A long-term study of the tooth replacement phenomenon in the young green iguana, *Iguana iguana*. J. Herpetol. 18, 176–185.

Kline, L.W., Cullum, D.R., 1985. Tooth replacement and growth in the young green iguana, *Iguana iguana*. J. Morphol. 186, 265–269.

Kupfer, A., Müller, H., Antoniazzi, M.M., Jared, C., Greven, H., Nussbaum, R.A., Wilkinson, M., 2006. Parental investment by skin feeding in caecilian amphibian. Nature (London) 440, 926–929.

Lawson, R., Wake, D.B., Beck, N.T., 1971. Tooth replacement in the redbacked salamander, *Plethodon cinereus*. J. Morphol. 134, 59–269.

Lawson, R., 1966. Tooth replacement in the frog *Rana temporaria*. J. Morphol. 119, 233–240.

Lubkin, S.R., 1997. On pattern formation in reptilian dentition. J. Theor. Biol. 186, 145–157.

Luer, C.L., Blum, P.C., Gilbert, P.W., 1990. Rate of tooth replacement in the nurse shark *Ginglymostoma cirratum*. Copeia 182–191.

McCullum, M., Sharpe, P.T., 2001. Evolution and development of teeth. J. Anat. Lond. 199, 153–159.

McDowell, S.B., Bogert, C.M., 1954. The systematic position of *Lanthanotus* and the affinities of the anguinomorph lizards. Bull. Mus. Comp. Zool. Harv. 105, 1–142.

Miller, W.A., Radnor, C.J.P., 1970. Tooth replacement patterns in young *Caiman scleros*. J. Morphol. 130, 501–510.

Miller, W.A., Radnor, C.J.P., 1973. Tooth replacement patterns in the bowfin (*Amia calva*: Holostei). J. Morphol. 140, 381–395.

Miller, W.A., Rowe, D.J., 1973. Preliminary investigation of variations in tooth replacement in adult *Necturus maculosus*. J. Morphol. 140, 63–76.

Mochizuki, K., Fukui, S., 1983. Development and replacement of upper jaw teeth in gobiid fish *Siryopterus japonicus*. Jap. J. Ichthyol. 30, 27–36.

Mochizuki, K., Fukui, S., Gultneh, S., 1991. Development and replacement of teeth on jaws and pharynx in a gobiid fish *Sitydium plumieri* from Puerto Rico, with comments on resorption of upper jaw teeth. Nat. Hist. Res. 1, 41–52.

Morgan, E.C., King, W.K., 1983. Tooth replacement in king mackerel, *Scomberomorus cavalla* (Pisces: Scombridae). Southwest Nat. 28, 261–269.

Moss, S.A., 1967. Tooth replacement in the lemon shark, *Negaprion brevirostris*. In: Gilbert, P.W., Mathewson, R.F., Rall, D.P. (Eds.), Sharks, Skates and Rays. Johns Hopkins University Press, Baltimore, pp. 319–329.

Moss, S.A., 1972. Tooth replacement and body growth rates in the smooth dogfish, *Mustelus canis* (Mitchill). Copeia 808–811.

Nakajima, T., 1979. The development and replacement pattern of the pharyngeal dentition in a Japanese cyprinid fish, *Gnathopogon coerulescens*. Copeia 22–28.

Nakajima, T., 1984. Larval vs adult pharyngeal dentition in some Japanese cyprinid fishes. J. Dent. Res. 63, 1140–1146.

Nakajima, T., 1987. Development of pharyngeal dentition in the cobitid fishes, *Misgurnus anguillicaudatus* and *Cobitis biwae* with a consideration of evolution of cypriniform dentitions. Copeia 208–213.

Norris, K.S., Prescott, J.H., 1959. Jaw structure and tooth replacement in the opeleye, *Girella nigricans* (Ayers) with notes on other species. Copeia 275–283.

Osborn, J.W., 1970. New approach to Zahnreihen. Nature (London) 225, 343–346.

Osborn, J.W., 1971. The ontogeny of tooth succession in *Lacerta vivipara* Jacquin (1787). Proc. R. Soc. Lond. B179, 261–289.

Osborn, J.W., 1972. On the biological improbability of Zahnreihen as embryological units. Evolution 26, 601–607.

Osborn, J.W., 1974. On the control of tooth replacement in reptiles and its relationship to growth. J. Theor. Biol. 46, 509–527.

Osborn, J.W., 1975. Tooth replacement: efficiency, patterns and evolution. Evolution 29, 80–186.

Osborn, J.W., 1993. A model simulating tooth morphogenesis without morphogens. J. Theor. Biol. 65, 429–445.

Parker, H.W., Dunn, E.R., 1964. Dentitional metamorphosis in the amphibia. Copeia 75–86.

Pflugfelder, O., 1930. Das Gebiss der Gymnodonten. Ein Beitrag zur Histogenese des Dentins Z. Anat. EntwGesch 93, 543–566.

Poole, D.F.G., Shellis, R.P., 1976. Eruptive tooth movements in non-mammalian vertebrates. In: Poole, D.F.G., Stack, M.V. (Eds.), The Eruption and Occlusion of Teeth (Colston Papers No. 27). Butterworths, London, pp. 65–69.

Porntaveetus, T., Ohazama, A., Choi, H.Y., Herz, J., Sharpe, P.T., 2012. Wnt signaling in the murine diastema. Eur. J. Orthod. 34, 518–524.

Reif, W.-E., McGill, D., Motta, P., 1978. Tooth replacement rates of the sharks *Triakis semifasciata* and *Ginglymostoma cirratum*. Zool. Jahrb. Anat. 99, 151–156.

Reif, W.-E., 1976. Morphogenesis, pattern formation and function of the dentition of *Heterodontus* (Selachii). Zoomorphologie 83, 1–47.

Reif, W.-E., 1982. Evolution of dermal skeleton and dentition in vertebrates: the odontode-regulation theory. Evol. Biol. 15, 287–368.

Reif, W.-E., 1984. Pattern regulation in shark dentitions. In: Malacinski, G.M. (Ed.), Pattern Formation a Primer in Developmental Biology. Macmillan, New York, pp. 603–621.

Richman, J.M., Handrigan, G.R., 2011. Reptilian tooth development. Genesis 49, 247–260.

Rieppel, O., 1978. Tooth replacement in anguimorph lizards. Zoomorphologie 91, 77–90.

Roberts, T.R., 1967. Tooth formation and replacement in characoid fishes. Stanf. Ichthyol. Bull. 4, 231–247.

Sharpe, P.T., 2001. Neural crest and tooth morphogenesis. Adv. Dent. Res. 15, 4–7.

Shaw, J.P., 1979. The time scale of tooth development and replacement in *Xenopus laevis* (Daudin). J. Anat. 129, 323–342.

Shaw, J.P., 1985. Tooth replacement in adult *Xenopus laevis* (Amphibia: Anura). J. Zool. Lond. 207, 171–179.

Shaw, J.P., 1989. Observations on the polyphyodont dentition of *Hemiphractus proboscideus* (Anura: Hylidae). J. Zool. Lond. 217, 499–510.

Shellis, R.P., Berkovitz, B.K.B., 1976. Observations on the dental anatomy of piranhas (Characidae), with special reference to tooth structure. J. Zool. Lond. 180, 69–84.

Sire, J.-Y., Davit-Béal, T., Delgado, S., van der Heyden, C., Huysseune, A., 2002. First-generation teeth in nonmammalian lineages: evidence for a conserved ancestral character? Micr. Res. Tech. 59, 408–434.

Slaughter, B., Springer, S., 1968. Replacement of rostral teeth in sawfishes and sawsharks. Copeia 499–506.

Sneyd, J., Atri, A., Ferguson, M.J., Lewis, M.A., Seward, W., Murray, J.D., 1993. A model for the spatial patterning of tooth primordia in the alligator: initiation of the dental determinant. J. Theor. Biol. 165, 633–658.

Springer, S., 1960. Natural history of the sandbar shark *Eulamia milberti*. U.S. Fish. Wildl. Serv. Fish. Bull. 61, 1–38.

Strasburg, D.W., 1963. The diet and dentition of *Isistius brasiliensis*, with remarks on tooth replacement in other sharks. Copeia 33–40.

Streelman, J.T., Webb, J.F., Albertson, R.C., Kocher, T.D., 2003. The cusp of evolution and development: a model of cichlid tooth shape diversity. Evol. Dev. 5, 600–608.

Trapani, J., Yamamoto, Y., Stock, D.W., 2005. Ontogenetic transition from unicuspid to multicuspid oral dentition in a teleost fish: *Astyanax mexicanus*, the Mexican tetra (Ostariophysi: Characidae). Zool. J. Linn. Soc. 145, 523–538.

Trapani, J., 2001. Position of developing replacement teeth in teleosts. Copeia 35–51.

Tucker, A., Fraser, G.J., 2014. Evolution and developmental diversity of tooth regeneration. Semin. Cell Dev. Biol. 25, 71–80.

Tucker, A.S., Matthews, K.L., Sharpe, P.T., 1998. Transformation of tooth type induced by inhibition of BMP signaling. Science 282, 1136–1138.

Tuisku, P., Hildebrand, C., 1994. Evidence for a neural influence on tooth germ generation in a polyphyodont species. Dev. Biol. 165, 1–9.

Underwood, C.J., Johanson, Z., Welten, M., Metscher, B., Rasch, L.J., Fraser, G.J., Smith, M.M., 2015. Development and evolution of dentition pattern and tooth order in the skates and rays (Batoidea; Chondrichthyes). PLoS One 10, e0122553 doi:10.1371.

Van der Heyden, C., Huysseune, A., 2000. Dynamics of tooth formation and replacement in the zebrafish (*Danio rerio*) (Teleostei, Cyprinidae). Dev. Dynam 219, 486–496.

Vari, R.P., 1989. A phylogenetic study of the neotropical characiform family Curimatidae (Pisces: Ostariophysi). Smithson. Contr. Zool. 471, 1–71.

Vassilieva, A.B., Smirnov, S.V., 2001. Development and morphology of the dentition in the Asian salamander, *Ranodon sibiricus* (Urodela: Hynobiidae). Russ. J. Herpetol. 8, 105–116.

Wake, M.H., 1976. The development and replacement of teeth in viviparous caecilians. J. Morphol. 148, 33–64.

Wake, M.H., 1977. Fetal maintenance and its evolutionary significance in the Amphibia: Gymnophiona. J. Herpetol. 11, 379–386.

Wake, M.H., 1980. Fetal tooth development and adult replacement in *Dermophis mexicanus* (Amphibia: Gymnophiona): fields versus clones. J. Morphol. 166, 203–216.

Wakita, M., Itoh, K., Kobayashi, S., 1977. Tooth replacement in the teleost fish *Prionurus rnicrolepidotus* Lacépède. J. Morphol. 153, 129–142.

Wang, X.P., O'Connell, D.J., Lund, J.J., Saad, I., Kuraguchi, M., Turbe-Doan, A., Cavallesco, R., Kim, H., Park, P.J., Harada, H., Kucherlapati, R., Maas, R.L., 2009. Apc inhibition of Wnt signaling regulates supernumerary tooth formation during embryogenesis and throughout adulthood. Development 136, 1939–1949.

Wass, R.C., 1973. Size, growth, and reproduction of the sandbar shark, *Carcharhinus milberti*, in Hawaii. Pacific. Sci. 27, 305–318.

Weishampel, D.B., 1991. A theoretical morphological approach to tooth replacement in lower vertebrates. In: Schmidt-Kitter, N., Vogel, K. (Eds.), Constructional Morphology and Evolution. Springer, Berlin, pp. 295–310.

Welten, M., Smith, M.M., Underwood, C., Johanson, Z., 2015. Evolutionary origins and development of saw-teeth on the sawfish and sawshark rostrum (Elasmobranchii; Chondrichthyes). R. Soc. Open Sci. 2, 150189.

Westergaard, B., Ferguson, M.W.J., 1986. Development of the dentition in *Alligator mississippiensis*. Early embryonic development in the lower jaw. J. Zool. Lond. 210, 575–597.

Westergaard, B., Ferguson, M.W.J., 1987. Development of the dentition in *Alligator mississippimsis*. later development in the lower jaws of hatchlings and young juveniles. J. Zool. Lond. 212, 191–222.

Westergaard, B., Ferguson, M.W.J., 1990. Development of the dentition in *Alligator mississippiensis*: upper jaw dental and craniofacial development in embryos. hatchlings, and young juveniles, with a comparison to lower jaw development. Am. J. Anat. 187, 393–421.

Westergaard, B., 1980. Evolution of the mammalian dentition. Soc. Geol. Fr. Mem. N. S. 139, 191–200.

Westergaard, B., 1983. A new detailed model for mammalian dentitional evolution. Z. Zool. Syst. Evol. Forsch. 21, 68–78.

Westergaard, B., 1986. The pattern of embryonic tooth initiation in reptiles. Teeth revisited (Proc VIIth Int Symp Dent Morphol) In: Russel, D.E., Santori, J.-P., Sigogneau-Russel, D. (Eds.), Mém Mus. Hist. Nat. Ser. C. 53, 55–63.

Westergaard, B., 1988. Early dentition development in the lower jaws of *Anguis fragilis* and *Lacerta agilis*. Soc. Fauna Flora Fenn. Mem. 64, 148–151.

Whitlock, J.A., Richman, J.M., 2013. Biology of tooth replacement in amniotes. Int. J. Oral Sci. 5, 66–70.

Woerdeman, M.W., 1919. Beitrage zur Entwicklungsgeschichte von Zähnen und Gebiss der Reptilien I. Die Anlage und Entwicklung des embryonalen Gebisses als Ganzes und seine Beziehung zur Zahnleiste. Arch. Mikr. Anat. 92, 104–192.

Woerdeman, M.W., 1921. Beitrage zur Entwicklungsgeschichte von Zähnen und Gebiss der Reptilien IV. Uber die Anlage und Entwicklung der Zähne. Arch. Mikr. Anat. 95, 265–395.

Wu, P., Wu, X., Jiang, T.X., Elsey, R.M., Temple, B.L., Divers, S.J., Glenn, T.C., Yuan, K., Chen, M.H., Widelitz, R.B., Chuong, C.M., 2013. Specialized stem cell niche enables repetitive renewal of alligator teeth. Proc. Natl. Acad. Sci. U.S.A. 110, E2009–E2018.

Zahradniček, O., Horacek, I., Tucker, A.S., 2008. Viperous fangs: development and evolution of the venom canal. Mech. Dev. 125, 786–796.

Zahradniček, O., Horacek, I., Tucker, A.S., 2012. Tooth development in a model reptile: functional and null generation teeth in the gecko *Paroedura picta*. J. Anat. 221, 195–208.

Zhao, Z., Weiss, K.M., Stock, D.W., 2000. Development and evolution of dentition patterns and their genetic basis. In: Teaford, M.F., Smith, M.M., Ferguson, M.W.J. (Eds.), Development, Function and Evolution of Teeth. University Press Cambridge, pp. 152–172.

Chapter 11

Dentine and Dental Pulp

Both dentine and dental pulp are derivatives of the dental papilla, the ectomesenchymal component of the tooth germ. Dentine is produced by the odontoblasts at the periphery of the dental papilla and the dental pulp is derived from the central region of the papilla.

In the functional mammalian tooth, the dental pulp is crucial to maintaining the integrity of the tooth by supporting and providing nutrients for the odontoblasts, by mounting reparative responses to injury or damage, and by immune responses to infection. This support is essential because, as the tooth cannot be replaced more than once, it has to be viable for most of the life of a mammal. As the dental pulp and the dentine are so interdependent in mammalian teeth, they are considered to form a single functionality entity, the **dentine–pulp complex**. In non-mammalian vertebrates, the functions of the dental pulp, other than dentine formation, remain largely obscure. In these vertebrates, the teeth are usually replaced at intervals throughout life, so it is possible that the support functions of the pulp are much more limited than in mammals or even absent, but this question can be answered only through future research.

Dentine is a composite material with two main components. The major component (about 70% by weight of the tissue) is a mineral phase consisting of finely divided crystals of calcium phosphate. The second component (about 20% by weight of the tissue) is an organic matrix of which 90% consists of a protein, Type I collagen, which is organized into fibers. The remainder of dentine (about 10% by weight) is water, which fills the pores within the tissue. There are several types of dentine but the differences between them are, as far as is known, histological rather than compositional. By far the most widely distributed type of dentine is **orthodentine**, which forms a substantial layer surrounding an undivided pulp (Fig. 11.1) and is permeated by enormous numbers of minute tubules containing cytoplasmic processes of the odontoblasts (Figs. 11.1 and 11.2). In the other varieties of dentine, the dentine layer can be folded (**plicidentine**), subdivided (**osteodentine**), or permeated by blood capillaries (**vasodentine**). Tooth structure has been studied in only a fraction of vertebrate genera, so it is not possible to make fully accurate statements about the distribution of each type of dentine.

In this chapter, we first describe the process of dentine formation and the functions of dental pulp. We then describe the composition, mechanical properties, and structure of orthodentine and the histology of the remaining types of dentine. This chapter concludes with descriptions of two hypermineralized tissues which form major components of the toothplates of Holocephali and Dipnoi. Most of our knowledge of dentine and pulp has come from the study of human tissue and the reader is referred to Berkovitz et al. (2009) for detailed reviews.

DENTINOGENESIS

Odontoblast Differentiation

The peripheral cells of the dental papilla differentiate into odontoblasts under the influence of signals from the dental epithelium (Smith and Lesot, 2001). These epithelial/mesenchymal interactions are continuous and reciprocal. Odontoblast differentiation first occurs at the tips of unicuspid teeth or at the sites of the future cusps of multicuspid teeth and then extends laterally. A number of transcription factors and signaling molecules are involved in odontoblast differentiation, including members of the TGFβ, BMP, FGF, and IGF families (Smith and Lesot, 2001). It is also clear that the basement membrane separating the dental epithelium and papilla plays a central role and acts as a reservoir and also as a modulating and potentiating agent for signaling molecules (Smith and Lesot, 2001).

As they differentiate into odontoblasts the peripheral cells of the dental papilla enlarge by expansion of the distal cytoplasm toward the dental epithelium and form a closely packed, pseudo-epithelial layer (Fig. 11.3). Adjacent odontoblasts are connected by three types of junctions: desmosomes, tight junctions and gap junctions (Fig. 11.4). The distal region of the cytoplasm (toward the epithelium) expands and acquires the suite of organelles concerned with synthesis and secretion of protein: rough endoplasmic reticulum and a well-developed Golgi apparatus (Linde and Goldberg, 1993; Figs. 11.5 and 11.6). A unique characteristic of the mature odontoblast is the possession of an **odontoblast process**, a single cytoplasmic process extending from

The Teeth of Non-Mammalian Vertebrates. http://dx.doi.org/10.1016/B978-0-12-802850-6.00011-4

the distal cell surface into a tubule within the predentine and dentine. The process is formed as the odontoblast begins to lay down extracellular matrix and is left behind as the cell body retreats toward the center of the papilla in advance of the forming matrix. Typically, the odontoblast process contains microtubules, microfibrils and vesicles, and occasional mitochondria, but not the synthetic and secretory organelles found in the adjacent distal cytoplasm (Fig. 11.7). The odontoblast process allows for the transport of molecules synthesized in the cell body along the dentinal tubule. The initial diameter of the tubules is about 3 μm as seen at the pulp–dentine surface. As the odontoblasts migrate centripetally toward the center of dental pulp, the tubules become more crowded, so that, eg, in humans the density of tubules in inner dentine may be as high as 50,000 per mm² as compared with 20,000 per mm² in outer dentine.

Formation of Dentine

Dentine formation follows the same pattern as odontoblast differentiation. It comprises two overlapping processes: secretion of matrix by the odontoblasts, and its subsequent mineralization. During active dentinogenesis, a thin layer of unmineralized matrix (**predentine**) is always present between the odontoblast layer and the mineralized dentine, indicative of a time lag between deposition of the organic matrix and its mineralization (Figs. 11.3 and 11.7).

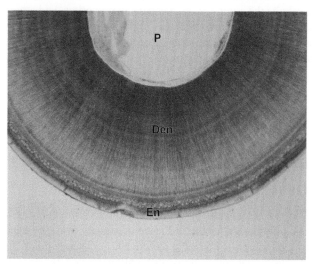

FIGURE 11.1 Horizontal ground section of tooth from the common dentex (*Dentex dentex*: Teleostei), showing pulp cavity (P) surrounded by orthodentine (Den) with radially oriented tubules and an outer layer of collar enameloid (En). Image width = 2.11 mm. *Courtesy of Royal College of Surgeons. Tomes Slide Collection, Cat. no. 399.*

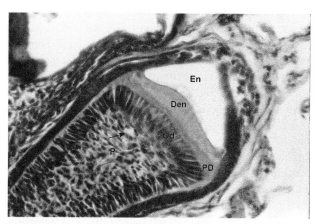

FIGURE 11.3 Tooth germ of European eel (*Anguilla anguilla*) at stage when enameloid is fully mineralized and is represented by a space left after demineralization (En). Dentine (Den) forming beneath enameloid and separated from odontoblast layer (Od) by a layer of unmineralized predentine (PD). *P*, pulp. Note small blood vessel (*arrow*) beneath odontoblast layer. Image width = 180 μm.

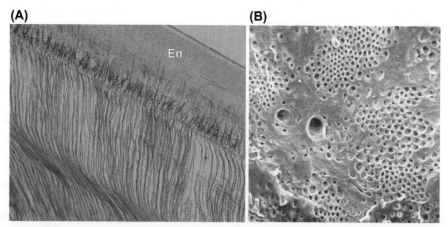

FIGURE 11.2 Dentine in the piranha (*Serrasalmus*). (A) Longitudinal ground section showing orthodentine (Den) permeated by large numbers of dentinal tubules, which also project into the enameloid (En) *(Courtesy Royal College of Surgeons. Tomes Slide Collection. Cat. no. 279.).* Field width = 520 μm. (B) Scanning electron micrograph of hypochlorite-treated wall of pulp, showing tubular appearance of dentinal tubules. The tubules show variation in distribution and diameter *(From Shellis, R.P., Berkovitz, B.K.B., 1976. Observations on the dental anatomy of piranhas (Characidae) with special reference to tooth structure. J. Zool. Lond. 180, 69–84, Courtesy Editors of the Journal of Zoology, London.).*

After synthesis, matrix protein is packaged in the Golgi apparatus into vesicles containing cross-striated, fibrous material, which indicates the presence of collagen (Figs. 11.6 and 11.7). The vesicles discharge their contents at the base of the odontoblast process into the space at the inner aspect of the predentine, where the collagen molecules are assembled into fibers which add to the thickness of the predentine (Figs. 11.3 and 11.7).

Although some calcium and phosphorus ions may pass between the odontoblasts, the bulk of calcium and phosphate ions required for mineralization of dentine are probably transferred through the odontoblast cell bodies. As free calcium is toxic at more than micromolar concentrations, protective transport mechanisms exist for calcium transport within the odontoblast. The restriction of mineral-ion transport to the intracellular route allows the odontoblasts to regulate mineral ion transport, which ensures orderly mineral deposition.

Mineralization of Dentine

During dentinogenesis, the first mineral appears in the matrix of the outermost dentine as clusters of crystals. The clusters enlarge until they become confluent and mineralization then spreads inwards toward the pulp. Because the apposition of matrix fibers at the inner boundary roughly equals the inward advance of mineralization, the thickness of the predentine layer remains approximately constant (about 20 µm in human enamel but less in smaller species, such as rodents, where the thickness is about 4 µm). The transition between predentine and mineralized dentine is abrupt (Figs. 11.3 and 11.7), suggesting that the mineralization process is very rapid. Mineral deposition also occurs at isolated sites within the predentine in advance of the main mineralization front (see Fig. 11.29).

The mineral of dentine is finely divided and consists of numerous very small, poorly crystallized crystals of a calcium phosphate (hydroxyapatite) which are plate-like in form, about 30-nm wide and 3-nm thick (Verbeeck, 1986; Märten et al., 2010). The mineral crystals in large mammalian teeth are oriented in two patterns. Some are aligned closely with the collagen fibers and about 20% are actually located within the fibers (Kinney et al., 2003a). The remaining crystals appear randomly oriented by some methods (Märten

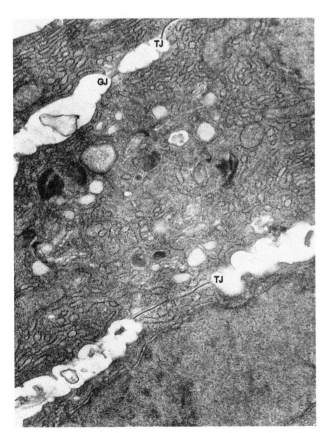

FIGURE 11.4   Three odontoblasts, showing a gap junction (GJ) and two tight junctions (TJ). European eel (*Anguilla anguilla*: Teleostei), transmission electron micrograph. Image width = 5.7 µm.

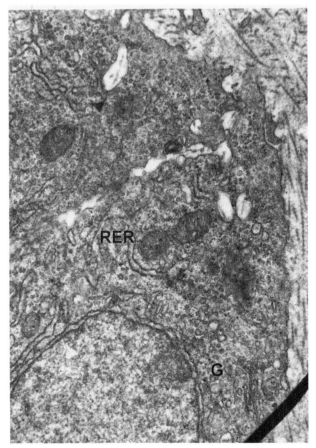

FIGURE 11.5   Preodontoblast, showing first stages of enlargement of distal cytoplasm, with appearance of Golgi apparatus (G) and rough endoplasmic reticulum (RER). At this stage, dentine formation has not commenced and the cell does not possess a process. European eel (*Anguilla anguilla*; Teleostei), transmission electron micrograph. Image width = 4.2 µm.

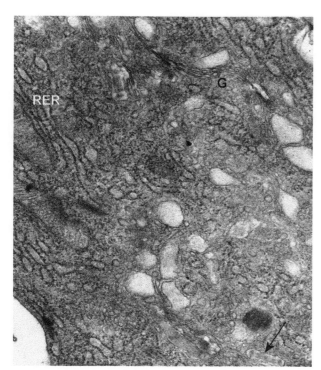

FIGURE 11.6 Cytoplasm of fully functional odontoblast distal to nucleus, showing rough endoplasmic reticulum (RER), Golgi apparatus (G) and numerous vesicles. Newly formed vesicles near the Golgi saccules are large, with amorphous contents. As the vesicles mature they become smaller and the contents appear more fibrous. In mature vesicles (bottom right, *arrow*), cross-striations are visible. European eel (*Anguilla anguilla*; Teleostei), transmission electron micrograph. Image width=5.7 μm.

FIGURE 11.7 Odontoblast process extending from odontoblast body (lower right) into dentine (upper left). Between the mineralized dentine and the odontoblast is a layer of unmineralized dentine containing a number of extracellular matrix vesicles. The process itself is devoid of organelles except for microtubules but at its base are several secretory vesicles (*arrows*). Image width=3.8 μm. Inset: a single secretory vesicle with fibrous contents showing cross-striations. European eel (*Anguilla anguilla*; Teleostei), transmission electron micrograph. Image width=0.6 μm.

et al., 2010) but in fact are arranged radially in arcade-shaped domains known as **calcospherites** (Fig. 11.8A and B), which result from growth of crystals outward from focal points (Fig. 11.9; Schmidt and Keil, 1971; Shellis, 1983) and intersect with the crystals co-oriented with the collagen fibers. Calcospherites are easily identified in large teeth, including those of some non-mammalian vertebrates such as crocodylians (Schmidt and Keil, 1971), but are much less obvious or not detectable in teeth of small vertebrates, whether mammalian or not. In larger teeth, calcospherites tend to be smaller and more numerous in the outer and inner dentine than in the intermediate dentine where, because they are more widely spaced, they can grow to a larger size. Some areas of dentine have few calcospherites (Fig. 11.8B).

The processes by which mineral ions are deposited in dentine from solution are not yet resolved. One theory suggests that the first clusters of mineral in dentine are formed within small, extracellular, membrane-bound, **matrix vesicles**. It is hypothesized that mineral ions are accumulated within the vesicles by the pumping action of membrane-bound enzymes until eventually precipitation occurs and the resulting crystals grow to the point where they rupture the vesicle membrane. Similar matrix vesicles have been implicated in the initial mineralization

of bone and calcified cartilage. However, it is implausible that deposition of mineral crystals co-oriented with the matrix collagen fibers could be established by matrix vesicles, which generate randomly oriented crystals. Therefore attention has also focused on the possible role of matrix components in mineralization.

The noncollagenous portion of dentine matrix consists largely of a variety of proteins and proteoglycans (Goldberg et al., 2011). Most of the noncollagenous proteins in dentine, including osteopontin, dentin sialophosphoprotein (DSPP), dentine matrix protein 1, matrix extracellular phosphoglycoprotein, belong to the small integrin-binding ligand N-linked glycoproteins (SIBLING) family of proteins (Goldberg et al., 2011; Staines et al., 2012). These proteins are extremely acidic because of their high content of aspartic and glutamic acids and numerous serines that are 90% phosphorylated. They adopt an open and random

(A)

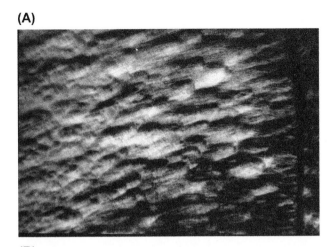

(B)

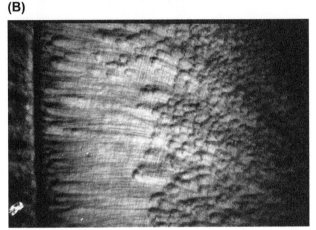

FIGURE 11.8 (A) Longitudinal ground section of human molar, viewed in polarized light, showing group of large calcospherites in the middle dentine. (B) Similar section (enamel to left). In this section, calcospherites are sparse in the middle region of dentine. Image width=2.3 mm. Image width=0.9 mm.

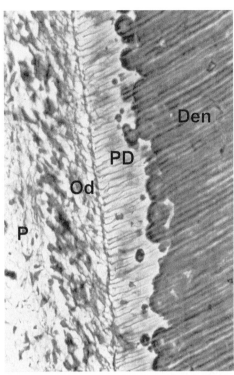

FIGURE 11.9 Demineralized section of developing human tooth. Pulp to left, mineralized dentine to right. Large, spherical calcospherites forming in predentine layer. Silver stain. Image width=0.9 mm. *Den*, dentine; *Od*, odontoblast layer; *P*, pulp; *PD*, predentine. *From Berkovitz, B.K.B., 2013. Nothing but the Tooth. Elsevier, London; Berkovitz, B.K.B., Holland, G.R., Moxham, B.J., 2009. Oral Anatomy, Histology and Embryology, fourth ed. Elsevier, London. Courtesy of london.*

configuration in solution, providing freedom for interaction with other molecules. Accumulating evidence has implicated SIBLING proteins in matrix mineralization in dentine (and bone). It is likely that one or more of these strongly anionic molecules can bind calcium in an ordered pattern to form nuclei for formation of mineral crystals on the matrix collagen fibers. The prime candidates for this role are probably **dentine phosphoprotein** (DPP), which is derived from DSPP by enzymic hydrolysis and is the major noncollagenous protein in dentine, and **dentine matrix protein 1** (Prasad et al., 2010; Deshpande et al., 2011). Both are highly anionic proteins, especially when phosphorylated, and can adhere to collagen as well as bind crystals. DPP appears to bind to the collagen fibers at the sites of the "gap regions," where there exists a space between consecutive collagen molecules. These gaps in turn provide sites for crystal deposition within the fiber (Landis, 1996). It is likely that more than one mechanism of mineralization is operative in dentine, especially given the complex pattern of crystal orientation (Shellis, 1983).

The preceding description refers to the dentine formed between the dentinal tubules (**intertubular dentine**), which makes up by far the largest fraction of the tissue. Soon after mineralization of the intertubular dentine, additional mineral, in the form of closely packed, small particles, plates, or needles, begins to be deposited between the odontoblast process and tubule wall. With time, this material, termed **peritubular dentine** (strictly speaking, intratubular dentine), narrows the tubule or even occludes it completely. Peritubular dentine lacks collagen, is rich in dentine matrix protein 1, and lacks DPP (Linde and Goldberg, 1993). The process of mineralization is therefore different from that of intertubular dentine. Peritubular dentine is more highly mineralized (85% mineral by weight) than the intertubular dentine (70% mineral by weight) (Figs. 11.10 and 11.11). It is also harder than intertubular dentine (1.3–4.7 vs 0.52–1.2 GPa) and has a higher elastic modulus (29–45 vs 12–25 GPa) (Zhang et al., 2014).

Most of the dentine is formed before eruption is completed; this tissue is known as **primary dentine**. Once the crown and root of the tooth are completed in mammals, the activity of the odontoblasts is reduced but does not cease; they continue to form **secondary dentine** by apposition at the inner surface of the primary dentine, but much more slowly than primary dentine. In mammals, therefore, a

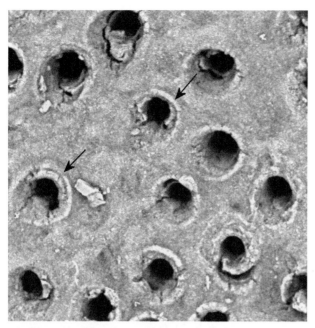

FIGURE 11.10 Ground and polished surface of human dentine, parallel with the outer root surface, prepared by the method of Shellis and Curtis (2010). The peritubular dentine (*arrows*) surrounds each dentinal tubule as a hypermineralized layer. Scanning electron micrograph. Image width=25 μm.

recognizable odontoblast layer is usually present at the periphery of the pulp and there is always a thin layer of predentine lining the pulpal surface of the dentine.

DENTAL PULP

As stated previously, nearly all research on dental pulp has been carried out on teeth of humans and a few other mammals. Therefore, the following account deals mainly with the mammalian pulp. The fragmentary knowledge of the pulp in non-mammalian vertebrates is summarized at the end of this section.

Organization of Pulp

The dental pulp of a functional mammalian tooth is a vascular, innervated connective tissue (Fig. 11.12). As in any connective tissue, the pulp comprises a scaffold of collagen fibers embedded in ground substance, which are products of the main connective-tissue cells, the fibroblasts. Because the pulp is protected by the dentine from displacing forces and is a low-compliance tissue, its collagen framework consists of just a fine network of thin fibers. Judging from the quantity and quality of intracellular organs present in its fibroblasts, the turnover of the extracellular matrix of the pulp is slow. It is richly supplied by nerves and blood vessels, perhaps a reflection of its low compliance. The pulp tissue contains immune cells (T-lymphocytes, macrophages, and dendritic antigen-presenting cells). It also contains a population of **stem cells**. In response to external stimuli, such as

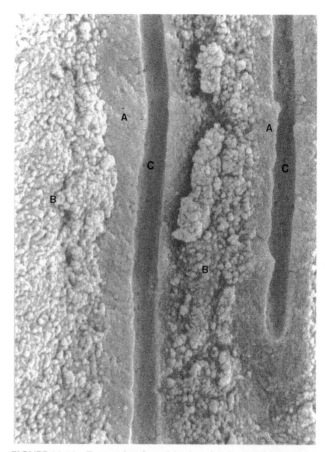

FIGURE 11.11 Fractured surface of dentine, showing two tubules which have split open longitudinally. Peritubular dentine (A) surrounding tubules (C) presents a flat, smooth fracture surface, while intertubular dentine (B) presents a rough fracture surface. The fracture surfaces of the two areas of dentine reflect the underlying structure: fine-grained and fibrous, respectively. Scanning electron micrograph. Image width=13 μm. *Image courtesy Professor M. M. Smith, From Berkovitz, B.K.B., Holland, G.R., Moxham, B.J., 2009. Oral Anatomy, Histology and Embryology, fourth ed. Elsevier, London, Courtesy of Elsevier.*

wear, these are capable of differentiating into new odontoblast-like cells which produce reparative dentine. Isolated pulpal stem cells have principally odontogenic, osteogenic, adipogenic, and neurogenic capabilities. Thus, these stem cells could potentially be utilized in regenerative dentistry, for instance in pulp therapy (Huang, 2009; Ulmer et al., 2010; Arthur et al., 2014)

About 70–90% of pulpal nerves consist of slow-conducting, unmyelinated C-type fibers and the rest of myelinated Aδ and Aβ fibers (Hildebrand et al., 1995). The main sensory nerve trunks enter the pulp via the root canals and branch to form a dense network around and beneath the odontoblast layer (Fig. 11.13). The dentine itself is innervated by penetration of unmyelinated nerve endings extending 100–200 μm into the tubules and lying alongside the odontoblast processes. Such nerve endings are found in all or most tubules of coronal dentine but in no more than 0.2% of tubules in root dentine. The pulp

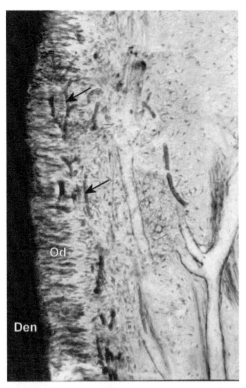

FIGURE 11.12 Low-power view of histological section of tooth, showing general histological structure of dental pulp. Left: dentine (Den) and odontoblast layer (Od). To the right, the central region of the pulp, with a connective-tissue matrix permeated by blood vessels. Note small capillaries (*arrows*) forming a plexus beneath and within the odontoblast layer. Masson's trichrome. Magnification approximately ×300. *From Berkovitz, B.K.B., Holland, G.R., Moxham, B.J., 2009. Oral Anatomy, Histology and Embryology, fourth ed. Elsevier, London.*

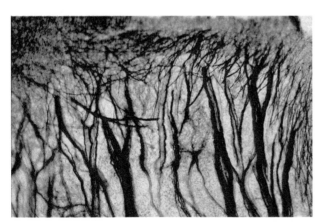

FIGURE 11.13 Histological section of periphery of dental pulp, showing terminal branches of nerve fibers. Silver stain (cells are not visible by this staining method). Magnification approximately ×300.

also contains sympathetic fibers, which are most abundant in the pulp horns and relatively sparse within the root canals, where the density of blood vessels is low. It is likely that parasympathetic nerves are also present in the pulp, although this remains somewhat controversial (Hildebrand et al., 1995; Caviedes-Buchali et al., 2008).

FIGURE 11.14 Longitudinal ground section of human molar. Pulp chamber below field. Primary dentine at top. At the bottom of the field is a layer of secondary dentine, demarcated from the primary dentine by an abrupt change in tubule orientation, producing an Owen's line (*arrows*). Image width=520 μm. *From Berkovitz, B.K.B., Holland, G.R., Moxham, B.J., 2009. Oral Anatomy, Histology and Embryology, fourth ed. Elsevier, London.*

The major blood vessels of the pulp consist of a few arterioles and veins which run up the roots of the tooth and branch profusely, especially in the coronal region, to form a plexus of small blood vessels beneath and between the odontoblasts (Fig. 11.12). Some capillaries are fenestrated and high tissue fluid pressures have been reported in mammalian dental pulps (Berkovitz et al., 2009). Blood flow through the pulp is controlled mainly by the sympathetic nerves, which cause constriction of blood vessels. However, C and Aδ fibers release various neuropeptides which have the opposite effect of dilating blood vessels and promoting flow (Hildebrand et al., 1995; Caviedes-Buchali et al., 2008).

Secondary and Tertiary Dentine Formation

After completion of the primary dentine, the odontoblasts become smaller and much less active but continue to produce **secondary dentine**. This has a regular, tubular structure and may be indistinguishable from primary dentine, except for a marked synchronous change in orientation of the dentinal tubules, known as an Owen's line (Fig. 11.14). Although proceeding very slowly (approximately 0.5 μm/day), apposition of secondary dentine over long periods causes a progressive reduction in the size of the pulp cavity and alterations in its shape and relative position.

In the event of injury to the tooth, active **tertiary dentine** formation can be rapidly initiated locally, at the pulpal ends of those tubules associated with the site of external damage. The main function of tertiary dentine formation is to block the inner openings of the dentinal tubules and hence to prevent ingress of bacteria, which would result in pulpal and systemic infection. There appear to be two levels of response, which depend on the severity of the injury. Severe injury can kill the odontoblasts. The eventual outcome here is the deposition of

reparative tertiary dentine. The lost odontoblasts are replaced by odontoblast-like cells which differentiate from stem cells within the pulp. Reparative dentine has an irregular structure and can be atubular. Any tubules present are not continuous with those of the primary dentine. When injury to the tooth is less severe, the odontoblasts are rapidly reactivated and initiate **reactive tertiary dentine**. Such tertiary dentine contains dentinal tubules but these are highly irregular in form and arrangement. An important example of this process is the formation of tertiary dentine beneath areas of attrition of the coronal enamel. Even if wear proceeds through the enamel and the primary and secondary dentine, exposure of the dental pulp to the oral environment is prevented by the presence of tertiary dentine (Fig. 11.15). Clearly, this is vital to tooth integrity in herbivorous mammals, in which wear is extensive.

The stimulus for reactivation of quiescent odontoblasts, or for differentiation of new odontoblast-like cells from stem/progenitor cells, is the release of growth factors in the pulp as the result of an inflammatory process. In addition, dentine itself contains a number of growth factors which are not normally available in the sound tissue but which can be released when the dentine is damaged and can diffuse to the pulp along the dentinal tubules (Smith and Lesot, 2001; Smith et al., 2012). Most research on tertiary dentine formation has been concerned with situations, eg, dental caries or erosion, where growth factors can be solubilized, but mechanical damage through wear or fracture presumably also releases growth factors. Reactivation of surviving odontoblasts in initiation of reactive tertiary dentine seems to resemble the process of

odontoblast differentiation during normal tooth ontogeny, with TGFβ-1, TGFβ-3, and BMP-7 playing important roles. The TGFβ family of signaling molecules also seems to be important in differentiation of new odontoblasts in reparative tertiary dentine formation. For reviews of the control of odontoblast activation, see Smith and Lesot (2001) and Smith et al. (2012).

Immune Responses

In the long-lived teeth of mammals, defense against bacterial infection resulting from damage to the dentine layer is vital and the pulp can mount effective immune responses, as described by Hahn and Liewehr (2007), Caviedes-Bucheli et al. (2008), and Bhingare et al., (2014). Capture of bacterial antigens by antigen-presenting dendritic cells, and their subsequent presentation to T-cell lymphocytes, sets off a complex cascade of cellular and cytokine responses. There is evidence that the odontoblasts themselves act as receptors that can detect molecules of bacterial origin, such as cell-wall constituents, and also secrete TGFβ and various chemokines, which stimulate the cellular immune response (Hahn and Liewehr, 2007; Bleicher, 2014).

Pulpal nerves have various roles in the responses of the pulp to external stimuli, such as infection or damage (Hildebrand et al., 1995; Caviedes-Bucheli et al., 2008). Stimulation of sympathetic nerves leads to vasoconstriction. Stimulation of sensory or parasympathetic nerves, on the other hand, causes vasodilatation and increased vascular permeability, and also modulates the activities of various immune cells. These changes can result in pulpal inflammation. However, at sites where external damage is manifested in the pulp, there may be extensive sprouting of nerve fibers expressing signaling agents. These agents eventually stimulate growth of fibroblasts and odontoblast-like cells and induce differentiation of new odontoblast-like cells.

Tooth Sensitivity

The sensitivity of teeth to noxious stimuli is well known; changes of temperature, mechanical contact, changes in osmotic pressure or damage through fracture or dental caries can produce pain of varying intensity and type. The adaptive significance of dental pain is obscure (Hildebrand et al., 1995), although it may result in an infected or damaged tooth being loaded less during mastication until the healing process has begun. Different stimuli evoke different types of pain (Hildebrand et al., 1995). C fibers are activated by stimuli which reach the pulp and cause a dull, aching type of pain. Stimulation of Aδ fibers within dentine produces sharp, intense pain. The latter type of pain has been intensively studied because of its manifestation as hypersensitivity of exposed root surfaces.

Dental hypersensitivity depends on the presence of patent dentinal tubules at the root surface. They may be developmental, when cementum has not been produced to cover

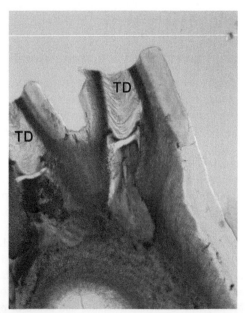

FIGURE 11.15 Longitudinal section of molars of **bank vole** (*Myodes glareolus*). Despite the complete loss of the primary dentine and enamel through wear at the cusps, the continuous formation of reactive tertiary dentine (TD) has ensured that the pulp remains isolated from the oral environment. Image width = 1.05 mm. *Courtesy of Royal College of Surgeons. Tomes Slide Collection, Cat. no. 1097.*

the dentine, or they may have been opened by mechanical action (wear or fracture) or by pathological processes, such as caries or erosion (West et al., 2013). According to a widely accepted theory, external stimuli such as changes in temperature or mechanical stimulation do not act directly on the receptors but indirectly, by inducing fluid flow within the dentinal tubules. The identity of the receptors remains unresolved but it is most widely accepted that the unmyelinated nerve fibers within the dentinal tubules and around the odontoblast layer act as nociceptors. There is also some evidence that the odontoblasts themselves are sensitive to mechanical and other stimuli. Although there are numerous nerve fibers between the odontoblasts, no specialized intercellular junctions between the two have been observed. Therefore, if the odontoblasts do perform a sensory function, the nerve fibers must be stimulated by chemical messengers released by the odontoblasts. For details of the possible pathways involved in dentine sensation, see Magloire et al. (2010), Chung et al. (2013), West et al. (2013), and Bleicher (2014).

Dental Pulp in Non-Mammalian Vertebrates

Possible functions of the dental pulp in non-mammalian vertebrates, other than dentine formation, have received very little research attention, so knowledge in this field is fragmentary. No studies on immune responses by the pulp of non-mammalian vertebrates have been made. There is some evidence that secondary or tertiary dentine can be formed in the mature tooth; a layer of atubular dentine at the pulpal surface of the dentine was reported in *Latimeria* (Shellis and Poole, 1978) and an example from an elasmobranch is illustrated in Fig. 11.16. Whether formation of this tissue is age-related (like secondary dentine) or has a reparative function (like tertiary dentine) is not known.

As regards innervation, Hildebrand et al. (1995) cited G. Retzius as having described pulpal nerve fibers in the teeth of a bony fish (goldfish) and in amphibian teeth, but were able only to cite textbooks to support the presence of nerves in teeth of elasmobranchs and reptiles. There is therefore much scope to establish more securely the presence of nerve fibers within non-mammalian teeth. However, Tuisku and Hildebrand (1995) confirmed the presence of nerve fibers in a teleost (*Tilapia mariae*; Cichlidae). The arrangement of fibers is similar to that in mammals. Myelinated fibers enter the basal part of the pulp and branch toward the tip of the pulp to form a plexus of unmyelinated fibers near the odontoblasts. A few nerve endings are present within tubules. Recently, Crucke et al. (2015) showed that the pharyngeal teeth of the zebrafish (*Danio rerio*: Cyprinidae) were innervated by a branch of the vagus nerve. Nerve fibers entered the dental pulp only during late development, when the attachment tissues were forming, and did not occur in teeth at earlier stages of development. No intratubular nerve fibers were observed. Crucke et al. suggested that the nerves functioned in vasomotor control.

In developing teeth of non-mammalian vertebrates, the dental pulp is very cellular. Blood vessels are present in larger teeth (Fig. 11.17) but are inconspicuous or absent in small teeth (Sire et al., 2002; Davit-Béal et al., 2006).

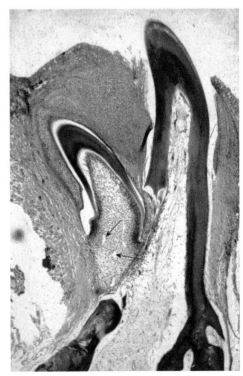

FIGURE 11.17 Rainbow trout (*Oncorhynchus mykiss*). Functional tooth in process of resorption, with adjacent replacement tooth. The pulp of the replacement tooth is highly cellular and contains blood vessels (*arrow*). The functional tooth in process of resorption contains few cells, although it remains vascular. Masson trichrome. Image width=1.07 mm. *From Berkovitz, B.K.B., 1977. Chronology of tooth development in rainbow trout* (Salmo gairdneri). *J. Exp. Zool. 200, 65–69, Courtesy Editors of the Journal of Experimental Zoology.*

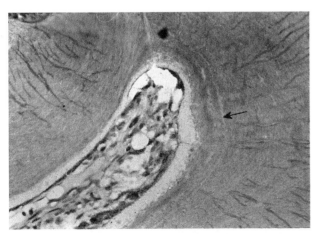

FIGURE 11.16 Functional tooth of **thornback ray** (*Raja clavata*), showing layer of atubular dentine at the pulpal surface of the orthodentine. *Arrow* indicates a boundary line between tubular and atubular tissue. The atubular tissue appears to be a form of secondary or tertiary dentine. Image width=0.26 mm.

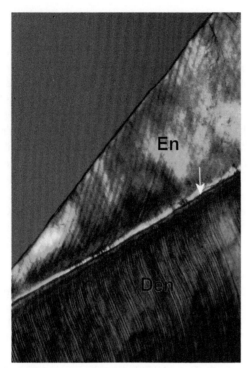

FIGURE 11.18 Junction between enamel (En) and dentine (Den) in human tooth, viewed in polarized light with first-order red plate. The circumpulpal dentine is blue, because the matrix fibers are highly oriented (approximately parallel with the horizontal axis), while the mantle dentine (*arrow*) appears reddish, because a large proportion of the fibers are oriented perpendicular to the enamel–dentine junction. Image width = 0.9 mm. *From Berkovitz, B.K.B., Holland, G.R., Moxham, B.J., 2009. Oral Anatomy, Histology and Embryology, fourth ed. Elsevier, London, Courtesy of Elsevier.*

TABLE 11.1 Composition of Human Enamel and Dentine by Volume (Shellis et al., 2014), With Information on Some Mechanical Properties (From Zhang et al., 2014 Unless Otherwise Indicated)

Constituent	Enamel	Dentine
Mineral (vol%)	91.4	48.0
Organic material (protein + lipid) (vol%)	5.0	30.6
Water (vol%)	3.4	21.4
Young's modulus (GPa)	80–105	20–25
Hardness (GPa)	3.4–4.6	0.5–1.0
Ultimate tensile strength (MPa)	12–42	50–100
Fracture toughness (MPa.m$^{0.5}$)	0.5–1.3[a]	2.2–3.4[b]

In both enamel and dentine, there is considerable variation with site, age, and with the orientation and degree of hydration of the specimen.
[a]He and Swain (2008).
[b]Ivancik and Arola (2013).

Presumably in the latter, the diffusion distances are short enough for gases, nutrients, and metabolic waste products to be exchanged without vascularization of the pulp. In older functional teeth, the cell density is much lower, the odontoblasts are very small and inactive, and a definite predentine layer is not evident (Fig. 11.17).

ORTHODENTINE

Mechanical Properties

Teeth owe their mechanical strength to the combination of enamel or enameloid, which are hard but brittle, with dentine, which is softer and less rigid and may reduce the stress on the hypermineralized tissues at the functional surface. Dentine shows little mechanical anisotropy (Kinney et al., 2003b), due in part to the dispersion of crystal orientation (Märten et al., 2010), and this may improve resistance to compressive stresses (Märten et al., 2010). Dentine is much more resistant to propagation of cracks than enamel or enameloid (higher fracture toughness) because of the intimate association of small apatite crystals with strong protein fibers (Kinney et al., 2003b). The junction between dentine and enamel (possibly also enameloid) appears to be crucial to limiting propagation of cracks through the tooth. Most cracks traveling inwards from the tooth surface terminate at this junction and those which cross the junction seem to be arrested in the outer few micrometers of the dentine. Table 11.1 uses the example of human dentine and enamel to illustrate the differences in mechanical properties.

From the limited comparative data available, it appears that among the vertebrates orthodentine has similar mineral contents, although the hardness and elastic modulus seem to be more variable (Table 11.2).

Mammalian Orthodentine

The fine structure of orthodentine varies between locations in an individual tooth and there also exist significant variations in dentine microstructure between taxa.

The shell of orthodentine surrounding the pulp cavity can be divided into two layers: an inner layer of **circumpulpal dentine**, which accounts for the greater part of the thickness of the orthodentine, and a thinner peripheral layer, underlying the enamel or cementum only a few micrometers thick in small teeth (eg, rat), but up to 30-μm thick in large teeth, such as those of humans. In the tooth crown, the outer layer of dentine, which represents the first-formed region of the tissue, is known as the **mantle dentine**. Here, a large proportion of the collagen fibers are oriented obliquely or at right angles to the outer dentine surface. In human mantle dentine, the numbers of fibers oriented parallel to the dentine surface and perpendicular to it are approximately equal (Fig. 11.18). In other species, however, the fibers at right angles to the dentine surface are very prominent (Schmidt

TABLE 11.2 Composition and Mechanical Properties of Dentine From Some Vertebrates

Taxon	Calcium Content, wt%	Phosphorus Content, wt%	Hardness, GPa	Elastic Modulus, GPa
Orthodentine				
Elasmobranchii (4 species)	24.3–30.9[a]	13.6–15.7[a]	0.7–2.1[a,b]	17.0–49.0[a,b]
Dipnoi (2 species)			0.43–0.58[c]	
Reptilia (*Crocodilus porosus*)	20.6[d]	12.7[d]	0.6[d]	
Nonhuman Mammalia (14 species, incl. 11 dolphins)			0.56–0.88[b,c,e]	7.8–21.6[b,c,e]
Nonhuman Mammalia: *Loxodonta* ivory	19.4[f]	11.5[f]	0.43[c]	7.7[c]
Nonhuman Mammalia: *Monodon* ivory			0.53	8.9
Homo sapiens	26.9[g]	13.2[g]	0.5–17[h,*]	16–22[h,*]
Osteodentine				
Elasmobranchii (*Carcharinus taurus*)			1.21[b]	28.4[b]

*Overall values.
[a]Enax et al. (2012).
[b]Whitenack et al. (2010).
[c]Currey and Abeysekera (2003).
[d]Enax et al. (2013).
[e]Loch et al. (2013).
[f]Raubenheimer (1999).
[g]Verbeeck (1986).
[h]Zhang et al., (2014).

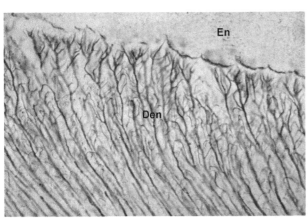

FIGURE 11.19 Enamel and outer dentine of human tooth. Mantle dentine contains numerous terminal branches of dentinal tubules. Image width=200 μm. *Den,* Dentine; *En,* enamel. *From Berkovitz, B.K.B., Holland, G.R., Moxham, B.J., 2009. Oral Anatomy, Histology and Embryology, fourth ed. Elsevier, London, Courtesy of Elsevier.*

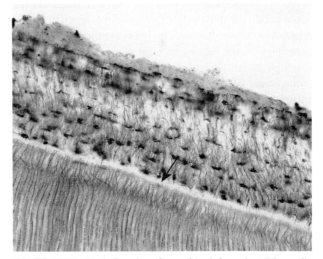

FIGURE 11.20 Vertical section of root of tooth from **civet** (Mammalia; Viverridae, species unknown). A thick layer of cellular cementum (top) is separated from the dentine (below). Between these layers is a thin atubular/acellular layer (*arrows*) which consists partly of acellular cementum and partly of the hyaline layer. There is no granular layer. Near their terminations, the tubules are branched. Image width=1.05 μm. *Courtesy of Royal College of Surgeons. Tomes Slide Collection, Cat. no. 1330.*

and Keil, 1971). Mantle dentine in human teeth is less well mineralized than the circumpulpal dentine (Märten et al., 2010). The mantle dentine is traversed by dentine tubules which tend to be extensively branched near their termini (Fig. 11.19).

In the roots of mammalian teeth, the layer corresponding to the mantle dentine is the **hyaline layer**, which lies between dentine and cementum (Fig. 11.20). The hyaline layer is so-called because it lacks dentine tubules. The tubules terminate in the immediately underlying region, which in larger mammals is distinguished as the **granular layer** (Fig. 11.21). In ground sections the granular layer appears dark and speckled because light is scattered by the numerous spaces created by

FIGURE 11.21   Longitudinal section of root of tooth from **sperm whale** (*Physeter macrocephalus*). From the outer surface inwards, there is a layer of cellular cementum overlying a thinner layer of acellular cementum, and then the orthodentine. The outermost part of the dentine is atubular: this is the hyaline layer. Beneath this is a dark speckled layer, which is the granular layer. Image width = 2.11 mm. *Courtesy of Royal College of Surgeons. Tomes Slide Collection, Cat. no. 795.*

FIGURE 11.22   High-power view of cross-section of root from **spring hare** (*Pedetes capensis*). Dentinal tubules run from upper right to lower left and are intersected by dark long-period (Andresen) lines. Image width = 1.05 mm. *Courtesy of Royal College of Surgeons. Tomes Slide Collection, Cat. no. 933.*

incomplete mineralization and by branching and dilatation of the tubule termini. A granular layer seems to be absent from teeth of smaller mammals (Fig. 11.20).

In the circumpulpal dentine, the collagen fibers are laid down parallel with the outer surface of the dental papilla in contrast to mantle dentine. The fibers are therefore oriented at an angle to the tubules, in most places about 90 degrees, although under the cusps the angle is more acute. This arrangement of the collagen fibers underlies an observed anisotropy in the elastic modulus of dentine, which is about 10% greater in the plane of the collagen fibers than perpendicular to it (Kinney et al., 2003b). Because dentine formation is initiated first beneath the cusps and then over a progressively wider area as the tooth germ expands, in longitudinal section the collagen fibers form a small angle with the outer surface of the dentine. In medium- and large-sized teeth, including those of humans, the fibers are arranged in layers and the fibers in one layer are oriented at about 50 degrees with respect to those in adjacent layers (Schmidt and Keil, 1971). Overall, therefore, the collagen fibers lie within ±25 degrees of the longitudinal direction.

Incremental Markings

Incremental markings, which mark the growth pattern of the tooth, can be observed in many teeth. In a variety of mammalian and non-mammalian vertebrates, the dentine displays incremental markings with an average periodicity of 16 μm/day (range 11.3–19.3 μm/day), although in primates, including humans, and some other animals, the periodicity is less (Dean, 1995).

In a number of mammals, the dentine contains two types of incremental line, one with a long and one with a short periodicity. The long-period lines (often known as **Andresen lines**) are a manifestation of periodic variations in the orientation of collagen fibers and are particularly clear under the polarizing microscope (Figs. 11.22 and 11.23B). The regular periodicity of the Andresen lines corresponds to increments of dentine apposition, over periods of 6–10 days in humans and great apes, 4–5 days in monkeys, and 13–14 days in proboscideans (Dean and Scandrett, 1996). In teeth of small animals, the fibers do not seem to be arranged in layers in the same way as in human teeth, and Andresen lines are either obscure or absent.

Between the long-period Andresen lines there occur short-period lines (known as **von Ebner lines**), which follow the mineralizing front at different stages during odontogenesis. They represent a daily increment of dentine mineralization (Dean, 1998). The appearance of von Ebner lines varies according to the regional distribution of calcospherites. In outer dentine, where calcospherites are smaller and more numerous, the von Ebner lines have a scalloped appearance (Fig. 11.23A), whereas in the inner dentine, which has fewer calcospherites, the von Ebner lines are smoother, suggesting a more continuous mineralizing front (Fig. 11.23B). Von Ebner lines are more difficult to visualize than Andresen lines and require the application of silver staining.

Both types of lines can be utilized in studies of tooth growth. Measurements of von Ebner lines show that the rate of mineralization of primary dentine in the crown of human teeth varies; it increases from 1.5 to 1.7 μm/day in the outer dentine to 4.0–4.6 μm/day in the inner dentine, although the rate may be higher in some parts of the crown (Dean, 1998). The rate seems to be similar in other primates (Dean, 1998).

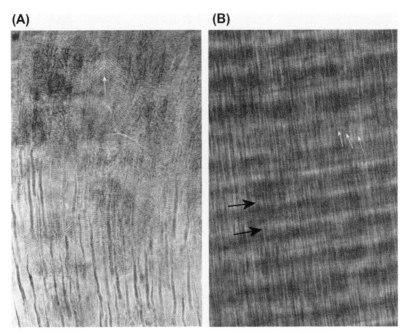

FIGURE 11.23   Short-period (von Ebner) lines in human dentine. (A) Outer root dentine. Lines follow course of successive positions of mineralization front around calcospherites (boundary between two calcospherites indicated by *small arrow*). Image width = 145 μm. (B) Both long-period lines and short-period lines in polarized light. Long-period lines appear as alternately light and dark bands (cf. Fig. 11.22): corresponding parts of successive lines indicated by *black arrows* (left). Short-period lines parallel with long-period lines in this region: three successive lines indicated by *small white arrows* (right). Image width = 145 μm. *Courtesy Professor M. C. Dean.*

In medium-to-large teeth, the dentinal tubules are not straight but exhibit curvature at two levels of scale. First, each tubule may show undulations (the **secondary curvatures**), with a wavelength of several micrometers, which are thought to result from oscillations of the odontoblasts within the dental papilla during dentine formation. At a larger scale, the tubules often demonstrate a sweeping S-shaped curve, the **primary curvature**, which reflects the fact that, as the dentine thickens during tooth formation, the inner surface gets smaller so the odontoblasts are pushed down the tooth.

Ivory

The largest mammalian teeth are the continuously growing tusks of several unrelated genera: elephants, hippopotamus, walrus, babirusa, and narwhal, together with the massive teeth of sperm whales. The dentine which makes up the greater part of these teeth is known as **ivory**. At least two types of ivory seem to be less well mineralized and softer than most types of orthodentine (Table 11.2). In the tusks of the narwhal (*Monodon monoceros*), walrus (*Odobenus rosmarus*), and sperm whale (*Physeter macrocephalus*), the histological structure of the ivory differs little from that of ordinary circumpulpal dentine. However, in the tusks of elephants (*Loxodonta africanus*, *Elephas indicus*) and of hippopotamus (*Hippopotamus amphibius*), the ivory has a much more complex structure.

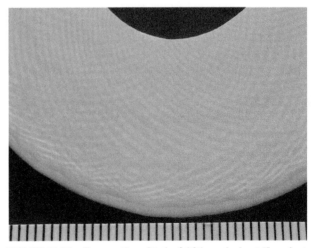

FIGURE 11.24   Cross-section of tusk of **African elephant** (*Loxodonta africana*), viewed by incident light, showing the curved checkerboard pattern in the ivory. Magnification unknown. *From Berkovitz, B.K.B., 2013. Nothing but the Tooth. Elsevier, London, Courtesy Dr. S. O'Connor.*

Cross-sections of elephant tusks viewed in incident light present a characteristic pattern. Dark and light patches are aligned in two sets of curving rows—one set intersecting the other—to form a distorted checkerboard-like pattern (Fig. 11.24). This effect depends on the characteristic morphology and arrangement of the dentinal tubules in the tissue. On their radial course through the dentine, the tubules undulate in the longitudinal place in a regular, synchronized fashion (Fig. 11.25). The crests of the undulations reflect

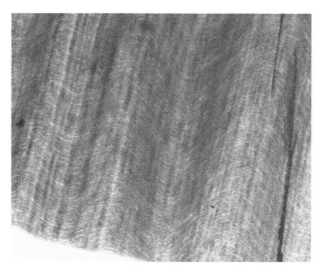

FIGURE 11.25 Longitudinal section of tusk of **African elephant** (*Loxodonta africana*), viewed by transmitted light. Note the synchronized, sinusoidal curvatures of the dentinal tubules. Image width=2.11 μm. *Courtesy of Royal College of Surgeons. Tomes Slide Collection, Cat. no. 850.*

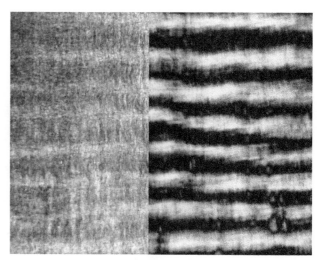

FIGURE 11.26 Transverse section of tusk of **hippopotamus** (*Hippopotamus amphibious*). Left: as viewed in ordinary light, showing synchronized undulations of tubules forming bands; Right: as viewed in polarized light. Changes in orientation of collagen between layers, are manifested as a pattern of black and white stripes, corresponding to tubule undulations in left. Image width=0.11 mm.

light back to a viewer of a transverse section whereas the troughs do not, so synchronized groups of tubules appear alternately dark and light. The checkerboard effect is generated by the fact that the tubules in alternate sectors are out of phase (Miles and White, 1960). There are correlated changes in orientation of the matrix collagen fibers and the three-dimensional structure of the matrix is very complex (Schmidt and Keil, 1971; Su and Cui, 1999).

In hippopotamus ivory, the dentinal tubules undulate markedly in the transverse plane and instead of the checkerboard pattern seen on elephant ivory, there is a pattern of circumferential stripes on transversely cut surfaces (Fig. 11.26). The stripes are not always complete, where the undulations of the tubules become irregular.

Non-Mammalian Orthodentine

The dentine closest in histological structure to that of mammals is that of **crocodylians**. It shows conspicuous incremental lines (Fig. 11.27). The short-period lines, approximately 16 μm apart, represent daily increments of growth (Erickson, 1996a,b). There are also major dense lines with a much longer period: these might reflect annual or seasonal periods, but the periodicities of these lines await clarification. The crocodylians are the only non-mammalian vertebrates that possess true roots consisting of dentine covered by cementum. As in mammals, the outermost layer of the root dentine consists of a hyaline layer with an underlying granular layer (Fig. 11.28). Dentine in other non-mammalian vertebrates shows little, if any, regional differentiation, and incremental lines have not been reported except in crocodylians.

Among amphibians and fish, a prominent feature is the presence of longitudinally oriented collagen fiber bundles

FIGURE 11.27 Transverse ground section of tooth of crocodile (*Crocodylus* sp.). Accentuated long-period lines separated by more numerous shorter-period lines. Image width=1.05 mm. *Courtesy of Royal College of Surgeons. Tomes Slide Collection, Cat. no. 639.*

(Fig. 11.29) that probably increase the resistance of the teeth, which are often elongated, to bending. At the functional tip of the tooth, the longitudinal fibers are continuous with similarly oriented crystals in the enameloid (Chapter 12), while at the basal extremity, the longitudinal fibers form part of the tooth attachment (see Chapter 4).

The presence of peritubular dentine in fish is suggested by the fact that dentine surrounding dentinal tubules has a higher electron density in electron micrographs (Shellis and Miles, 1976). This tissue has also been demonstrated by scanning electron microscopy in a **piranha** (*Serrasalmus rhombeus*; Shellis and Berkovitz, 1976).

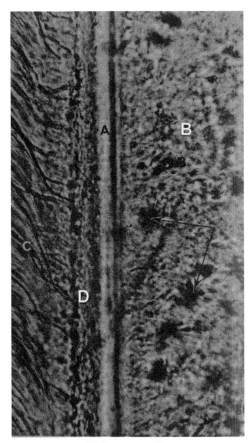

FIGURE 11.28 *Caiman sclerops* (Reptilia; Crocodylidae). Ground section of part of root showing acellular cementum (A), covered by a layer of cellular cementum (B), containing cementocyte lacunae (*arrowed*). (C) dentinal tubules and (D) granular layer. *From Berkovitz1, B.K.B., Sloan, P., 1979. Attachment tissues of the teeth in* Caiman sclerops *(Crocodilia). J. Zool. Lond. 187, 179–194, Courtesy Editors of the Journal of Zoology, London.*

OSTEODENTINE

The sole study of the mechanical properties of osteodentine (in elasmobranchs) suggests that it is similar in this respect to orthodentine (Table 11.2).

Osteodentine, the second most common form of dentine, is so-called because its histological structure superficially resembles osteonal bone, but each tubular element (**denteon**) consists of dentine, not bone, and has a central vascular canal surrounded by (ortho)dentine rather than cellular bone. During development, a network of fibers is laid down within the dental papilla and dentine is laid down on this scaffolding by odontoblasts.

Osteodentine occurs widely within all groups of **chondrichthyan fish**. It fills the pulp chamber of the tooth plates of chimaeras (Holocephali). In many sharks it forms the basal plate, which mediates attachment of the tooth. The pulp chamber of many sharks, eg, the mako shark, *Isurus oxyrinchus* (referred to as *Oxyrhina glauca* by Schmidt and Keil, 1971), the tiger shark, *Galeocerdo cuvier* (Enax et al., 2012), and the great white shark, *Carcharodon carcharias*, is filled with

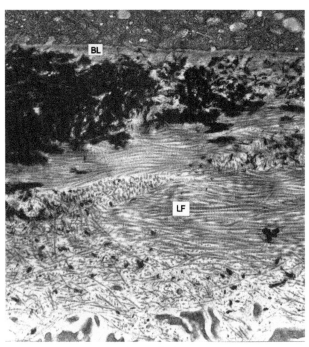

FIGURE 11.29 Parasagittal longitudinal section of developing tooth shaft of common eel (*Anguilla anguilla*). Top: inner dental epithelium separated from the tooth shaft (which is mineralizing at top left) by a basal lamina (BL). Within the predentine, prominent longitudinally oriented bundles of collagen fibers (LF) are present. Transmission electron micrograph. Image width=7.5 μm. *From Shellis, R.P., Miles, A.E.W., 1976. Observations with the electron microscope on enameloid formation in the common eel (*Anguilla anguilla; *Teleostei). Proc. R. Soc. Lond. B194, 253–269.*

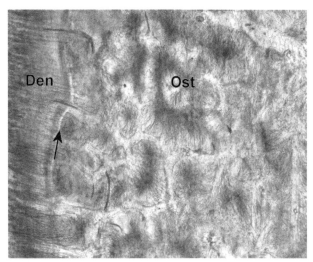

FIGURE 11.30 Longitudinal section through tooth of sand shark (*Carcharias* sp.). Junction (*arrow*) between orthodentine (Den), forming the shell of the tooth, and osteodentine (Ost), which fills the pulp cavity. Image width=2.11 mm. *Courtesy of Royal College of Surgeons. Tomes Slide Collection, Cat. no. 57.*

osteodentine: a condition referred to as **osteodont** (in contrast to **orthodont**, meaning a tooth with a connective-tissue pulp surrounded by a shell of orthodentine). The denteons in shark osteodentine have no preferred orientation and form a

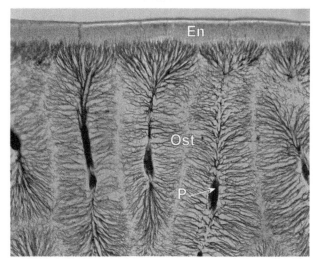

FIGURE 11.31 Vertical section of crushing tooth of eagle ray (*Myliobatis* sp.) The layer of enameloid (En) is supported by orthodentine overlying a layer of osteodentine (Ost). Each denteon surrounds a subdivision of the dental pulp (P), from which radiate dentinal tubules. Image width=2.11 mm. *Courtesy of Royal College of Surgeons. Tomes Slide Collection, Cat. no. 123.*

sponge-like structure (Fig. 11.30), adapted to countering the multidirectional forces acting on the tooth during use. The denteons that make up the osteodentine filling the pulp chamber of the block-shaped, crushing teeth of eagle rays (Myliobatidae) provide support for the flat, functional surface of orthodentine covered by enameloid (Fig. 11.31; Schmidt and Keil, 1971). This structure is adapted to resisting the strong vertical forces generated as shelled prey is crushed between the upper and lower teeth. Osteodentine also makes up the whole of the continuously growing "teeth" which project from the rostrum of **sawfish** (Elasmobranchii; Pristiformes). In these structures, the denteons are all oriented parallel with the long axis of the "tooth" and are bound together by a network of mineralized collagen fibers (Shellis and Berkovitz, 1980) (Fig. 11.32). Because the tensile strength of the denteons is considerably greater than that of the interstitial tissue, wear is more rapid at the edges of the tooth than at the tip and this helps to maintain the pointed tip and sharp edges of the "tooth."

The pulp chamber of the large fang-like teeth of several carnivorous **teleost fish** (eg, the pike, *Esox*, the barracuda, *Sphyraena* (Fig. 11.33), and the wolf-fish, *Anarhichas*) is also filled with osteodentine in which the denteons run longitudinally in the center of the tooth.

Finally, osteodentine is found in teeth of a few mammals, notably the **aardvark** (*Orycteropus afer*: Tubulidentata; Fig. 11.34). The presence of longitudinally oriented denteons in the osteodentine filling the pulp chamber gives the order its name.

VASODENTINE

This is a rare form of dentine in which blood capillaries become incorporated into the dentine as it is deposited. Blood

(A)

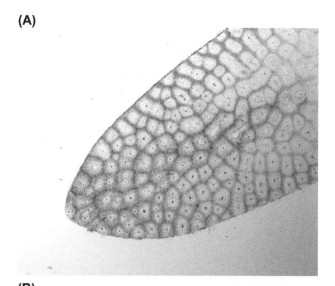

(B)

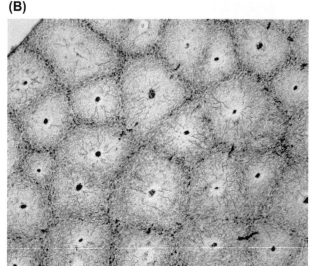

FIGURE 11.32 Sawfish (*Pristis* sp.). (A) Transverse section of rostral "tooth." Image width=4.28 mm; (B) (Inset). Detail of Fig. 11.32A. Image width=1.05 mm. *Courtesy of Royal College of Surgeons. Tomes Slide Collection, Cat. no. 104.*

continues to flow through these vessels in the living animal, so that the teeth appear red (see Fig. 4.61). From our own material and the slides in the Tomes Collection (Royal College of Surgeons), we have identified vasodentine in several **gadiform teleosts**: cod (*Gadus morhua*; Gadidae), haddock (*Melanogrammus aeglefinus*; Gadidae), hake (*Merluccius merluccius*; Merluccidae), and ling (*Molva molva*; Lotidae; Fig. 11.35). It is also present in the **pike-perch** (*Sander* sp.; Percidae), the **Atlantic wolf-fish** (*Anarhichas lupus*; Anarhichidae) and a **butterfly fish** (Chaetodontidae) (see Fig. 4.103). In all cases the outer dentine was free of capillaries and was a form of orthodentine. It has also been reported that vasodentine fills the pulp chamber of the three-toed sloth (*Bradypus*: Mammalia, Xenarthra; Schmidt and Keil, 1971), but this observation requires further confirmation. Schmidt

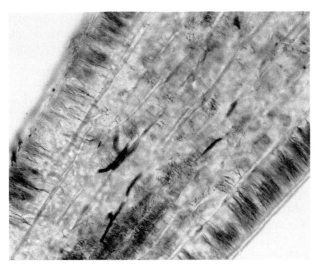

FIGURE 11.33 Longitudinal section of tooth from barracuda (*Sphyraena* sp.). The sides of the tooth are covered by a thin layer of enameloid. Inside this lies a layer of orthodentine in which tubules are oriented radially. The center of the tooth is occupied by osteodentine in which the denteons are aligned parallel with the long axis of the tooth. Image width=1.05 mm. *Courtesy of Royal College of Surgeons. Tomes Slide Collection, Cat. no. 341.*

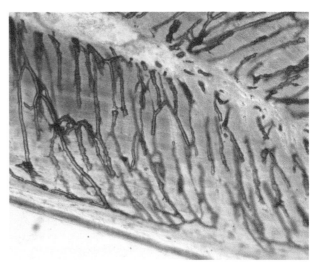

FIGURE 11.35 Longitudinal section of tooth from ling (*Molva molva*). At the bottom of the field is an outer layer of dentine lacking tubules. Most of the dentine thickness consists of vasodentine, which is permeated by blood capillaries that connect with the dental pulp (upper left). Image width=1.05 mm. *Courtesy of Royal College of Surgeons. Tomes Slide Collection, Cat. no. 376.*

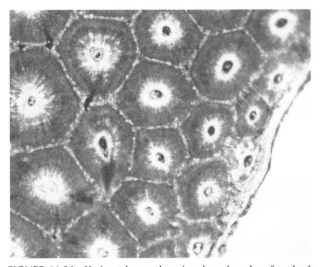

FIGURE 11.34 Horizontal ground section through molar of aardvark (*Orycteropus afer*), showing close-packed denteons of osteodentine. Image width=1.05 mm. *Courtesy of Royal College of Surgeons. Tomes Slide Collection, Cat. no. 776.*

and Keil reported that the tissue in question is formed as the teeth wear down, rather than as a primary tissue as in other species. Moreover, the histological structure resembles osteodentine rather than typical vasodentine.

PLICIDENTINE

As the name suggests, plicidentine is extensively folded vertically. This has the effect of increasing the bending resistance of the tooth by forming internal longitudinally oriented ribs. The formation of plicidentine has not been studied, but histological observations indicate that a scaffolding of fibrous tissue is laid down first, upon which dentine is then laid down by odontoblasts. The extinct amphibians known as **labyrinthodonts** were so named because all possessed teeth based on elaborately folded plicidentine. Within the extant vertebrates, plicidentine has a patchy distribution. Among Actinopterygii, the bases of the teeth are made of plicidentine in *Lepisosteus* (Holostei) and, among the teleosts, in *Arapaima* (Osteoglossiformes; Osteoglossidae) and several characiforms: *Hoplias* (Erythrinidae), *Hydrocynus* (Alestidae; see ridged tooth base in Fig. 4.30B), and possibly *Hydrolycus* (Cynodontidae; Meunier et al., 2015a). The bases of the fangs of *Latimeria* (Sarcopterygii) also consist of plicidentine (Meunier et al., 2015b). Among **reptiles**, plicidentine occurs in Helodermatidae, Lanthanotidae, and Varanidae (Fig. 11.36; Schultze, 1970; Kearney and Rieppel, 2006).

In all of these taxa, plicidentine forms the basal portion of the tooth shaft. The form of the infoldings is relatively simple, compared to that which occurred in many fossil amphibian and reptilian lineages, and the pulp chamber is not filled with osteodentine (Schultze, 1970). Rieppel and Labhardt (1979) suggested that the presence of plicidentine at the tooth base increases the strength of ankylosis.

HYPERMINERALIZED TISSUES OF TOOTH PLATES

The tooth plates of the chimaeroids (Chondrichthyes; Holocephali) and the dipnoans (Osteichthyes; Dipnomorpha)

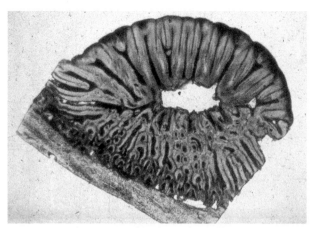

FIGURE 11.36 Ground transverse section of basal portion of the tooth of a monitor lizard (*Varanus* sp.) showing the folded appearance of plicidentine. Image width = 2.3 mm.

are not replaced, so are subject to considerable wear during the life of the fish, which can be of considerable duration. The enameloid or enamel layer covering the surface is too thin to withstand prolonged abrasion. Instead, the plates are reinforced by deposition of hypermineralized tissue, which forms a composite material with dentine. As the tooth is abraded by use, this tissue becomes exposed to form the wear-resistant functional surface and its loss through tooth function is compensated by continuing formation on the aspect away from the functional surface. The formative cells of these hypermineralized tissues are of mesenchymal origin and the initial matrix contains collagen fibers. Although collagen is present in the matrix, it either has no significant role in mineralization (pleromin) or, along with other organic components, it is removed during mineralization (petrodentine). The mineral content reaches a similar level to that of enamel or enameloid and the hypermineralization is achieved without contact with epithelial cells.

Pleromin

The tooth plates of chimaeroid fish are composed mainly of osteodentine. However, each plate carries on the oral surface a mass of hypermineralized tissue known as **pleromin** (Didier et al., 1994), which is used in crushing or cutting prey. The quantity of pleromin varies between the different tooth plates, there being least on the upper tooth plates. It also varies between taxa (Didier et al., 1994). In *Callorhinchus*, pleromin forms a relatively large mass known as the **tritor**. In *Chimaera*, the pleromin forms slender rods embedded in the bone making up the plates and is exposed at the cutting edges. Other taxa, such as *Rhinochimaera*, may lack pleromin.

Pleromin is formed by mesenchymal cells (**pleromoblasts**), apparently derived from osteoblasts (Ishiyama

et al., 1991). The cells are columnar in form and give off numerous cytoplasmic processes which penetrate the matrix. The matrix, which is formed completely before mineralization begins, contains collagen fibers, especially in the last-formed matrix adjacent to the layer of pleromoblasts, but the main component is a meshwork of membrane-bound tubular vesicles which are 25–250 nm in diameter. Mineralization begins by deposition of fine, needle-like crystals (possibly hydroxyapatite) in matrix vesicles located in the last-formed matrix. Such crystals are, however, deposited only in relation to the rather sparse collagen fibers. Most of the mineral is deposited as polygonal crystals within the tubular vesicles: a process reminiscent of enameloid mineralization of elasmobranchs (Chapter 12). Pleromin is unique in that the principal mineral is not hydroxyapatite, as in other dental tissues, but consists of coarsely granular crystals which have been identified (Ishiyama et al., 1984, 1991) as magnesium-containing β-tricalcium phosphate, β-$(Ca,Mg)_3(PO_4)_2$: a mineral which does not otherwise occur as a normal constituent of dental tissues.

Petrodentine

The tooth plates of dipnoans are composite structures: during their initiation and growth, odontodes are formed separately and then incorporated into the plate by continuing hard tissue formation. The complex histological structure of the plates has given rise to considerable disagreement about terminology (Smith, 1985). Here we will use as simple a terminology as possible. The outer shell of each odontode consists of a thin layer of enamel supported by a layer of mantle dentine. Petrodentine is first deposited within this shell to form an undivided core. Later, the petrodentine increases in thickness and in area and incorporates vertical columns of vascular tissue continuous with the pulp chamber. Dentine is laid down on the walls of these connective-tissue columns, so the tissue as a whole visually resembles osteodentine, but with widely spaced denteons within a matrix of hypermineralized petrodentine. Petrodentine is formed by mesenchymal cells (**petroblasts**) and it appears that these cells and those forming the denteons are both derived from the same population of dental pulp cells (Ishiyama and Teraki, 1990; Kemp, 2003). Petroblasts resemble odontoblasts in that the nucleus is located proximally and the expanded distal cytoplasm contains organelles indicative of protein synthesis and secretion. Ishiyama and Teraki (1990) detected several cytological variations among petroblasts and suggested that these represented stages in an episodic or cyclical process of petrodentine formation: (1) secretion of collagenous matrix by petroblasts with poorly developed cell processes; (2) breakdown of the matrix, mediated by elaborately branched and anastomosing processes; (3) mineralization, accompanied by disintegration of the petroblast processes. Many aspects of this hypothesis warrant further research, but it is certain that mineralization of petrodentine achieves much higher levels than in dentine and that this is accompanied by loss of nearly all the organic matrix (Ishiyama and Teraki, 1990; Kemp, 2003). Mature petrodentine has a very high mineral content, and consists of tightly interwoven, highly elongated crystals of hydroxyapatite which have cross-sectional dimensions of about 70×100 nm (Ishiyama and Teraki, 1990), which are similar to those of enamel (Verbeeck, 1986). Because of the high mineral content, the hardness of petrodentine (2.5–3.3 GPa) is much greater than that of dentine (Tables 11.1 and 11.2) and almost as great as that of enameloid (3.2–3.5 GPa) or mammalian enamel (3.0–3.9 GPa; Currey and Abeysekera, 2003; Whitenack et al., 2010).

ONLINE RESOURCES

US Fish and Wildlife Service Forensics Laboratory. "Identification Guide for Ivory and Ivory Substitutes." http://www.fws.gov/lab/ivory.php.

REFERENCES

Arthur, A., Shi, S., Gronthos, S., 2014. Dental pulp stem cells. In: Vishwakarma, A., Sharpe, P.T., Songtao, S., Ramalingam, M. (Eds.), Stem Cell Biology and Tissue Engineering in Dental Sciences. Elsevier, London, pp. 279–289.

Berkovitz, B.K.B., Sloan, P., 1979. Attachment tissues of the teeth in *Caiman sclerops* (Crocodilia). J. Zool. Lond. 187, 179–194.

Berkovitz, B.K.B., Holland, G.R., Moxham, B.J., 2009. Oral Anatomy, Histology and Embryology, fourth ed. Elsevier, London.

Berkovitz, B.K.B., 1977. Chronology of tooth development in rainbow trout (*Salmo gairdneri*). J. Exp. Zool. 200, 65–69.

Bhingare, A.C., Ohno, T., Tomura, M., Zhang, C., Aramaki, O., Otsuki, M., Tagami, J., Azuma, M., 2014. Dental pulp dendritic cells migrate to regional lymph nodes. J. Dent. Res. 93, 288–293.

Bleicher, F., 2014. Odontoblast physiology. Exp. Cell Res. 325, 65–71.

Caviedes-Bucheli, J., Muñoz, H.R., Azuero-Holguín, M.M., Ulate, E., 2008. Neuropeptides in dental pulp: the silent protagonists. J. Endod. 34, 773–788.

Chung, G., Jung, S.J., Oh, S.B., 2013. Cellular and molecular mechanisms of dental nociception. J. Dent. Res. 92, 948–955.

Crucke, J., van de Kelft, A., Huysseune, A., 2015. The innervation of the zebrafish pharyngeal jaws and teeth. J. Anat. 227, 62–71.

Currey, J.D., Abeysekera, R.M., 2003. The microhardness and fracture surface of the petrodentine of *Lepidosiren* (Dipnoi), and of other mineralised tissues. Arch. Oral Biol. 48, 439–447.

Davit Béal, T., Allizard, F., Sire, J.-Y., 2006. Morphological variations in a tooth family through ontogeny in *Pleuronectes waltl* (Lissamphibia, Caudata). J. Morphol. 257, 1048–1065.

Dean, M.C., Scandrett, A.E., 1996. The relation between long-period incremental markings in dentine and daily cross-striations in enamel in human teeth. Arch. Oral Biol. 41, 233–241.

Dean, M.C., 1995. The nature and periodicity of incremental lines in primate dentine and their relationship to periradicular bands in OH 16 (*Homo habilis*). In: Moggi-Cecchi, J. (Ed.), Aspects of Dental Biology: Palaeontology, Anthropology and Evolution. Florence, International Institute for the Study of Man, pp. 239–265.

Dean, M.C., 1998. Comparative observations on the spacing of short-period (von Ebner's) lines in dentine. Arch. Oral Biol. 43, 1009–1021.

Deshpande, A.S., Fang, P.-A., Zhang, X., Jayaraman, T., Sfeir, C., Beniash, E., 2011. Primary structure and phosphorylation of dentin matrix protein 1 (DMP1) and dentin phosphophoryn (DPP) uniquely determine their role in biomineralization. Biomacromolecules 12, 2933–2945.

Didier, D.A., Stahl, B.J., Zangerl, R., 1994. Development and growth of compound tooth plates. *Callorhinchus milii* (Chondrichthyes, Holocephali). J. Morphol. 222, 73–89.

Enax, J., Prymak, O., Raube, D., Epple, M., 2012. Structure, composition, and mechanical properties of shark teeth. J. Struct. Biol. 178, 290–299.

Enax, J., Fabritius, H.-O., Rack, A., Prymak, O., Raube, D., Epple, M., 2013. Characterization of crocodile teeth: correlation of composition, microstructure, and hardness. J. Struct. Biol. 184, 155–163.

Erickson, G.M., 1996b. Incremental lines of von Ebner in dinosaurs and the assessment of tooth replacement rates using growth line counts. Proc. Natl. Acad. Sci. U.S.A. 93, 14623–14627.

Erickson, G.M., 1996a. Daily deposition of dentine in juvenile *Alligator* and assessment of tooth replacement rates using incremental line counts. J. Morphol. 228, 189–194.

Goldberg, M., Kulkarni, A.B., Young, M., Boskey, A., 2011. Dentin: structure, composition and mineralization: the role of dentin ECM in dentin formation and mineralization. Front. Biosci. 3, 711–735.

Hahn, C.-L., Liewehr, F.R., 2007. Update on the adaptive immune responses of the dental pulp. J. Endod. 33, 773–781.

He, L.H., Swain, M.V., 2008. Understanding the mechanical behaviour of human enamel from its structural and compositional characteristics. J. Mech. Behav. Biomed. Mater 1, 18–29.

Hildebrand, C., Fried, K., Tuisku, F., Johansson, C.S., 1995. Teeth and tooth nerves. Progr. Neurobiol. 45, 165–222.

Huang, G.T.-J., Gronthos, S., Shi, S., 2009. Mesenchymal stem cells derived from dental tissues vs. those from other sources: their biology and role in regenerative medicine. J. Dent. Res. 88, 792–806.

Ishiyama, M., Teraki, Y., 1990. The fine structure and formation of hypermineralized petrodentine in the tooth plate of extant lungfish (*Lepidosiren paradoxa* and *Protopterus* sp.). Arch. Histol. Cytochem. 53, 307–321.

Ishiyama, M., Sasagawa, I., Akai, J., 1984. The inorganic content of pleromin in tooth plates of the living holocephalan, *Chimaera phantasma*, consists of a crystalline calcium phosphate known as β-$Ca_3(PO_4)_2$ (Whitlockite). Arch. Histol. Jap. 47, 89–94.

Ishiyama, M., Yoshie, S., Teraki, Y., Cooper, E.W.T., 1991. Ultrastructure of pleromin, a highly mineralized tissue comprizing crystalline calcium phosphate known as whitlockite, in holocephalan tooth plates. In: Suga, S., Nakahara, H. (Eds.), Mechanisms and Phylogeny of Mineralization in Biological Systems. Springer, Berlin, pp. 453–457.

Ivancik, J., Arola, D., 2013. The importance of microstructural variations on the fracture toughness of human dentin. Biomaterials 34, 864–874.

Kearney, M., Rieppel, O., 2006. An investigation into the occurrence of plicidentine in the teeth of squamate reptiles. Copeia 337–350.

Kemp, A., 2003. Ultrastructure of developing tooth plates in the Australian lungfish, *Neoceratodus forsteri* (Osteichthyes: Dipnoi). Tissue Cell 35, 401–426.

Kinney, J.H., Habelitz, S., Marshall, S.J., Marshall, G.W., 2003a. The importance of intrafibrillar mineralization of collagen on the mechanical properties of dentin. J. Dent. Res. 82, 957–961.

Kinney, J.H., Marshall, S.J., Marshall, G.W., 2003b. The mechanical properties of human dentine: a critical review and re-evaluation of the dental literature. Crit. Rev. Oral Biol. Med. 14, 13–29.

Landis, W.J., 1996. Mineral characterization in calcifying tissues: atomic, molecular and macromolecular perspectives. Conn. Tiss. Res. 34, 239–246.

Linde, A., Goldberg, M., 1993. Dentinogenesis. Crit. Rev. Oral Biol. Med. 4, 679–728.

Loch, C., Swain, M.V., van Vuuren, L.J., Kieser, J.A., Fordyce, R.E., 2013. Mechanical properties of dental tissues in dolphins (Cetacea: Delphinoidea and Inioidea). Arch. Oral Biol. 58, 773–779.

Magloire, H., Maurin, J.C., Couble, M.L., Shibukawa, Y., Tsumura, M., Thivichon-Prince, B., Bleicher, F., 2010. Topical review. Dental pain and odontoblasts: facts and hypotheses. J. Orofac. Pain 24, 335–349.

Märten, A., Fratzl, P., Paris, O., Zaslansky, P., 2010. On the mineral in collagen of human crown dentine. Biomaterials 31, 5479–5490.

Meunier, F.J., Mondéjar-Fernández, J., Goussard, F., Clément, G., Herbin, M., 2015b. Presence of plicidentine in the oral teeth of the coelacanth *Latimeria chalumnae* Smith 1939 (Sarcopterygii; Actinistia). J. Struct. Biol. 190, 31–37.

Meunier, F.J., De Mayrinck, D., Brito, P.M., 2015a. Presence of plicidentine in the labial teeth of *Hoplias aimara* (Erythrinidae; Ostariophysi; Teleostei). Acta Zool. 96, 174–180.

Miles, A.E.W., White, J.W., 1960. Ivory. Proc. R. Soc. Med. 53, 775–780.

Prasad, M., Butler, W.T., Qin, C., 2010. Dentin sialophosphoprotein in biomineralization. Conn. Tiss. Res. 51, 404–417.

Raubenheimer, E.J., 1999. Morphological aspect and composition of African elephant (*Loxodonta africana*) ivory. Koedoe 42, 57–64.

Rieppel, O., Labhardt, L., 1979. Mandibular mechanics in *Varanus niloticus* (Reptilia: Lacertilia). Herpetologica 35, 158–163.

Schmidt, W.J., Keil, A., 1971. Polarizing Microscopy of Dental Tissues (D.F.G. Poole, A.I. Darling, Trans.). Pergamon, Oxford.

Schultze, H.-P., November 1970. Folded teeth and the monophyletic origin of tetrapods. Am. Mus. 2408, 1–10.

Shellis, R.P., Berkovitz, B.K.B., 1976. Observations on the dental anatomy of piranhas (Characidae) with special reference to tooth structure. J. Zool. Lond. 180, 69–84.

Shellis, R.P., Berkovitz, B.K.B., 1980. Dentine structure in the rostral teeth of the sawfish *Pristis* (Elasmobranchii). Arch. Oral Biol. 25, 339–343.

Shellis, R.P., Miles, A.E.W., 1976. Observations with the electron microscope on enameloid formation in the common eel (*Anguilla anguilla*; Teleostei). Proc. R. Soc. Lond. B194, 253–269.

Shellis, R.P., Poole, D.F.G., 1978. The structure of the dental hard tissues of the coelacanthid fish *Latimeria chalumnae* Smith. Arch. Oral Biol. 23, 1105–1113.

Shellis, R.P., 1983. Structural organization of calcospherites in normal and rachitic human dentine. Arch. Oral Biol. 28, 85–95.

Shellis, R.P., Curtis, A.R., 2010. A minimally destructive technique for removing the smear layer from dentine surfaces. J. Dent. 38, 941–944.

Shellis, R.P., Featherstone, J.D.B., Lussi, A., 2014. Understanding the chemistry of dental erosion. In: Lussi, A., Ganss, C. (Eds.), Erosive Tooth Wear (Monographs in Oral Science 25). Karger, Base, pp. 163–179.

Sire, J.-Y., Davit-Béal, T., Delgado, S., van der Heyden, C., Huysseune, A., 2002. First-generation teeth in nonmammalian lineages: evidence for a conserved ancestral character? Microsc. Res. Tech. 59, 408–434.

Smith, A.J., Lesot, H., 2001. Induction and regulation of crown dentinogenesis: embryonic events as a template for dental tissue repair? Crit. Rev. Oral Biol. Med. 12, 425–437.

Smith, A.J., Scheven, B.A., Takahashi, Y., Ferracane, J.L., Shelton, R.M., Cooper, P.R., 2012. Dentine as a bioactive extracellular matrix. Arch. Oral Biol. 57, 109–121.

Smith, M.M., 1985. The pattern of histogenesis and growth of tooth plates in larval stages of extant lungfish. J. Anat. 140, 627–643.

Staines, K.A., MacRae, V.E., Farquharson, C., 2012. The importance of the SIBLING family of proteins on skeletal mineralisation and bone remodelling. J. Endocrinol. 214, 241–255.

Su, X.W., Cui, F.Z., 1999. Hierarchical structure of ivory: from nanometer to centimetre. Mat. Sci. Eng. C 7, 19–29.

Tuisku, F., Hildebrand, C., 1995. Immunohistochemical and electron microscopic demonstration of nerve fibres in relation to gingivae, tooth germs and functional teeth in the lower jaw of the cichlid *Tilapia mariae*. Arch. Oral Biol. 40, 513–520.

Ulmer, F.L., Winkel, A., Kohorst, P., Stiesch, M., 2010. Stem cells – prospects in dentistry. Schweiz Monatsschr. Zahnmed. 120, 860–872.

Verbeeck, R.M.H., 1986. Minerals in human enamel and dentine. In: Driessens, F.C.M., Wöltgens, J.H.M. (Eds.), Tooth Development and Caries. CRC Press, Boca Raton, pp. 95–152.

West, N.X., Lussi, A., Seong, J., Hellwig, E., 2013. Dentin hypersensitivity: pain mechanisms and aetiology of exposed cervical dentin. Clin. Oral Invest. 17 (Suppl. 1), S9–S19.

Whitenack, L.B., Simkins, D.C., Motta, P.J., Hirai, M., Kumar, A., 2010. Young's modulus and hardness of shark tooth biomaterials. Arch. Oral Biol. 55, 203–209.

Zhang, Y.-R., Du, W., Zhou, X.-D., Yu, H.-Y., 2014. Review of research on the mechanical properties of the human tooth. Int. J. Oral Sci. 6, 61–69.

Chapter 12

Enameloid and Enamel

COMPOSITION AND PROPERTIES

The functional surfaces of most teeth are covered by either **enamel** or **enameloid**, which are much more highly mineralized than the supporting dentine and are hence harder and more rigid. These properties enable teeth to overcome the mechanical resistance of foods. The available data (Table 12.1) show that the mineral content (calcium and phosphate contents) and porosity (water content) are similar in the two tissues. Although measurements of hardness and elastic modulus tend to vary according to method of measurement, age of the tooth, location within the tissue, and other factors, the data in Table 12.1 indicate that, overall, the mechanical properties of enamel and enameloid are similar.

These tissues are characterized, not only by high mineral contents, but by the large crystal size and high degree of crystallinity of the mineral phase. The mineral of enamel, and of the enameloid of urodele amphibians and actinopterygians, is a form of hydroxyapatite, $Ca_5(PO_4)_3OH$, which contains impurities, such as carbonate, magnesium, and sodium, but at much lower concentrations than the apatite in dentine or bone. The crystals are long and flattened. In human enamel the crystals are approximately 70-nm wide and 25-nm thick (Daculsi and Kérébel, 1978).

The enameloid of elasmobranchs is distinguished by a consistently high-fluoride content approximating that in fluorapatite, $Ca_5(PO_4)_3F$ (Suga et al., 1991): 2.33–3.1% (w/w) among rays and 3.1–4.9% (w/w) among sharks. The mineral was considered by Miake et al. (1991) to be a highly fluoridated form of carbonate-containing apatite. The morphology of the crystals varies during formation but in mature elasmobranch enameloid they have regular hexagonal cross-sections with a diameter of up to about 100 nm (Fig. 12.1; Sasagawa and Akai, 1992; Chen et al., 2014).

Among bony fish, the fluoride content of enameloid varies widely, from about 0.2% (w/w) to about 3% (w/w); the fluoride content is not related to diet or the fluoride concentration in the aquatic environment (Suga et al., 1991). Crystals from high-fluoride enameloid, eg, that of the red sea bream (*Pagrus major*), have similar morphology to those in elasmobranch enameloid, whereas low-fluoride enameloid has crystals with flattened hexagonal cross-sections up to 50–60 nm wide (Fig. 12.2; Shellis and Miles, 1976; Miake et al., 1991). The true length of crystals in

enamel and enameloid is difficult to determine, but by use of special techniques, lengths of over 100 μm have been observed in immature human enamel (Daculsi et al., 1984). Enameloid has only been studied in conventional ultrathin sections, in which crystals tend to fracture, but the crystals in eel enameloid are clearly more than 400-nm long (Shellis and Miles, 1976). For comparison, crystals of dentine mineral are only about 30-nm long and 3-nm thick (Daculsi et al., 1978).

The structure of enameloid is highly variable, but the enamel of non-mammalian vertebrates is much simpler, in that the enamel crystals are oriented fairly uniformly at right angles to the enamel surface. This is because the distal poles of the inner dental epithelium (IDE) cells are flat, and they form a smooth enamel surface. Because enamel crystals grow along the concentration gradients, which run from outside to inside, the smooth enamel surface is correlated with uniform, radial crystal orientation. This type of enamel is called **aprismatic** or **prismless**. In contrast, the enamel of mammals contains **prisms** or **rods**: elongated domains which have a one-to-one relationship with the IDE cells and which are distinguished by abrupt and regular changes in crystal orientation. The formation of prisms is correlated with the presence of a projection at the distal pole of the IDE cells (the **Tomes process**).

GENERAL ASPECTS OF FORMATION

Some aspects of the histology of formation of enamel and enameloid are described in Chapter 9.

Enamel

Enamel is a product solely of the IDE and is laid down on the first-formed layer of dentine. It forms in two stages: matrix production and maturation. Both stages are under the control of the IDE cells, which pass through two corresponding stages of differentiation (Smith and Nanci, 1995).

During **matrix formation**, the IDE cells elongate and become polarized, with the nucleus located proximally (away from the tooth). The distal cytoplasm (toward the tooth) acquires the organelles needed for synthesis and secretion of protein: rough endoplasmic reticulum, an active Golgi apparatus and secretory vesicles containing the matrix proteins destined for export. The organic matrix

The Teeth of Non-Mammalian Vertebrates. http://dx.doi.org/10.1016/B978-0-12-802850-6.00012-6

TABLE 12.1 Mineral Content (Calcium and Phosphorus Concentrations), Porosity (Water Concentration), Hardness, and Elastic Modulus of Enamel and Enameloid

Class	Species	%Ca (w/w)	%P (w/w)	%H$_2$O (w/w)	Hardness (GPa)	Reduced Elastic Modulus (GPa)	References
Enamel							
Mammalia	*Homo sapiens*	36.6	17.7	1.15	3.1–5.7	73–105	Verbeeck (1986), Dibdin and Poole (1982), and Zhang (2014)
Reptilia	*Sphenodon punctatus* (indentation measurements)				4.0	73.2	Kieser et al. (2011)
Reptilia	*S. punctatus* (nanobeam measurements)					30 normal / 42 parallel	Yilmaz et al. (2014)
Reptilia	*Crocodilus porosus*	38.8	14.4	3.3	3.2		Enax et al. (2013)
Enameloid							
Chondrichthyes	*Sphyrna tiburo*				3.5	68.9	Whitenack et al. (2010)
Chondrichthyes	*Carcharinus taurus*				3.2	72.6	Whitenack et al. (2010)
Chondrichthyes	*Isurus oxyrhinchus*	37.8	17.7	3.08	6.2–7.5	99–131	Enax et al. (2012)
Chondrichthyes	*Galeocerdo cuvier*	31.2	17.0	3.13	5.8–6.1	94–100	Enax et al. (2012)
Chondrichthyes	*Carcharodon carcharias*				2.6, 4.1[a]	77.2,84.4[a]	Chen et al. (2012)
Osteichthyes	*Serrasalmus manueli*				3.1.4.1[a]	81.9,86.5[a]	Chen et al. (2012)

[a]*First value=the mean value recorded in dry tissue; second=the mean value for hydrated tissue.*

of immature mammalian enamel consists of a number of unique proteins, mainly amelogenin (which makes up 90% of the matrix), enamelin, ameloblastin, and a number of proteases (Moradian-Oldak, 2013). The enamel matrix begins to mineralize soon after secretion. Slender crystals are deposited near the enamel–dentine junction and then grow outward toward the outer surface as the matrix grows in thickness.

Once the enamel has reached its full thickness, **maturation** begins. The synthesis of matrix protein by the IDE cells ceases, but the cells secrete proteases, such as matrix metalloproteinase 20 and kallikrein-4, which break down matrix protein. Not only is there a quantitative reduction in the amount of protein present in enamel, but there is also a qualitative change, as amelogenin is preferentially removed. The products of matrix hydrolysis are removed by the IDE

and the process completed within lysosomes. At the same time the IDE cells pump calcium and phosphate ions into the enamel. A feature of maturation-stage IDE cells in nearly all vertebrates which form enamel is elaborate folding of the cell membrane facing the enamel to become a ruffled border. This border has a large surface area, which facilitates both inward and outward transport. The distal cell membrane also becomes enriched with ion-transporting enzymes such as ATPases. In mammals, the cellular changes in the IDE during maturation are more complex (Smith and Nanci, 1995) but less is known about the process in non-mammalian vertebrates.

As the basic features of enamel formation and of IDE cell differentiation are shared by all vertebrates which have enamel-covered teeth, they will not be repeated in the following account.

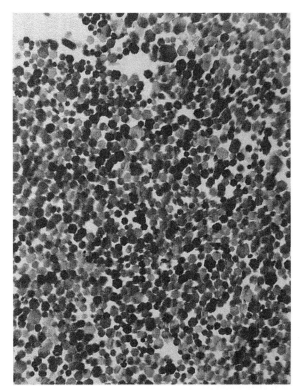

FIGURE 12.1 Crystals from enameloid of lesser spotted catfish (*Scyliorhinus canicula*; Elasmobranchii). Note the symmetrical hexagonal cross-sections. Transmission electron micrograph. Image width approximately 4 μm.

FIGURE 12.2 Crystals from mature enameloid of a teleost fish, the European eel (*Anguilla anguilla*; Teleostei), in an ultrathin section which has fractured. The crystals are plate-like (cf. Fig. 12.1) and elongated but, as they fracture easily during sectioning, their true length is difficult to determine. Transmission electron micrograph. Image width = 900 nm.

Enameloid

In contrast to enamel, enameloid matrix is laid down between the IDE and the ectomesenchymal cells of the dental papilla; before dentine apposition begins, its formation is accompanied by signs of differentiation in both tissue layers. Enameloid matrix contains collagen and other proteins, although the nature and origin of the matrix differ considerably between elasmobranchs on the one hand, and actinopterygians and urodeles on the other. As in enamel formation, mineralization involves removal of most of the matrix and its replacement by considerable quantities of mineral.

CHONDRICHTHYAN ENAMELOID

Structure

Among chondrichthyans, the structure of enameloid differs between sharks and rays (Schmidt and Keil, 1971; Reif, 1973; Enault et al., 2013). In shark enameloid four variations in the form and arrangement of crystals can be distinguished (Reif, 1973) and, in any given species, the enameloid contains two or three structurally differentiated regions.

The enameloid on the crushing posterior teeth of hornsharks (*Heterodontus*) consists of two layers (Reif, 1973). The outer layer (about 100 μm thick) consists of **single-crystal enameloid**, which is made up of thick, elongated crystals oriented perpendicular to the functional surface. This structural variant does not occur among other neoselachians but was the dominant form of enameloid in nonneoselachian sharks (Reif, 1979; Gillis and Donoghue, 2007). In *Heterodontus*, the outer layer of single-crystal enameloid rests upon an underlying layer of enameloid in which the crystals are thinner than in single-crystal enameloid and are grouped into bundles which show a variety of orientations (Fig. 12.3A). This tissue was termed **woven enameloid** by Reif (1979). Both single-crystal and woven enameloid are adapted to resisting compressive forces (Reif, 1973; Preuschoft et al., 1974).

Among other extant sharks, which have teeth with tall cusps or sharp cutting edges, the structure of enameloid is more complex than in *Heterodontus* and comprises three layers (Fig. 12.3B). The inner layer consists of woven enameloid. Overlying this is a layer of **parallel-structured enameloid**. Like woven enameloid, this consists of crystals organized into bundles. However, most of the fiber bundles are longitudinally oriented and form "palisades" (Garant, 1970), between which run radially oriented crystal bundles. The longitudinal bundles of the parallel-structured enameloid are structurally continuous with the inner, woven layer and the radial bundles run through both the outer and inner layers, tying them together. The longitudinal bundles of parallel-structured enameloid are adapted to resisting bending forces (Reif, 1973; Preuschoft et al., 1974). The combination of this layer with an inner woven

FIGURE 12.3 Diagrams of elasmobranch enameloid structure. (A) *Heterodontus*, showing outer layer of single-crystal enameloid (SCEn) overlying an inner layer of woven enameloid (WEn); (B) Part of transverse section of shark tooth, showing the three-layered enameloid structure: from the inside out, woven enameloid (WEn), parallel-structured enameloid, in which the crystals are cut and are represented by dots (PSEn) and the outer parallel-structured enameloid (OPSEn), which also forms the cutting edge. The parallel-structured enameloid is traversed by radial crystal bundles (RCB). The outer surface of the tooth is covered by the thin shiny layer (SL). An osteodont tooth, with the pulp cavity filled with osteodentine (OstDen) is represented. Orthodentine is shown black in both (A) and (B). *After Schmidt W.J., Keil A., 1971. Polarizing Microscopy of Dental Tissues; Poole DFG, Darling AI (Transl). Oxford, Pergamon and Reif W.-E., 1973. Morphologie und Ultrastruktur des Hai-Schmelzes. Zool. Scr. 2, 231–250.*

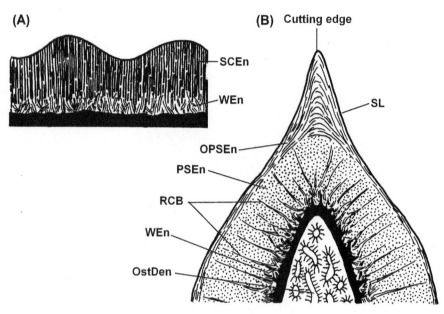

layer therefore confers great strength on the enameloid layer as a whole. The images in Schmidt and Keil (1971) show that the relative thickness of the inner woven layer varies considerably: it is virtually absent in the mako shark *Isurus oxyrhinchus* [*Oxyrhina glauca*] but occupies about 25% of the enameloid layer in the sand tiger shark *Carcharias*. The mechanical significance of this variation has not been explored, but it is worth noting that *Isurus* has pointed teeth with a fairly rounded cross-section, while *Carcharias* has triangular, flattened teeth with sharp cutting edges.

The cutting edges of shark teeth have a structure distinct from that of the rest of the tooth. The enameloid at the edges is composed of crystal bundles originating in the thin outer layer of the parallel-structured enameloid. The crystal bundles contributing to outer parallel-fibered enameloid and the cutting edges lie parallel with the tooth surface but are oriented approximately perpendicular to the cutting edge. They may be continuous with the vertical bundles of the outer enameloid layer or may be superimposed on the latter (Fig. 12.3B).

The outermost layer of shark teeth is composed of a **shiny layer**, 1–4 µm thick, which consists of randomly oriented large, elongated crystals lying in the plane of the tooth surface. These large crystals are embedded in a fine-grained acid-soluble matrix so become evident only after brief etching (Reif, 1973).

The enameloid of rays lacks the structural differentiation seen in sharks. Although there exist only a very small number of studies, these suggest that, among myliobatids and rajiforms, the constituent crystals are typically oriented perpendicular to the functional surface (Schmidt and Keil, 1971; Enault et al., 2013). The enameloid crystals in the myliobatoids studied by Sasagawa and Akai (1992, 1999) have the morphology typical for elasmobranchs, with regular hexagonal cross-sections. In the mobulids and gymnurids, which are plankton feeders, the enameloid layer is reduced in thickness. The resolution of available scanning electron microscopy observations (Enault et al., 2013) is not adequate to discern whether the enameloid has a single-crystal structure or not.

Formation

The formation of enameloid matrix is preceded by differentiation of the peripheral cells of the dental papilla into odontoblasts and of the IDE into a layer of low-columnar cells. The enameloid matrix contains odontoblast processes and collagen fibers, which originate between the odontoblasts (Fig. 12.4). Fibers are finer and more widely spaced in the outer part of the matrix layer than in the inner part. Electron microscopy shows that the fibers are of two types: fibers of ordinary thickness (15–35 nm) and a cross-banding periodicity of 55 nm, and much thicker "giant fibers" (up to 300 nm diameter) with a cross-banding periodicity of about 18 nm. (Sasagawa and Akai, 1992) However, most of the volume of the matrix is filled with a network of tubular vesicles, about 15–20 nm in diameter (Fig. 12.5), of which the walls consist of unit cell membranes (Garant, 1970; Sasagawa and Akai, 1992). There is strong evidence that the vesicles originate from the cell membranes of the odontoblasts (Sasagawa and Akai, 1992). During matrix deposition, the IDE cells continue to elongate and their cytoplasm accumulates large quantities of glycogen. The cells acquire organelles associated with protein synthesis and secretion, but mostly after the matrix has been formed. Interpretations of the ultrastructure of enameloid formation have varied widely. Nanci et al. (1983) and Kemp (1985) concluded that the IDE is the major source of matrix components, while others (Prostak and Skobe, 1988; Sasagawa and Akai, 1992) thought that it contributed little or nothing.

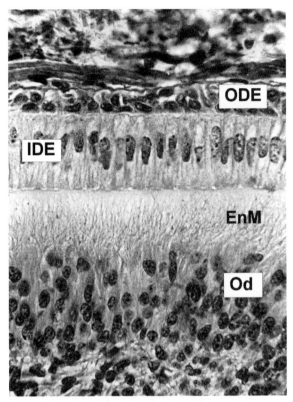

FIGURE 12.4   Enameloid matrix formation in thornback ray (*Raja clavata*). Outer dental epithelium (ODE). Inner dental epithelial (IDE) cells elongated but not polarized, with nuclei toward center of cells. The only visible components of the enameloid matrix (EnM) are collagen fibers projecting radially into the layer. Odontoblasts (Od) are polarized, with abundant distal cytoplasm and form a layer several cells deep. Demineralized section, Masson trichrome. Image width = 170 μm.

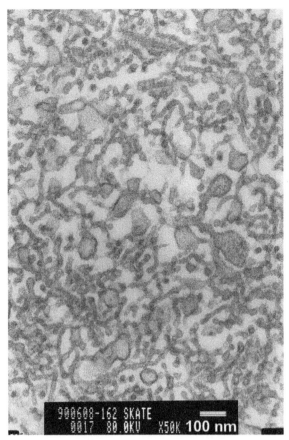

FIGURE 12.5   Round ray (*Urolophus aurantiacus*): high-power transmission electron micrograph of the enameloid matrix during the stage of matrix formation, showing a network of tubular vesicles. Demineralized with a 2.5% solution of EDTA-2Na, stained with uranyl acetate and lead citrate. Specimen no. 12217. *Courtesy of Prof. I. Sasagawa. © I. Sasagawa.*

There is other evidence that the IDE cells synthesize and secrete protein but the activity seems to be very low (Shellis, 1978; Graham, 1984). The enameloid matrix does not react differently from predentine, either with histological stains or with histochemical tests for amino acids or carbohydrate groups (Shellis, 1978), which implies that any epithelial component of the matrix is present at low concentrations.

After reaching its full thickness, the enameloid layer begins to mineralize and, at the same time, dentine begins to form on its pulpal aspect (Fig. 12.6). Enameloid mineralization begins at the cutting edge and spreads laterally. Crystals seem to be initiated exclusively within the tubular vesicles (Fig. 12.7). They then grow until they reach full thickness (Fig. 12.8; Sasagawa and Akai, 1992). During enameloid mineralization, the matrix fibers and tubular vesicles are reduced to fragments but many fibers persist. The basal lamina associated with the IDE is intact during early mineralization but becomes fragmented during growth of the mineral crystals. The glycogen which has accumulated in the IDE cells is lost during early enameloid mineralization, possibly because it provides a source of energy for the processes occurring during

mineral deposition. The IDE cells acquire Golgi complexes, smooth endoplasmic reticulum, vesicles, vacuoles, and granules of electron-dense material. A large number of vesicles are present at the distal pole next to the mineralizing enameloid. The cell membranes demonstrate activity of enzymes, such as alkaline phosphatase, acid phosphatase, and Ca-dependent ATPase. These features suggest that the IDE cells are involved in transport of mineral ions into the enameloid and the breakdown and removal of organic matrix fragments (Sasagawa, 2002).

Sawada and Inoue (2003) observed that the outermost layer of enameloid, which has a fine-grained texture and corresponds to the shiny layer, seems to form by mineralization of the inner portion of the lamina densa component of the basement membrane.

ACTINOPTERYGIAN ENAMELOID

Among bony fish, enameloid is formed in two locations. **Cap enameloid** forms a hollow cap or sheath at the tip of the tooth, while **collar enameloid** forms a layer on the surface of the tooth shaft.

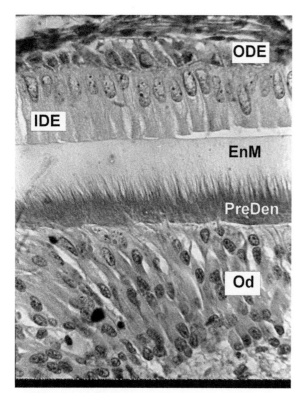

FIGURE 12.6 Early enameloid mineralization in thornback ray (*Raja clavata*). Outer dental epithelium (ODE). Inner dental epithelium (IDE) cells taller than during matrix formation (Fig. 12.4) and are polarized, with nuclei located proximally. Inner layer of collagen fibers persist in enameloid matrix (EnM). Odontoblasts (Od) still form a multicell layer of cells actively forming predentine (PreDen). Demineralized section, Masson's trichrome. Image width = 173 μm.

Cap Enameloid

Structure

The thickness of cap enameloid is greatest at the tip of the cap and diminishes laterally. On grasping/piercing teeth, the cap has the shape of a cone, usually recurved, which rests upon a relatively short "peg" at the tip of the dentine (Fig. 12.9). In carnivorous fish the cap often has a lancet shape, with a cross-section like a biconvex lens, with mesial and distal cutting edges, as in pikes (*Esox*), barracuda (*Sphyraena*), or the viper fish *Chauliodus* (Schmidt and Keil, 1971; Greven et al., 2009). In some carnivorous fish, eg, piranhas (Serrasalmidae), the teeth are still more compressed and the cutting edges are very sharp (Shellis and Berkovitz, 1976). Enameloid caps with cutting edges are usually hollow and enclose the supporting dentine in a similar way as enamel on a mammalian cusp.

The cap enameloid on the cutting teeth of piranhas and barracudas consists of an outer parallel-structured layer, which in turn overlies a woven layer. This structure is convergent with that of sharks (Shellis and Berkovitz, 1976; Reif, 1979), although the parallel-structured layer is less regular than in sharks. Reif (1979) considered that actinopterygian

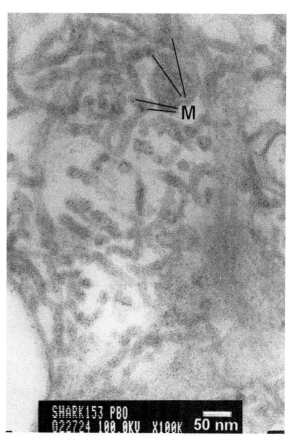

FIGURE 12.7 Starspotted smooth-hound (*Mustelus manazo*). High-power transmission electron micrograph of the enameloid matrix during the early stage of mineralization. Electron-dense mineral particles (M) are found in many tubular vesicles. Note the absence of crystals between the vesicles. Nondemineralized, stained with Pb alone. Specimen no. 11588. *Courtesy of Prof. I. Sasagawa. © I. Sasagawa.*

enameloid mostly has a woven structure. However, in piercing teeth the cap enameloid contains prominent longitudinal bundles of crystals (Schmidt and Keil, 1971) and may have a structure that is a simplified version of the structure seen in piranhas. Thus, in the European eel (*Anguilla anguilla*), crystals curve inward from the longitudinal bundles and in the central zone cross from one side of the cap to the other, while radial crystal bundles intersect the inner enameloid (Fig. 12.10) (Shellis and Miles, 1976).

A woven structure is characteristic of the thick enameloid on the crushing teeth of durophagous bony fish (Figs. 12.11 and 12.14B), although the crystal bundles in the outer region tend to be oriented perpendicular to the cap surface (Reif, 1979; Schmidt and Keil, 1971). This structure is obviously adapted to resisting the high compressive forces exerted by the crushing action of the teeth.

The pharyngeal teeth of the parrot fish *Scarus rivulatus* are arranged in rows on the upper and lower pharyngeal plates, forming a grinding mill which reduces food scraped from the surface of coral. The enameloid structure is asymmetrical and is adapted to resisting the complex pattern of

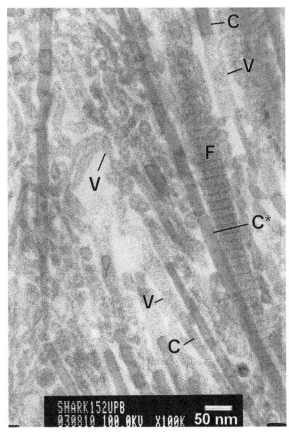

FIGURE 12.8 Starspotted smooth-hound (*Mustelus manazo*). High-power transmission electron micrograph of the enameloid matrix during the stage of mineralization. Crystals (C) are growing within tubular vesicles (V). One well-formed crystal (C*) runs diagonally across the field. Note the lack of mineralization on the cross-banded collagen fibers (F). Nondemineralized, stained with uranyl acetate and lead citrate. Specimen no. 11712. *Courtesy of Prof. I. Sasagawa. © I. Sasagawa.*

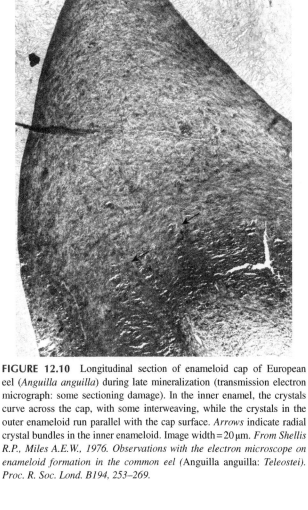

FIGURE 12.10 Longitudinal section of enameloid cap of European eel (*Anguilla anguilla*) during late mineralization (transmission electron micrograph: some sectioning damage). In the inner enamel, the crystals curve across the cap, with some interweaving, while the crystals in the outer enameloid run parallel with the cap surface. *Arrows* indicate radial crystal bundles in the inner enameloid. Image width=20 μm. *From Shellis R.P., Miles A.E.W., 1976. Observations with the electron microscope on enameloid formation in the common eel* (Anguilla anguilla: *Teleostei). Proc. R. Soc. Lond. B194, 253–269.*

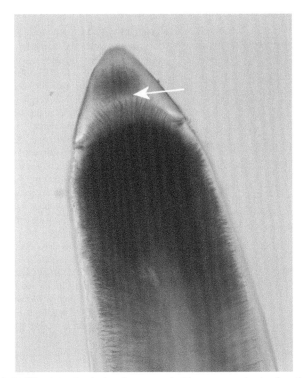

FIGURE 12.9 Tooth tip of conger eel (*Conger conger*), showing the enameloid cap resting on a short projection of dentine. Note the odontoblast processes crossing from the dentine into the enameloid (*arrow*). Image width=420 μm. *Courtesy of Royal College of Surgeons. Tomes Slide Collection, Cat. no. 307.*

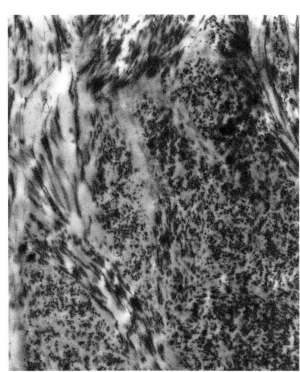

FIGURE 12.11 Ballan wrasse (*Labrus bergylta*). Inner enameloid during mineralization, showing woven structure, with crossing bundles of crystals. Transmission electron micrograph. Image width=10.4 μm.

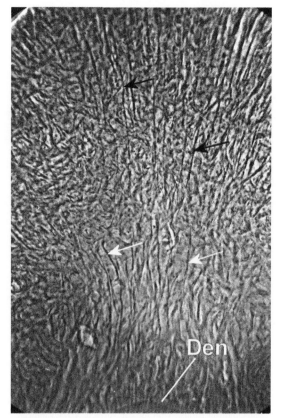

FIGURE 12.12 Wrasse enameloid (late mineralization stage), showing dentinal tubules (*white arrows*), entering the tissue from the dentine (Den) at the bottom of the field, and enameloid tubules (*black arrows*), entering from the outer surface (top) and branching toward the inside of the tooth. Demineralized: phase-contrast. Image width=91 μm.

shearing forces exerted on the teeth. On the leading and trailing edges of the teeth, most of the enameloid crystals lie perpendicular to the tooth surface and thus resist abrasion (Carr et al., 2006). These crystals are intersected by other crystals, oriented horizontally or vertically, which are aligned with the tensile stresses, bind the structure together, and inhibit crack propagation. Next to the enameloid–dentine junction the crystal bundles have a woven arrangement, which Carr et al. suggested acts as a crack-stopping layer.

Because its organic matrix is partly derived from odontoblasts, actinopterygian cap enameloid nearly always contains odontoblast processes continuous with those in the underlying dentine (Fig. 12.9: see also Fig. 11.2A). A feature of cap enameloid in many crushing teeth is the presence of **enameloid tubules**, which are wider than dentinal tubules and, in newly completed teeth, are covered by a thin layer of enameloid (Figs. 12.12 and 12.18; Isokawa et al., 1964). Enameloid tubules run into the cap from the outer surface, tapering and branching toward the interior and interlacing with the extremities of odontoblast processes. These features suggest that the tubules must either penetrate the enameloid before mineralization or may even be left behind by IDE cells if these deposit material into the outer enameloid as Prostak et al. (1993) suggested. The function of these tubules is unknown.

Formation

The formation of cap enameloid is very consistent among cladistia, holosteans, and teleosts, although there are a

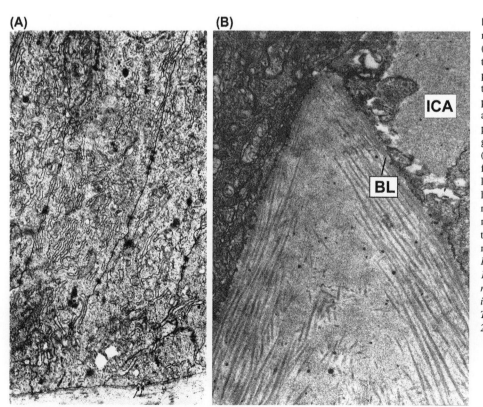

FIGURE 12.13 (A) Enameloid matrix formation in European eel (*Anguilla anguilla*; transmission electron micrograph). The IDE cells are polarized, with the nuclei located at the proximal pole. The distal cytoplasm contains well-developed Golgi apparatus, abundant rough endoplasmic reticulum and some dense granules. Image width=8.1 μm; (B) Tip of cap. IDE cells separated from enameloid matrix by basal lamina (BL). Intercellular accumulations (ICA) of amorphous material between IDE cells, with similar material between collagen fibers of the matrix. Transmission electron micrograph. Image width=2.9 μm. *From Shellis R.P., Miles A.E.W., 1976. Observations with the electron microscope on enameloid formation in the common eel (Anguilla anguilla: Teleostei). Proc. R. Soc. Lond. B194, 253–269.*

(A)   **(B)**

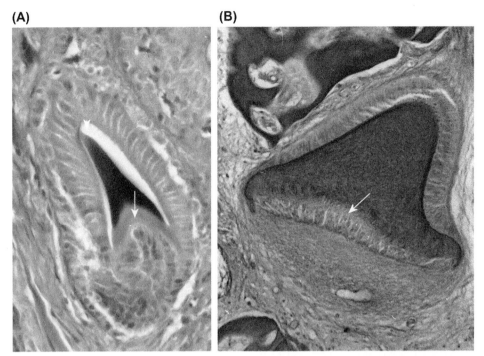

FIGURE 12.14   Histochemical reactions of enameloid matrix during early mineralization. Note low reactivity of predentine (*arrow*) for both tests. (A) European eel (*Anguilla anguilla*). DDD reaction for SS and SH groups (as in the amino acids cysteine and cystine). Image width=100 μm; (B) Ballan wrasse (*Labrus bergylta*). Positive reaction to PAS+Alcian blue staining for neutral and acidic glycosyl groups. Image width=780 μm.

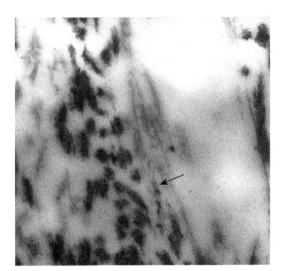

FIGURE 12.15 Early enameloid mineralization in Ballan wrasse (*Labrus bergylta*). TEM. Thin collagen fibers (*arrow*) and associated slender crystals. Image width=1.2 μm.

few differences between species (Shellis and Miles, 1976; Meinke, 1982; Prostak et al., 1989, 1993, Sasagawa and Ishiyama, 2004; Sasagawa et al., 1995, 1997, 2005). The matrix of cap enameloid is laid down between the IDE and a layer of odontoblasts differentiated from the outer cells of the dental papilla. It has a framework of collagen fibers, much more substantial than enameloid matrix in elasmobranchs and the constituent fibers have a similar density to predentine. The fibers are arranged in a specific pattern and during mineralization act as a template for mineral

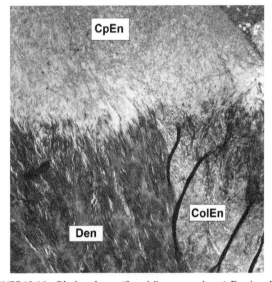

FIGURE 12.16   Black sea bream (*Spondyliosoma cantharus*). Demineralized section of unerupted tooth, showing cap enameloid (CpEn), dentine (Den), and collar enameloid (ColEn). TEM. Matrix of cap enameloid reduced to small fragments, while the matrix of collar enameloid persists and consists of an inner layer of circumferentially oriented and woven fibers and an outer layer of obliquely oriented fibers. Image width=20 μm.

deposition so that, in the mature tissue, the arrangement of crystals is the same as that of the fibers in the early matrix. Once the cap has been completely formed, dentine deposition commences on the internal aspect of the cap and mineralization of enameloid and dentine begin shortly thereafter.

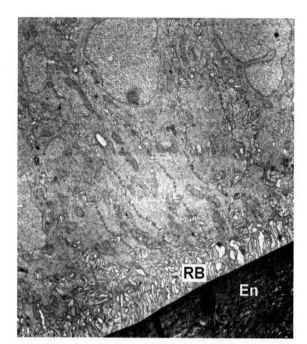

FIGURE 12.17 European eel (*Anguilla anguilla*). IDE cells during mineralization of cap enameloid (En). Distal cytoplasm contains a Golgi apparatus, numerous mitochondria, and a few lysosome-like bodies. Distal cell membrane folded into a ruffled border (RB), with which coated vesicles are connected. Transmission electron micrograph. Image width=16.4 μm. *From Shellis R.P., Miles A.E.W., 1976. Observations with the electron microscope on enameloid formation in the common eel (*Anguilla anguilla: Teleostei). Proc. R. Soc. Lond. B194, 253–269.*

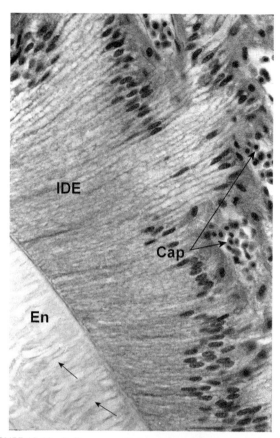

FIGURE 12.18 Ballan wrasse (*Labrus bergylta*). IDE cells during mineralization of cap enameloid. Note blood capillaries (Cap) extending into the IDE layer. In the enameloid, enameloid tubules are visible (*arrow*). Ehrlich's hematoxylin and eosin. Image width=110 μm.

Both the odontoblasts and the IDE contribute to the matrix of cap enameloid (Andreucci and Blumen, 1971; Shellis and Miles, 1974). The odontoblasts lay down collagen fibers and leave processes behind in the cap. The IDE cells begin to differentiate into protein synthesizing and secreting cells during enameloid matrix formation and secrete protein into the matrix up to the point when it starts to mineralize (Fig. 12.13). Originally it was assumed that the IDE cells secreted proteins similar to those found in enamel (Shellis and Miles, 1974; Shellis, 1975) but subsequent studies have identified procollagen vesicles, identical with those seen in odontoblasts in the IDE cells (Fig. 11.7) and it seems clear that some of the matrix collagen is of ectodermal origin (Prostak et al., 1991, 1993; Sasagawa, 1995). Prostak (1993) considered that, in the relatively large lamellae making up the "beak" of tetraodonts, most of the enameloid matrix collagen originates from the IDE, whereas Sasagawa (1995) believed that the contribution was much less in the cichlid *Tilapia* and procollagen vesicles were not observed in the IDE of the European eel by Shellis and Miles (1976). It seems possible that the IDE secretes a larger fraction of the enameloid collagen in large teeth. This is supported by the fact that odontoblast processes extend only into the inner enameloid of the cap enameloid of crushing teeth (Fig. 12.12).

The only chemical analysis of cap enameloid matrix (Kawasaki et al., 1987) found that 79% of the protein of "soft" enameloid, which is at an early stage of mineralization, is soluble in cold 0.5 M acetic acid and has an amino acid composition similar to that of dentine collagen. The fact that this collagen fraction is acid-soluble indicates that it is incompletely cross-linked. The remaining 21% of the matrix is acid-insoluble and was considered to consist of noncollagenous proteins. The latter may include the proteins secreted by the IDE cells and are probably responsible for the greater reactivity of enameloid matrix toward histochemical stains for amino acids compared with dentine (Fig. 12.14A; Shellis, 1975; Meinke, 1982). Moreover, enameloid matrix stains strongly with histochemical methods for sulfated glyconjugates, including chondroitin sulfates (Fig. 12.14B; Kakizawa et al., 1976; Kogaya, 1989).

Mineralization of actinopterygian teeth begins at the junction between the cap enameloid and the underlying dentine. In this region there occur numerous matrix vesicles which generate clusters of mineral crystals. Mineralization of enameloid spreads outwards from the initial site and proceeds by deposition of mineral on the matrix collagen fibers (Fig. 12.15), so that the crystals are highly cooriented

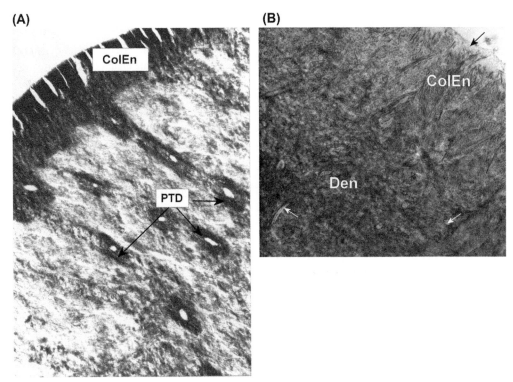

FIGURE 12.19   Collar enameloid (ColEn) of European eel (*Anguilla anguilla*). (A) Surface of mature tooth shaft, showing hypermineralized outermost layer. Image width=7.8 μm; (B) Similar region, demineralized. Note the radial arrangement of the matrix collagen fibers. At the shaft surface, where the collar enameloid is most highly mineralized (Fig. 12.19A), the collagen fibers are partly broken down. Image width=12.4 μm. *Den*, dentine; *PTD*, peritubular dentine. *Both figures from Shellis R.P., Miles A.E.W., 1976. Observations with the electron microscope on enameloid formation in the common eel (Anguilla anguilla: Teleostei). Proc. R. Soc. Lond. B194, 253–269.*

with the fibers (Inage, 1975; Prostak et al., 1991, 1993; Sasagawa, 1997, 1998a). It seems implausible that crystals forming on the collagen fibers are derived from the randomly oriented crystals formed within matrix vesicle, and it is more likely that nucleation occurs directly on the collagen fibers. As mineralization progresses, the matrix collagen fibers break down and are resorbed, so that the mature tissue contains only a small residue of organic material (Fig. 12.16). In a sea bream (*Pagrus major*), the protein content fell from 19% by weight to less than 1% during mineralization (Kawasaki et al., 1987). During mineralization, the acid-soluble fraction seemed to be removed preferentially, as it made up a decreasing proportion of total protein as mineral content increased. The matrix protein is broken down by proteolytic enzymes (Kawasaki et al., 1987; Shimoda et al., 1989).

As the enameloid cap mineralizes, the IDE cells switch from matrix production to resorbing organic matrix and secreting mineral. These activities are characterized by cytological changes similar to those seen during maturation of enamel: acquisition of a ruffled border and, within the cytoplasm, numerous mitochondria and lysosomes (Fig. 12.17). At this stage the basal lamina breaks down. The cell membranes of the IDE cells display enzyme activities associated with transfer of materials into and out of the cells

(Sasagawa, 1998a,b). In large tooth germs, the dental epithelium may be invaded by blood capillaries, which clearly have the function of facilitating transport, especially of mineral (Fig. 12.18).

Collar Enameloid and Enamel

Structure

In most actinopterygians, collar enameloid is only a few micrometers thick, but in the stout crushing teeth of durophagous species it can be a few hundred micrometers thick. The collar enameloid is usually more highly mineralized overall than the underlying dentine but the degree of mineralization varies and is highest at the outermost surface.

The most obvious organic component of the collar enameloid consists of fine collagen fibers which are laid down around the outside of the longitudinal bundles and run obliquely outward toward the IDE (Figs. 12.16 and 12.19B). The collagen fibers are probably formed mainly by the odontoblasts (Sasagawa and Ishiyama, 1988). During formation of the shaft, the IDE cells are active in protein synthesis and secrete protein into the collar enameloid and subsequently they seem to be involved in mineralization: both at a lower level than in cap enameloid formation

(Shellis and Miles, 1974; Sasagawa and Ishiyama, 1988). During mineralization of collar enameloid, crystals are deposited on the collagen fibers (Sasagawa and Ishiyama, 1988). The mineral content of the outer layer of collar enameloid can be very high (Fig. 12.19A) and the matrix collagen fibers are partly broken down (Fig. 12.19B; Herold, 1974; Shellis and Miles, 1976; Sasagawa and Ishiyama, 1988). It has been suggested that this thin outer layer is a form of enamel (Herold, 1974; Shellis and Miles, 1976), but this is probably incorrect in view of the fact that teleosts lack the genes for enamel matrix proteins, and amelogenin has not been detected in collar enameloid of teleosts (Table 12.3).

In contrast, a thin layer of enamel (approximately 1–2-μm thick) has been positively identified as enamel on the tooth shaft of the cladistian *Polypterus* (Sasagawa et al., 2012) and the holostean *Lepisosteus* (Prostak et al., 1989; Ishiyama et al., 1999; Sasagawa et al., 2014). The identification as enamel was substantiated by the detection of amelogenin in the layer (Table 12.2). In *Lepisosteus* the enamel layer is confined to the tooth shaft, while in *Polypterus* it extends over the cap enameloid as well (Sasagawa et al., 2012).

SARCOPTERYGIAN ENAMEL

The teeth of both the coelacanth and lungfish are covered with a thin layer of true enamel and enameloid does not occur in this group. The enamel layer is thin: 1–2 μm in young *Neoceratodus*, about 10 μm in adult *Neoceratodus*, *Lepidosiren*, and *Latimeria* (Shellis and Poole, 1978; Smith, 1979; Barry and Kemp, 2007). The enamel crystals in *Latimeria* are oriented perpendicular to the outer surface (Smith, 1978; Shellis and Poole, 1978). In *Neoceratodus*, the crystals form small, radially oriented bundles (0.2–0.25-μm thick; Barry and Kemp, 2007), in which there is a fan-like arrangement of crystals and which originate from specific points at the enamel–dentine junction during development (Kemp, 2003). Barry and Kemp (2007) referred to these bundles as protoprisms but, as this term suggests that they are precursors of prisms, it is inappropriate (Sander, 2000). One or more incremental lines are present (Smith, 1978, 1979; Shellis and Poole, 1978; Barry and Kemp, 2007). The orientation of these lines, sloping gently from the outer enamel in the apical region to the inner enamel in the basal region, demonstrates that the pattern of enamel deposition is like that in tetrapods.

TABLE 12.2 Distribution of Enamel Proteins in Enamel, as Determined by Immunohistochemistry

Group	Species	Amelogenin	Enamelin	Tuftelin	MMP20*	EMSP1**	References
Actinopterygii	*Lepisosteus* sp.[a]	+					Ishiyama et al. (1994, 1999)
	Polypterus senegalus[a]	+					Sasagawa et al. (2012)
Sarcopterygii	*Protopterus* sp.	+	−				Ishiyama et al. (1994)
	Neoceratodus forsteri	+	−	+	+	+	Satchell et al. (2002) and Diekwisch et al. (2002)
Amphibia	*Ambystoma maculatum* (adult)	+	+				Herold et al. (1989)
	Triturus pyrrhogaster (larva)	+					Kogaya (1999)
	Salamander	+					Ishiyama et al. (1998b)
	Hyla cinerea	+	−	−	+	−	Satchell et al. (2002)
	Rana pipiens	+		−			Diekwisch et al. (2002)
Reptilia	Crocodile	+					Ishiyama et al. (1994)
	Gekko gecko	+	+	+			Herold et al. (1989)
	Hemidactylus turcicus	+		+	+	+	Satchell et al. (2002) and Diekwisch et al. (2002)
	Iguana iguana	+		+			Diekwisch et al. (2002)
	Elaphe trivirgatus	+					Ishiyama et al. (1998b)
Mammalia	*Monodelphis domestica* (Marsupialia)	+		+			Diekwisch et al. (2002)
	Opossum (Marsupialia)	+					Ishiyama et al. (1994)
	Mus musculus	+	+	+	+	+	Herold et al. (1989), Satchell et al. (2002), and Diekwisch et al. (2002)
	Rattus norvegicus	+					Ishiyama et al. (1994)
	Bos taurus (calf)	+	+				Herold et al. (1989)

*Matrix metalloproteinase 20; **enamel matrix serine protease.
[a]Collar enamel. Green=present; red=absent.

Sarcopterygian enamel is formed by the IDE after the first layer of dentine has been deposited (Miller, 1969; Grady, 1970; Shellis and Poole, 1978; Satchell et al., 2000; Kemp, 2003; Barry and Kemp, 2007). In *Neoceratodus*, secretory IDE cells, although shorter than in other taxa, contain rough endoplasmic reticulum and numerous vesicles, while the maturation-stage IDE cells do not develop a ruffled border but have numerous resorptive vesicles at the surface next to the enamel (Kemp, 2003). Enamel formation has not been studied at the ultrastructural level in *Latimeria*.

AMPHIBIAN ENAMELOID AND ENAMEL

The teeth of adult urodeles and anurans are covered by a layer of true enamel, which is thickest over the two cusps and tapers off down the shaft of the tooth. The crystals tend to be smaller than in mammalian enamel. In amphibian enamel they are usually oriented uniformly perpendicular to the outer surface, as in the tree frog (*Hyla japonica*) (Sato et al., 1990), but may be more irregular, as in the urodeles *Ambystoma maculatum* and *Salamandra salamandra* (Sato et al., 1993).

In the embryonic and larval stages of urodeles, the apex of the tooth consists of enameloid (Meredith Smith and Miles, 1971; Davit-Béal et al., 2007). This in turn is covered by a thin layer of true enamel (Davit-Béal et al., 2007). Davit-Béal et al. (2007) showed that, in *Pleurodeles waltl*, the enameloid cap is largest and the enamel layer thinnest in the embryo, but during the following two generations of larval teeth the enameloid is progressively reduced and then lost so that, at metamorphosis, the teeth are covered with enamel. The enameloid cap contains collagen fibers (Meredith Smith and Miles, 1971; Kogaya, 1999; Davit-Béal et al., 2007), together with sulfated glycoconjugates (Kogaya, 1999; Wistuba et al., 2003a). The enameloid forms between odontoblasts and the IDE, which shows features associated with protein synthesis during matrix formation, although Davit-Béal et al., (2007) considered that this is not evidence that the IDE supplies matrix components. The enamel layer forms as the enameloid cap mineralizes. Mature enameloid is more highly mineralized than the dentine but crystal size is similar, rather than enlarged as in actinopterygian cap enameloid (Meredith Smith and Miles, 1971). The reduction of enameloid during ontogeny and the apparently earlier appearance of enamel, were considered by Davit-Béal et al. (2007) to be due to a progressive slowing of odontoblast activity. This resulted in production of thicker collagen fibers embedded in an interstitial matrix, which favored dentine formation rather than enameloid formation.

The differentiation of the IDE cells during matrix formation and maturation follows the usual stages and the pattern is broadly similar in both enameloid and enamel (Meredith Smith and Miles, 1971; Sato et al., 1990; Wistuba et al., 2002, 2003b; Davit-Béal et al., 2007).

IRON AND CUTICLES

In many actinopterygian and urodele teeth the enameloid or enamel covering the tip contains high concentrations of ferric iron, up to 11.5% at the outer surface in *Gnathonemus petersii* (Teleostei; Mormyridae). When the iron concentration is high enough, the tooth tip appears yellow, red, or brown, although apparently colorless teeth can prove to contain iron when tested. The concentration of iron increases from the enamel(oid)–dentine junction to the surface; in some species the concentration rises steadily, while in others it increases more rapidly near the outer surface (Suga et al., 1991, 1997, 1993; Sato et al., 1993; Richter et al., 2010). It has been suggested that iron substitutes for calcium in the hydroxyapatite lattice (Sparks et al., 1990), although when the iron concentration is very high, some may be adsorbed to crystal surfaces (Suga et al., 1997).

Among actinopterygians, the concentration of iron in enameloid is correlated with neither the iron concentration in the aquatic environment nor with diet (Suga et al., 1991). Variations in enameloid iron concentration can, however, be related to phylogeny. The ability to concentrate iron in enameloid is present among basal actinopterygians and this has been enhanced or reduced in different teleost lineages (Suga et al., 1993). Suggestions as to the function of iron in enameloid and enamel include: increased hardness; inhibition of crack propagation; and elimination of excess iron from the body.

At the outer tooth surface, iron is often incorporated in a distinct layer, usually referred to as a cuticle, which was found to contain small, densely packed crystals in one fish (Sparks et al., 1990). The cuticle is resistant to both alkali and acid, and it can be removed from the tooth cap by sequential treatment with these agents (Fig. 12.20; Schmidt and Keil, 1971). The cuticle is firmly bonded to the enamel or enameloid surface and probably acts to inhibit formation of cracks (Shellis and Berkovitz, 1976). The cuticle is produced after the formation of enameloid or enamel (Fig. 12.21) and is entirely a product of the IDE (Meredith Smith and Miles, 1971; Davit-Béal et al., 2007).

REPTILIAN ENAMEL

The enamel layer of reptilian teeth is thickest at the tip of the tooth and tapers off toward the base. Relief on the outer tooth surface, such as longitudinal ridges or wrinkles, is due to variations in enamel thickness and not to undulations of the enamel–dentine junction, which is flat (Sander, 2000). In most reptiles, the enamel is not prismatic (Fig. 12.22; Schmidt and Keil, 1971): the constituent crystals are parallel and oriented perpendicular to the outer surface. However, in some species divergence or convergence of the crystals creates structural differentiation, which is described in detail and classified by Sander (2000). Sometimes, the

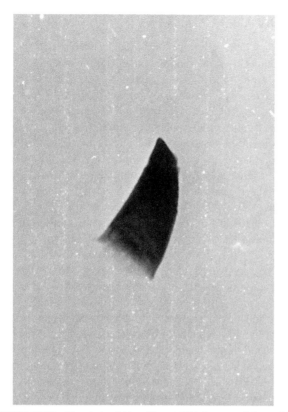

FIGURE 12.20 European eel (*Anguilla anguilla*). Iron-containing cuticle isolated from cap enameloid by boiling in potassium hydroxide-glycerol to remove organic material, followed by treatment with Perls reagent, which is acidic and dissolves the tooth mineral, while simultaneously staining the iron in the cuticle. Image width=330 μm.

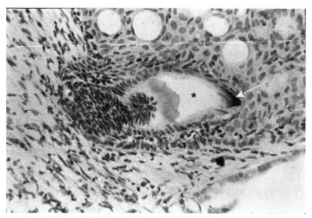

FIGURE 12.21 European eel (*Anguilla anguilla*). Tooth germ after complete mineralization of cap enameloid, represented by a space (*) in this demineralized section. The cuticle is present at the surface of the (*cap arrow*). Harris's hematoxylin. Image width=245 μm.

enamel is composed of linear arrays, in which there is a fan-like arrangement of crystals, and which vary in size from columns extending most or all of the way through the enamel down to smaller structural units only a few crystals wide and a few micrometers long. This structure is seen in the enamel of the tuatara (*Sphenodon punctatus*)

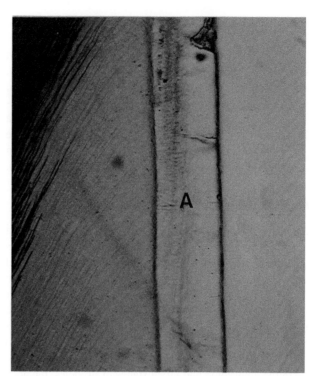

FIGURE 12.22 Nile crocodile (*Crocodilus niloticus*). Longitudinal ground section of tooth. Enamel (A) is nonprismatic, so appears homogeneous in this section. Image width=460 μm.

(Yilmaz et al., 2014). Prismatic enamel has been identified in only one extant reptile, the agamid (*Uromastyx hardwickii*; Fig. 12.23; Cooper and Poole, 1973).

The formation of enamel in reptiles follows the sequence of matrix formation, mineralization, and maturation described earlier, with a few differences from mammals (Delgado et al., 2005). First, the secretory surfaces of the ameloblasts are flat, as in sarcopterygians and amphibians; the Tomes process, which is associated with prism formation in mammals, is absent. Second, at least in the Canarian skink (*Chalcides viridanus*), the full thickness of the enamel matrix is secreted before it starts to mineralize, in contrast to mammals, where enamel matrix starts to mineralize almost immediately after secretion. Third, while the ameloblasts lose the organelles associated with protein secretion as the enamel mineralizes, and their cell membranes acquire numbers of folds and processes, they do not form a ruffled border as in mammals.

EVOLUTION OF HYPERMINERALIZED TISSUES

One approach to elucidating the relationship between dentine, enamel, and enameloid has been to investigate the distribution of matrix proteins among the three tissues. Studies utilizing immunohistochemistry to determine whether enamel proteins are present in the teeth of different vertebrates are summarized in Tables 12.2 and 12.3. There are

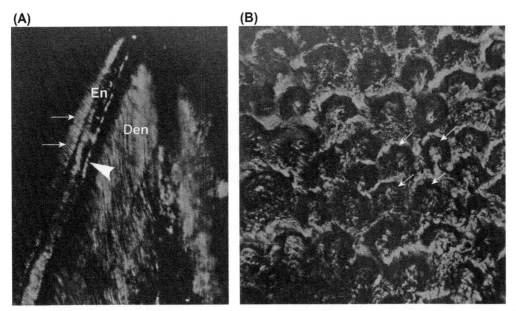

FIGURE 12.23 Hardwicke's spiny-tailed lizard (*Uromastyx hardwickii*). (A) Longitudinal ground section of tooth, viewed in polarized light. Junction between enamel (En) and dentine (Den) marked by *large arrowhead*. The enamel layer shows radially oriented striations (*small arrows*) which are due to the presence of prisms: crystal bundles, about 5 μm in diameter. Image width = 1.2 mm; (B) Surface which has been ground into the enamel parallel with the outer surface, polished and etched lightly with acid. The etching creates horseshoe-shaped clefts (*arrows*) at the sites of discontinuities of crystal orientation, which define the boundaries of enamel prisms. Image width = 41 μm. *From Cooper J.S., Poole D.F.G., 1973. Dentition and dental tissues of the agamid lizard, Uromastyx. J. Zool. 169, 85–100. Courtesy of the Editors Journal of Zoology.*

TABLE 12.3 Distribution of Enamel Proteins in Enameloid, as Determined by Immunohistochemistry

Group	Species	Amelogenin	Enamelin	Tuftelin	MMP20*	EMSP1**	References
		Enameloid					
Chondrichthyes	*Heterodontus francisci*	+	+	+	+	+	Satchell et al. (2002)
	Heterodontus francisci	+		–			Diekwisch et al. (2002)
	Squalus acanthias	–	+				Herold et al. (1989)
	Mustelus manazo	–	–				Ishiyama et al. (1994)
Actinopterygii	*Lepisosteus* sp.[a]	–	–				Ishiyama et al. (1994, 2001)
	Polypterus senegalus[a]	–					Satchell et al. (2002)
	Pomatomus saltatrix	–	+				Diekwisch et al. (2002)
	Esox lucius	–	+				Herold et al. (1989)
	Cynoscion nothus	–	+				Ishiyama et al. (1994)
	Prionotus carolinus	–	+				Ishiyama et al. (1994, 2001)
	Poecilia reticulata	–		+			Kogaya (1999)
	Tilapia sp.	–	–				Herold et al. (1989)
	Fugu niphobles	–	–				Herold et al. (1989)
	Halichoeres poecilopterus[b]	–					Herold et al. (1989)
Amphibia	*Ambystoma maculatum* (larva)	–	+				Herold et al. (1989)
	Ambystoma mexicanum (larva)	–		+			Diekwisch et al. (2002)
	Triturus pyrrhogaster (larva)	–					Ishiyama et al. (1994)

*Matrix metalloproteinase 20; **enamel matrix serine protease.
[a]Cap enameloid; [b]Both cap and collar enameloid. Green = present; red = absent.

uncertainties because of problems such as cross-reactivity and differences in reactivity with the antibodies to the mammalian enamel proteins that have been used as probes. Some reports do not describe the particular stage of development of the tooth germs studied. The staining patterns sometimes seem anomalous. For instance, the reactions for amelogenin reported for *Heterodontus* differ markedly between Satchell et al. (2002) and Diekwisch et al. (2002). Nevertheless, these data are valuable because they include a wide variety of species.

In recent years, genetic methods have clarified the origins of the matrices of hard tissues and hence the relationships between them.

Both the noncollagenous small integrin-binding ligand N-linked glycoproteins (SIBLING) of dentine (and bone) and the matrix proteins of enamel are coded by genes belonging to the secretory calcium-binding phosphoprotein (*SCPP*) family (Kawasaki, 2009). This gene family is believed to have originated ultimately from a single gene (SPARC, which codes for osteonectin) by two cycles of gene duplication. The first cycle, due to whole-genome duplication, seems to have occurred after the divergence of jawless and jawed vertebrates (Venkatesh et al., 2014) and produced the SPARC-like 1 gene (*SPARCL1*). In the second cycle, which is believed to have occurred after the divergence of Osteichthyes from Chondrichthyes (Venkatesh et al., 2014), tandem duplication of *SPARCL1* generated the family of *SCPP* genes (Kawasaki, 2009, 2011). In humans, the *SCPP* gene family comprises two subfamilies. The acidic *SCPP* subfamily codes for SIBLING proteins of bone and dentine (Chapter 11). The proline/glutamine (P/Q)-rich *SCPP* subfamily codes for the enamel matrix proteins enamelin and ameloblastin, as well as amelogenin, which is, however, located on a different chromosome. Considerable changes have occurred within the *SCPP* gene family during evolution.

Enamel

Genomic analysis has established that the genes for the enamel matrix proteins amelogenin, enamelin, and ameloblastin are expressed in several tetrapods (humans, the lizard *Anolis*, the frog *Xenopus*) and also in the sarcopterygian *Latimeria* (Kawasaki, 2009; Kawasaki and Amemiya, 2014). Reptilian and amphibian homologues of eutherian amelogenin, ameloblastin, and enamelysin genes have been characterized (Ishiyama et al., 1998a; Toyosawa et al., 1998; Shintani et al., 2003, 2006). These genes often show large differences from their mammalian counterparts; for instance, Ishiyama et al. (1998a) found only 45% homology between the amelogenin genes from snake and humans.

The IDE also expresses genes for two proteins which are not associated with the matrix: odontogenic ameloblast-associated protein (ODAM) and (except in *Xenopus*) amelotin (Kawasaki, 2009; Kawasaki and Amemiya, 2014). ODAM

and amelotin are associated with enamel maturation: they participate in removal of degraded matrix and in transporting calcium ions to the hard tissue surface while preventing excessive supersaturation (Kawasaki, 2013).

The presence of amelogenin in enamel matrix has been confirmed by immunohistochemistry, not only in tetrapods, but also in the lungfish *Protopterus* and *Neoceratodus* (Dipnomorpha), and in *Polypterus* (Cladistia) and *Lepisosteus* (Holostei; see Table 12.2). In *Lepisosteus*, the amelogenin gene is missing from its usual location (Qu et al., 2015). The proteins detected by immunohistochemistry are cross-reactive with the C-terminal and middle portions of porcine amelogenin, but not the N-terminal region. The two immunoreactive *Lepisosteus* proteins also have higher molecular weights than mammalian amelogenins (65 and 78 kDa versus 25–27 kDa; Sasagawa et al., 2014).

In immunohistochemical studies, enamelin has been identified in enamel of tetrapods but not in that of the tree frog *Hyla cinerea* or in lungfish (Table 12.2).

Actinopterygian and Urodele Enameloid

In contrast to enamel, the genes for the three enamel matrix proteins were not identified in enameloid of two teleosts: fugu (*Tokifugu rubripes*) and zebra fish (*Danio rerio*), but the *odam* gene was expressed by the IDE (Kawasaki, 2009; Kawasaki and Amemiya, 2014). During enameloid formation, expression of a number of genes is common to the IDE and the odontoblasts. Collagen Type 1 (*col1*) is expressed by both cells. This supports the suggestion by Prostak et al. (1991, 1993) that the IDE is a source of enameloid collagen, and confirms the finding by Huysseune et al. (2008) that *col1α1* (the gene coding the α1 chain of type I collagen) is expressed by IDE cells of salmon tooth germs. Huysseune et al. (2008) observed that *col1α1* expression was not confined to the cells surrounding the cap enameloid but occurred in the IDE covering the forming tooth shaft, where secretion of protein was observed in the great crested newt (*Triturus cristatus*; Amphibia) by Meredith Smith and Miles (1969) and in teleosts by Shellis and Miles (1974). Besides collagen, both the IDE and the odontoblasts of teleosts also express the acidic *SCPP* genes *sparc* (osteonectin), *scpp1*, and *scpp5*.

The IDE cells also secrete *col1α1* during formation of enameloid on the teeth of larval urodeles (Assaraf-Weill et al., 2014). As the amount of enameloid formed decreases during successive generations of teeth (Davit-Béal et al., 2007) the expression of *col1α1* diminishes and ceases in the adult teeth, which instead express *amel* (amelogenin).

Immunohistochemical studies agree as to the absence of amelogenin from actinopterygian and urodele enameloid but have produced contradictory results with regard to enamelin (Table 12.3).

It is clear that, in teleosts (and, probably, other actinopterygians and in urodeles), enameloid and dentine share a common set of extracellular matrix proteins. However, the higher reactivity of the teleost enameloid matrix to histochemical tests for amino acids and glycosyl groups indicates a difference in composition from dentine matrix. The difference could be related to differences in posttranslational modification of the known matrix proteins; to the presence of proteases; or to the presence of (glyco) proteins that have not yet been characterized.

In both dentine, and actinopterygian/urodele enameloid, mineralization is initiated in matrix vesicles and on the matrix collagen fibers. The hypermineralization of actinopterygian enameloid is due to proteolysis of the matrix by proteases and the activity of the IDE cells. The proteases which break down the matrix (Kawasaki et al., 1987) have not been characterized, although *mmpf20l*, a gene related to *MMP20*, which is important in enamel maturation, is expressed in teleosts (Kawasaki and Suzuki, 2011). *Odam* is also expressed in the distal portions of the IDE cells of teleosts during enameloid mineralization and probably fulfills the same functions in enameloid mineralization as it does in enamel maturation (Kawasaki, 2009, 2013).

Chondrichthyan Enameloid

Venkatesh et al. (2014) analyzed the genome of the holocephalian elephant shark (*Callorhinchus milii*) and two elasmobranchs: the little skate (*Leucoraja erinacea*) and the lesser spotted catshark (*Scyliorhinus canicula*). Their results showed that *sparc* and *sparcl1* were both present but *scpp* genes were absent. Thus, Chondrichthyes lack both enamel matrix proteins and genes for acidic proteins associated with enameloid and dentine in actinopterygians and urodeles. Positive immunohistochemical reactions reported for amelogenin and enamelin in elasmobranchs enameloid (Table 12.3) may therefore be artifactual. However, several authors have reported that enamelin-like proteins can be isolated from elasmobranch enameloid (Slavkin et al., 1983a,b; Graham, 1985; Deutsch et al., 1991), and the amino acid composition of mature shark enameloid matrix resembles that of mature enamel (Levine et al., 1966). In the light of the reported absence of enamel matrix genes, the nature of these proteins requires further investigation. The absence of acidic SCPP genes raises the question of how dentine, which is structurally and mechanically indistinguishable from that of other vertebrates, is formed in elasmobranch teeth. The most likely explanation is that mineralization of chondrichthyan dentine is mediated by *sparc*, *sparcl1*, or both, as Venkatesh et al. (2014) suggested in the case of dermal bone.

The roles of proteolytic enzymes, and of molecules such as ODAM and amelotin in elasmobranch mineralization have not been ascertained. Nevertheless, it is clear that maturation of elasmobranch enameloid, as in actinopterygians, depends on active inward transport of mineral ions and removal of organic material by the IDE cells (Sasagawa, 2002).

In both elasmobranch enameloid and pleromin, mineralization takes place within tubular vesicles instead of in matrix vesicles or on collagen fibers (Sasagawa, 2002; Chapter 11). However, while the dental epithelium is important in mineralization of enameloid (Sasagawa, 2002), pleromin mineralizes without participation by epithelium (Ishiyama et al., 1991). Thus, elasmobranch enameloid is distinct from both actinopterygian enameloid and pleromin and Sasagawa (2002) proposed the term **adameloid** for elasmobranchs enameloid to emphasize these differences.

Relationships Between Dentine, Enameloid, and Enamel

There are uncertainties about the detailed composition of enameloid matrices. However, it seems clear that the dental hard tissues form a sequence of grades of organization with increasing genetic complexity (Kawasaki, 2009): dentine—(actinopterygian) enameloid—enamel. Adameloid may represent a branch from this sequence, which occurred before the genome duplication that occurred before chondrichthyans and osteichthyans diverged. The main steps in the sequence are the development of a mechanism for hypermineralization, based on ODAM, and the appearance of novel P/Q-SCPP genes coding for enamel matrix proteins. At present, pleromin and petrodentine do not fit into this scheme, as they become hypermineralized without participation by epithelium. Hypermineralization is thus achieved by some process different from the ODAM-based system located in the IDE.

The correlation of the evolutionary sequence of dental tissues with a sequence of increasing genetic complexity, brought about by tandem duplication, allows previous hypotheses on the relationships between enameloid to be tested. First, the suggestion (Smith, 1995) that enameloid evolved after enamel may be incorrect, as it implies that a process of enlarging the SCPP gene family was followed by its reduction. However, genomic analysis in more fish species is required to elucidate the evolutionary relationships between acctinopterygian enameloid and the enamel in Sarcopterygii, *Polypterus* and *Lepisosteus*. The hypothesis (Shellis and Miles, 1974) that a simple heterochronic shift in the timing of IDE activity could account for the evolution of enamel from actinopterygian-type enameloid appears to be untenable, since it rested on the assumptions that the protein(s) secreted by the IDE were similar to enamel proteins and that they had a role in enameloid mineralization distinct from their role in enamel formation. It is clear from the molecular evidence that enamel-producing IDE cells secrete matrix molecules that are not produced by enameloid-producing IDE cells.

REFERENCES

Andreucci, R.D., Blumen, G., 1971. Radioautographic study of *Spheroides testudineus* denticles (checkered puffer). Acta Anat. 79, 76–83.

Assaraf-Weill, N., Gasse, B., Silvent, J., Bardet, C., Sire, J.-Y., Davit-Béal, T., 2014. Ameloblasts express Type I collagen during amelogenesis. J. Dent. Res. 93, 502–507.

Barry, J.C., Kemp, A., 2007. High resolution transmission electron microscopy of developing enamel in the Australian lungfish, *Neoceratodus forsteri* (Osteichthyes: Dipnoi). Tissue Cell 39, 387–398.

Carr, A., Kemp, A., Tibbetts, I., Truss, R., Drennane, J., 2006. Microstructure of pharyngeal tooth enameloid in the parrotfish *Scarus rivulatus* (Pisces: Scaridae). J. Micr. 221, 8–16.

Chen, C., Wang, Z., Saito, M., Tohei, T., Takano, Y., Ikuhara, Y., 2014. Fluorine in shark teeth: its direct atomic-resolution imaging and strengthening function. Angew. Chem. Int. Ed. 53, 1543–1547.

Chen, P., Schirer, J., Simpson, A., Nay, R., Lin, Y.S., Yang, W., Lopez, M.I., Li, J., Olevsky, E.A., Meyers, M.A., 2012. Predation versus protection: fish teeth and scales evaluated by nanoindentation. J Mate. Res. 27, 100–112.

Cooper, J.S., Poole, D.F.G., 1973. Dentition and dental tissues of the agamid lizard, *Uromastyx*. J. Zool. 169, 85–100.

Daculsi, G., Kérébel, B., 1978. High-resolution electron microscope study of human enamel crystallites: size, shape, and growth. J. Ultrastruct. Res. 65, 163–172.

Daculsi, G., Kérébel, B., Verbaere, A., 1978. Méthode de mesure des cristaux d'apatite de la dentine humaine en microscopie électronique à transmission de haute résolution. C. R. Acad. Sci. Ser. D. 286, 1439–1442.

Daculsi, G., Menanteau, J., Kérébel, L.M., Mitre, D., 1984. Length and shape of enamel crystals. Calcif. Tiss. Int. 36, 550–555.

Davit-Béal, T., Allizard, F., Sire, J.-Y., 2007. Enameloid/enamel transition through successive tooth replacements in *Pleurodeles waltl* (Lissamphibia, Caudata). Cell Tiss. Res. 328, 167–183.

Delgado, S., Davit-Béal, T., Allizard, F., Sire, J.-Y., 2005. Tooth development in a scincid lizard, *Chalcides viridanus* (Squamata), with particular attention to enamel formation. Cell Tiss. Res. 319, 71–89.

Deutsch, D., Palmon, A., Dafni, L., Shenkman, A., Sherman, J., Fisher, L., Termine, J.D., Young, M., 1991. Enamelin and enameloid. In: Suga, S., Nakahara, H. (Eds.), Mechanisms and Phylogeny of Mineralization in Biological Systems. Springer, Berlin, pp. 73–80.

Dibdin, G.H., Poole, D.F.G., 1982. Surface area and pore size analysis for human enamel and dentine by water vapour sorption. Arch. Oral. Biol. 27, 235–241.

Diekwisch, T.G.H., Berman, B.J., Anderton, X., Gurinsky, B., Ortega, A.J., Satchell, P.G., Williams, M., Arumugham, C., Luan, X., McIntosh, J.E., Yamane, A., Carlson, D.S., Sire, J.-Y., Shuler, C.F., 2002. Membranes, minerals, and proteins of developing vertebrate enamel. Microsc. Res. Tech. 59, 373–395.

Enault, S., Cappetta, H., Adnet, S., 2013. Simplification of the enameloid microstructure of large stingrays (Chondrichthyes: Myliobatiformes): a functional approach. Zool. J. Linn. Soc. 169, 144–155.

Enax, J., Prymak, O., Raabe, D., Epple, M., 2012. Structure, composition, and mechanical properties of shark teeth. J. Struct. Biol. 178, 290–299.

Enax, J., Fabritius, H.-O., Rack, A., Prymak, O., Raabe, D., Epple, M., 2013. Characterization of crocodile teeth: correlation of composition, microstructure, and hardness. J. Struct. Biol. 184, 155–163.

Garant, P.R., 1970. An electron microscopic study of the crystal-matrix relationship in the teeth of the dogfish *Squalus acanthias* L. J. Ultrastruct. Res. 441–449.

Gillis, J.A., Donoghue, P.C.J., 2007. The homology and phylogeny of chondrichthyan tooth enameloid. J. Morphol. 268, 33–49.

Grady, J.E., 1970. Tooth development in *Latimeria chalumnae* (Smith). J. Morphol. 132, 377–388.

Graham, E.E., 1984. Protein biosynthesis during spiny dogfish (*Squalus acanthias*) enameloid formation. Arch. Oral. Biol. 29, 821–825.

Graham, E.E., 1985. Isolation of enamelinlike proteins from blue shark (*Prionace glauca*) enameloid. J. Exp. Zool. 234, 185–191.

Greven, H., Walker, Y., Zanger, K., 2009. On the structure of teeth in the viperfish *Chauliodus sloani* Bloch & Schneider, 1801 (Stomiidae). Bull. Fish. Biol. 11, 87–98.

Herold, R.C., Rosenbloom, J., Granovsky, M., 1989. Phylogenetic distribution of enamel proteins: immunohistochemical localization with monoclonal antibodies indicates the evolutionary appearance of enamelins prior to amelogenins. Calcif. Tiss. Res. 45, 88–94.

Herold, R.C.B., 1974. Ultrastructure of odontogenesis in the pike (*Esox lucius*). Role of dental epithelium and formation of enameloid layer. J. Ultrastruct. Res. 48, 435–454.

Huysseune, A., Takle, H., Soenens, M., Taerwe, K., Witten, P.E., 2008. Unique and shared gene expression patterns in Atlantic salmon (*Salmo salar*) tooth development. Dev. Genes Evol. 218, 427–437.

Inage, T., 1975. Electron microscopic study of early formation of the tooth enameloid of a fish (*Hoplognathus fasciatus*). I. Odontoblasts and matrix fibers. Arch. Histol. Jpn. 38, 209–227.

Ishiyama, M., Inage, T., Shimokawa, H., Yoshibe, S., 1994. Immunocytochemical detection of enamel proteins in dental matrix of certain fish. Bull. Inst. Océanogr. Monaco 14 (Suppl. 1), 175–182 Numéro special.

Ishiyama, M., Mikami, M., Shimokawa, H., Oida, S., 1998a. Amelogenin protein in tooth germs of the snake *Elaphe quadrivirgata*, immunohistochemistry, cloning and cDNA sequence. Arch. Histol. Cytol. 61, 467–474.

Ishiyama, M., Inage, T., Shimokawa, H., 1998b. Phylogenetic distribution of amelogenin proteins and amelogenin genes in vertebrates. Conn. Tiss. Res. 39, 219 Abstract.

Ishiyama, M., Inage, T., Shimokawa, H., 1999. An immunocytochemical study of amelogenin proteins in the developing tooth enamel of the gar-pike, *Lepisosteus oculatus* (Holostei, Actinopterygii). Arch. Histol. Cytol. 62, 191–197.

Ishiyama, M., Inage, T., Shimokawa, H., 2001. Abortive secretion of an enamel matrix in the inner enamel epithelial cells during an enameloid formation in the gar-pike, *Lepisosteus oculatus* (Holostei, Actinopterygii). Arch. Histol. Cytol. 64, 99–107.

Ishjiyama, M., Yoshie, S., Teraki, Y., Cooper, E.W.T., 1991. Ultrastructure of pleromin, a highly mineralized tissue comprising crystalline calcium phosphate known as whitlockite, in holocephalian tooth plates. In: Suga, S., Nakahara, H. (Eds.), Mechanisms and Phylogeny of Mineralization in Biological Systems. Springer, Berlin, pp. 453–458.

Isokawa, S., can Huysen, G., Kosakai, T., 1964. Historadiography of tubular enamel and dentin of osseous piscine teeth. J. Nihon Univ. Sch. Dent. 6, 79–87.

Kakizawa, Y., Kasuya, K., Kojima, N., Nagai, S., Takagi, H., Fukui, K., 1976. An histochemical study on acid mucopolysaccharides in the enameloid formation stages of fish (*Oplegnathus fasciatus*). J. Nihon Univ. Sch. Dent. 18, 105–113.

Kawasaki, K., Amemiya, C.T., 2014. SCPP genes in the coelacanth: tissue mineralization genes shared by sarcopterygians. J. Exp. Zool. Mol. Dev. Evol. 322B, 390–402.

Kawasaki, K., Suzuki, T., 2011. Molecular evolution of matrix metalloproteinase 20. Eur. J. Oral. Sci. 119 (Suppl. 1), 247–253.

Kawasaki, K., Shimoda, S., Fukae, M., 1987. Histological and biochemical observations of developing enameloid of the sea bream. Adv. Dent. Res. 1, 191–195.

Kawasaki, K., 2009. The SCPP gene repertoire in bony vertebrates and graded differences in mineralized tissues. Dev. Genes Evol. 219, 147–157.

Kawasaki, K., 2011. The SCPP gene family and the complexity of hard tissues in vertebrates. Cells Tissues Organs 194, 108–112.

Kawasaki, K., 2013. Odontogenic ameloblast-associated protein (ODAM) and amelotin: major players in hypermineralization of enamel and enameloid. J. Oral. Biosci. 55, 85–90.

Kemp, N.E., 1985. Ameloblastic secretion and calcification of the enamel layer in shark teeth. J. Morphol. 184, 215–230.

Kemp, A., 2003. Ultrastructure of developing tooth plates in the Australian lungfish, *Neoceratodus forsteri* (Osteichthyes: Dipnoi). Tissue Cell 35, 401–426.

Kieser, J.A., He, L.H., Dean, M.C., Jones, M.E., Duncan, W.J., Swain, M.V., Nelson, N.J., 2011. Structure and compositional characteristics of caniniform dental enamel in the tuatara *Sphenodon punctatus* (Lepidosauria: Rhynchocephalia). N.Z. Dent. J. 107, 44–50.

Kogaya, Y., 1989. Histochemical properties of sulphated glycoconjugates in developing enameloid matrix of the fish *Polypterus senegalus*. Histochemistry 91, 185–190.

Kogaya, Y., 1999. Immunohistochemical localisation of amelogenin-like proteins and type I collagen and histochemical demonstration of sulphated glycoconjugates in developing enameloid and enamel matrices of the larval urodele (*Triturus pyrrhogaster*) teeth. J. Anat. 195, 455–464.

Levine, P.T., Glimcher, M.J., Seyer, J.M., Huddleston, J.I., Hein, J.W., 1966. Noncollagenous nature of the proteins of shark enamel. Science 154, 1192–1194.

Meinke, D.K., 1982. A histological and histochemical study of developing teeth in *Polypterus* (Pisces, Actinopterygii). Arch. Oral. Biol. 27, 197–206.

Meredith Smith, M., Miles, A.E.W., 1969. An autoradiographic investigation with the light microscope of proline-H^3 incorporation during tooth development in crested newt (*Triturus cristatus*). Arch. Oral. Biol. 14, 479–490.

Meredith Smith, M., Miles, A.E.W., 1971. Ultrastructure of odontogenesis in larval and adult urodeles. Differentiation of the dental epithelial cells. Z Zellforsch Mikr Anat. 121, 470–498.

Miake, Y., Aoba, T., Moreno, E.C., Shimoda, S., Prostak, K., Suga, S., 1991. Ultrastructural studies on crystal growth of enameloid minerals in elasmobranch and teleost fish. Calcif. Tiss. Res. 48, 204–217.

Miller, W.A., 1969. Tooth enamel in *Latimeria chalumnae* (Smith). Nature Lond. 221, 1244.

Moradian-Oldak, J., 2013. Protein-mediated enamel mineralization. Front. Biosci. 17, 1996–2023.

Nanci, A., Bringas, P., Samuel, N., Slavkin, H.C., 1983. Selachian tooth development: III. Ultrastructural features of secretory amelogenesis in *Squalus acanthias*. J. Craniofac. Dev. Biol. 3, 53–73.

Preuschoft, H., Reif, W.-E., Müller, W.H., 1974. Funktionsanpassungen von Haifisch-Zähnen in Form und Struktur. Z Anat. Entwicklungsges 143, 315–344.

Prostak, K.S., Skobe, Z., 1988. Ultrastructure of odontogenic cells during enameloid matrix synthesis in tooth buds from an elasmobranch, *Raja erinacea*. Am. J. Anat. 182, 59–72.

Prostak, K.S., Seifert, P., Skobe, Z., 1989. Ultrastructure of the developing teeth in the gar pike (*Lepisosteus*). In: Fearnhead, R.W. (Ed.), Tooth Enamel V. Yokohama, pp. 188–192 Florence.

Prostak, K., Seifert, P., Skobe, Z., 1991. Tooth matrix formation and mineralization in extant fishes. In: Suga, S., Nakahara, H. (Eds.), Mechanisms and Phylogeny of Mineralization in Biological Systems. Springer, Berlin, pp. 465–469.

Prostak, K.S., Seifert, P., Skobe, Z., 1993. Enameloid formation in two tetraodontiform fish species with high and low fluoride contents in enameloid. Arch. Oral. Biol. 38, 1031–1044.

Qu, Q., Haitina, T., Zhu, M., Ahlberg, P.E., 2015. New genomic and fossil data illuminate the origin of enamel. Nat. Lond 526, 108–112.

Reif, W.-E., 1973. Morphologie und Ultrastruktur des Hai-„Schmelzes". Zool. Scr. 2, 231–250.

Reif, W.-E., 1979. Structural convergence between enameloid of actinopterygian teeth and of shark teeth. Scanning Electron Microsc. II, 547–554.

Richter, H., Kierdorf, H., Kierdorf, U., Clemen, G., Greven, H., 2010. A microscopic and microanalytical study (Fe, Ca) of the teeth of the larval and juvenile *Ambystoma mexicanum* (Amphibia: Urodela: Ambystomatidae). Vert. Zool. 60, 27–35.

Sander, P.M., 2000. Prismless enamel in amniotes: terminology, function and evolution. In: Teaford, M.F., Smith, M.M., Ferguson, M.W.J. (Eds.), Development, Function and Evolution of Teeth. Cambridge University Press, Cambridge, pp. 92–106.

Sasagawa, I., Akai, J., 1992. The fine structure of the enameloid matrix and initial mineralization during tooth development in the sting rays, Dasyatis akajei and *Urolophus aurantiacus*. J. Electron Micr 41, 242–252.

Sasagawa, I., Akai, J., 1999. Ultrastructural observations of dental epithelial cells and enameloid during enameloid mineralization and maturation stages in stingrays, *Urolophus aurantiacus*, an elasmobranch. J. Electron Micr. 48, 455–463.

Sasagawa, I., Ishiyama, M., 1988. The structure and development of the collar enameloid in two teleost fishes, *Halichoeres poecilopterus* and *Pagrus major*. Anat. Embryol. 178, 499–511.

Sasagawa, I., Ishiyama, M., 2004. Fine structural and cytochemical observations on Ihe dental epithelial cells during cap enameloid formation stages in *Polypterus senegalus*, a bony fish (Actinopterygii). Conn. Tiss. Res. 46, 33–52.

Sasagawa, I., Ishiyama, M., 2005. Fine structural and cytochemical mapping of enamel organ during the enameloid formation stages in gars, *Lepisosteus oculatus*, Actinopterygii. Arch. Oral. Biol. 50, 373–391.

Sasagawa, I., Yokosuka, H., Ishiyama, M., Mikami, M., Shimokawa, H., Uchida, Y., 2012. Fine structural and immunohistochemical detection of collar enamel in the teeth of *Polypterus senegalus*, an actinopterygian fish. Cell Tiss. Res. 347, 369–381.

Sasagawa, I., Ishiyama, M., Yokosuka, H., Mikami, M., Shimokawa, H., Uchida, T., 2014. Immunohistochemical and Western blot analyses of collar enamel in the jaw teeth of gars, *Lepisosteus oculatus*, an actinopterygian fish. Conn Tiss. Res. 55, 225–233.

Sasagawa, I., 1995. Fine structure of tooth germs during the formation of enameloid matrix in *Tilapia nilotica*, a teleost fish. Arch. Oral. Biol. 40, 801–814.

Sasagawa, I., 1997. Fine structure of the cap enameloid and of the dental epithelial cells during enameloid mineralisation and early maturation stages in the tilapia, a teleost fish. J. Anat. 190, 589–600.

Sasagawa, I., 1998a. Mechanisms of mineralization in the enameloid of elasmobranchs and teleosts. Conn. Tiss. Res. 39, 207–214.

Sasagawa, I., 1998b. Activity of alkaline and acid phosphatases in dental epithelial cells and enameloid during odontogenesis in two teleost fish, *Oreochromis niloticus* and *Tilapia buttikoferi*. Eur. J. Oral. Sci. 106 (Suppl.), 513–518.

Sasagawa, I., 2002. Mineralization patterns in elasmobranch fish. Microsc. Res. Techn. 59, 396–407.

Satchell, P.G., Shuler, C.F., Diekwisch, T.G.H., 2000. True enamel covering in teeth of the Australian lungfish *Neoceratodus forsteri*. Cell Tiss. Res. 299, 27–37.

Satchell, P.G., Anderton, X., Ryu, O.H., Luan, X., Ortega, A.J., Opamen, R., Berman, B.J., Witherspoon, D.E., Gutmann, J.L., Yamane, A., Zeichner-David, M., Simmer, J.P., Shuler, C.F., Diekwisch, T.G.H., 2002. Conservation and variation in enamel protein distribution during vertebrate tooth development. J. Exp. Zool. Mol. Dev. Evol. 294, 91–106.

Sato, I., Shimada, K., Sato, T., Kitagawa, T., 1990. Fine structure and histochemistry of the teeth of the tree frog (*Hyla japonica*). Okajimas Folia Anat. Jpn. 67, 11–20.

Sato, I., Shimada, K., Sato, T., Etherige, R., 1993. Morphological study of the teeth of *Ambystoma maculatum, Salamandra salamandra* and *Aneides lugubris*: fine structure and chemistry of enamel. Okajimas Folia Anat. Jpn. 70, 157–164.

Sawada, T., Inoue, S., 2003. Ultrastructure of basement membranes in developing shark tooth. Calcif. Tiss. Int. 72, 65–73.

Schmidt, W.J., Keil, A., 1971. Polarizing Microscopy of Dental Tissues; Poole DFG, Darling AI (Transl). Pergamon, Oxford.

Shellis, R.P., Berkovitz, B.K.B., 1976. Observations on the dental anatomy of piranhas (Characidae) with special reference to tooth structure. J. Zool. Lond. 180, 69–84.

Shellis, R.P., Miles, A.E.W., 1974. Autoradiographic study of the formation of enameloid and dentine matrices in teleost fishes using tritiated amino acids. Proc. R. Soc. Lond. B185, 51–72.

Shellis, R.P., Miles, A.E.W., 1976. Observations with the electron microscope on enameloid formation in the common eel (*Anguilla anguilla*: Teleostei). Proc. R. Soc. Lond. B194, 253–269.

Shellis, R.P., Poole, D.F.G., 1978. The structure of the dental hard tissues of the coelacanthid fish *Latimeria chalumnae* Smith. Arch. Oral. Biol. 23, 1105–1113.

Shellis, R.P., 1975. A histological and histochemical study of the matrices of enameloid and dentine in teleost fishes. Arch. Oral. Biol. 20, 183–187.

Shellis, R.P., 1978. The role of the inner dental epithelium in the formation of the teeth in fishes. In: Butler, pm, Joysey, K.S. (Eds.), Development, Function and Evolution of Teeth. Academic Press, London, pp. 32–42.

Shimoda, S., 1989. [Histogenesis of sea bream (Pagrus major) enameloid]. Tsurumi Shigaku Tsurumi Univ. Dent. J. 15, 267–284 (Japanese, English abstract).

Shintani, S., Kobata, M., Toyosawa, S., Ooshima, T., 2003. Identification and characterization of ameloblastin gene in an amphibian, *Xenopus laevis*. Gene 318, 125–136.

Shintani, S., Kobata, M., Toyosawa, S., Ooshima, T., 2006. Expression of ameloblastin during enamel formation in a crocodile. J. Exp. Zool. Mol. Dev. Evol. 306B, 126–133.

Slavkin, H.C., Samuel, N., Bringas, P., Nanci, A., Santos, V., 1983a. Selachian tooth development. II. Immunolocalization of amelogenin polypeptides in epithelium during secretory amelogenesis in *Squalus acanthias*. J. Craniofac. Genet. Dev. Biol. 3, 43–52.

Slavkin, H.C., Graham, E., Zeichner-David, M., Hildemann, W., 1983b. Enamel-like antigens in hagfish: possible evolutionary significance. Evolution 37, 404–412.

Smith, C.E., Nanci, A., 1995. Overview of morphological changes in enamel organ cells associated with major events in amelogenesis. Int. J. Dev. Biol. 39, 153–161.

Smith, M.M., 1978. Enamel in the oral teeth of *Latimeria chalumnae* (Pisces: Actinistia): a scanning electron microscope study. J. Zool. Lond 185, 355–369.

Smith, M.M., 1979. SEM of the enamel layer in oral teeth of fossil and extant crossopterygian and dipnoan fishes. Scanning Electron Microsc. II, 483–490.

Smith, M.M., 1995. Heterochrony in the evolution of enamel in vertebrates. In: McNamara, K.J. (Ed.), Evolutionary Change and Heterochrony. Wiley, Chichester, pp. 125–150.

Sparks, N.H.C., Motta, P.J., Shellis, R.P., Wade, V.J., Mann, S., 1990. An analytical electron microscopy study of iron-rich teeth from the butterflyfish (*Chaetodon ornatissimus*). J. Exp. Biol. 151, 371–385.

Suga, S., Taki, Y., Wada, K., Ogawa, M., 1991. Evolution of fluoride and iron concentrations in the enameloid of fish teeth. In: Suga, S., Nakahara, H. (Eds.), Mechanisms and Phylogeny of Mineralization in Biological Systems. Springer, Berlin, pp. 439–446.

Suga, S., Taki, Y., Ogawa, M., 1993. Fluoride and iron concentrations in the enameloid of lower teleostean fish. J. Dent. Res. 72, 912–922.

Suga, S., Taki, Y., Ogawa, M., 1997. Iron in the enameloid of perciform fish. J. Dent. Res. 71, 1316–1325.

Toyosawa, S., O'Huigin, C., Figueroa, F., Tichy, H., Klein, J., 1998. Identification and characterization of amelogenin genes in monotremes, reptiles, and amphibians (tooth formation/evolutionary innovations). Proc. Natl. Acad. Sci. U. S. A. 95, 13056–13061.

Venkatesh, B., Lee, A.P., Ravi, V., Maurya, A.K., Lian, M.M., Swann, J.B., Ohta, Y., Flajnik, M.F., Sutoh, Y., Kasahara, M., Hoon, S., Gangu, V., Roy, S.R., Irimia, M., Korzh, V., Kondrychyn, I., Lim, Z.W., Tay, B.H., Tohari, S., Kong, K.W., Ho, S., Lorente-Galdos, B., Quilez, J., Marques-Bonet, T., Raney, B.J., Ingham, P.W., Tay, A., Hillier, L.W., Minx, P., Boehm, T., Wilson, R.K., Brenner, S., Warren, W.C., 2014. Elephant shark genome provides unique insights into gnathostome evolution. Nature Lond. 505, 174–179.

Verbeeck, R.M.H., 1986. Minerals in human enamel and dentine. In: Driessens, F.C.M., Wöltgens, J.H.M. (Eds.), Tooth Development and Caries. CRC Press, Boca Raton, pp. 95–152.

Whitenack, L.B., Simkins, D.C., Motta, P.J., Hirai, M., Kumar, A., 2010. Young's modulus and hardness of shark tooth biomaterials. Arch. Oral. Biol. 55, 203–209.

Wistuba, J., Greven, H., Clemen, G., 2002. Development of larval and transformed teeth in *Ambystoma mexicanum* (Urodela, Amphibia): an ultrastructural study. Tiss. Cell 34, 14–27.

Wistuba, J., Völker, W., Ehmcke, J., Clemen, G., 2003a. Characterization of glycosaminoglycans during tooth development and mineralization in the axolotl, Ambystoma mexicanum. Tissue Cell 35, 353–361.

Wistuba, J., Ehmcke, J., Clemen, G., 2003b. Tooth development in *Ambystoma mexicanum*: phosphatise activities, calcium accumulation and cell proliferation in tooth-forming tissues. Ann. Anat. 185, 239–245.

Yilmaz, E.D., Bechtle, S., Özcobana, H., Kieser, J.A., Swain, M.V., Schneider, G.A., 2014. Micromechanical characterization of prismless enamel in the tuatara, *Sphenodon punctatus*. J. Mech. Behav. Biomed. Mater 39, 210–217.

Zhang, Y.-R., Du, W., Zhou, X.-D., Yu, H.-Y., 2014. Review of research on the mechanical properties of the human tooth. Int. J. Oral. Sci. 6, 61–69.

Index

Note: Page numbers followed by f and t indicate figures and tables respectively

A

Aardvark (*Orycteropus afer*), 306, 307f
Acanthosaura armata, 190, 191f
Acris gryllus blanchardi, 146
Acrodont position, 154
Actinopterygii
  dentitions
    Cladista, 54, 54f
    Holostei, 54–55, 55f
    Teleostei. *See* Teleostei
  food capture
    biting, 47
    ram, 46–47
    suction, 46–47
    suspension-feeding fishes, 47
  intraoral transport and prey manipulation,
    47–48, 48f
  intrapharyngeal transport, 48–49, 48f
  oral jaws, 44–46, 45f–46f
  tooth attachment
    ankylosis, 50, 50f
    direct fibrous attachment, 50, 51f
    hinged teeth, 52, 52f
    pedicellate attachment, 50–52, 51f
    thecodont attachment, 53, 53f
    "unattached" teeth, 53
  tooth formation, 49–50
African characins, 60
African cichlids
  feeding method, 82–83
  inhibition, 82
  oral and pharyngeal dentitions, 83, 83f–84f
African clawed frog (*Xenopus laevis*), 113–114,
  144–145, 145f, 265, 274–276
African elephant (*Loxodonta africana*),
  303–304, 303f–304f
African savannah monitor (*Varanus
  exanthematicus*), 188f
Agalychnis callydrias (Hylidae), 113
Agama agama, 189, 189f
Agama agricollis, 189, 190f
Agamid (*Uromastyx hardwickii*), 323–324
Agamidae, 189–191, 189f–192f
Agnatha, 1
Alethinophidia
  Aniliidae, 209, 210f
  Boidae, 210, 210f–211f
  Colubridae
    aglyphous and nonvenomous, 218
    Colubrinae, 220–221, 220f–223f, 222t
    Dipsadinae, 220, 220f
    Homalopsinae, 221
    Natricinae, 218–220, 219f

    number of teeth, 218
    venom delivery, 218
  Elapidae, 216–218, 216f–218f
  Lamprophiidae, 214, 214t
    Atractaspidinae, 215–216, 215f
    Lamprophiinae, 214–215, 215f
  Pythonidae, 210–211, 211f
  Viperidae, 211–212, 211f
    Crotalinae, 213–214, 214f–215f
    Viperinae, 212–213, 212f–213f
Alligator gar (*Atractosteus spatula*), 55, 55f
Alligatoridae, 225, 227–228
  American alligator, 228, 229f
  black caiman, 228, 229f
  Chinese alligator, 228, 229f
  Cuvier's dwarf caiman, 228, 230f
  spectacled caiman, 228, 229f–230f
Alligator mississippiensis, 226
Allison's anole (*Anole allisoni*), 197,
  197f–198f, 250f
Alluaud's haplo (*Astatoreochromis alluaudi*),
  86, 87f
Alveolar bone, 226
Alytes obstetricans (Alytidae), 113
Amblyrhynchus cristatus, 193–194, 195f
Amboina sailfin lizard (*Hydrosaurus
  amboinensis*), 190, 191f
Ambystoma, 130–131
Ambystoma annulatum, 126
Ambystoma mabeei (Ambystomatidae), 113
Ambystoma maculatum, 323
Ambystomatidae, 130–133, 131f–134f
Ameloblasts, 241
American alligator (*Alligator mississippiensis*),
  228, 229f, 267
American bullfrog (*Rana* (*Lithobates*)
  catesbeiana), 143f, 148
American crocodile, 230, 230f–231f
Amphibians
  adult stage, 113
  adult teeth, 113
  Anurans. *See* Anurans
  feeding, 113
  Gymnophiona. *See* Caecilians
    (Gymnophiona)
  larval stage, 113
  ontogeny of dentition, 272–276,
    275f–277f
  prey, 113
  subclass, 113
  tongues, 113
  tooth replacement, 114
    Anurans, 265–266

    caecilians, 264–265
    Urodeles, 265
  Urodela. *See* Caudata (Urodela)
Amphignathodontidae, 145–146, 145f–146f0
Amphisbaenia
  Amphisbaenidae, 178–181, 180f–181f
  Bipedidae, 179, 179f
  diet, 178
  families, 177
  Iberian worm lizard, 177, 177f
  morphology, 177–178
  Rhineuridae, 178–179, 178f
  skull types, 178
  tooth numbers, 178
  Trogonophidae, 178–179, 180f
Amphistylic jaw suspension, 5
Anarhichadidae, 105, 105f–106f
Anaspid condition, 153
Andresen lines, 302
Angel sharks (*Squatina*), 23, 25f
Anglerfishes, 95–97
Angler or monkfish (*Lophius piscatorius*), 97,
  98f
Anguilliformes, 56–57, 56f–58f
Ankylosed position, 154
Ankylosis, 50, 50f, 243–244
Anolis carolinensis, 197, 197f
Anoxypristis cuspidata, 35
Anurans
  characteristics, 137–138
  dentitions of adult
    bicuspid teeth, 144, 144f
    Guenther's marsupial frog, 143–144, 143f
    marine toad, 142–143
    pair of bony projections, 144
    prey hunt at night, 142–143
    role in food capture, 143–144
  Neobatrachia, 145–150, 146f–149f
  Nonneobatrachian, 144–145, 145f
  suborders, 137–138
  tadpole dentitions, 138–142, 140f–142f
Aparasphenodon brunoi, 146
Aprismatic enamel, 311
Aquatic salamanders, 128
Arapaima (*Arapaima* sp.), 47–48, 48f
Arboreal blind snake (*Typhlops anguisticeps*),
  201
Arboreal salamander (*Aneides lugubris*), 137,
  138f
Armored catfish, 278
Asian clawed salamander (*Onychodactylus
  japonicus*), 134–135
Asian salamander (*Ranodon sibiricus*), 128, 275f

Printed and bound by CPI Group (UK) Ltd, Croydon, CR0 4YY

08/05/2025

01865031-0001